AF885080

SIGMA SERIES

STRENGTH OF MATERIALS

About the Author

L S Negi is the Principal of Guru Nanak Dev Co-educational Polytechnic, Rohini, Delhi. An M Tech in Structural Engineering from the Indian Institute of Technology, Delhi, he has over thirty years of teaching experience. He is also the author of *Design of Steel Structures*, 2e, and co-author of *Structural Analysis*, both published by Tata McGraw-Hill.

STRENGTH OF MATERIALS

L. S. Negi

Principal
Guru Nanak Dev Polytechnic, Delhi

Tata McGraw-Hill Publishing Company Limited
NEW DELHI

McGraw-Hill Offices

New Delhi New York St Louis San Francisco Auckland Bogotá Caracas
Kuala Lumpur Lisbon London Madrid Mexico City Milan Montreal
San Juan Santiago Singapore Sydney Tokyo Toronto

Tata McGraw-Hill

Published by Tata McGraw-Hill Publishing Company Limited,
7 West Patel Nagar, New Delhi 110 008.

Copyright © 2008, by Tata McGraw-Hill Publishing Company Limited.
No part of this publication may be reproduced or distributed in any form or by any means, electronic, mechanical, photocopying, recording, or otherwise or stored in a database or retrieval system without the prior written permission of the publishers. The program listings (if any) may be entered, stored and executed in a computer system, but they may not be reproduced for publication.

This edition can be exported from India only by the publishers,
Tata McGraw-Hill Publishing Company Limited.

ISBN: (13): 978-0-07-070222-6
ISBN: (10): 0-07-070222-5

General Manager: Publishing—SEM & Tech Ed: *Vibha Mahajan*
Asst. Sponsoring Editor—SEM & Tech Ed: *Shukti Mukherjee*
Executive—Editorial Services: *Sohini Mukherjee*
General Manager: Marketing—Higher Education and School: *Michael J. Cruz*
Product Manager—SEM & Tech Ed: *Biju Ganesan*
Controller—Production: *Rajender P Ghansela*
Asst. General Manager—Production: *B L Dogra*
Senior Production Executive: *Anjali Razdan*

Information contained in this work has been obtained by Tata McGraw-Hill, from sources believed to be reliable. However, neither Tata McGraw-Hill nor its authors guarantee the accuracy or completeness of any information published herein, and neither Tata McGraw-Hill nor its authors shall be responsible for any errors, omissions, or damages arising out of use of this information. This work is published with the understanding that Tata McGraw-Hill and its authors are supplying information but are not attempting to render professional services. If such services are required, the assistance of an appropriate professional should be sought.

Typeset at Bukprint India, B-180A, Guru Nanak Pura, Laxmi Nagar, Delhi-110 092

Cover Design: Kapil Gupta

RALLCRQYRARRY

The McGraw·Hill Companies

Contents

Preface

Strength of materials, or mechanics of solids, as it is commonly called, forms one of the integral topics of mechanical engineering. The vastness of this subject and its application in different fields like design and analysis of structures, and theory of machines, prompted me to present this subject in a compact form. An effort has been made to provide something extra to take the reader beyond the existing courses being taught in various places.

This book is aimed at students of undergraduate courses of mechanical, civil, production and industrial engineering and also all those preparing for various competitive examinations in engineering. It endeavors to cover all the fundamental topics in a clear, concise and lucid manner. Practicing engineers will also find it to be a handy reference material.

The book is divided into 16 chapters. Hooke's law for three-dimensional state of stress and differential equations of equilibrium in Cartesian and polar coordinates are presented in Chapter 1, which are later used for analysis of thick cylinders in Chapter 14. Details into theory of elasticity is avoided due to space constraints. Matrix formulations have been used for Hooke's law in Chapter 1, and also for analysis of stresses and strains in chapters 2 and 3 and trusses in Appendix D. Torsion has been explained in Chapter 4 while Chapter 5 deals with shearing forces and bending moments. Chapter 6 deals with centroid, second moment of area and product moment of area. Chapter 7 is on stresses in beams. Chapter 8 handles trusses. Chapter 9 describes about compression members and Chapter 10 is on combined direct and bending stress. Chapter 11 explains the theories of failure. Chapter 12 is on the geometrical methods of deflection. Chapters 13 and 14 deal with thin shells and thick-walled cylinders respectively. Chapter 15 explains the concept of plastic theory of structures, and finally Chapter 16 deals with springs.

Suitable simple computer programs in BASIC, FORTAN and C++ languages have been provided which can be further modified by the reader, e.g., the computer program in Example 5.11 calculates the shear force and bending moments at various nodes besides maximum value of bending moments and its location within two nodes. This program can be extended to plot shear force and bending moment diagrams. In Appendix D the computer program can analyze a determinate truss but it does not check if the truss is a determinate truss. Readers should also check the computer program for bugs. These programs can be downloaded from http://www.mhhe.com/negi/som.

I sincerely thank Dr V Soami Das, Professor, Dept. of Mechanical Engineering, Dayalbagh Educational Institute, Agra, Mr P C Gope, Professor, Department of Mechanical Engineering, College of Technology, G. B. Pant University, Pantnagar, Prof. Akshoy Ranjan Paul, Member of Faculty, Mechanical Engineering, Kalyani Govt. Engineering College, Kalyani, West Bengal, Prof. Deepak Sengupta, Professor, Dept. of Applied Mechanics, Bengal Engineering College, Howrah, Dr J V Gurumurthy, Professor, Dept. of Mechanical Engineering, Vidyawardhini's College of Engineering and Technology, Thane, and Dr Srinivasa Raghavan, Head, Dept. of Civil Engineering, Crescent Engineering College, Chennai. They have helped improve the book to a great extent by reviewing it thoroughly and giving me valuable feedback.

I also wish to express my regards to all those teachers without whom this book would not have been complete. I am also thankful to the staff of Tata McGraw-Hill, including Mrs Vibha Mahajan, Ms Shukti Mukherjee, Ms Sohini Mukherjee and Ms Anjali Razdan for their dedicated effort in publishing this book. Last but not the least, I am grateful to my family for their help, support and patience during the preparation of this book.

Suggestions for further improvement will always be welcome.

L S NEGI

1

Stress and Strain

CONCEPT REVIEW

1.1 INTRODUCTION

This book mainly deals with the deformation of elastic materials in engineering field. Most engineering materials are considered to be homogeneous. The properties of a homogeneous material are uniform throughout its body. However a homogeneous material may be isotropic or anisotropic. A material is said to be isotropic if its properties are same in all directions. Many engineering materials like bricks, concrete, etc. are isotropic. Materials like timber are anisotropic since their properties differ significantly for directions parallel and perpendicular to the grains. Metals can also acquire directional properties through rolling or extruding process because these processes tend to orient the crystals in certain direction.

1.1.1 Stress vs. Strain

Load applied on a rigid body may be axial tension or compression, shear or volumetric as shown in Fig. 1.1. Resistance of a body to external load is called stress and is governed by Hooke's law within elastic zone: "Stress in an elastic material is proportional to strain". This can be written mathematically for three cases of deformation (Fig. 1.1) as follows:

(i) For axial deformation (compression or tension)

$$\frac{\text{Axial stress}}{\text{Axial strain}} = \frac{\sigma}{\varepsilon} = E \tag{1.1}$$

where,

$$\text{Axial stress, } \sigma = \frac{\text{Axial force}}{\text{Area}} = \frac{F}{A} \tag{1.2}$$

$$\text{Axial strain, } \varepsilon = \frac{\text{Elongation}}{\text{Initial length}} = \frac{\Delta L}{L} \tag{1.3}$$

and, E = Young's modulus, constant for each elastic material.

(ii) For shear deformation

$$\frac{\text{Shear stress}}{\text{Shear strain}} = \frac{\tau}{\gamma} = G \tag{1.4}$$

where,

$$\text{Shear stress, } \tau = \frac{\text{Shear force}}{\text{Area}} = \frac{F}{A} \tag{1.5}$$

Shear strain,

γ = Angular deformation in radians,

G = Modulus of rigidity or shear modulus, constant for each elastic material.

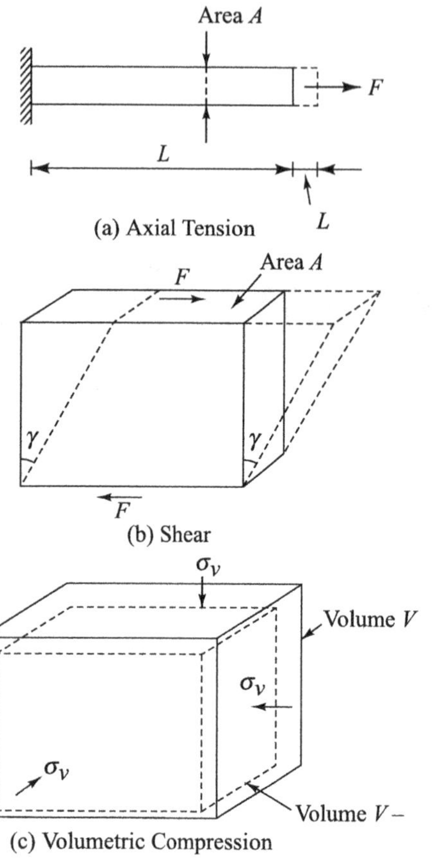

Fig. 1.1

(iii) For volumetric deformation (compression or tension)

$$\frac{\text{Volumetric stress}}{\text{Volumetric strain}} = \frac{\sigma_v}{\varepsilon_v} = K \tag{1.6}$$

where, Volumetric strain,

$$\varepsilon_v = \frac{\text{Change in volume}}{\text{Initial volume}} = \frac{\Delta V}{V} \tag{1.7}$$

and, K = Bulk modulus

It should be noted that the magnitude of stress and external pressure will be equal at equilibrium (i.e., when there is no movement of the body under the applied load).

Sign Convention: Tensile stresses and strains are normally considered +ve.

1.2 AXIAL TENSILE TEST

All axial load test are conducted at a constant rate of loading or at a constant rate of strain, the former being more common in use. Figure 1.2 shows stress–strain plot for mild steel. It is linear from starting point O to the limit of proportionality point P, up to which Hooke's law remains valid.

After point P the stress–strain curve deviates slightly from the straight line towards elastic limit E. Beyond point E, the stress increases up to the upper yield point Y_1 and then there is an unstable

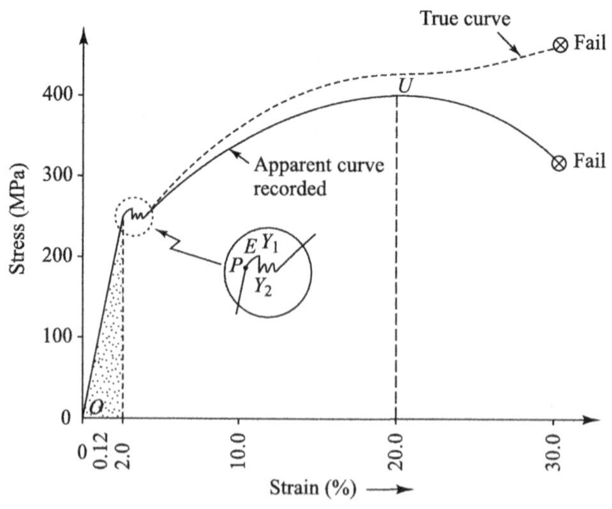

Fig. 1.2 *Stress–Strain Curve for Mild-steel*

region from upper yield point Y_1 to lower yield point Y_2. Mild steel remains elastic up to point E i.e., unloading will trace the loading curve. Beyond yield point Y_1 there is a small plastic zone where steel deforms at a constant stress. Thereafter stress increases with strain but the relation is no more linear. Unloading in this semi-plastic zone will be almost parallel to the initial stress–strain curve and a permanent strain ε_P is stored in the material. Points P, E, Y_1, Y_2 are very close and may merge into a single point Y as shown in idealized stress–strain curve of Fig. 1.3. It is also seen that yield stress increases to Y' after unloading. This phenomenon is called strain hardening or work hardening. This is due to rearrangement of randomly oriented grains in the direction of tension.

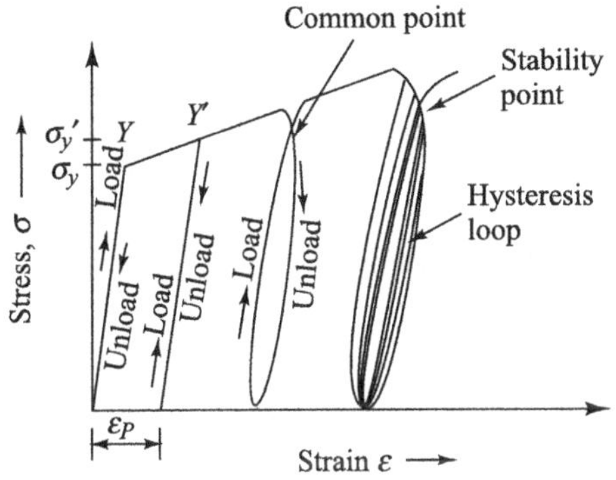

(a) Loading/unloading

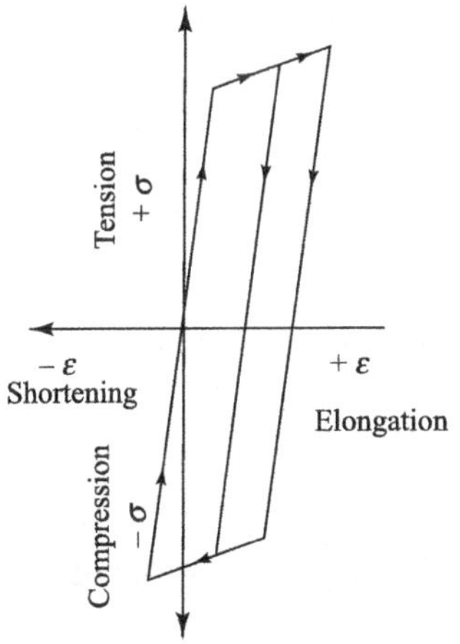

(b) Tension/Compression

Fig. 1.3 *Cyclic Loading*

The stress–strain curve beyond yield point is flatter and rises to an ultimate stress U. Beyond point U, the strain tends to be localized forming a neck in the test piece where cross-sectional area decreases rapidly till breaking of sample takes place at rupture stress.

1.2.1 Proof Stress

Figure 1.4 shows the stress–strain diagram of mild steel in comparison to other materials. Many metals like high carbon do not show a definite yield point and their yield point is defined corresponding to a proof stress which causes a permanent strain of 0.2% upon unloading as shown in Fig. 1.5.

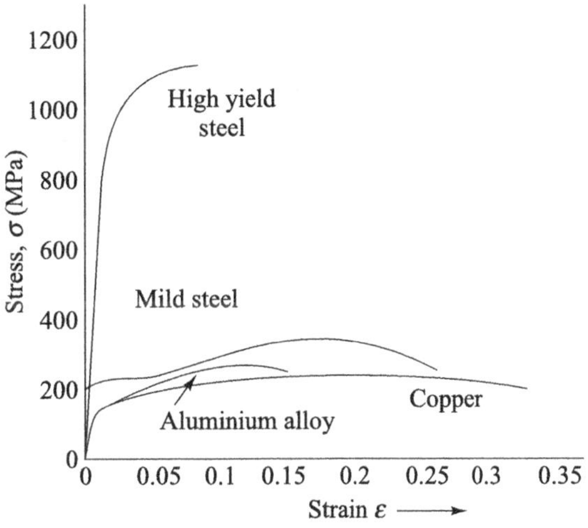

Fig. 1.4 *Typical Stress–strain Diagrams for Different Materials*

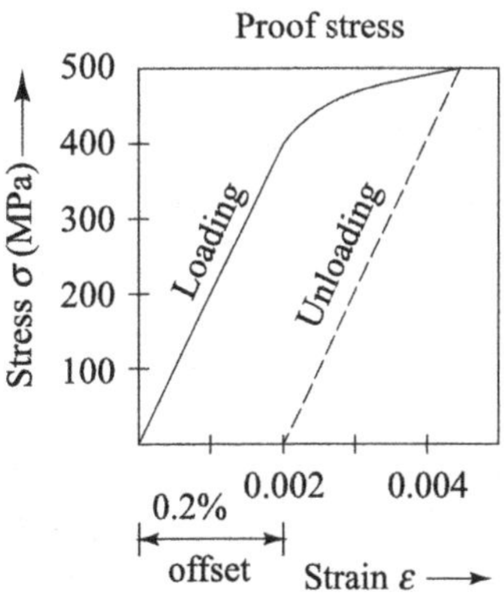

Fig. 1.5 *Stress–Strain Curve for Cold Worked Steel*

1.2.2 True Stress–True Strain

There is an error in the computation of stress and strain if they are calculated based on initial area and length since both are changing during the test. In Fig.1.2 firm line shows this apparent stress–strain relation. True stress–strain curve (shown by dotted line in Fig.1.2) can be obtained as follows.

Let initial cross-sectional area and length of the sample be A_0 and L_0, respectively (I.S., 1608–1972 recommends $L_0 = 5.65/A_0$ for axial tension test)

At any stage cross-sectional area and length of the sample be A and L, respectively,

Therefore at any stage,

$$\text{Apparent stress, } \sigma = \frac{\text{Force}}{\text{Area}} = \frac{F}{A_0} \quad (1.8)$$

$$\text{Apparent strain, } \varepsilon = \frac{L - L_0}{L_0} \quad (1.9)$$

$$\text{True stress, } \bar{\sigma} = \frac{F}{A} \quad (1.10a)$$

$$\text{True strain, } \bar{\varepsilon} = \int_{L_0}^{L} \frac{dL}{L} = \log_e \frac{L}{L_0} \quad (1.11a)$$

If the volume changes are neglected, then

$$A_0 \times L_0 = A \times L$$

$$\text{True stress, } \bar{\sigma} = \frac{F}{A_0} \times \frac{L}{L_0} = \sigma(1 + \varepsilon) \quad (1.10b)$$

$$\text{True strain, } \bar{\varepsilon} = \log_e \frac{L}{L_0} = \log_e(1 + \varepsilon) \quad (1.11b)$$

The difference between true stress–strain and apparent stress–strain is significant only at higher strain beyond elastic limit.

1.2.3 Poisson's Ratio

The cross-section of the specimen reduces under the axial load which causes transverse or lateral strain ε_l.

$$\text{Lateral strain, } \varepsilon_l = \frac{d - d_0}{d_0} \quad (1.12)$$

where, d = diameter of specimen at any stage,

d_0 = initial diameter of specimen.

The ratio of lateral strain $|\varepsilon_l|$, and longitudinal strain $|\varepsilon|$ under axial load is called Poisson's ratio.

$$\text{Poisson's ratio, } \mu = \frac{-\varepsilon_l}{\varepsilon} \quad (1.13)$$

Minus sign indicates that lateral and axial strains are of opposite signs. Poisson's ratio is constant for most of the materials up to proportionality limit and its value ranges from 0.25 to 0.35. The theoretical maximum value of μ is 0.5 for incompressible substance.

1.2.4 Modulus of Resilience and Strain Energy

The modulus of resilience is the strain energy stored in a material per unit volume when loaded up to elastic limit.

$$\text{Strain energy} = \frac{\text{Average force} \times \text{distance moved}}{\text{Volume}}$$

$$\text{Modulus of resilience} = \left(\frac{0 + \sigma_y}{2}\right) \times \frac{A \times \Delta L}{A \times L}$$

$$= \frac{1}{2} \times \sigma_y \times \varepsilon_y = \frac{\sigma_y^2}{2E} \quad (1.14)$$

= area under stress-strain plot shown shaded in Fig. 1.2.

1.2.5 Working Stress

If a structure is to work within elastic limit, the design stress or working load should always be less then the yield stress with a factor of safety.

$$\text{Working stress} = \frac{\text{Yield stress}}{\text{Factor of safety}} \quad (1.15)$$

1.2.6 Modulus of Toughness

It is the measure of total energy a material can absorb prior to fracture. Modulus of toughness is the total area under stress-strain plot.

1.2.7 Creep, Relaxation and Fatigue

These three phenomena are associated with the plastic deformation of solids. The stress-strain plot in Fig.1.2 is drawn by increasing stress or strain continuously and no time is allowed for one to develop due to increases in other. However, at higher stress level strain will increases with time at constant load. This phenomenon is called creep and corresponding strain is creep strain.

Relaxation is somewhat reverse of creep. If a material is strained at a constant level, the stress in the material reduces with time. This phenomenon is called relaxation. The materials fail at a fatigue stress much smaller than their rupture stress if they are subjected to a specified number of cycles of alternating tensile/compression loading or loading/unloading. The endurance limit or fatigue limit is the maximum reversed stress below which an infinite number of cycles may be applied without rupturing the specimen. The endurance limit for most engineering materials is less than the elastic limit. Figure 1.4 shows loading/unloading curves meet at a common point which is decreasing with successive cycles upto a stability point. The area of hysteresis loop shown is the energy dissipated with each cycle. Similar hysteresis loop is obtained in complete reversal of load cycles. These phenomena are now getting special attention since the basis of structural design is now shifted from working load theory to plastic design.

1.2.8 Volume Change

Consider a cube of initial volume $V_0 = a \times b \times c$, of the tension specimen. If the strain parallel to the x-axis is ε, the strains parallel to y-and z-axis are,

$$\varepsilon_Y = \varepsilon_Z = \mu \times \varepsilon$$

The deformed volume,

$$V = a(1+\varepsilon) \times b(1-\mu\varepsilon) \times c(1-\mu\varepsilon)$$
$$= a \times b \times c\,(1+\varepsilon)(1-\mu \times \varepsilon)^2$$
$$= a \times b \times c\,[1 + (1-2\mu)\,\varepsilon]$$

(neglecting higher order terms of ε)

Thus change in volume,

$$\Delta V = a \times b \times c\,(1-2\mu)\,\varepsilon \qquad (1.16)$$

If the volume does not change during the test, Eq. (1.16) indicates the maximum possible value of μ is 0.5.

1.2.9 Shear Stress-Strain Diagram

Ordinarily, the important properties of metals are obtained from axial tension test, provided Poisson's ratio is determined during the test. The modulus of rigidity can then be calculated from the formula, $G = E/2\,(1+\mu)$. Mechanical properties such as shearing proportional limit, shearing yield strength and shear ultimate strength are defined in a manner similar to the axial tension test. The mechanical properties of a material are occasionally determined from shear test. The most common shear test uses a thin walled tube under torsion. The shear stress–strain relationship is then established using Eq. (4.1) through (4.4) if r, l, θ and T are known by measurements.

1.2.10 Compression Stress-Strain Diagram

For most ductile materials, true stress–strain curves are practically the same in tension and compression. The two curves are quite different when engineering strains are used, since the specimen tend to bulge in compression. Figure 1.6 shows stress–strain curve for ductile and brittle materials in compression and tension.

1.3 GENERAL 3-DIMENSIONAL STATE OF STRESS

A body can be subjected to varying stress in all the three directions as shown in Fig.1.7. Body forces F_x, F_y, F_z may also be acting alongwith direct stresses and shear stresses. The equations for equilibrium in three directions can then be written as follows.

$$\sum F_x = 0, \quad \frac{\partial \sigma_x}{\partial x} + \frac{\partial \tau_{yx}}{\partial y} + \frac{\partial \tau_{zx}}{\partial z} + F_x = 0 \qquad (1.17a)$$

$$\sum F_y = 0, \quad \frac{\partial \tau_{xy}}{\partial x} + \frac{\partial \sigma_y}{\partial y} + \frac{\partial \tau_{zy}}{\partial z} + F_y = 0 \qquad (1.17b)$$

$$\sum F_z = 0, \quad \frac{\partial \tau_{xz}}{\partial x} + \frac{\partial \tau_{yz}}{\partial y} + \frac{\partial \sigma_z}{\partial z} + F_z = 0 \qquad (1.17c)$$

Now consider equilibrium of moment of the forces about an axis parallel to the z-axis and passing through center of the face $ABCD$,

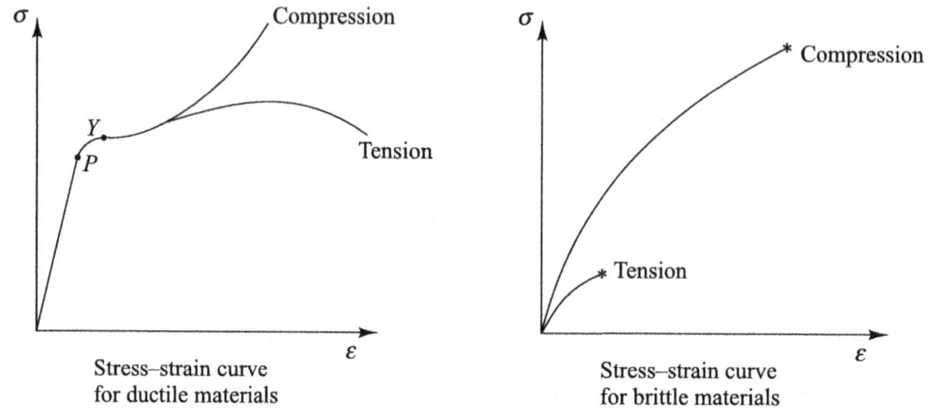

Fig. 1.6

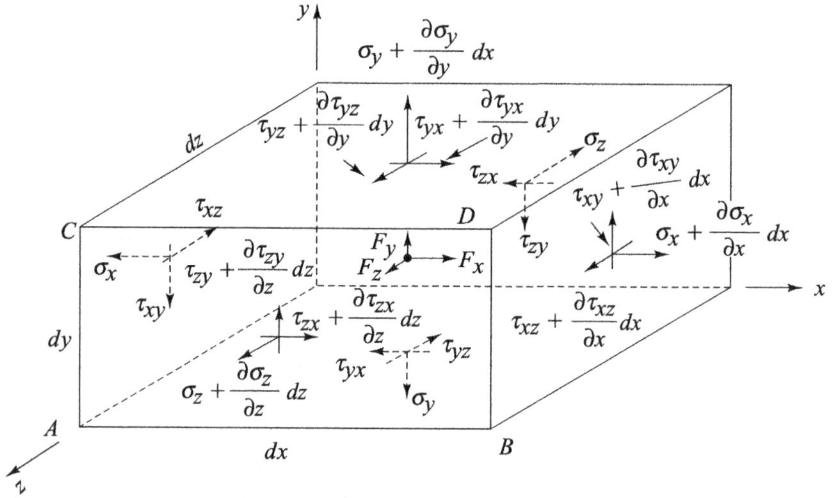

Fig. 1.7 *General 3-Dimensional State of Stress*

$$\left(\tau_{xy} + \frac{\partial \tau_{xy}}{\partial x} dx\right) dzdy \frac{dx}{2} + \tau_{xy} dzdy \frac{dx}{2}$$

$$-\left(\tau_{yx} + \frac{\partial \tau_{yx}}{\partial y} dy\right) dzdy \frac{dy}{2} - \tau_{yx} dzdx \frac{dy}{2} = 0$$

Neglecting product of four differential terms in comparison to three differential terms gives,

$$\tau_{xy} = \tau_{yx} \qquad (1.17\text{d})$$

Similarly it can be shown that,

$$\tau_{yz} = \tau_{zy} \quad \text{and} \quad \tau_{zx} = \tau_{xz} \qquad (1.17\text{e,f})$$

The differential equations for equilibrium in polar coordinates are derived as follows (Example 1.25)

$$\frac{\partial \sigma_r}{\partial r} + \frac{1}{r}\frac{\partial \tau_{r\theta}}{\partial \theta} + \frac{\sigma_r - \sigma_\theta}{r} + F_r = 0 \qquad (1.18\text{a})$$

$$\frac{1}{r}\frac{\partial \sigma_\theta}{\partial \theta} + \frac{\partial \tau_{r\theta}}{\partial r} + 2\frac{\tau_{r\theta}}{r} + F_\theta = 0 \qquad (1.18\text{b})$$

1.3.1 Strain–Displacement Relations

Consider an element *ABCD* unit thickness undergoes strains and displaced to *A′ B′ C′ D′* as shown in Fig. 1.8. If u and v represent displacements in x- and y-directions then axial strain in x direction,

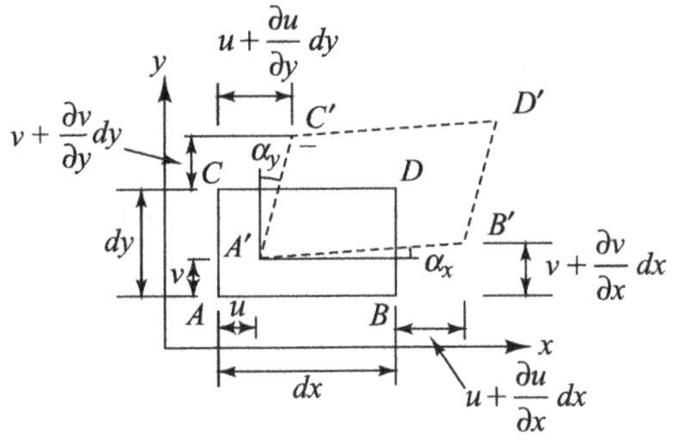

Fig. 1.8 *Displacement for Strains*

$$\varepsilon_x = \frac{u + \frac{\partial u}{\partial x}dx - u}{dx} = \frac{\partial u}{\partial x} \tag{1.19a}$$

Similarly, $$\varepsilon_y = \frac{\partial v}{\partial y} \tag{1.19b}$$

The shear strain in xy plane, γ_{xy} is given by the angles of rotation α_x and α_y as as follows

$$\gamma_{xy} = \alpha_x + \alpha_y \approx \tan\alpha_x + \tan\alpha_y$$

$$= \frac{\frac{\partial v}{\partial x}dx}{dx} + \frac{\frac{\partial u}{\partial y}dy}{dy}$$

or, $$\gamma_{xy} = \frac{\partial u}{\partial y} + \frac{\partial v}{\partial x} \tag{1.19c}$$

For three-dimensional elements of size $dx\,dy\,dz$, the normal and shearing strains are similarly related to the displacements u, v, w in x, y and z directions as follows

$$\varepsilon_x = \frac{\partial u}{\partial x},\ \varepsilon_y = \frac{\partial v}{\partial y},\ \varepsilon_z = \frac{\partial w}{\partial z} \tag{1.20a,b,c}$$

$$\gamma_{xy} = \frac{\partial u}{\partial y} + \frac{\partial v}{\partial x} \tag{1.20d}$$

$$\gamma_{yz} = \frac{\partial v}{\partial z} + \frac{\partial w}{\partial y} \tag{1.20e}$$

$$\gamma_{zx} = \frac{\partial w}{\partial x} + \frac{\partial u}{\partial z} \tag{1.20f}$$

Since the angular change in x to y or y to x direction is same,

$$\gamma_{xy} = \gamma_{yx},\quad \gamma_{yz} = \gamma_{zy} \text{ and } \gamma_{zx} = \gamma_{xz} \tag{1.20g, h, i}$$

The strain-displacement relations in polar coordinates shown as follows are derived later in Example 1.26.

$$\varepsilon_r = \frac{\partial u}{\partial r} \tag{1.21a}$$

$$\varepsilon_\theta = \frac{u}{r} + \frac{1}{r}\frac{\partial v}{\partial \theta} \tag{1.21b}$$

and, $$\gamma_{r\theta} = \frac{1}{r}\frac{\partial u}{\partial \theta} + \frac{\partial v}{\partial r} - \frac{v}{r} \tag{1.21c}$$

Compatibility Condition The strains are not independent of one another and the condition of compatibility is given by double differentiation of Eq.(1.19) as follows

$$\frac{\partial^2 \varepsilon_x}{\partial y^2} = \frac{\partial^2 u}{\partial x \partial y^2},\quad \frac{\partial^2 \varepsilon_y}{\partial x^2} = \frac{\partial^2 v}{\partial x^2 \partial y},$$

$$\frac{\partial^2 \gamma_{xy}}{\partial x \partial y} = \frac{\partial^2 u}{\partial x \partial^2 y} + \frac{\partial^2 v}{\partial x^2 \partial y}$$

$$\therefore \quad \frac{\partial^2 \varepsilon_x}{\partial y^2} + \frac{\partial^2 \varepsilon_y}{\partial x^2} = \frac{\partial^2 \gamma_{xy}}{\partial x \partial y} \tag{1.22a}$$

Similarly, it can be shown that Eq. (1.21a–c) gives the compatibility equation in polar coordinates as follows.

$$\frac{\partial^2 \varepsilon_\theta}{\partial r^2} + \frac{1}{r^2}\frac{\partial^2 \varepsilon_r}{\partial \theta^2} + \frac{2}{r}\frac{\partial \varepsilon_\theta}{\partial r} - \frac{1}{r}\frac{\partial \varepsilon_r}{\partial r} = \frac{1}{r}\frac{\partial^2 \gamma_{r\theta}}{\partial r \partial \theta} + \frac{1}{r^2}\frac{\partial \gamma_{r\theta}}{\partial \theta} \tag{1.22b}$$

1.3.2 Generalized Hooke's Law

Axial stress in x-direction causes (i) axial strain in x-direction and (ii) lateral strain in y and z directions.

When axial stress is acting in all three x, y and z directions, strain in one direction is caused by stress in all three directions as follows

$$\varepsilon_x = \frac{1}{E}\left[\sigma_x - \mu(\sigma_y + \sigma_z)\right] \tag{1.23a}$$

$$\varepsilon_y = \frac{1}{E}\left[\sigma_y - \mu(\sigma_z + \sigma_x)\right] \tag{1.23b}$$

$$\varepsilon_z = \frac{1}{E}\left[\sigma_z - \mu(\sigma_y + \sigma_x)\right] \tag{1.23c}$$

In case of small strains within proportional limit, shear strains do not affect the strains in other directions. Therefore, Eq. (1.4) for shear strains in three directions is written as follows

$$\gamma_{xy} = \frac{1}{G}\tau_{xy}, \quad \gamma_{yz} = \frac{1}{G}\tau_{yz}, \quad \gamma_{xz} = \frac{1}{G}\tau_{xz} \tag{1.23d, e, f}$$

The generalized Hooke's law for the three-dimensional stress-strain relation is written by rearranging the above equations as equation 1.23 g and h.

$$\{\sigma\} = [E] \times \{\varepsilon\} \tag{1.23g}$$

or,

$$\begin{bmatrix}\sigma_x\\ \sigma_y\\ \sigma_z\\ \tau_{xy}\\ \tau_{yz}\\ \tau_{zx}\end{bmatrix} = \frac{E}{(1+\mu)(1-2\mu)}\begin{bmatrix}1-\mu & \mu & \mu & 0 & 0 & 0\\ \mu & 1-\mu & \mu & 0 & 0 & 0\\ \mu & \mu & 1-\mu & 0 & 0 & 0\\ 0 & 0 & 0 & \frac{1-2\mu}{2} & 0 & 0\\ 0 & 0 & 0 & 0 & \frac{1-2\mu}{2} & 0\\ 0 & 0 & 0 & 0 & 0 & \frac{1-2\mu}{2}\end{bmatrix}\begin{bmatrix}\varepsilon_x\\ \varepsilon_y\\ \varepsilon_z\\ \gamma_{xy}\\ \gamma_{yz}\\ \gamma_{zx}\end{bmatrix} \tag{1.23h}$$

The above equations show that the strain can exist without stress e.g., if $\sigma_y = \sigma_z = 0$, even then $\varepsilon_y = \varepsilon_z = -\mu\ \sigma_x/E$. It should also be noted that stress is not proportional to strain when stresses exist in more than one direction, so modified modulus of elasticity can be defined as,

$$E_{m,x} = \frac{E\sigma_x}{\sigma_x - \mu(\sigma_y + \sigma_z)} \tag{1.24a}$$

$$E_{m,y} = \frac{E\sigma_y}{\sigma_y - \mu(\sigma_z + \sigma_x)} \tag{1.24b}$$

$$E_{m,z} = \frac{E\sigma_z}{\sigma_z - \mu(\sigma_y + \sigma_x)} \tag{1.24c}$$

1.3.3 Plane Stress

For two-dimensional state of stress such as stresses in a thin plate wherein σ_z, τ_{xz}, τ_{yz} are zero, Eq. 1.23 reduces to,

$$\varepsilon_x = \frac{1}{E}(\sigma_x - \mu\sigma_y) \tag{1.23i}$$

$$\varepsilon_y = \frac{1}{E}(\sigma_y - \mu\sigma_x) \tag{1.23j}$$

$$\gamma_{xy} = \frac{1}{G}\tau_{xy} \tag{1.23k}$$

This after rearrangement can be written as,

$$\begin{bmatrix}\sigma_x\\ \sigma_y\\ \tau_{xy}\end{bmatrix} = \frac{E}{(1-\mu^2)}\begin{bmatrix}1 & \mu & 0\\ \mu & 1 & 0\\ 0 & 0 & \frac{1-\mu}{2}\end{bmatrix}\begin{bmatrix}\varepsilon_x\\ \varepsilon_y\\ \tau_{xy}\end{bmatrix} \tag{1.23l}$$

For plane stress case ε_z is given by,

$$\varepsilon_z = \frac{-\mu}{E}(\sigma_x + \sigma_y).$$

1.3.4 Plane Strain

If physical constrains prevent axial deformation in one direction, then such a state of strain is called plane strain. For example a vertical thin slice in a long wall will not deform along the length of the wall due to its bulk. Thus if strain are prevented in z-direction i.e., $\varepsilon_z = \gamma_{xz} = \gamma_{zy} = 0$ then Eq. (1.23c) gives,

$$\sigma_z = \mu(\sigma_y + \sigma_x)$$

Substituting σ_z in Eq. (1.23a, b) yield the equations for plane strain problem as follows:

$$\varepsilon_x = \frac{1+\mu}{E}[\sigma_x(1-\mu) - \mu\sigma_y] \qquad (1.23m)$$

$$\varepsilon_y = \frac{1+\mu}{E}[\sigma_y(1-\mu) - \mu\sigma_z] \qquad (1.23n)$$

$$\gamma_{xy} = \frac{1}{G}\tau_{xy} \qquad (1.23o)$$

Inversion of these equations or substituting $\varepsilon_z = \gamma_{xz} = \gamma_{zy} = 0$ in Eq. (1.23h) yields,

$$\begin{bmatrix} \sigma_x \\ \sigma_y \\ \tau_{xy} \end{bmatrix} = \frac{E}{(1+\mu)(1-2\mu)} \times \begin{bmatrix} 1-\mu & \mu & 0 \\ \mu & 1-\mu & 0 \\ 0 & 0 & \frac{1-2\mu}{2} \end{bmatrix} \begin{bmatrix} \varepsilon_x \\ \varepsilon_y \\ \tau_{xy} \end{bmatrix} \qquad (1.23p)$$

Table 1.1 shows the stress-strain condition in plane stress/plane strain conditions.

Table 1.1

	Stress direction	*Zero stress*	*Zero strain*
Plane stress	x, y	$\sigma_z, \tau_{xz}, \tau_{yz}$	γ_{xz}, γ_{yz}
Plane strain	x, y, z	τ_{xz}, τ_{yz}	$\varepsilon_z, \gamma_{xz}, \gamma_{yz}$

1.4 RELATION BETWEEN ELASTIC CONSTANTS *E*, *G* AND *K*

Normally, Young's modulus E and Poisson's ratio μ are considered as independent constants for a material and the constants G and K are derived as dependent constants.

1.4.1 *E* vs. *G*

Consider a square plate *ABCD* unit thickness and size $b \times b$ subjected to shear force τ as shown in Fig. 1.9(a). Its deformed shape $AB'C'D'$ is redrawn as $ABC'D'$ in Fig. 1.9(b) for convenience in calculations.

$$\text{Shear strain, } \gamma = \frac{CC'}{BC}$$

Strain along diagonal AC,

$$\varepsilon_{AC} = \frac{AC' - AC}{AC} \approx \frac{FC'}{AC} = \frac{CC'\cos 45^\circ}{BC\sec 45^\circ} = \frac{\gamma}{2}$$

or,
$$\varepsilon_{AC} = \frac{\gamma}{2} = \frac{\tau}{2G} \qquad (1.24a)$$

Now considering the equilibrium of the free body *ABD* along the diagonal *AC* in Fig. 1.9(c),

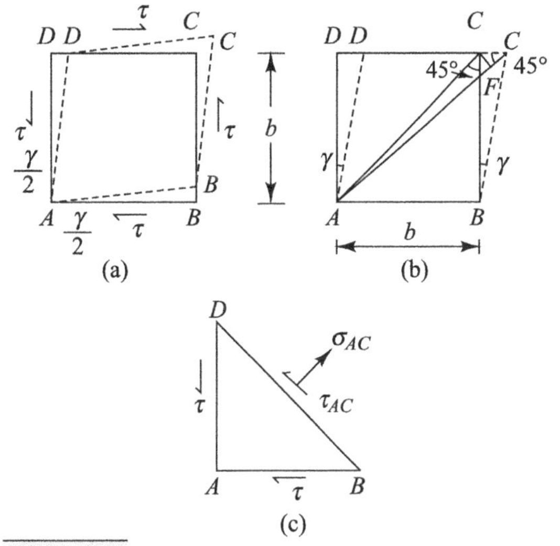

Fig. 1.9

$$\sigma_{AC} \times (\sqrt{2}b \times 1) = 2 \times \tau \times (b \times 1)\cos 45^\circ$$

or,
$$\sigma_{AC} = \tau$$

Similarly, the equilibrium condition of the free body *ABD* along the diagonal *BD* gives,

$$\tau_{AC} = 0$$

Thus the square plate is under axial tension $\sigma_{AC} = \tau$, along the diagonal AC. Similarly, it can be derived that the plate is under axial compressive stress $\sigma_{BD} = \tau$, along the diagonal BD. The strain along

diagonal AC is due to tension along AC and compression along diagonal BD.
From Eq. (1.23a)

$$\varepsilon_{AC} = \frac{1}{E}(\sigma_{AC} + \mu\sigma_{BD}) = \frac{\tau}{E}(1+\mu) \quad (1.24b)$$

From Eqs (1.24a and 1.24b),

$$G = \frac{E}{2(1+\mu)} \quad (1.24c)$$

1.4.2 *E* vs. *K*

Let a cube of initial volume $dx\ dy\ dz$ be subjected to the stresses σ_x, σ_y and σ_z. Its final volume is given by,

$$V + \Delta V = (1+\varepsilon_x)\,dx\,(1+\varepsilon_y)\,dy\,(1+\varepsilon_z)\,dz$$

$$= (1 + \varepsilon_x + \varepsilon_y + \varepsilon_z)\,dx\,dy\,dz$$

or, $$\Delta V = (\varepsilon_x + \varepsilon_y + \varepsilon_z)\,dx\,dy\,dz$$

or, volumetric strain, $\varepsilon_v = \dfrac{\Delta V}{V} = \varepsilon_x + \varepsilon_y + \varepsilon_z$

Substituting Eqs 1.23, $\varepsilon_v = \dfrac{1-2\mu}{E}(\sigma_x + \sigma_y + \sigma_z)$

If the cube is under volumetric stress,

$$\sigma_v = \sigma_x = \sigma_y = \sigma_z$$

Then, $$\varepsilon_v = \frac{3(1-2\mu)\sigma_v}{E}$$

Substituting in Eq. (1.6),

$$K = \frac{E}{3(1-2\mu)} \quad (1.25)$$

Since K is a +ve quantity, $1 - 2\mu > 0$ or $\mu < 0.5$. A material is incompressible if K is infinite or if $\mu = 0.5$.

1.5 STATICALLY INDETERMINATE PROBLEMS

If unknown forces in a system of forces applied on a body under equilibrium can be determined by the equilibrium equations i.e., $\sum$ Forces = 0, and $\sum$ Moments = 0, then it is said to be a statically determinate system. However, in some cases this is not possible and additional equations are required for determining the unknown forces, then they are called statically indeterminate problems. Additional equations are provided by the compatibility of displacements.

1.6 SAINT VENANT PRINCIPLE

The stress-force relation $\sigma = F/A$ is valid very near to application of the load or at the change of shape of section, e.g. Fig. 1.10 shows the stress distribution across the sections a–a and b–b in an axially loaded bar. The region over which the force-stress relation is not valid tends to be localized as expressed in Saint Venant Principle: " Except in the vicinity of the load, the method of application of load has no effect on stresses as long as the resultant load remains the same". These localized effects become negligible roughly at a distance equal to width of the member. The correction factor for localized effect may be determined experimentally or by empirical formulae.

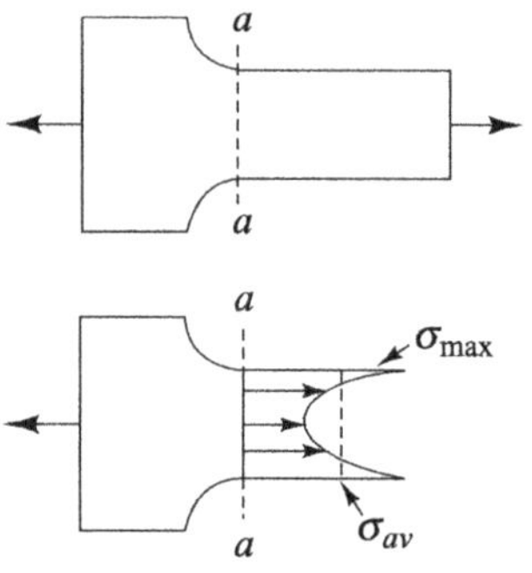

(a) Stress distribution near a fillet

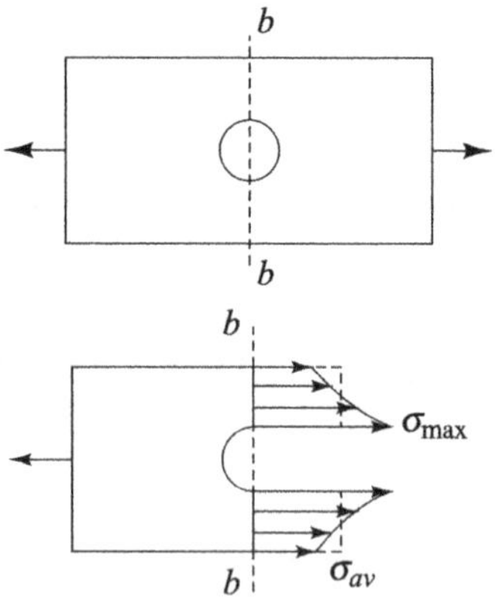

(b) Stress distribution near a hole

Fig. 1.10

1.7 IMPACT OR DYNAMIC LOAD

Till now, the analysis was for gradually applied load. If a weight W talls through a height h on a column of length L and cross-sectional area A as shown in Fig. 1.11, the maximum stress generated at the time of impact is much greater than the static stress which would occur when the weight W just rest on the head of the column. The behaviour of structures under dynamic load is complicated and an approximate analysis is obtained by assuming that the maximum stress σ_{max} produced at the time of impact is below the elastic limit and there is no loss of energy during impact due to plastic deformation, heat and sound. Thus the work done by the falling weight is equal to the strain energy of the column (Eq. 1.14),

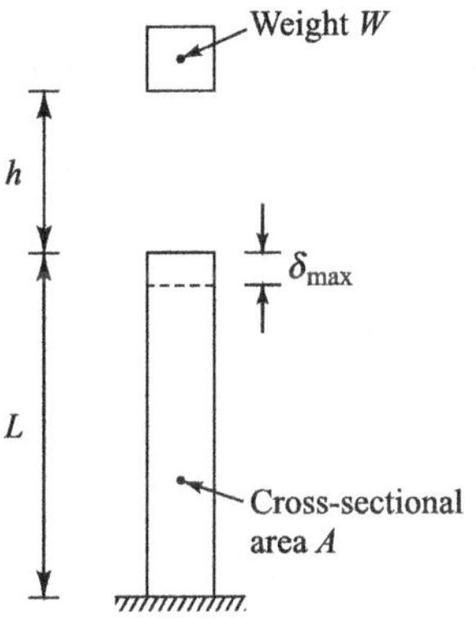

Fig. 1.11

$$W(h + \delta_{max}) = \frac{\sigma_{max}^2}{2E} \times AL \qquad (1.26)$$

where δ_{max} = maximum displacement corresponding to maximum stress σ_{max} at the time of impact

$$= \frac{\sigma_{max} L}{E}$$

$$\therefore \quad W\left(h + \frac{\sigma_{max} L}{E}\right) = \frac{\sigma^2_{max}}{2E} AL \qquad (1.27)$$

or $$\sigma^2_{max} - 2\frac{W}{A}\sigma_{max} - \frac{2WhE}{AL} = 0 \qquad (1.28)$$

or, $$\sigma_{max} = \frac{W}{A}\left(1 + \sqrt{1 + \frac{2AEh}{WL}}\right) \qquad (1.29)$$

The quantity within brackted is called as *impact factor* and is equal to the ratio of the maximum dynamic deflection δ_{max} and the static deflection δ_{st} as follow.

$$\frac{\delta_{max}}{\delta_{st}} = 1 + \sqrt{1 + \frac{2h}{\delta_{st}}} \qquad (1.30)$$

where, $$\delta_{st} = \frac{WL}{AE}$$

When $h >> \delta_{max}$ in Eq. 1.28, then

$$\sigma_{max} = \sqrt{2WhE / AL} \text{ and } \delta_{max} = \sqrt{2\delta_{st} h}$$

Suddenly Applied Load When $h = 0$, i.e. load is on the top of the column but support in such a way that their contact pressure is zero. Now if the weight is released suddenly then $\sigma_{max} = 2\sigma_{st}$ in Eq. 1.29.

Horizontal Impact When weight W moving with a horizontal velocity v strikes an elastic body, then the work done by weight W is $Wv^2/2g$ in place of $W(h + \delta_{max})$ in Eq. 1.26 and the maximum stress and deflection are as follow.

$$\sigma_{max} = \frac{W}{A}\sqrt{\frac{v^2}{g\,\delta_{st}}} \qquad (1.31)$$

and $$\delta_{max} = \delta_{st}\sqrt{\frac{v^2}{g\,\delta_{st}}} \qquad (1.32)$$

EXAMPLES

Example 1.1 In an axial tension test on mild steel specimen of 12 mm diameter and 60 mm gauge length, the data of load/extension was obtained as follows.

Load (kN)	*Extension (mm)*	*Load (kN)*	*Extension (mm)*
9.1	0.025	32.2	0.5
18.7	0.05	40.5	1.5
28.5	0.075	45.2	2.5
33.5	0.1	51.5	5.0
32.6	0.125	53.6	10.0
32.6	0.15	49.3	15.0
32.2	0.3	38.2	rupture

Determine (i) proportionality limit, (ii) yield stress, (iii) ultimate stress, (iv) modulus of toughness, (v) % elongation, (vi) % reduction of area, if the gauge length and diameter at failure are 77 mm and 9 mm, respectively.

Solution Initial cross-sectional area,

$$A_0 = \frac{\pi}{4} \times 12^2 = 113.1 \text{ mm}^2$$

Initial gauge length, $L_0 = 60$ mm

Load, P(kN)	*Stress, P/A_0 (MPa)*	*Extension, ΔL (mm)*	*Strain, $\Delta L/L_0 \times 10^{-4}$ (mm/mm)*
9.1	80.5	0.025	4.2
18.7	165.3	0.05	8.3
28.5	252.0	0.075	12.5
33.5	296.2	0.1	16.7
32.6	288.2	0.125	20.8
32.6	288.2	0.15	25.0
32.2	284.7	0.3	50.0
32.2	284.7	0.5	83.3
40.5	358.1	1.5	250.0
45.2	399.6	2.5	416.7
51.5	444.7	5.0	833.3
53.6	473.9	10.0	1666.7
49.3	424.4	15.0	2500.0
38.2	337.7	Rupture	–

Stress-strain relation is drawn in Fig. 1.12. It is seen from the plot,

Proportionality limit = 280 MPa at 0.14% strain,

Yield stress, Upper σ_{y1} = 296.2 MPa,

Lower σ_{y2} = 284.7 MPa,

Ultimate stress, σ_u = 473.9 MPa

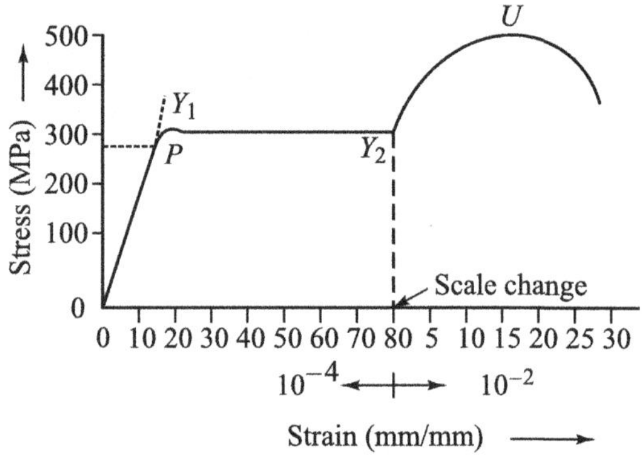

Fig. 1.12

Modulus of elasticity,

$$E = \frac{\Delta\sigma}{\Delta\varepsilon} = \frac{252 - 165.3}{(12.5 - 8.3) \times 10^{-4}}$$

$$= 2.06 \times 10^5 \text{ MPa}$$

Modulus of resilience = Area below stress–strain plot up to proportionality limit

$$= \frac{280 \times 14 \times 10^{-4}}{2} = 0.196 \text{ N mm/mm}^3$$

Modulus of toughness

= Total area below stress–strain plot

$= 125.5$ N mm/mm^3

% Energy absorbed up to proportionality limit

$$= \frac{0.196}{125.5} \times 100 = 0.15\%$$

Energy absorbed upto proportionality limit is very small in case of ductile materials. Therefore it is economical to design them for loads upto ultimate load with a load factor for safety.

$$\% \text{ reduction in area} = \frac{(\pi/4)\times(12^2-9^2)}{(\pi/4)\times 12^2}\times 100$$

$$= 43.75\%$$

$$\% \text{ elongation} = \frac{\Delta L}{L}\times 100 = \frac{(7.7-6)}{6}\times 100$$

$$= 28.33\%$$

Example 1.2 In an axial tensile test on 15 mm diameter bar of gauge length 200 mm, the load at proportionality limit is found to be 25 kN and the corresponding changes in length and diameter are 0.25 mm and 0.00625 mm, respectively. Determine (i) modulus of elasticity, (ii) Poisson's ratio and (iii) the % volume change.

Solution At proportionality limit,

$$\text{Stress, } \sigma = \frac{25000}{(\pi/4)\,15^2} = 141.47 \text{ MPa}$$

$$\text{Strain, } \varepsilon = \frac{0.25}{200} = 1.25\times 10^{-3} \text{ mm/mm}$$

Modulus of elasticity,

$$E = \frac{\sigma}{\varepsilon} = \frac{141.47}{1.25\times 10^{-3}} = 113176 \text{ MPa}$$

Lateral strain,

$$\varepsilon_1 = \frac{0.00625}{15} = 4.17\times 10^{-4} \text{ mm/mm}$$

Poisson's ratio,

$$\mu = \frac{\varepsilon_1}{\varepsilon} = \frac{4.17\times 10^{-4}}{1.25\times 10^{-3}} = 0.333$$

Volume change

Initial volume = $(\pi/4)\times 15^2\times 200 = 35343$ mm^3

Final volume = $(\pi/4)\times(15-0.00625)^2\times 200.25$

$= 35357.6$ mm^3

% change in volume

$$= \frac{(35357.6-35343)\times 100}{35343} = 0.041\%$$

Example 1.3 A mild steel bar of cross-sectional area 500 mm^2 is subjected to the forces shown in Fig. 1.13 (a). Determine the total elongation in the bar if E = 200 GPa.

Solution Free bodies $A-B$, $A-C$ and $A-D$ are considered to make load diagram in Fig. 1.13(b) – (e). 50 kN, 35 kN and 45 kN (all tensile) at B, C and D are required for the equilibrium of the free bodies.

$$\frac{\text{Stress}}{\text{Strain}} = \frac{F/A}{\Delta L/L} = E$$

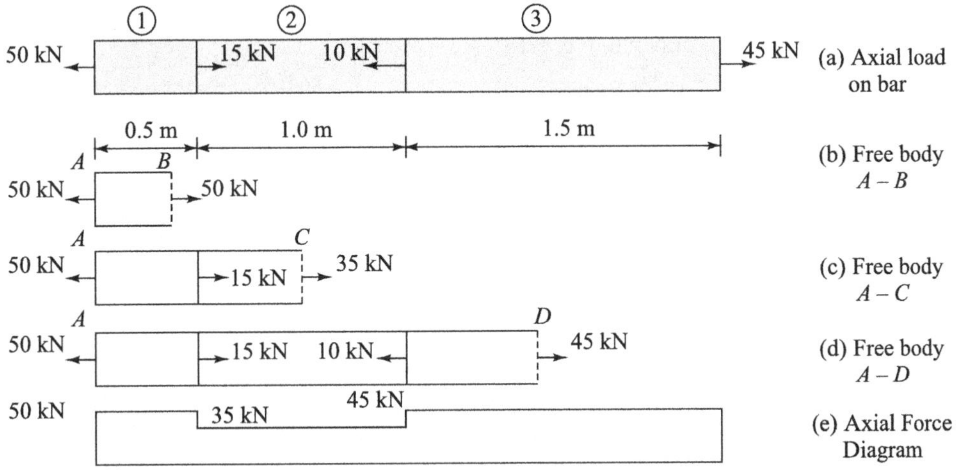

Fig. 1.13

Elongation of a section, $\Delta L = FL/AE$

For section 1, elongation

$$\Delta L_1 = \frac{50000 \times 500}{500 \times 2 \times 10^5} = 0.25 \text{ mm}$$

For section 2, elongation

$$\Delta L_2 = \frac{35000 \times 1000}{500 \times 2 \times 10^5} = 0.35 \text{ mm}$$

For section 3, elongation

$$\Delta L_3 = \frac{45000 \times 1500}{500 \times 2 \times 10^5} = 0.675 \text{ mm}$$

Total elongation $\Delta L = 1.275$ mm

Example 1.4 Determine the total elongation of the stepped steel bar under loads as shown in Fig. 1.14(a), if the modulus of elasticity, $E = 210$ GPa.

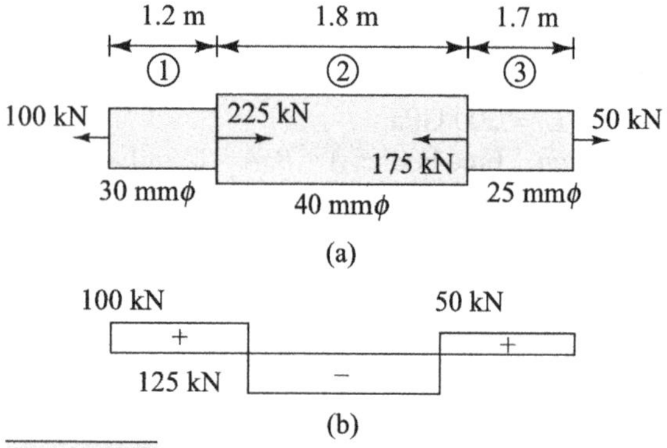

Fig. 1.14

Soluton Axial load diagram is drawn in Fig. 1.14.

Elongation of a section, $\Delta L = FL/AE$

For section 1,

$$\Delta L_1 = \frac{+100000 \times 1200}{(\pi/4) \times 30^2 \times 210000} = +\,0.808 \text{ mm}$$

For section 2,

$$\Delta L_2 = \frac{-125000 \times 1800}{(\pi/4) \times 40^2 \times 210000} = -\,0.853 \text{ mm}$$

For section 3,

$$\Delta L_3 = \frac{+50000 \times 1700}{(\pi/4) \times 25^2 \times 210000} = +\,0.825 \text{ mm}$$

Total elongation, $\Delta L = 0.78$ mm

Example 1.5 A 12 mm diameter steel bar is welded to a rigid plate which is supported by a brass pipe of outside diameter 30 mm and inside diameter 20 mm. The system is loaded as shown in Fig. 1.15. Determine the displacement of lower end of the steel bar and the stress in the steel and brass if the modulii of elasticity of the steel and brass are 210 GPa and 105 GPa, respectively.

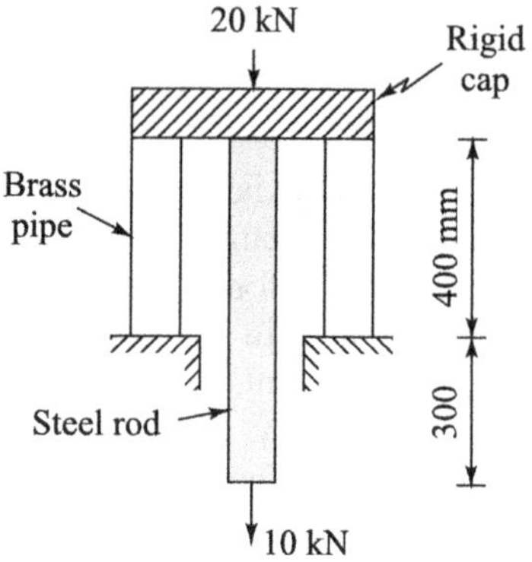

Fig. 1.15

Solution Total load on brass pipe = 10 + 20 = 30 kN

Displacement of brass pipe

$$= \frac{FL}{AE} = \frac{30000 \times 400}{(\pi/4)\,(30^2 - 20^2) \times 105000}$$

$$= 0.291 \text{ mm}$$

Load on steel bar = 10 kN

Displacement in steel bar

$$= \frac{10000 \times 700}{(\pi/4)12^2 \times 210000} = 0.295 \text{ mm}$$

Total displacement of the lower end = 0.586 mm

$$\text{Stress in steel} = \frac{10000}{(\pi/4)12^2} = 88.42 \text{ MPa}$$

$$\text{Stress in brass} = \frac{30000}{(\pi/4)\,(30^2 - 20^2)}$$

$$= 76.39 \text{ MPa}$$

Example 1.6 Determine the elongation of a circular tapered bar of length L under axial load P

if the diameter increases uniformly from D_1 to D_2 as shown in Fig. 1.16.

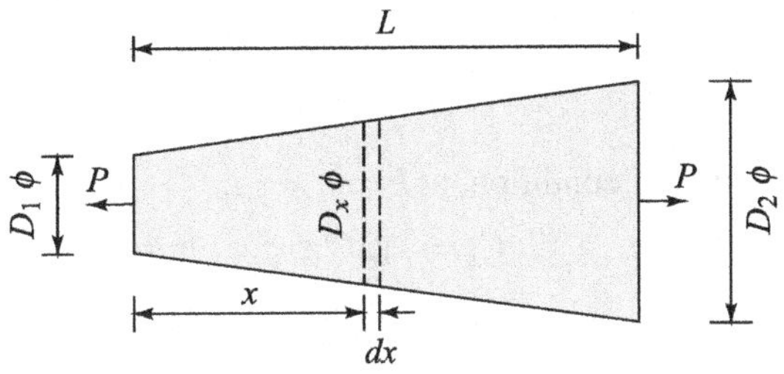

Fig. 1.16

Solution Elongation ds of an elementary section dx is given by,

$$ds = \frac{P\,dx}{E \times A} = \frac{P\,dx}{E(\pi/4)D_x^2}$$

where $D_x = D_1 + (D_2 - D_1) \times x/L = D_1 + k \times x$

Total elongation,

$$\Delta L = \int_0^L ds = \frac{4P}{\pi E}\int_0^L D_x^{-2}dx$$

$$= \frac{4P}{\pi E}\int_0^L (D_1 + k \times x)^{-2}dx$$

$$= \frac{4P}{\pi E}\left|\frac{-(D_1 + k \times x)^{-1}}{k}\right|_0^L = \frac{4PL}{\pi E D_1 D_2}$$

Example 1.7 Determine the total elongation of a uniform bar of area A and length L hanging freely under its self-weight w per unit volume as shown in Fig. 1.17.

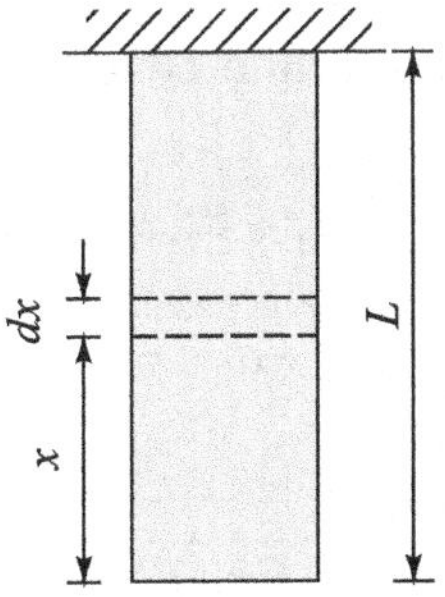

Fig. 1.17

Solution The elongation ds of an elementary section dx under self-weight is given by,

$$ds = \frac{(A \times x \times w)\,dx}{EA}$$

Total elongation, $\Delta L = \int_0^L ds = \frac{w}{E}\int_0^L x\,dx$

or, $$\Delta L = \frac{wL^2}{2E}$$

Example 1.8 Determine the deflection of the free end due to self-weight of vertically hung tapering bar having larger and smaller radius r_2 and r_1, respectively, and length L as shown in Fig. 1.18.

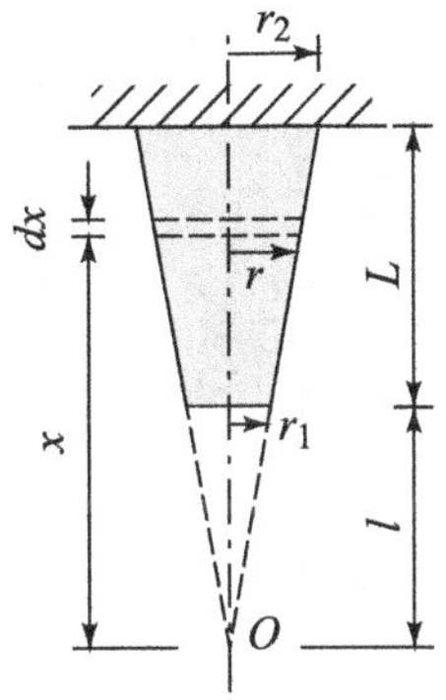

Fig. 1.18

Solution If point O is the vertex of the truncated cone, then self-weight P_x acting on any elementary section d_x is given by,

$$P_x = \frac{w \times \pi}{3}\left(r^2 x - r_1^2 l\right)$$

Extension $d\delta$ of elementary section d_x under self-weight P_x,

$$d\delta = \frac{P_x dx}{EA}$$

where $$A = \pi \times r^2$$

and $$r = \frac{r_1 \times x}{l}$$

Total elongation,

$$\Delta L = \int d\delta = \frac{w \times \pi}{3E}\int_l^{L+l}\frac{(r^2 x - r_1^2 l)\,dx}{\pi \times r^2}$$

or, $$\Delta L = \frac{wL^2 (L + 3l)}{6E (L + l)}$$

but, $$l = \frac{r_1 L}{r_2 - r_1}$$

therefore $$\Delta L = \frac{wL^2 (r_2 + 2r_1)}{6Er_2}$$

Example 1.9 A body having the form of a solid of revolution is under a vertical load P as shown in Fig. 1.19. The radius of a body at top is r_0 and the unit weight of the body is w. If the compressive stress at all horizontal sections are desired to be constant, determine the variation of radius along the axis.

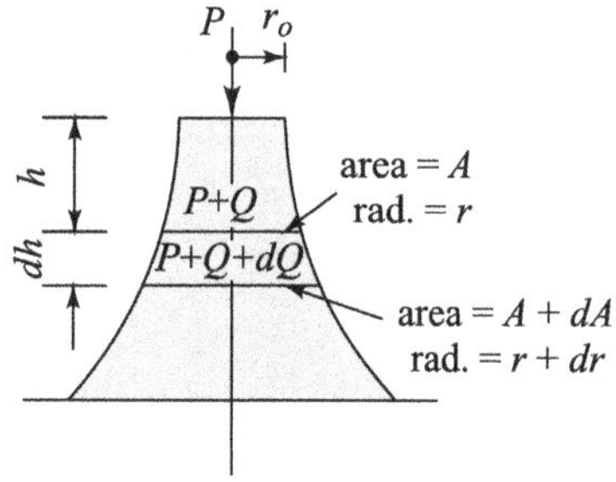

Fig. 1.19

Solution Consider an elementary section of thickness dh at a depth h, having area of cross-section A.

If, Q = self-weight on upper face of elementary section,

and $Q + dQ$ = self-weight on lower face of elementary section,

Then, $$\frac{P + Q}{A} = \frac{P + Q + dQ}{A + dA} = \frac{P}{\pi r_0^2}$$

or, $$\frac{dQ}{dA} = \frac{P}{\pi r_0^2}$$

or, $$\frac{(\pi r^2\ dh)\ w}{2\pi r\ dr} = \frac{P}{\pi r_0^2}$$

Integrating for entire body,

$$\int dh = \frac{2P}{\pi r_0^2 w} \int \frac{dr}{r}$$

$$h = \frac{2P}{\pi r_0^2 w} \ln r + C_1$$

Boundary condition, at $h = 0$, $r = r_0$,

Therefore, $$C_1 = -(2P/\pi r_0^2\ w) \ln r_0$$

or, $$h = (2P/\pi r_0^2\ w) \ln (r/r_0)$$

or, $$r = r_0 e^{\pi r_0^2 wh/2P}$$

Statically Indeterminate Problems—Examples 1.10 through 1.21

Example 1.10 A circular bar of diameter 50 mm and length 3.3 m is fixed rigidly between two supports. An axial load of 216 kN is applied at 0.3 m from one support as shown in Fig. 1.20. Find the stresses in two regions of the bar.

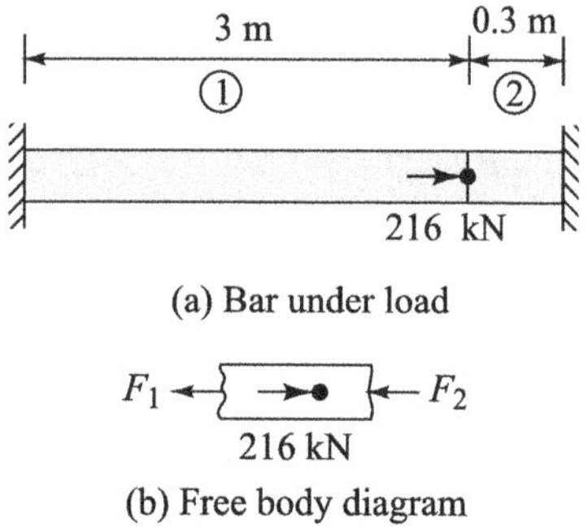

Fig. 1.20

Solution Equilibrium condition of free body diagram gives,

$$F_1 + F_2 = 216 \text{ kN}$$

or, $$\frac{\pi}{4} \times 50^2 \times (\sigma_1 + \sigma_2) = 216000 \text{ N} \qquad \text{(i)}$$

For compatibility,

Extension of region 1 = Compression of region 2

or, $$\Delta L_1 = \Delta L_2$$

$$\text{or,} \quad \frac{(\sigma_1 \times A) \times 3000}{E \times A} = \frac{(\sigma_2 \times A) \times 300}{E \times A}$$

or, $$10\sigma_1 = \sigma_2 \qquad \text{(ii)}$$

From Eqs (i) and (ii),

$$\sigma_1 = 10 \text{ MPa}$$

and $$\sigma_2 = 100 \text{ MPa}$$

Example 1.11 Determine the stresses induced in aluminium and steel portions of the composite bar due to a force of 300 kN applied at the junction as shown in Fig. 1.21. The cross-sectional areas of steel and aluminium are 1250 mm^2 and 2500 mm^2, respectively. Modulii and steel and aluminium are 210 GPa and 70 GPa, respectively.

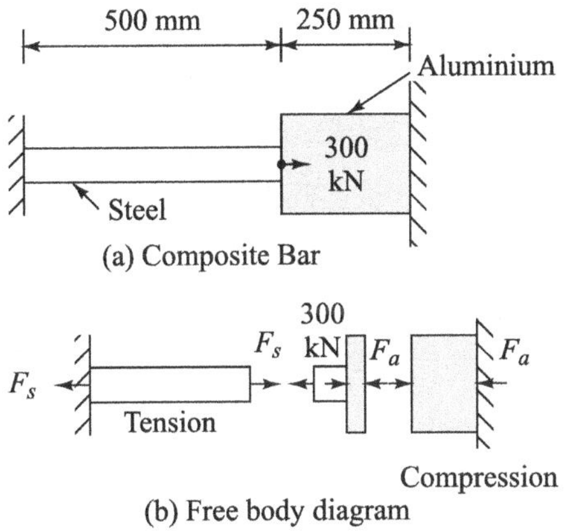

Fig. 1.21

Solution **Equilibrium equation:** From free body diagram,

$$F_s + F_a = 300 \text{ kN} \qquad \text{(i)}$$

Compatibility equation:

Extension in steel = Compression in aluminium

or, $$\Delta L_s = \Delta L_a$$

Substituting stress–strain relation,

$$\frac{F_s \times L_s}{A_s \times E_s} = \frac{F_a \times L_a}{A_a \times E_a}$$

$$\text{or,} \quad \frac{F_s \times 500}{1250 \times 210000} = \frac{F_a \times 250}{2500 \times 70000}$$

or, $$F_s = 0.75\, F_a \qquad \text{(ii)}$$

For Eqs (i) and (ii)

$$F_s = 171.4 \text{ kN}$$

and $$F_a = 128.6 \text{ kN}$$

Stress in steel, $\sigma_s = 102.88$ MPa (Tension)

Stress in aluminium,

$$\sigma_a = 68.56 \text{ MPa (Compression)}$$

Since stresses in steel and aluminium are below their yield stresses, the analysis carried out on the basis of elastic stress-strain relation is valid.

Example 1.12 For composite bar of Example 1.11, calculate the stresses due to increase in temperature alone by 5°C if the coefficients of thermal expansion for steel and aluminium 12 × 10^{-6} mm/mm/°C and and 23.2 × 10^{-6} mm/mm/°C respectively.

Solution **Equilibrium equation:** From free body diagram (Fig. 1.22)

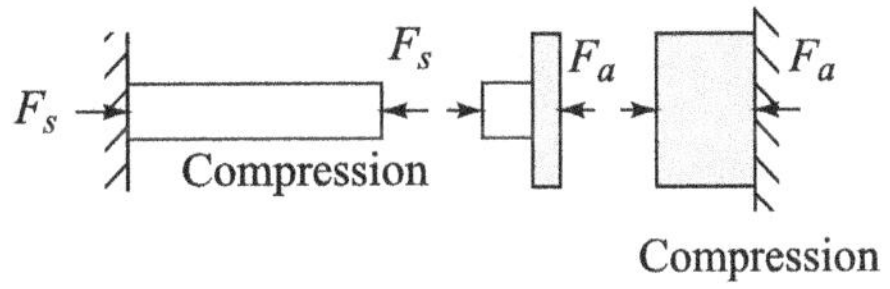

Fig. 1.22

$$F_s = F_a \qquad \text{(i)}$$

Compatibility equation:

Unconstrained elongation of the composite bar

$$= \alpha_s \times L_s \times \Delta t + \alpha_a \times L_a \times \Delta t$$

$$= 12 \times 10^{-6} \times 500 \times 5 + 23.2 \times 10^{-6} \times 250 \times 5$$

$$= 0.03 + 0.029 = 0.059 \text{ mm}$$

Since the bar is restrained, this will cause internal stress and strain. For compatibility of displacements,

$$\Delta L_s + \Delta L_a - 0.059 = 0$$

Substituting stress–strain relation,

$$\frac{F_s \times 500}{1250 \times 210000} + \frac{F_a \times 250}{2500 \times 70000} = 0.059 \qquad \text{(ii)}$$

From Eqs (i) and (ii), $F_s = F_a = 17.7$ kN

Stress in steel, $\sigma_s = 17700/1250$

$= 14.16$ MPa (Compression)

Stress in aluminium,

$\sigma_a = 17700/2500$

$= 7.08$ MPa (Compression)

Example 1.13 A bar of length L and cross-sectional area A is fixed to a rigid support at one end and to a spring of stiffness k at the other as shown in the Fig. 1.23(a). Derive a formula for stress in the bar caused by an increase in temperature Δt if coefficient of thermal expansion is α for the bar.

Solution **Equilibrium equation:** From free body diagram Fig. 1.23(b),

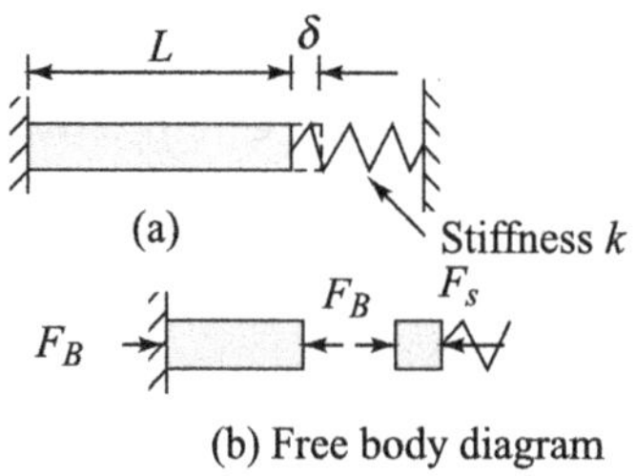

Fig. 1.23

$$F_{\text{bar}} = F_{\text{spring}}$$

or, $$\sigma \times A = k \times \delta \qquad \text{(i)}$$

Compatibility condition:

Total strain = Temperature strain–strain relieved

$$\frac{\sigma}{E} = \frac{\alpha L \Delta t}{L} - \frac{\delta}{L} \qquad \text{(ii)}$$

From Eqs (i) and (ii),

$$\sigma = \frac{E\alpha\Delta t}{1 + \dfrac{EA}{kL}}$$

Example 1.14 A steel bolt of cross-sectional area 600 mm^2 is fitted snugly in an 900 mm^2 aluminium collar. The pitch of the bolt is 2 mm. Determine the stresses in steel bolt and aluminium collar if the bolt is given 1/8th turn. Take moduli of steel and aluminium as 210 GPa and 70 GPa, respectively.

Solution **Equilibrium condition:** Since no external force is acting at the final position of the nut, the equilibrium of internal force in free diagram [Fig. 1.24(d)] gives,

$$F_s = F_a \qquad \text{(i)}$$

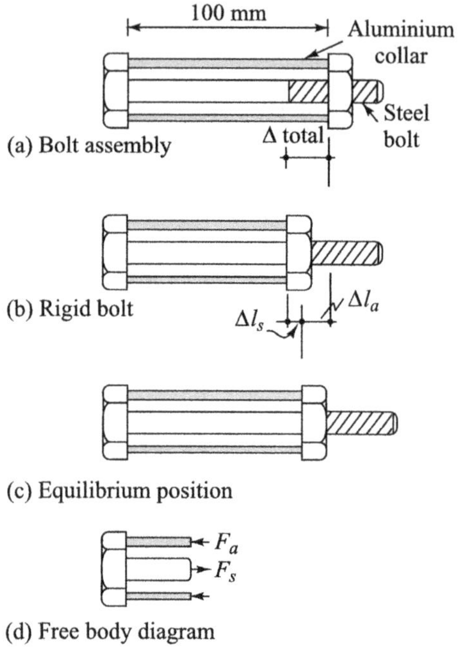

Fig. 1.24

Compatibility condition: First consider the bolt is rigid and only the aluminium collar gets compressed [Fig.1.24(b)]. Next consider the bolt is also elastic and the collar and bolt move together to the final position as shown in Fig. 1.24(c) such that,

$$\Delta L_s + \Delta L_a = \Delta L_{\text{total}} \qquad \text{(ii)}$$

Substituting stress–strain relation in Eq. (ii),

$$\frac{F_s \times L_s}{A_s \times E_s} + \frac{F_a \times L_a}{A_a \times E_a} = \frac{2}{8} \text{ mm} \qquad \text{(iii)}$$

or, $$\frac{F_s \times 100}{600 \times 210000} + \frac{F_a \times 100}{900 \times 70000} = \frac{2}{8}$$

or, $$F_s = F_a = 105 \text{ kN}$$

Stress in steel, σ_s = 105000/600

= 175 MPa (Tension)

Stress in aluminium,

σ_a = 105000/900

= 116.7 MPa (Compression)

Example 1.15 Determine the stresses induced in steel bolt and aluminium collar assembly of Example 1.14 if the temperature also increases by 50 °C. Take the coefficient of thermal expansion for steel and aluminium to be $\alpha_s = 18 \times 10^{-6}$ mm/mm/°C and $\alpha_s = 23 \times 10^{-6}$ mm/mm/°C, respectively.

Solution First consider the temperature stresses only.

Equilibrium condition: Since no external force is acting at the final position of the assembly, the equilibrium of internal forces in free body diagram [Fig. 1.25(d)] gives,

$$F_s = F_a \qquad \text{(i)}$$

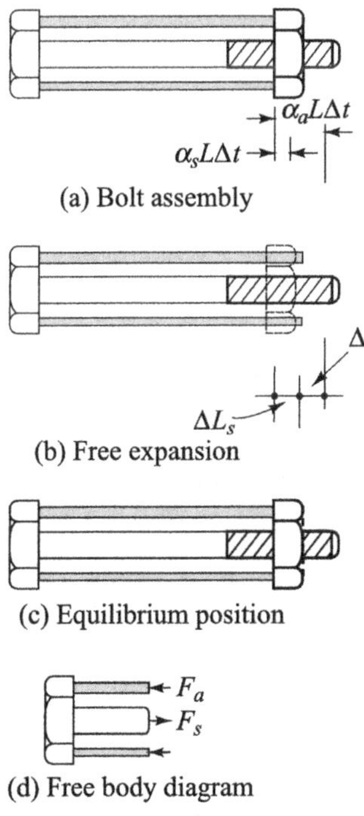

(a) Bolt assembly

(b) Free expansion

(c) Equilibrium position

(d) Free body diagram

Fig. 1.25

Compatibility condition: Fig. 1.25(b) shows the free expansion of the steel bolt and aluminium collar. The expansion of aluminium collar will be more than steel bolt since its coefficient of thermal expansion is higher. Thus at equilibrium aluminium collar will shorten by an amount ΔL_a and steel bolt will extend by an amount ΔL_s as shown in Fig. 1.25(c) such that,

$$\Delta L_s + \Delta L_a = (\alpha_a - \alpha_s) \times L \times \Delta t \qquad \text{(ii)}$$

Substituting stress-strain relation in Eq. (ii),

$$\frac{F_s \times L_s}{A_s \times E_s} + \frac{F_a \times L_a}{A_a \times E_a} = (\alpha_a - \alpha_s) \times L \times \Delta t \qquad \text{(iii)}$$

or,
$$\frac{F_s}{600 \times 210000} + \frac{F_a}{900 \times 70000} = (23 - 18) \times 10^{-6} \times 50$$

or, $F_s = F_a$ = 10500 kN

Stress in steel, σ_s = 10500/600

= 17.5 MPa (Tension)

Stress in aluminium,

σ_a = 10500/900

= 11.67 MPa (Compression)

Combined stresses: If the external loads are also acting with rise in temperature, the mechanical and thermal stress will superimpose as follows.

Stress in steel, $\sigma_s = \sigma_{s,\text{mech}} + \sigma_{s,\text{therm}}$

= 175 + 17.5

= 192.5 MPa (Tension)

Stress in aluminium,

$\sigma_a = \sigma_{a,\text{mech}} + \sigma_{a,\text{therm}}$

= 116.67 + 11.67

= 128.34 MPa (Compression)

Example 1.16 If an external compressive load of 60 kN is applied on the same bolt and aluminium collar assembly of Example 1.15, calculate the stresses in steel bolt and aluminium collar.

Solution First consider the external load alone.

Equilibrium condition: Consider the free body diagram and assuming the direction of force as shown in Fig. 1.26,

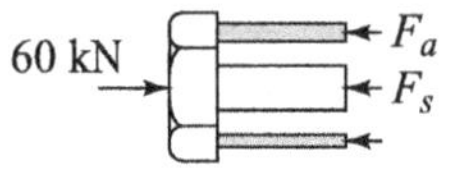

Fig. 1.26

$$F_s + F_a = 60 \times 10^3 \text{ N} \quad \text{(i)}$$

Compatibility condition: The displacement of steel bolt and aluminium collar will be equal under external load.

$$\Delta L_s = \Delta L_a \quad \text{(ii)}$$

Substituting stress–strain relation in Eq. (ii),

$$\frac{F_s \times L_s}{A_s \times E_s} = \frac{F_a \times L_a}{A_a \times E_a} \quad \text{(iii)}$$

or,
$$F_s = \frac{600 \times 210 \times 10^3}{900 \times 70 \times 10^3} F_a = 2F_a$$

Substituting in Eq. (i)

$$F_a = 60 \times 10^3/3 = 2 \times 10^4 \text{ N}$$

and
$$F_s = 4 \times 10^4 \text{ N}$$

Therefore, stresses due to external load of 60 kN acting alone,

$$\sigma_s = 4 \times 10^4/600 = 66.67 \text{ MPa (Compression)}$$
$$\sigma_s = 2 \times 10^4/900 = 22.22 \text{ MPa (Compression)}$$

Total stresses due to tightening of nut + rise in temperature + external load,

$$\sigma_s = 192.5 - 66.67$$
$$= 125.83 \text{ MPa (Tension)}$$
$$\sigma_s = 128.34 + 22.22$$
$$= 150.56 \text{ MPa (Compression)}$$

Example 1.17 A reinforced concrete column of 450 mm square has four steel bars of 30 mm diameter as shown in Fig. 1.27. It carries as load of 600 kN. Find the stresses in steel and concrete if their modulii of elasticity are 210 GPa and 22 GPa, respectively. Also calculate the adhesive force concrete and steel.

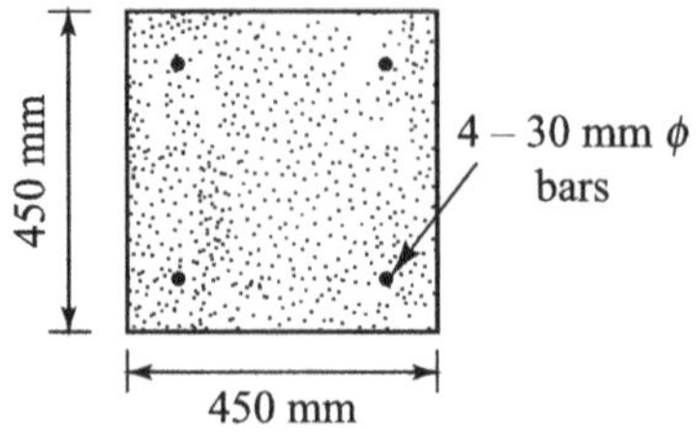

Fig. 1.27

Solution Area of steel bars,

$$A_s = 4 \times (\pi/4) \times 30^2 = 2827.4 \text{ mm}^2$$

Area of concrete,

$$A_c = 450 \times 450 - 2827.4 = 199672.6 \text{ mm}^2$$

Equilibrium condition: External load is shared by load in steel bars F_s and load in concrete F_c as follows.

$$F_s + F_c = 600 \times 10^3 \text{ N} \quad \text{(i)}$$

Compatibility condition: Steel and concrete will have equal extension under the load.

$$\Delta L_s = \Delta L_c \quad \text{(ii)}$$

Substituting stress–strain relation (assuming to be linear) in Eq. (ii),

$$\frac{F_s \times L_s}{A_s \times E_s} = \frac{F_c \times L_c}{A_c \times E_c}$$

or,
$$F_s = \frac{210 \times 2827.4}{22 \times 199672.6} F_c$$
$$F_s = 0.135 \ F_c$$

Substituting in Eq. (i),

$$F_c = 528557.4 \text{ N}$$

and
$$F_s = 71442.6 \text{ N}$$

Stress in steel, $\sigma_s = 25.27$ MPa

Stress in concrete, $\sigma_c = 2.65$ MPa

External load/unit area

$$= \frac{600 \times 10^3}{450 \times 450} = 2.96 \text{ MPa}$$

Additional stress in steel

$$= 25.27 - 2.96 = 22.31 \text{ MPa}$$

Therefore, adhesive force required between steel and concrete = $22.31 \times 2827.4 = 63 \times 10^3$ N

Example 1.18 Determine the central transverse load W which will produce central deflection δ in the initially straight bar AB of area A and length $2L$ hinged at both ends as shown in Fig. 1.28.

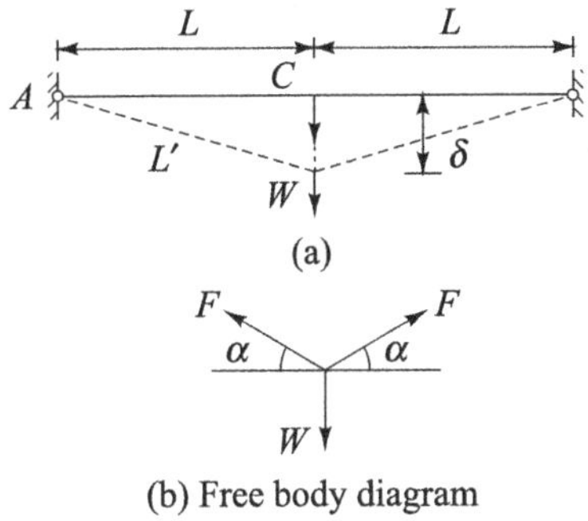

Fig. 1.28

Solution **Equilibrium condition:** The forces in section AC and CB will be equal due to symmetry. Therefore, from free body diagram,

$$W = 2F \sin \alpha = 2F \times \frac{\delta}{L'} \quad \text{(i)}$$

Compatibility condition: Elongation of section AC is given by,

$$\Delta L = L' - L = \frac{FL}{AE} \quad \text{(ii)}$$

From Eqs (i) and (ii),

$$W = \frac{2(L' - L)AE}{L} \times \frac{\delta}{L'} = \frac{2\delta AE}{L}\left(1 - \frac{L}{L'}\right)$$

$$= \frac{2\delta AE}{L}\left(1 - \frac{1}{\sqrt{L^2 + \delta^2}}\right)$$

Example 1.19 Three rods are hinged together to support a vertical load W as shown in Fig. 1.29. Find the force carried by each rod if A, L and E denote their respective cross-sectional area, length and modulus of elasticity. Outer rods AD and CD are identical.

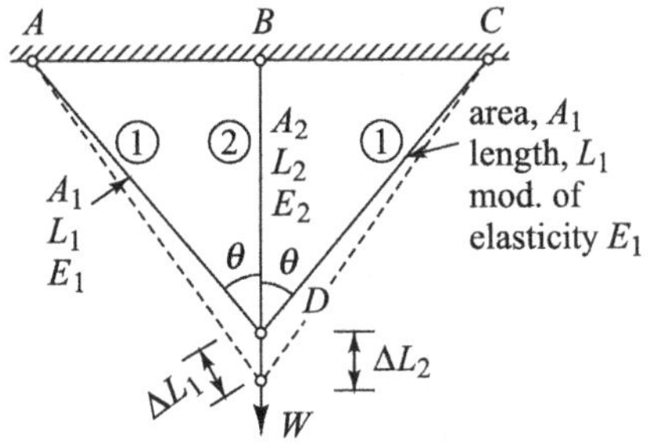

Fig. 1.29

Solution **Equilibrium condition:** For static equilibrium of joint D,

$$\sum F_x = 0: \quad F_{AD} \sin \theta = F_{CD} \sin \theta$$

or,
$$F_{AD} = F_{CD} = F_1$$

$$\sum F_y = 0: \quad 2F_1 \cos \theta + F_2 = W \quad \text{(i)}$$

Compatibility condition: The deformed shape of three members is shown by dotted line. It is assumed that the angle θ does not change appreciably, then the compatibility equation,

$$\Delta L_1 = \Delta L_2 \cos \theta \quad \text{(ii)}$$

or,
$$\frac{F_1 L_1}{A_1 E_1} = \frac{F_2 L_2}{A_2 E_2} \cos \theta$$

But,
$$L_2 = L_1 \cos \theta$$

Therefore,
$$F_1 = F_2 \frac{A_1 E_1}{A_2 E_2} \cos^2 \theta \quad \text{(iii)}$$

From Eqs (i) and (ii),

$$F_2 = \frac{W}{1 + \frac{2A_1 E_1}{A_2 E_2} \cos^3 \theta}$$

and
$$F_1 = \frac{W \cos^2 \theta}{\frac{A_2 E_2}{A_1 E_1} + 2\cos^3 \theta}$$

Example 1.20 A rigid weightless horizontal bar AB is supported by three bars of steel, brass and copper as shown in Fig. 1.30. The temperature of the bars is increases by 15 K and an external load of 100 kN is also applied such that the bar remains horizontal. Determine the stresses in the bars and the location of 100 kN force if the properties of the bars are as follows.

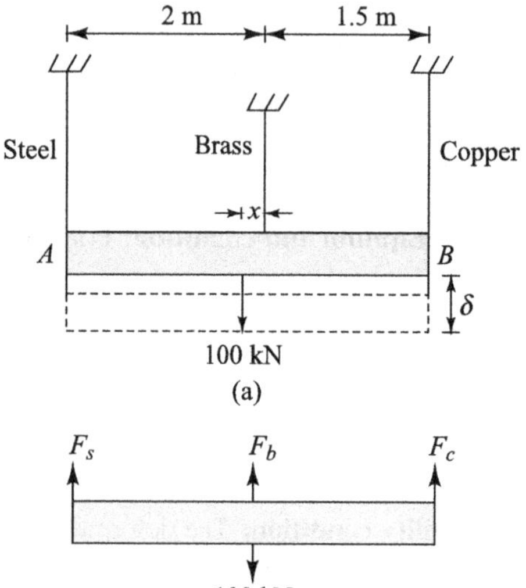

(b) Free body diagram

Fig. 1.30

Bar	*Area (mm²)*	*Length (mm)*	*Modulus of elasticity (GPa)*	*Coefficient of expansion (mm/mm/K)*
Steel	100	2000	200	12×10^{-6}
Brass	300	1500	90	20×10^{-6}
Copper	200	1750	120	18×10^{-6}

Solution **Equilibrium condition:** Let the location of 100 kN load be at a distance x from the brass bar, static equilibrium of free body diagram gives,

$$\sum F = 0: \quad F_s + F_b + F_c - 100 \times 10^3 = 0 \quad \text{(i)}$$

$\sum M = 0$: Taking moment about end B,

$$F_s \times 3500 + F_b \times 1500 - 100 \times 10^3 \times (x + 1500) = 0 \quad \text{(ii)}$$

Compatibility condition: Total elongation of all bars due to external load and temperature rise is equal and is denoted by δ.

For steel bar,

$$\frac{F_s \times 2000}{100 \times 200 \times 10^3} + 12 \times 10^{-6} \times 15 \times 2000 = \delta \quad \text{(iii)}$$

For brass bar,

$$\frac{F_b \times 1500}{300 \times 90 \times 10^3} + 20 \times 10^{-6} \times 15 \times 1500 = \delta \quad \text{(iv)}$$

For copper bar,

$$\frac{F_c \times 1750}{200 \times 120 \times 10^3} + 18 \times 10^{-6} \times 15 \times 1750 = \delta \quad \text{(v)}$$

Solving five unknowns in five Eq. (i) through (v),

$\delta = 2.83$ mm, $F_s = 24730$ N,

$F_b = 42894$ N, $F_c = 32372.6$ N and

$x = 8.96$ mm.

Stress in steel bar = 247.3 MPa,

Stress in brass bar = 142.98 MPa,

Stress in copper bar = 161.86 MPa.

Example 1.21 A rigid horizontal bar is supported by three equally spaced bars of same cross-sectional area as shown in Fig. 1.31 (a). The length of middle bar is half of the outer bars. Find the ultimate load P_u on the horizontal rigid bar if the load is applied at the center. Draw load–deflection curve.

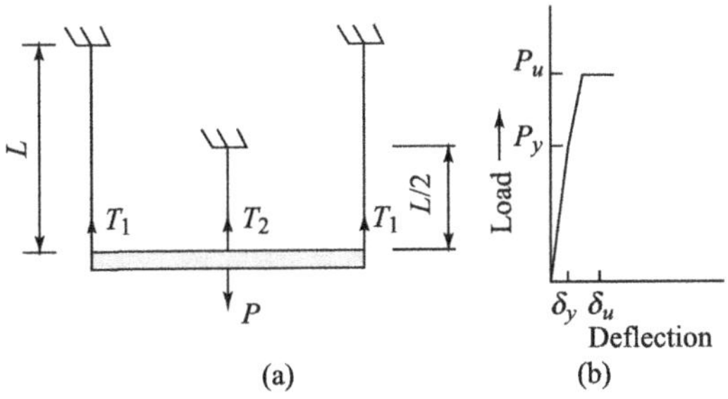

Fig. 1.31

Solution Let T_1 and T_2 = tension in outer and inner bars, respectively

Equilibrium condition:

$$P = 2 \times T_1 + T_2 \qquad \text{(i)}$$

(1) *Elastic stage:*

Compatibility condition: At any stage, the elonga-tions of all the bars are equal,

or,
$$\frac{T_1 L_1}{AE} = \frac{T_2 L_2}{AE} \qquad \text{(ii)}$$

From Eqs (i) and (ii),

$$P = 2 \times T_2 = 4 \times T_1 \qquad \text{(iii)}$$

Since $T_2 > T_1$, the middle bar will reach the yield stress earlier than the outer and the corresponding load P is called the yield load P_y.
Thus at yield stage Eq. (iii) gives,

$$P_y = 2T_2 = 2\sigma_y A$$

And load deflection at yield stage,

$$\delta_y = \frac{T_2 L_2}{AE} = \frac{\sigma_y AL/2}{AE} = \frac{\sigma_y L}{2E}$$

(2) *Plastic stage:*

At plastic stage stress in all three bars are at yield stress and the corresponding load P is called the ultimate load P_u.
Equilibrium condition at plastic stage gives,

$$P_u = 2 \times T_1 + T_2 = 2\sigma_y A + \sigma_y A = 3\sigma_y A$$

Load deflection at plastic stage

$$\delta = \frac{\sigma_y L}{E}$$

Figure 1.31(b) shows load–deflection curve.

Generalized Hooke's Law—Examples 1.22 to 1.26

Example 1.22 A thin rectangular steel plate 100 mm by 50 mm undergoes deformations 0.05 mm and 0.01 mm in x and y directions due to stresses σ_x and σ_y as shown in Fig. 1.32. Determine the magnitude of σ_x and σ_y if modulus of elasticity is 200 GPa and Poisson's ratio is 0.3.

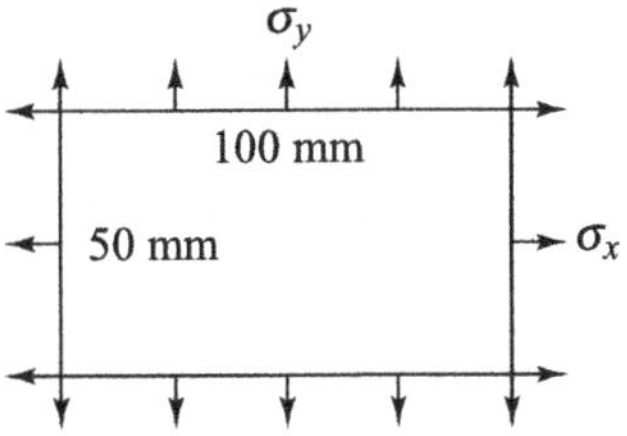

Fig. 1.32

Solution It is plane stress problem where,

$$\varepsilon_x = \frac{0.05}{100} = 0.005 \text{ mm/mm, and}$$

$$\varepsilon_y = \frac{0.01}{50} = 0.0002 \text{ mm/mm}$$

Substituting in Eq. (1.231),

$$\sigma_x = \frac{E}{1-\mu^2}\left(\varepsilon_x + \mu\varepsilon_y\right)$$

$$= \frac{200000}{1-0.3^2}(0.0005 + 0.3 \times 0.0002)$$

$$= 123.07 \text{ MPa}$$

$$\sigma_y = \frac{E}{1-\mu^2}(\varepsilon_y + \mu\varepsilon_x)$$

$$= \frac{200000}{1-0.3^2}(0.0002 + 03 \times 0.0005)$$

$$= 76.92 \text{ MPa}$$

Example 1.23 Figure 1.33 shows metal cubes A, B and C of side 100 mm in direct contact resting on rigid base and confined in x-direction between two rigid-end plates. Young's modulus of elasticity and Poisson's ratio for cubes A and C are 150 GPa and 0.25 respectively and for cube B are 200 GPa and 0.3 respectively.

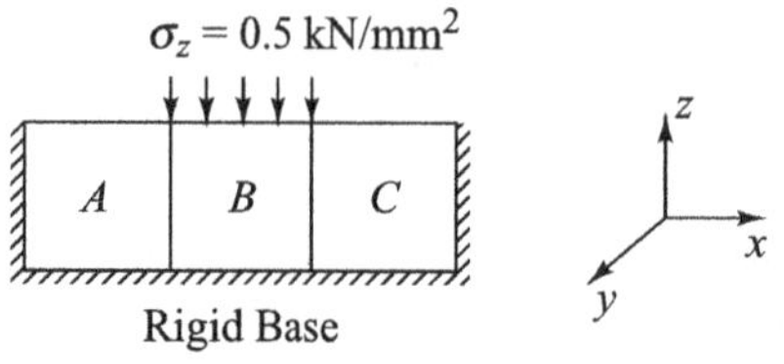

Fig. 1.33

If the upper face of the centre cube B is subjected to uniform compressive stress of 0.5 kN/mm^2, compute for cube B, the following:

(i) the direct stress σ_x in x-direction,
(ii) the direct strain in x, y and z direction,
(iii) the volumetric strain.

Solution Using suffix 1 for blocks A and C and suffix 2 for block B,

Equilibrium condition: $\sigma_{x2} = \sigma_{x1} = \sigma_x$ (comp.)

Compatibility condition:

Displacement of block B

= Total displacement of blocks A and C

or, $\Delta L_{x2} = 2\Delta L_{x1}$

or, $\varepsilon_{x2} = 2\varepsilon_{x1}$ (i)

Assuming that there is no friction at any interface, Hooke's law gives extension for block B,

$$\varepsilon_{x2} = \frac{1}{E_2}(-\sigma_x + \mu \times \sigma_z)$$

Contraction for blocks A and C,

$$\varepsilon_{x1} = \frac{\sigma_x}{E_1}$$

Substituting values of ε_{x1} and ε_{x2} in Eq. (i),

$$\frac{1}{E_2}(-\sigma_x + \mu \times \sigma_z) = 2\frac{\sigma_x}{E_1}$$

or, $$\frac{1}{200}(-\sigma_x + 0.3 \times 0.5) = \frac{2\sigma_x}{150}$$

or, $$\sigma_x = 0.09 \text{ kN/mm}^2 \text{ (comp.)}$$

For block B,

$$\varepsilon_x = \frac{1}{E}(-\sigma_x + \mu \times \sigma_z)$$

$$= \frac{1}{200}(-0.09 + 0.3 \times 0.5) = 0.0003$$

$$\varepsilon_y = \frac{-\mu}{E}(\sigma_x + \sigma_z) = -\frac{0.3}{200}(-0.09 - 0.5)$$
$$= 0.000885$$

$$\varepsilon_z = \frac{1}{E}(-\sigma_z + \mu \times \sigma_x)$$

$$= \frac{1}{200}(-0.5 + 0.3 \times 0.09) = -0.002365$$

and volumetric strain,

$$\varepsilon_v = \varepsilon_x + \varepsilon_y + \varepsilon_z$$
$$= 0.0003 + 0.000885 - 0.002365 = -0.00118$$

Example 1.24 A steel bar 50 mm × 50 mm × 300 mm is under compressive load of 100 kN applied in the direction of its length. Find the changes in length and volume if all lateral strains are prevented. Also find the apparent modulus of elasticity if $E = 200$ GPa and $\mu = 0.3$.

Solution σ_x, σ_y and σ_z are compressive and therefore taken –ve in Eq. (1.23),

$$\varepsilon_x = \frac{1}{E}[-\sigma_x + \mu(\sigma_y + \sigma_z)]$$

$$\varepsilon_y = \frac{1}{E}[-\sigma_y + \mu(\sigma_z + \sigma_x)] = 0$$

$$\varepsilon_z = \frac{1}{E}[-\sigma_z + \mu(\sigma_y + \sigma_x)] = 0$$

By symmetry, $\sigma_y = \sigma_z = \sigma$

or, $$-\sigma + 0.3(\sigma_x + \sigma) = 0$$

or, $$\sigma = \frac{0.3\sigma_x}{0.7} = \frac{0.3 \times 100000}{0.7 \times 50 \times 50} = 17.14 \text{ N/mm}^2$$

which gives,

$$\varepsilon_x = \frac{1}{200000}(-40 + 0.3 \times 2 \times 17.14)$$
$$= -0.00014858 \text{ mm/mm}$$

or, $\Delta L = -0.00014858 \times 300 = -0.044574$ mm

Apparent modulus of elasticity,

$$E_{m,x} = \frac{\sigma_x}{\varepsilon_x} = \frac{40}{0.00014858} = 269215 \text{ N/mm}^2$$

Change in volume

$$= \Delta L \times A = -0.0446 \times 50 \times 50$$
$$= 111.5 \text{ mm}^3$$

Example 1.25 Derive the differential equation for equilibrium in polar coordinates.

Solution Figure *E*. 1.34 shows an element under radial stress σ_r and tangential stress σ_θ, respectively. The shear stress $\tau_{\theta r}$ is equal to $\tau_{r\theta}$ as derived earlier in Eq. (1.17). Considering equilibrium of forces in radial direction,

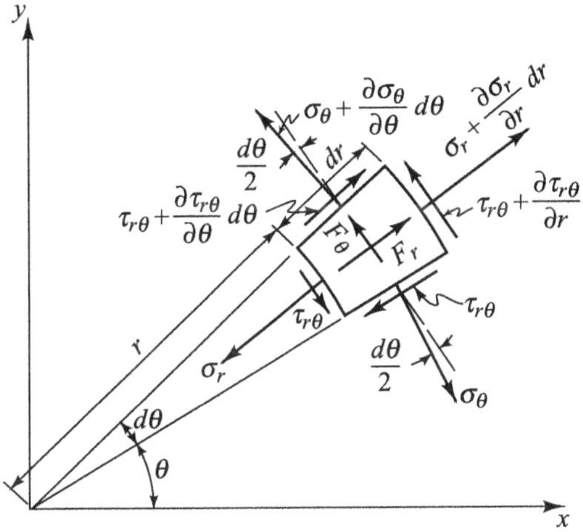

Fig. 1.34

$$-\sigma_r r\, d\theta + \left(\sigma_r + \frac{\partial \sigma_r}{\partial r} dr\right)(r + dr)d\theta - \sigma_\theta\, dr$$
$$\sin\frac{d\theta}{2} - \left(\sigma_\theta + \frac{\partial \sigma_\theta}{\partial \theta} d\theta\right) dr \sin\frac{d\theta}{2} - \tau_{r\theta}\, dr$$
$$\cos\frac{d\theta}{2} + \left(\tau_{r\theta} + \frac{\partial \tau_{r\theta}}{\partial \theta} d\theta\right) dr \cos\frac{d\theta}{2} + F_r\, r\, dr\, d\theta$$
$$= 0$$

Neglecting higher order differential terms, and taking,

$$\sin\frac{d\theta}{2} \approx \frac{d\theta}{2} \text{ and } \cos\frac{d\theta}{2} \approx 1,$$

$$\frac{\partial \sigma_r}{\partial r} + \frac{1}{r}\frac{\partial \tau_{r\theta}}{\partial \theta} + \frac{\sigma_r - \sigma_\theta}{r} + F_r = 0 \qquad \text{(i)}$$

Now consider equilibrium of forces in tangential direction,

$$-\tau_{r\theta} r\, d\theta + \left(\tau_{r\theta} + \frac{\partial \tau_{r\theta}}{\partial r} dr\right)(r + dr)\, d\theta -$$
$$\sigma_\theta\, dr \cos\frac{d\theta}{2} + \left(\sigma_\theta + \frac{\partial \sigma_\theta}{\partial \theta} d\theta\right) dr \cos\frac{d\theta}{2} +$$
$$\tau_{r\theta}\, dr \sin\frac{d\theta}{2} + \left(\tau_{r\theta} + \frac{\partial \tau_{r\theta}}{\partial \theta} d\theta\right) dr \sin\frac{d\theta}{2} +$$
$$F_\theta r\, dr\, d\theta = 0$$

which reduces to,

$$\frac{1}{r}\frac{\partial \sigma_\theta}{\partial \theta} + \frac{\partial \tau_{r\theta}}{\partial r} + 2\frac{\tau_{r\theta}}{r} + F_\theta = 0 \qquad \text{(ii)}$$

Example 1.26 Derive strain-displacement relations in polar coordinates.

Solution An element *ABCD* shown in Fig. 1.35, deforms to *A′ B′ C′ D′*. The radial and tangential components of displacements are represented by *u* and *v* respectively.

Considering side *AB*, the radial strain ε_r is given by,

$$\varepsilon_r = \frac{\left(u + \frac{\partial u}{\partial r} dr - u\right)}{dr} = \frac{\partial u}{\partial r} \qquad \text{(i)}$$

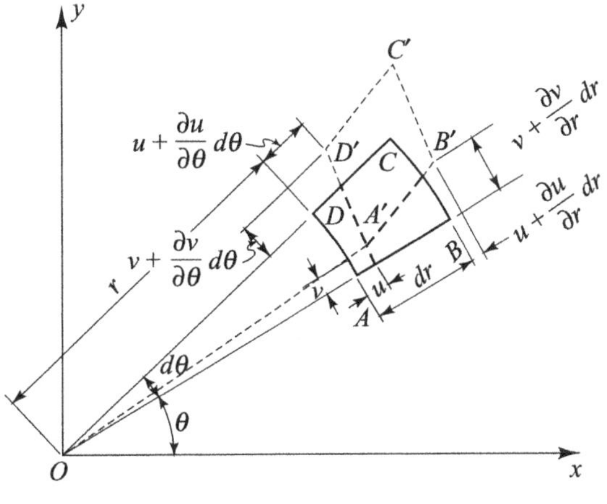

Fig. 1.35

For tangential strain ε_θ, one part due to radial displacement is given by,

$$\frac{(r+u)\,d\theta - r\,d\theta}{r\,d\theta} = \frac{u}{r}$$

Another part of tangential strain ε_θ, due to radial displacement of D to D' is given by,

$$\frac{v + \dfrac{\partial v}{\partial \theta} d\theta - v}{r\,d\theta} = \frac{1}{r}\frac{\partial v}{\partial \theta}$$

Therefore the total tangential strain,

$$\varepsilon_\theta = \frac{u}{r} + \frac{1}{r}\frac{\partial v}{\partial \theta} \qquad \text{(ii)}$$

For shearing, angular displacement of side AD,

$$\frac{\dfrac{\partial u}{\partial \theta} d\theta}{r\,d\theta} = \frac{1}{r}\frac{\partial u}{\partial \theta}$$

Similarly angular displacement of side AB,

$$\frac{\dfrac{\partial v}{\partial r} dr}{dr} = \frac{\partial v}{\partial r}$$

However side $A'B'$ has an initial rigid body displacement v/r, which does not contribute to deformation. Thus the total shearing deformation is given by,

$$\gamma_{r\theta} = \frac{1}{r}\frac{\partial u}{\partial \theta} + \frac{\partial v}{\partial r} - \frac{v}{r} \qquad \text{(iii)}$$

Example 1.27 Determine the maximum height from which a weight of 25 N shown in Fig. 1.36 can be dropped onto the flange attached to a 10 mm diameter rod of length 1 m if the maximum stress is not to exceed 150 MPa. Take $E = 200$ GPa.

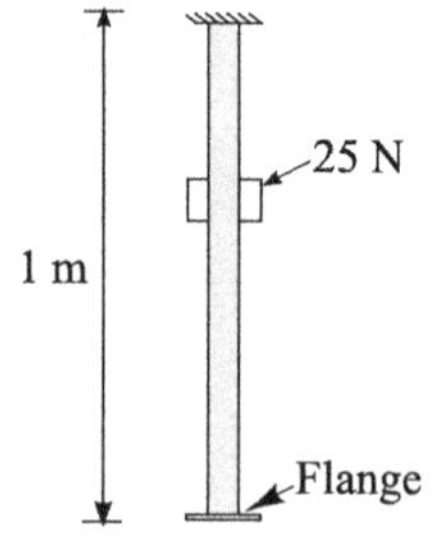

Fig. 1.36

Solution Using Eq. 1.28,

$$W\left(h + \frac{\sigma_{max} L}{E}\right) = \frac{\sigma_{max}^2}{2E} \times AL$$

where, $\sigma_{max} = 150$ MPa, $W = 25$ N

$L = 1000$ mm, $E = 200000$ MPa

$$A = \frac{\pi}{4} \times 10^2 = 78.54 \text{ mm}^2$$

or, $$h = \frac{150^2 \times 78.54 \times 1000}{2 \times 200000 \times 25} - \frac{150 \times 1000}{200000}$$

$= 175.96$ mm

Example 1.28 A concrete pile 6 m long has a diameter 300 mm. It is driven into the ground using a hammer of 5 kN. Calculate the height of fall of the hammer if the maximum instantaneous stress the concrete can withstand is 20 N/mm^2. Take the modulus of elasticity of concrete as 25000 N/mm^2.

Solution Using Eq. (1.28)

$$W\left(h + \frac{\sigma_{max} L}{E}\right) = \frac{\sigma_{max}^2}{2E} \times AL$$

$$h = \frac{20^2}{2 \times 25000} \times \frac{\pi}{4} \times \frac{300^2 \times 6000}{5000} - \frac{20 \times 6000}{25000}$$

$= 673.8$ mm

Example 1.29 A weight of 200 N is dropped at mid span of a simply supported beam from a height of 0.5 m. Determine the instantaneous maximum deflection and maximum stress if the span of the beam is 3 m and the cross-section of beam is 10 cm wide and 20 cm deep. Take $E = 200$ GPa.

Solution The deflection at midspan due to statically applied load is as follows.

$$\delta_{st} = \frac{WL^2}{48EI} = \frac{200 \times 3000^2}{48 \times 200000 \times 100 \times 200^3/12}$$

$= 0.0844$ mm

Impact factor

$$= 1 + \sqrt{1 + \frac{2h}{\delta_{st}}} = 1 + \sqrt{1 + \frac{2 \times 500}{0.0844}}$$

$$= 109.85$$

$$\delta_{max} = 109.85 \times 0.0844 = 9.27 \text{ mm}$$

Maximum bending moment at mid span due to static load 200 N at midspain,

$$M = \frac{WL}{4} = \frac{200 \times 3000}{4} = 150{,}000 \text{ N. mm,}$$

Maximum bending stress at mid span due to static load,

$$\delta_{st} = \frac{My}{I} = \frac{150000 \times 100}{100 \times 200^3 / 12} = 0.225 \text{ MPa}$$

Maximum stress due to dynamic load,

$$\delta_{max} = 0.225 \times 109.85 = 24.7 \text{ MPa}$$

Example 1.30 Redo Example 1.29 if the beam is supported on springs of stiffness, $k = 200$ kN/m.

Solution Static deflection of the beam,

$$\sigma_{st} = 0.0844 + \frac{100}{200} = 0.5844 \text{ mm}$$

$$\text{Impact factor} = 1 + \sqrt{1 + \frac{2 \times 500}{0.5844}} = 42.38$$

$$\delta_{max} = 0.5844 \times 42.38 = 24.76 \text{ mm}$$

$$\sigma_{max} = 0.225 \times 42.38 = 9.53 \text{ MPa}$$

EXERCISES

PROBLEMS

1.1 In an axial tension test on a circular steel specimen of diameter 17.5 mm and gauge length 200 mm, the load–deflection observations are as follows.

Plot the stress–strain diagram and determine (i) the Young's modulus of elasticity, (ii) the modulus of resilience, (iii) the modulus of toughness, (iv) percentage elongation and (v) percentage reduction in area if the gauge length at fracture is 256 mm and the diameter at fractured section is 12 mm.

Table 1.2 *Sample*

Load (kN)	*Deflection (mm)*	*Load (kN)*	*Deflection (mm)*	*Load (kN)*	*Deflection (mm)*
13.0	0.05	72.0	0.3	112.5	25.0
25.0	0.1	69.5	0.325	114.0	30.0
37.0	0.15	68.5	0.35	114.0	35.0
43.0	0.175	68.5	0.5	105.0	50.0
49.0	0.2	86.0	2.5	81.5	55.0
60.5	0.25	100.5	5.0	79.0	Fracture
66.5	0.275	107.5	10.0		

1.2 Figure 1.37 shows the stress–strain relation from a tensile test on a specimen of diameter 12 mm and gauge length 60 mm. Determine (i) Young's modulus of elasticity, (ii) proportionality limit, (iii) Proof stress at 0.2% offset, (iv) the ultimate stress, and (v) the modulus of resilience.

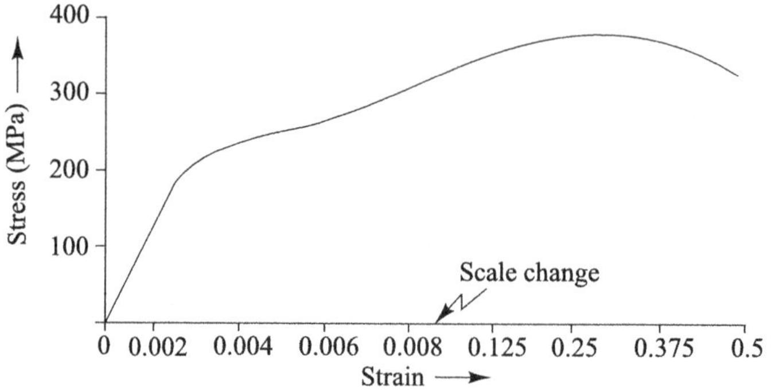

Fig. 1.37

1.3 A 50 mm square steel bar is subjected to an axial tensile load of 250 kN. Determine the decrease in lateral dimension if the modulus of elasticity for steel 200 GPa and Poisson's ratio is 0.3.

1.4 Determine the final volume of a square bar under axial strain of 0.01 if the initial unloaded sectional area and length of the bar are 2500 mm^2 and 250 mm, respectively. Take Poisson's ratio = 0.3.

1.5 Figure 1.38 shows a steel rod carrying 20 kN at *A* and 25 kN at *E*. It is supported at *C*. Find the change in length of the rod. Take Young's modulus of elasticity as 210 GPa.

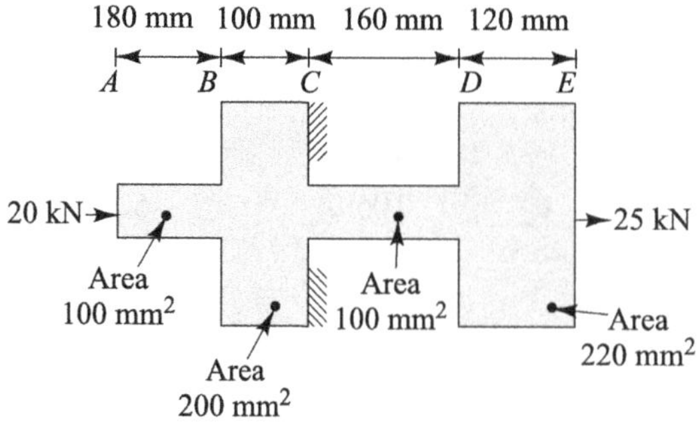

Fig. 1.38

1.6 A trapezoidal plate of thickness 10 mm, length 500 mm and width varying uniformly from 50 to 100 mm is under axial tension of 200 kN as shown in Fig. 1.39. Determine the elongation of the plate if the modulus of elasticity is 200 GPa.

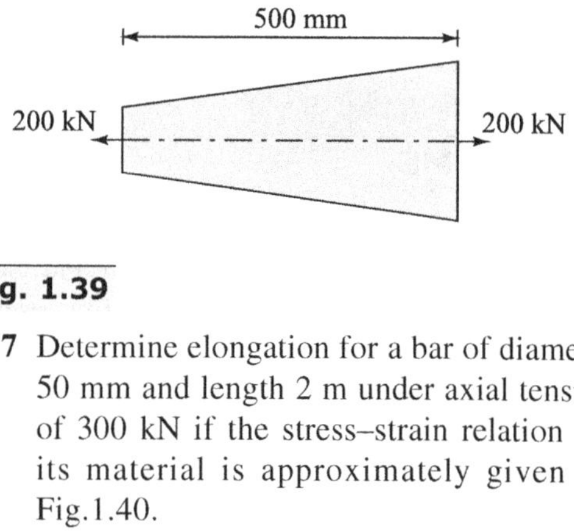

Fig. 1.39

1.7 Determine elongation for a bar of diameter 50 mm and length 2 m under axial tension of 300 kN if the stress–strain relation for its material is approximately given by Fig.1.40.

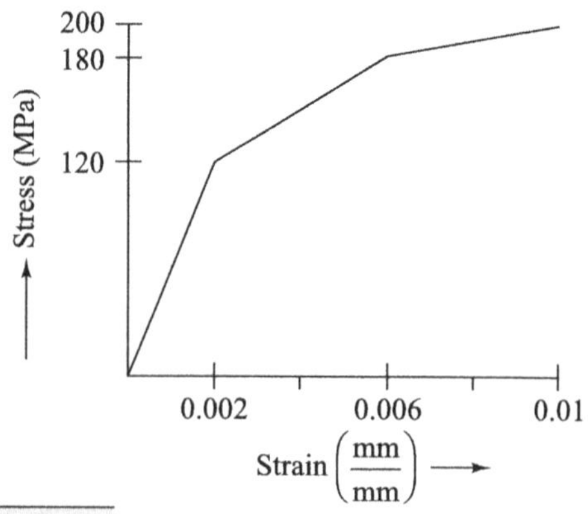

Fig. 1.40

1.8 A 50 mm square steel bar is fixed rigidly at both the ends. Determine the force in the bar if the temperature rises by 14°C if $E = 200$ GPa and $\alpha = 11.7 \times 10^{-6}$ mm/mm/°C.

1.9 A railway signal is operated by a steel wire rope of diameter 5 mm and length 600 m. Determine the movement to be given at the switch if the movement required at the signal is 17.5 cm under a pull of 1.5 kN. Take modulus of elasticity as 200 GPa.

1.10 Determine the maximum load P that can be applied to the axially loaded composite bar shown in Fig. 1.41 if the maximum permissible elongation is 0.025 mm. Take Young's modulus of elasticity for steel as 210 and 70 GPa for aluminium.

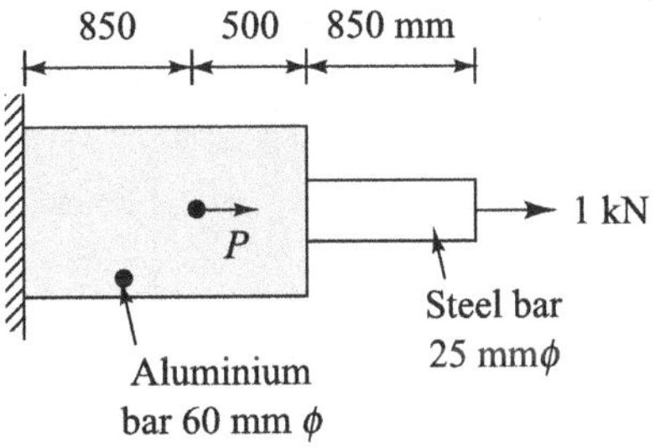

Fig. 1.41

1.11 A trapezoidal plate of thickness t is suspended from a rigid support as shown in Fig. 1.42. Determine the displacement of the free end if the unit weight of the bar is w and Young's modulus is E.

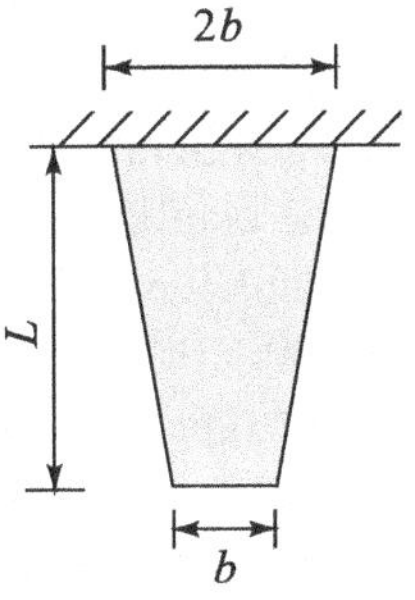

Fig. 1.42

1.12 Figure 1.43 shows an axially loaded composite bar of brass and steel fixed to a rigid support. The brass bar has a hollow section of 20 mm diameter. Determine the displacement of the free end and the maximum stress in steel and brass sections if their modulii of elasticity are 210 and 105 GPa, respectively.

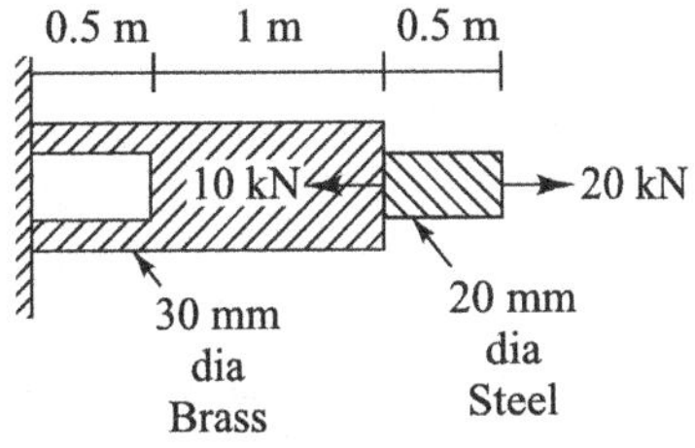

Fig. 1.43

1.13 Determine the temperature change that will cause a compressive stress of 36 MPa in the composite bar shown in Fig.1.44 if the modulus of elasticity and coefficient of thermal expansion for steel and aluminium are 210 GPa, 70 GPa, 12×10^{-6} mm/mm/°C and 23×10^{-6} mm/mm/°C, respectively.

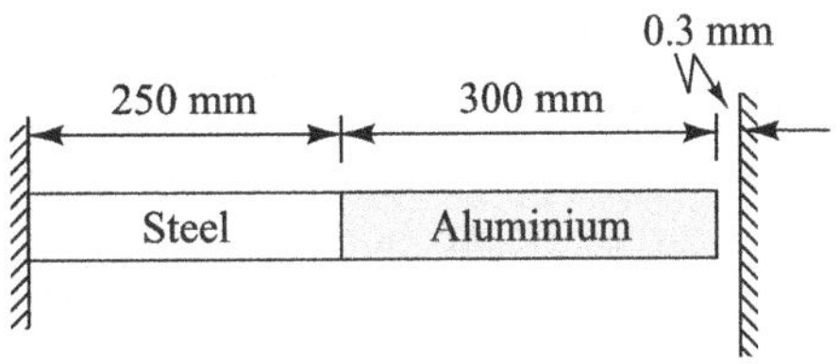

Fig. 1.44

1.14 A horizontal weightless rigid bar AB is supported by two bars CD and BE and a roller as shown in Fig. 1.45. The corss-sectional area of bars CD and BE are 500 mm^2 and 250 mm^2, respectively. Determine the stresses in the bars CD and BE if a load of 100 kN is applied. Take modulus of elasticity for bar CD 120 GPa and for bar $BE = 200$ GPa.

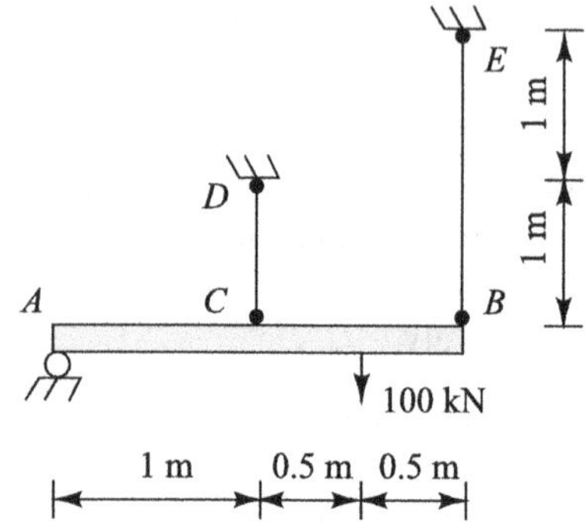

Fig. 1.45

1.15 A conical bar 1 m long tapers uniformly from 100 mm diameter to 200 mm diameter. The bar is rigidly fixed at both ends. Determine the maximum stress caused by 20°C rise in temperature if Young's modulus and coefficient expansion are 210 GPa and 12×10^{-6} mm/mm/°C, respectively.

1.16 A steel bolt of diameter 25 mm passes through a steel tube of inside and outside diameters 50 mm and 62.5 mm, respectively. The steel tube has a thick flange and the bolt is tightened between them such that the bolt has a tension of 30 kN. The length of bolt between nuts is 50 cm and the length of tube between thick flange is 40 cm. Determine the forces in the bolt and the tube if an additional tensile load of 30 kN is also applied on the flange.

1.17 Three short struts of equal length carry together a compressive load of 130 kN as shown in Fig. 1.46. Determine (i) the stresses in each strut, (ii) the changes in temperature which will cause zero stress in aluminium strut if Young's modulus of steel and aluminium are 210 GPa and 70 GPa, respectively and coefficient of thermal expansion for them are 12×10^{-6} mm/mm/°C and 23.4×10^{-6} mm/mm/°C, respectively.

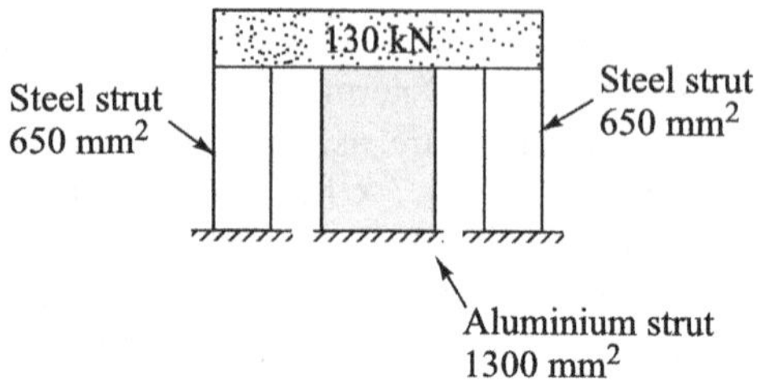

Fig. 1.46

1.18 Two mild steel bars of diameters 5 mm and 8 mm are supporting a horizontal rigid bar which is hinged at one end as shown in Fig. 1.47. Determine the magnitude of the ultimate load P_u acting at 2.5 m from the hinged end that will cause yielding of both the bars. Yield stress of the material is 250 MPa.

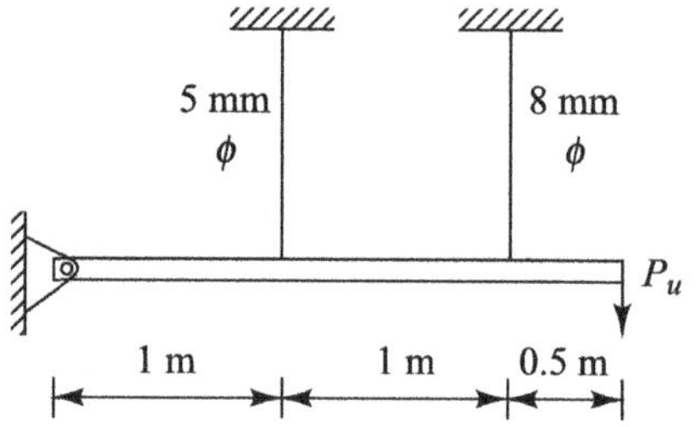

Fig. 1.47

1.19 Two bridge trusses are connected together by 10 m long steel bars of 25 mm diameter. The bars are heated to 100°C and the ends are hammered to form rivet heads on both ends. When a bar cools to 25°C, its end connections yield relatively by 5 mm. Determine the force in the bar if the coefficient of expansion and Young's modulus for steel is 12×10^{-6} mm/mm/°C and 200 GPa, respectively.

1.20 An aluminium bolt of 22 mm diameter is fitted snugly in a 3 mm thick tube of internal diameter 25 mm. Determine the stresses in the bolt and the tube when the temperature falls by 20°C if for steel and aluminium,

Young's modulii are 200 GPa and 70 GPa, respectively and the coefficient of thermal expansion are 12×10^{-6} mm/mm/°C and 23×10^{-6}mm/mm/°C, respectively.

1.21 A reinforced concrete column 300 mm square has four bars of 20 mm diameter. The length of the column is 4 m. Determine stresses in steel and concrete due to 25°C rise in temperature if the coefficients of thermal expansions are 23×10^{-6}/°C for steel and 10.8×10^{-6}/°C for concrete and the modulii of elasticity of steel and concrete are 210 GPa and 22 GPa, respectively.

1.22 Plot axial force diagram for the steel bar subjected to axial loads as shown in Fig. 1.48.

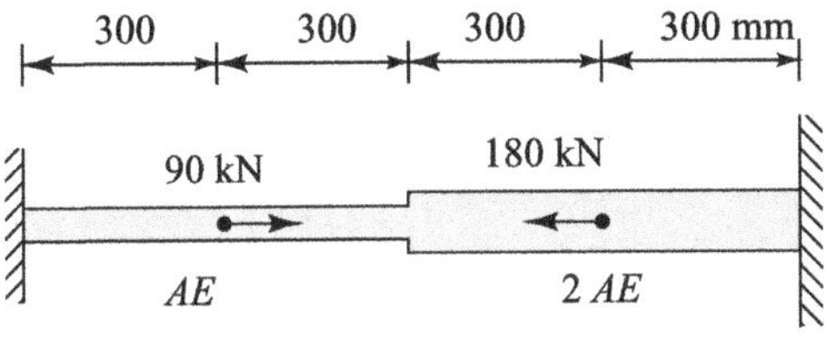

Fig. 1.48

1.23 Three 8 mm diameter high tensile steel bars of length 1.5 m each, are connected by hinges to carry a load of W as shown in Fig. 1.49. Assuming that the material of each member is elastic in tension upto a yield stress of 415 MPa and yields continuously thereafter at the yield stress. Calculate (i) the value of W at which yield stress is attained in any member, and (ii) the final collapse load W. Also sketch the load displacement curve for W. Take E = 200 GPa.

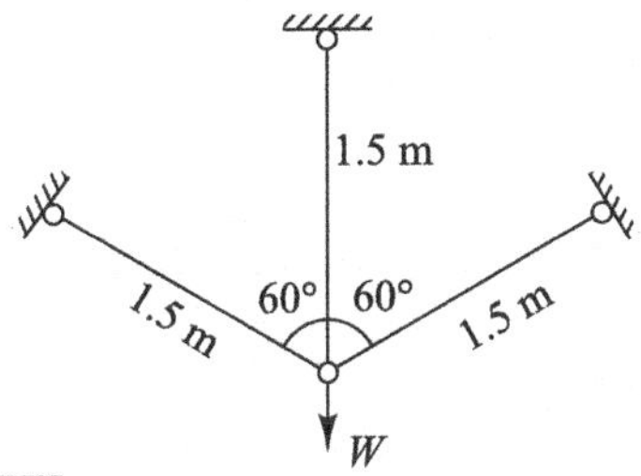

Fig. 1.49

1.24 A brass cone of height 500 mm and base diameter 300 mm is immersed in sea to a depth of 10 kN. Determine the changes in the dimensions of the cone if Young's modulus and Poisson's ratio for brass are 100 GPa and 0.35 GPa, respectively. Take the density of sea water to be 1050 kg/m^3.

1.25 A rubber pad is provided at bearing to support a bridge structure. The bond between rigid supports and the bearing restricts the strains in the plane of the bearing pad. Determine the ratio of apparent and true Young's modulii if Poisson's ratio for rubber is equal to 0.49.

1.26 A steel bar of cross-sectional area A has length L. It is hung vertically and a downward load of W is applied at its lower end. Determine the change in volume if the unit weight of the bar is w.

1.27 A square bar of length L is under axial compressive load F acting along its length. The material of the bar is linearly elastic with modulus of elasticity E and Poisson's ratio μ. Determine the change in volume due to applied load if the bar is restricted to have lateral deformations.

1.28 A steel plate 100 mm by 80 mm is subjected to plane strain and the dimensions have changed to 100.04 mm by 80.03 mm after deformations. Determine the stresses acting on the plate if E = 200 GPa and μ = 0.3.

1.29 Derive compatibility equation for three-dimensional state of strains.

1.30 A weight of 10 kN falls on the top of a follow cylindrical steel column 3 m high, outer diameter 250 mm and inner diameter 200 mm. If the height of fall is 1 m, determine the maximum stress and displacement due to impact if the modulus of elasticity is 200 GPa.

1.31 A steel bar has diameter 20 mm for 1 m length and 10 mm diameter for lower 1 m length. It is fixed rigidly at the top and a rigid plate is fixed at the bottom. Determine the maximum deflection if a 30 N weight is released from the rest position as shown in Fig. 1.50.

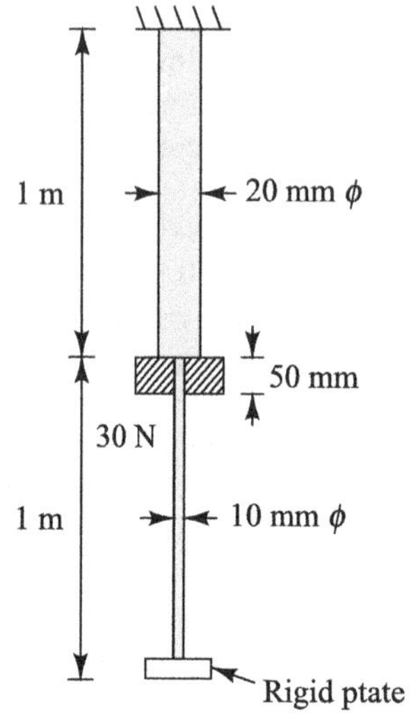

Fig. 1.50

1.32 A high yield steel bolt of nominal diameter 30 mm has diameter 26 mm at the root of the threads. If a proof load of 450 kN is applied on the bolt, determine the proof resilience of the bolt. Take modulus of elasticity, $E = 210$ GPa and the lengths of threaded and unthreaded portions of bolt are 50 mm and 80 mm, respectively.

1.33 A diver weighing 650 N jumps from a height of 800 mm on a wooden diving board 500 mm wide and 50 mm thick as shown in Fig.1.51. Determine the impact load and maximum stress developed in the diving board if Young's modulus for wood is 9.6 GPa.

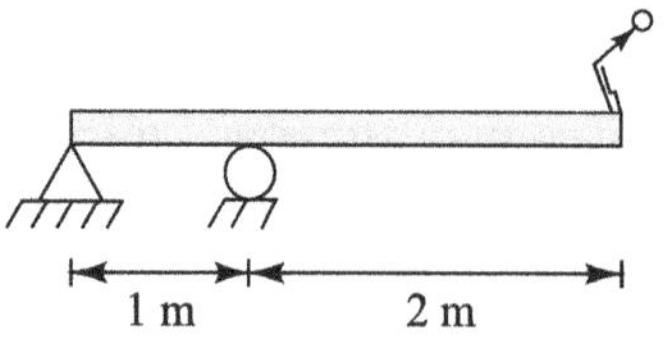

Fig. 1.51

1.34 A weight of 100 N falls on the end of rigid massless bar shown in Fig. 1.52 from a height of 500 mm. Determine the impact factor for the spring if the stiffness of spring is 20 MN/m.

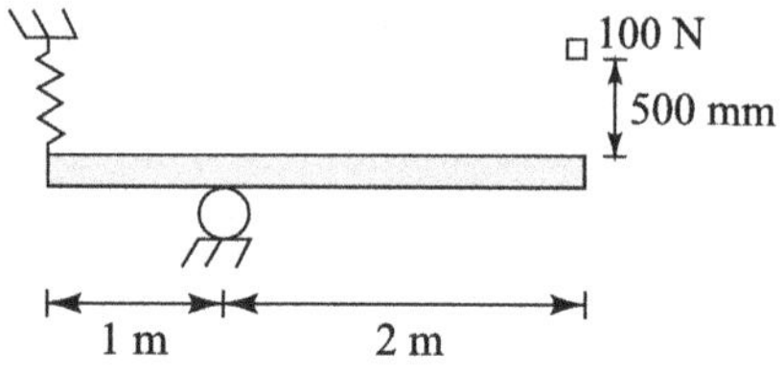

Fig. 1.52

1.35 A weight of 25 N falls from a height of 100 mm onto a rigid massless bar supported by springs as shown in Fig. 1.53. Determine the stiffness of springs, k if the maximum deflection is not to exceed 10 mm.

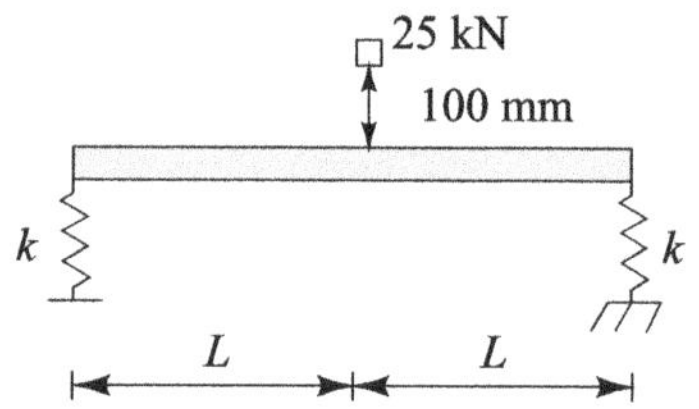

Fig. 1.53

1.36 A mass of 2 kg hits the end of the rod shown in Fig. 1.54 with a velocity of 10 m/s. Determine the maximum stress in the rod if $E = 200$ GPa. Also determine the maximum stress if the rod had a uniform diameter of 50 mm.

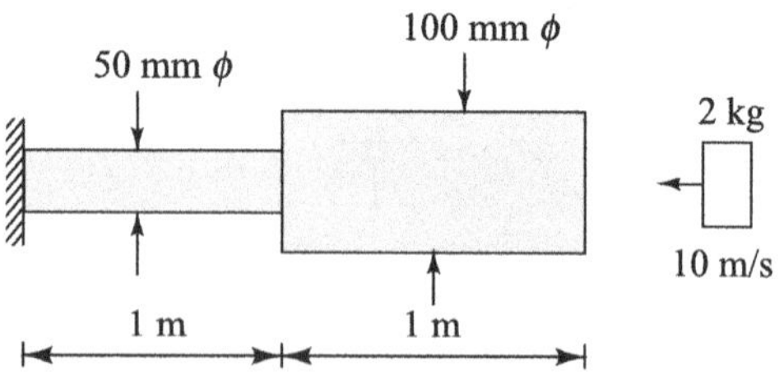

Fig. 1.54

1.37 A weight of 50 N talls through 100 mm on to a rigid weightless flange attached to the lower end of a inner 20 mm dia steel bar enclosed by a copper tube of inner diameter 25 mm and external diameter 30 mm as shown in Fig.1.55. Determine the maximum stress in copper tube and steel rod if the modulii of elasticity of steel and copper are 210 GPa and 70 GPa, respectivity.

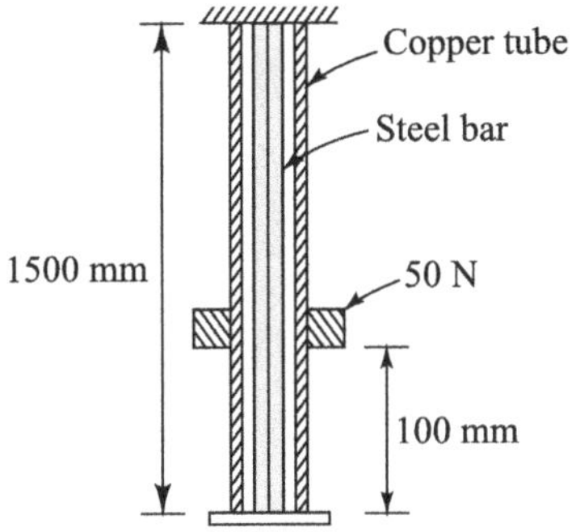

Fig. 1.55

OBJECTIVE QUESTIONS

1.38 In an axial tensile test on mild steel bar, axial strain ε develops under axial load F. If the corresponding lateral strain is ε_1 and the modulus of elasticity is E, then lateral stress is equal to,

(a) $\varepsilon_1 \times E$ (b) $(\varepsilon_1/\varepsilon) \times E$
(c) $(\varepsilon/\varepsilon_1) \times E$ (d) None of the above

1.39 Strain hardening of mild steel increase
(a) Modulus of elasticity
(b) Modulus of resilience
(c) Modulus of rigidity
(d) Modulus of toughness

1.40 Every elastic material does not show
(a) Logarithmic strain
(b) Modulus of toughness
(c) Yield point
(d) Fatigue

1.41 Poisson's ratio is the ratio of
(a) Axial stress to lateral stress
(b) Axial strain to lateral strain
(c) Lateral strain to axial strain
(d) Lateral deformation to axial deformation

1.42 To remove strain hardening, a process is used which is called
(a) Relaxation (b) Creep
(c) Work hardening (d) Annealing

1.43 A rigid weightless horizontal beam is supported by two identical steel bars *CD* and *EF* and a hinge at one end as shown in Fig. 1.56. Determine the load in the bars *CD* if a load of 40 kN is applied at end *B*.
(a) 13.33 kN (b) 26.67 kN
(c) 24 kN (d) 48 kN

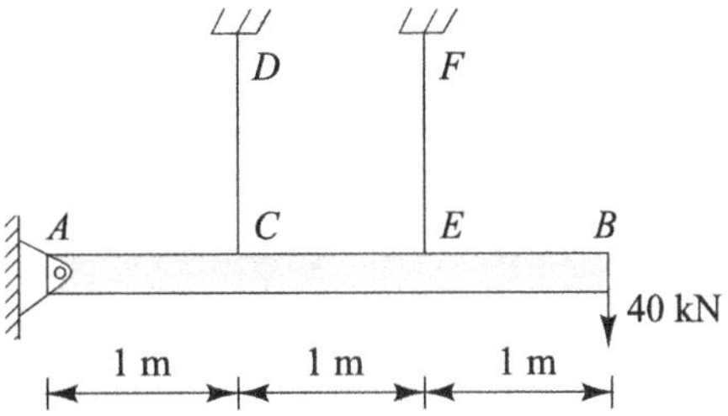

Fig. 1.56

1.44 A bar of length L is built fixed rigidly at both the ends as shown in Fig. 1.57. The reaction at end A due to axial load P is given by,

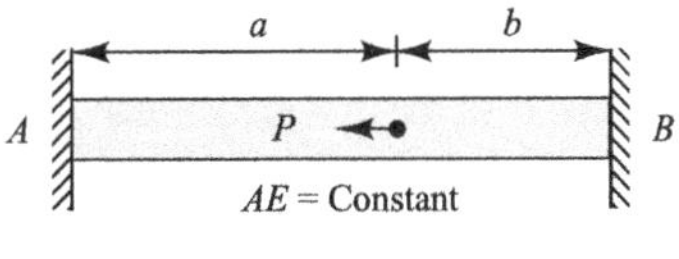

Fig. 1.57

(a) bP/L (b) aP/L
(c) $aP/(2a + b)$ (d) $bP/(2a + b)$

1.45 The ratio of modulus of rigidity and modulus of elasticity (G/E) for any elastic isotropic material is always

(a) less than 1/2 (b) less than 1/3
(c) more than 1/3 (d) *a* and *c* above

1.46 Which of the following strain fields describe a state of plane stress in a continuous elastic medium?

(a) $\begin{vmatrix} x^2 + c & xy \\ xy & cy \end{vmatrix}$ (b) $\begin{vmatrix} x + y^2 & 2xy \\ 2xy & x^2 + y \end{vmatrix}$

(c) $\begin{vmatrix} y^2 + c & xy \\ xy & cx \end{vmatrix}$ (d) $\begin{vmatrix} cy^2 & 2cxy \\ 2cxy & x + y^2 \end{vmatrix}$

1.47 A structure designed to withstand an impact load effectively should have
(1) Large volume
(2) High yield strength
(3) High modulus of elasticity
Out of these statements,
(a) 1 and 2 are correct
(b) 2 and 3 are correct
(c) 1 and 3 are correct
(d) All are correct

2

Analysis of Stress

CONCEPT REVIEW

2.1 INTRODUCTION

Different parts of a body under stress respond to external loads in different ways. In case of a thin strip *ABCD* under axial tension alone, as shown in Fig. 2.1, as section *A′ B′ C′ D′* will carry shear stress in addition to axial tension. Materials have different values of failure loads under different type of loads. Ductile materials are weak in shear and fail by shearing along 45° to the axis of the axial load forming cup and cone. The purpose of this chapter is to investigate the stresses on any plane in a body subjected to several loads.

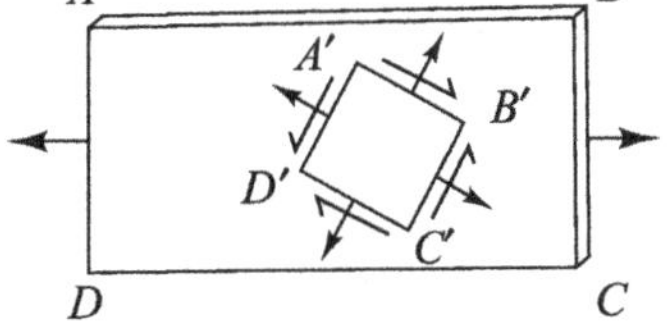

Fig. 2.1

2.2 TRANSFORMATION OF PLANE STRESSES

Consider a plane section *ABCD* of unit thickness, subject to normal stresses σ_x and σ_y along with shear stresses τ_{xy} and τ_{yx} as shown in Fig. 2.2. If the section is under equilibrium under applied loads then,

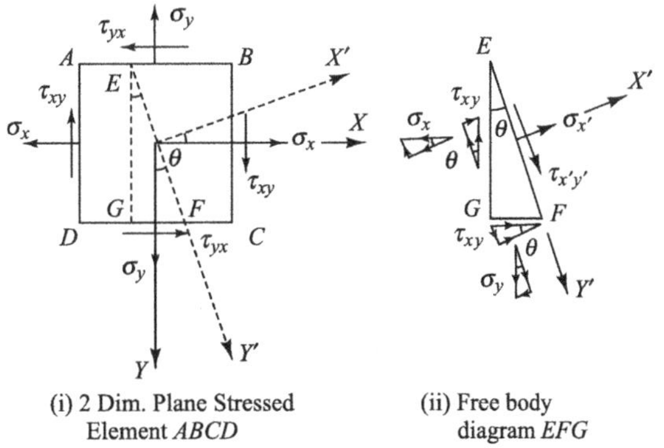

(i) 2 Dim. Plane Stressed Element *ABCD* (ii) Free body diagram *EFG*

Fig. 2.2

$$\tau_{xy} = \tau_{yx}$$

In order to investigate normal stress $\sigma_{x'}$ on a plane EF inclined at an angle θ to y-axis, consider the equilibrium of the free body EFG along normal x'-direction.

$$\Sigma F_{x'} = 0,\ \sigma_{x'} \times EF - \sigma_x \times EG \cos\theta + \tau_{xy} \times EG \sin\theta - \sigma_y \times FG \sin\theta + \tau_{xy} \times FG \cos\theta = 0$$

$$\text{or, } \sigma_{x'} = \sigma_x \frac{EG}{EF}\cos\theta - \tau_{xy}\frac{EG}{EF}\sin\theta + \sigma_y \frac{FG}{EF}\sin\theta - \tau_{xy}\frac{FG}{EF}\cos\theta$$

$$\text{or, } \sigma_{x'} = \sigma_x \cos^2\theta + \sigma_y \sin^2\theta - \tau_{xy}\sin 2\theta \quad (2.1a)$$

$$\text{or, } \sigma_{x'} = \frac{\sigma_x + \sigma_y}{2} + \frac{\sigma_x - \sigma_y}{2}\cos 2\theta - \tau_{xy}\sin 2\theta \quad (2.1b)$$

Similarly it can be seen that,

$$\sigma_{y'} = \frac{\sigma_x + \sigma_y}{2} - \frac{\sigma_x - \sigma_y}{2}\cos 2\theta + \tau_{xy}\sin 2\theta \quad (2.1c)$$

From above equations,

$$\sigma_x + \sigma_y = \sigma_{x'} + \sigma_{y'} \quad (2.1d)$$

Similarly for shear stress $\tau_{x'y'}$ consider the equilibrium of the free body EFG along tangential y'-direction,

$$\Sigma F_{y'} = 0,\ \tau_{x'y'} EF - \sigma_x EG \sin\theta - \tau_{xy} EG \cos\theta + \sigma_y FG \cos\theta + \tau_{xy} FG \sin\theta = 0$$

$$\text{or, } \tau_{x'y'} = \sigma_x \frac{EG}{EF}\sin\theta + \tau_{xy}\frac{EG}{EF}\cos\theta - \sigma_y \frac{FG}{EF}\cos\theta - \tau_{xy}\frac{FG}{EF}\sin\theta$$

$$\text{or, } \tau_{x'y'} = \frac{\sigma_x - \sigma_y}{2}\sin 2\theta + \tau_{xy}\cos 2\theta \quad (2.2)$$

Sign Convention

Usually +ve direction forces on +ve faces and –ve direction forces on –ve faces are taken +ve as shown in Fig. 2.3. According to this sign convention, all tensile stresses are +ve and the shear stresses τ_{xy}, τ_{yx} and $\tau_{x'y'}$ shown in Fig. 2.2 are +ve. Equations 2.1 and 2.2 are derived by anticlockwise rotation of $x - y$ axis and therefore anticlockwise rotation of $x - y$ axis to $x' - y'$ axis is considered as +ve.

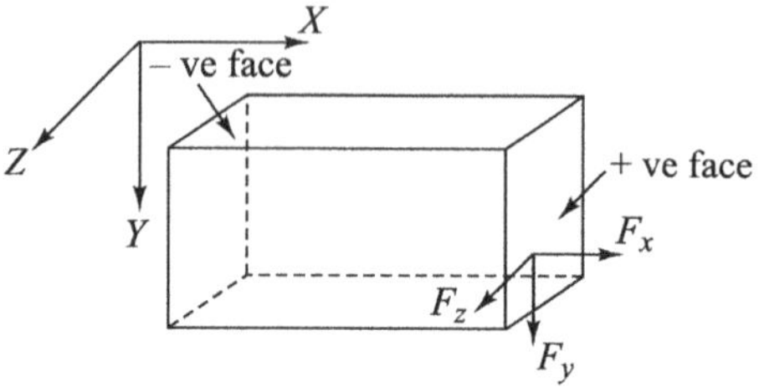

Fig. 2.3

2.2.1 Principal Stresses

Maximum or minimum value of normal stress $\sigma_{x'}$ are called principal stresses and the corresponding planes are called principal planes. Differentiate $\sigma_{x'}$ w.r.t. θ for maximum or minimum value of normal stress $\sigma_{x'}$,

$$\frac{d\sigma_{x'}}{d\theta} = 0$$

$$\text{or, } -\left(\frac{\sigma_x - \sigma_y}{2}\right) 2\sin 2\theta - 2\tau_{xy}\cos 2\theta = 0$$

$$\text{or, } \tan 2\theta_p = \frac{-\tau_{xy}}{\left(\frac{\sigma_x - \sigma_y}{2}\right)} \quad (2.3)$$

This gives two solutions differing by 90° for principal plane angle θ_p as shown in Fig. 2.4. Substituting values of sin $2\theta_p$ and cos $2\theta_p$ from Fig. 2.4 in Eq. (2.1b), the maximum and minimum principal stresses are given by,

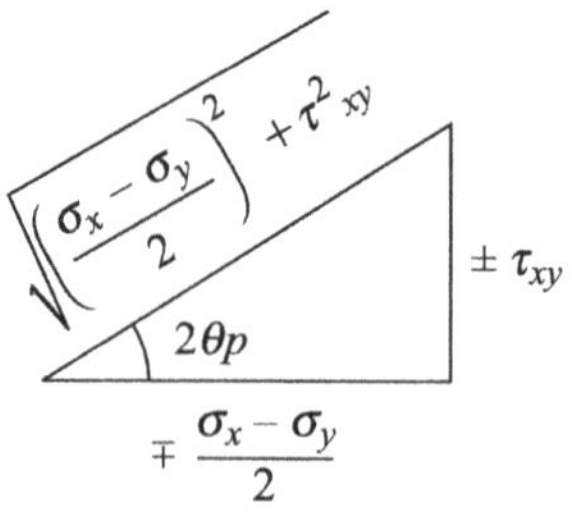

Fig. 2.4 *Principal Plane Angle θ_p*

$$\sigma_{x', \text{ max/min}} = \frac{\sigma_x + \sigma_y}{2} \pm \sqrt{\left(\frac{\sigma_x - \sigma_y}{2}\right)^2 + \tau_{xy}^2} \quad (2.4)$$

Shear stress at principal planes is zero.

2.2.2 Maximum Shear Stress

Differentiate $\tau_{x'y'}$ w.r.t. θ for maximum or minimum value of shear stress $\tau_{x'y'}$,

$$\frac{d\tau_{x'y'}}{d\theta} = 0 = \left(\frac{\sigma_x - \sigma_y}{2}\right) 2 \cos 2\theta - 2\tau_{xy} \sin 2\theta$$

$$\text{or,} \quad \tan 2\theta_s = \frac{\left(\dfrac{\sigma_x - \sigma_y}{2}\right)}{\tau_{xy}} \quad (2.5)$$

This also gives two values of $\sin 2\theta_s$ and $\cos 2\theta_s$ as per Fig. 2.5 for maximum value of shear stress as follows:

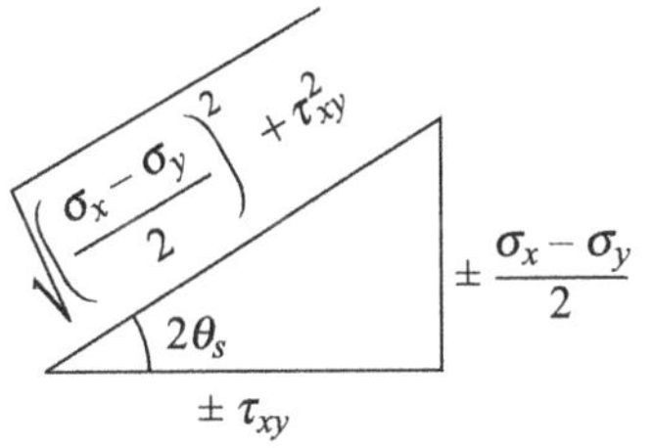

Fig. 2.5 *Maximum Shear Angle θ_s*

$$\tau_{x'y', \text{ max}} = \pm \sqrt{\left(\frac{\sigma_x - \sigma_y}{2}\right)^2 + \tau_{xy}^2} \quad (2.6)$$

Normal stress at the plane of maximum shear stress,

$$\sigma_{x'} = \frac{\sigma_x + \sigma_y}{2} \quad (2.7)$$

$2\theta_s$ and $2\theta_s$ differ by 90°, i.e. planes of maximum shear stress are at 45° to principal planes.

2.3 MOHR'S CIRCLE

All the equations derived so far can be represented graphically by Mohr's circle as shown in Fig. 2.6. Normal stresses σ_x, σ_y and shear stress τ_{xy} are

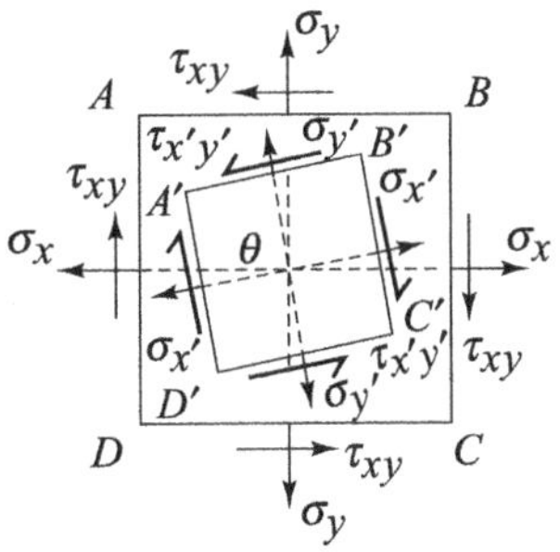

(i) 2 Dim. Plane Stressed element *ABCD*

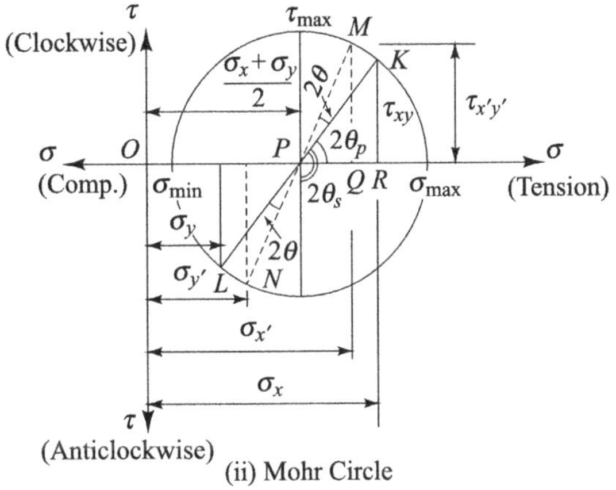

(ii) Mohr Circle

Fig. 2.6

plotted to scale and a circle is drawn through these points with center at horizontal axis.

Diameter *KL* represents the stress conditions of element *ABCD*. Horizontal diameter represents principal planes and vertical diameter represents planes of maximum shear stress. $2\theta_p$ and $2\theta_s$ are the angles between *KL* with horizontal and vertical diameter, respectively.

Diameter *MN* rotated by 2θ with respect to *KL* represents stress condition of element *A′ B′ C′ D′* where,

$$\sigma_{x'} = OQ = OP + PQ$$

$$= \frac{\sigma_x + \sigma_y}{2} + PM \cos(2\theta + 2\theta_p)$$

$$\text{or,} \quad \sigma_{x'} = \frac{\sigma_x + \sigma_y}{2} + PK(\cos 2\theta \cos 2\theta_p - \sin 2\theta \sin 2\theta_p)$$

Substituting,

$$PK \cos 2\theta_p = PR = \frac{\sigma_x - \sigma_y}{2}$$

and, $$PK \sin 2\theta_p = KR = \tau_{xy}$$

$$\sigma_{x'} = \frac{\sigma_x + \sigma_y}{2} + \frac{\sigma_x - \sigma_y}{2} \cos 2\theta - \tau_{xy} \sin 2\theta$$

Similarly,

$$\tau_{x'y'} = MQ = PM \sin (2\theta + 2\theta_p)$$
$$= PK (\sin 2\theta \cos 2\theta_p + \cos 2\theta \sin 2\theta_p)$$
$$= \frac{\sigma_x - \sigma_y}{2} \sin 2\theta + \tau_{xy} \cos 2\theta$$

Sign Convention

For Mohr's circle, the shear stresses which tend to rotate the element clockwise are considered +ve. Except this sign convention is same as in Art. 2.2.

2.4 GENERAL STATE OF STRESS *IN 3D*

Slightly different notations are used in this section for the sake of simplicity in presenting the results. Consider a cuboid of material shown in Fig. 2.7(a), under three-dimensional stress as follows.

$$[\tau_{ij}] = \begin{bmatrix} \tau_{xx} & \tau_{yx} & \tau_{zx} \\ \tau_{xy} & \tau_{yy} & \tau_{zy} \\ \tau_{xz} & \tau_{yz} & \tau_{zz} \end{bmatrix}$$

Cut out an elementary tetrahedron *KLMO* from one corner as shown in Fig. 2.7(b). *ON* is normal to the surface *KLM*. If the area of face *KLM* is *A*, then,

Area of face $KLO = A \times c_{nz} = A \cos \gamma$ (2.8a)

Area of face $LMO = A \times c_{nx} = A \cos \alpha$ (2.8b)

Area of face $KMO = A \times c_{ny} = A \cos \beta$ (2.8c)

where c_{nx}, c_{ny} and c_{nz} are called the direction cosines of angles α, β and γ which the normal *ON* makes with *x*, *y*, *z* directions.

If *X*, *Y* and *Z* are the forces acting on the inclined surface *KLM*, then equating the forces in *x*, *y* and *z* directions.

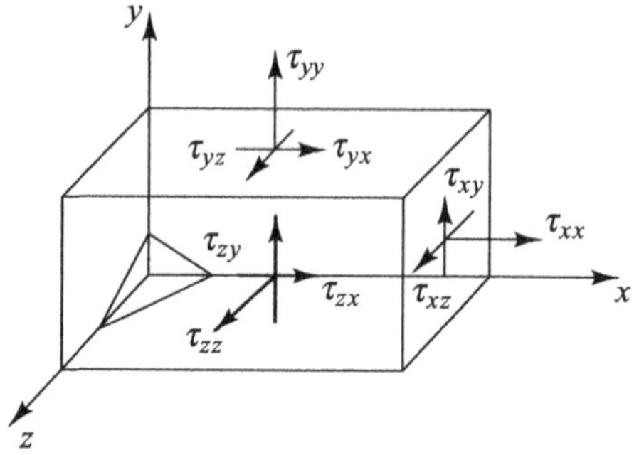

(a) 3 Dimensional Stresses

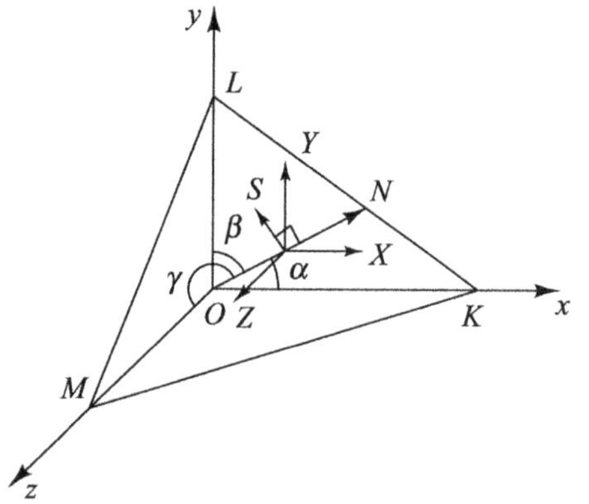

(b) Elementary Tetrahedron

Fig. 2.7

$$X \times A = \tau_{xx} (A \times c_{nx}) + \tau_{yx} (A \times c_{ny}) + \tau_{zx} \times (A \times c_{nz}) \quad (2.9a)$$

$$Y \times A = \tau_{xy} (A \times c_{nx}) + \tau_{yy} (A \times c_{ny}) + \tau_{zy} \times (A \times c_{nz}) \quad (2.9b)$$

$$Z \times A = \tau_{xz} (A \times c_{nx}) + \tau_{yz} (A \times c_{ny}) + \tau_{zz} \times (A \times c_{nz}) \quad (2.9c)$$

Resolving all forces in the direction of the normal *ON*,

$$\tau_{nn} \times A = X \times A \times c_{nx} + Y \times A \times c_{ny} + Z \times A \times c_{nz} \quad (2.10a)$$

or, $$\tau_{nn} = \tau_{xx} \times (c_{nx})^2 + \tau_{yy} \times (c_{ny})^2 + \tau_{zz} \times (c_{nz})^2 + 2\tau_{xy} (c_{nx} c_{ny}) + 2\tau_{xz} (c_{nx} c_{nz}) + 2\tau_{yz} (c_{ny} c_{nz}) \quad (2.10b)$$

This equation can be written in a condensed form as follows.

$$\tau_{nn} = \sum_i \sum_j \times \tau_{ij} \times c_{ni} \times c_{nj} \qquad (2.10c)$$

where $i, j = x, y$ or z.

Similarly a generalized form of Eq. (2.10c), can be derived for two perpendicular directions n and s (s in the plane KLM) as follows.

$$\tau_{ns} = \sum_i \sum_j \times \tau_{ij} \times c_{ni} \times c_{sj} \qquad (2.10d)$$

Equation (2.10c) and (d) can be written in matrix form as in Eq. (3.11b).

2.4.1 Principal Normal Stress

It was derived for plane stress two-dimensional problems that the shear stress at principal planes is zero. This remains true for three-dimensional problems. Assume that the oblique plane KLM is a principal plane, then the shear stress, $\tau_{ns} = 0$, and the forces X, Y and Z in Eq. (2.9) will be the function of τ_{nn} alone, i.e.,

$$X = \tau_{nn} \times c_{nx}$$
$$= \tau_{xx} \times c_{nx} + \tau_{yx} \times c_{ny} + \tau_{zx} \times c_{nz}$$
$$Y = \tau_{nn} \times c_{ny}$$
$$= \tau_{xy} \times c_{nx} + \tau_{yy} \times c_{ny} + \tau_{zy} \times c_{nz}$$
$$Z = \tau_{nn} \times c_{nz}$$
$$= \tau_{xz} \times c_{nx} + \tau_{yz} \times c_{ny} + \tau_{zz} \times c_{nz}$$

or, $$(\tau_{xx} - \tau_{nn}) \times c_{nx} + \tau_{yx} \times c_{ny} + \tau_{zx} \times c_{nz} = 0 \qquad (2.11a)$$

$$\tau_{xy} \times c_{nx} + (\tau_{yy} - \tau_{nn}) \times c_{ny} + \tau_{zy} \times c_{nz} = 0 \qquad (2.11b)$$

$$\tau_{xz} \times c_{nx} + \tau_{yz} \times c_{ny} + (\tau_{zz} - \tau_{nn}) \times c_{nz} = 0 \qquad (2.11c)$$

c_{nx}, c_{ny} and c_{nz} cannot be zero simultaneously because of the relation,

$$(c_{nx})^2 + (c_{ny})^2 + (c_{nz})^2 = 1 \qquad (2.12)$$

The non-zero solution for c_{nx}, c_{ny} and c_{nz} exists only if the determinate of their coefficients is zero, i.e.,

$$\begin{bmatrix} (\tau_{xx} - \tau_{nn}) & \tau_{yx} & \tau_{zx} \\ \tau_{xy} & (\tau_{yy} - \tau_{nn}) & \tau_{zy} \\ \tau_{xz} & \tau_{yz} & (\tau_{zz} - \tau_{nn}) \end{bmatrix} = 0 \qquad (2.13a)$$

Expansion of this determinate gives,

or, $$\tau_{nn}^3 - I_1\, \tau_{nn}^2 + I_2\, \tau_{nn} - I_3 = 0 \qquad (2.13b)$$

where

$$I_1 = \tau_{xx} + \tau_{yy} + \tau_{zz}$$
$$I_2 = \tau_{xx}\tau_{yy} + \tau_{yy}\tau_{zz} + \tau_{zz}\tau_{xx} - \tau_{xy}^2 - \tau_{yz}^2 - \tau_{zx}^2$$

and,

$$I_3 = \begin{vmatrix} \tau_{xx} & \tau_{yx} & \tau_{zx} \\ \tau_{xy} & \tau_{yy} & \tau_{zy} \\ \tau_{xz} & \tau_{yz} & \tau_{zz} \end{vmatrix}$$

I_1, I_2 and I_3 are called stress invariants since they are independent of the direction cosines c_{nx}, c_{ny} and c_{nz}.

The roots of Eq. (2.13b), give three values of principal stresses say, σ_I, σ_{II} and σ_{III} and a set of direction cosines for them is given by Eqs (2.11) and (2.12).

2.4.2 Octahedral Normal and Shearing Stress

If x, y and z planes are principal planes and the plane KLM makes equal angles with them, then the stresses on plane KLM are called octahedral stresses. An octahedron is a solid having eight equilateral triangular faces which are inclined equally to x, y and z directions. For such a plane, the direction cosines are equal, i.e. $c_{nx} = c_{ny} = c_{nz} = 1/3$ because $(c_{nx})^2 + (c_{ny})^2 + (c_{nz})^2 = 1$.

Substituting these values in Eq. (2.10b), along with $\tau_{xy} = \tau_{xz} = \tau_{yz} = 0$, since shears on principal x, y, z planes are zero.

Normal octahedral stress,

$$\tau_{nn,\text{oct}} = \frac{\tau_{xx} + \tau_{yy} + \tau_{zz}}{3} = \frac{\sigma_I + \sigma_{II} + \sigma_{III}}{3} \qquad (2.14)$$

If τ_{oct} is the total resultant stress on octahedral plane, then,

$$\tau_{oct}^2 = X^2 + Y^2 + Z^2 \quad (2.15)$$

Substituting $\tau_{xy} = \tau_{xz} = \tau_{yz} = 0$, in Eq. (2.9)

$$X = \sigma_I \times c_{nx}, \quad Y = \sigma_{II} \times c_{ny} \text{ and}$$
$$Z = \sigma_{III} \times c_{nz} \quad (2.16\text{ a–c})$$

From Eqs. (2.15) and (2.16),

$$\tau_{oct}^2 = \frac{\sigma_I^2 + \sigma_{II}^2 + \sigma_{III}^2}{3} \quad (2.17)$$

Octahedral shear stress is given by,

$$\tau_{ns,\,oct}^2 = \tau_{oct}^2 - \tau_{nn,\,oct}^2$$

$$\text{or, } \tau_{ns,\,oct}^2 = \frac{\sigma_I^2 + \sigma_{II}^2 + \sigma_{III}^2}{3} - \frac{(\sigma_I + \sigma_{II} + \sigma_{III})^2}{9}$$

$$\text{or, } \tau_{ns,oct} = \frac{1}{3}\sqrt{(\sigma_1 - \sigma_{II})^2 + (\sigma_{II} - \sigma_{III})^2 + (\sigma_{III} - \sigma_1)^2} \quad (2.18)$$

This is an important equation and is used in failure theories of materials.

Lame's stress ellipsoid Eq.(2.16) can be written as,

$$c_{nx} = \frac{X}{\sigma_I}, \; c_{ny} = \frac{Y}{\sigma_{II}}, \text{ and } c_{nz} = \frac{Z}{\sigma_{III}} \quad (2.19\text{a–c})$$

Substituting Eq. (2.19a-c) in the relation,

$$(c_{nx})^2 + (c_{ny})^2 + (c_{nz})^2 = 1$$

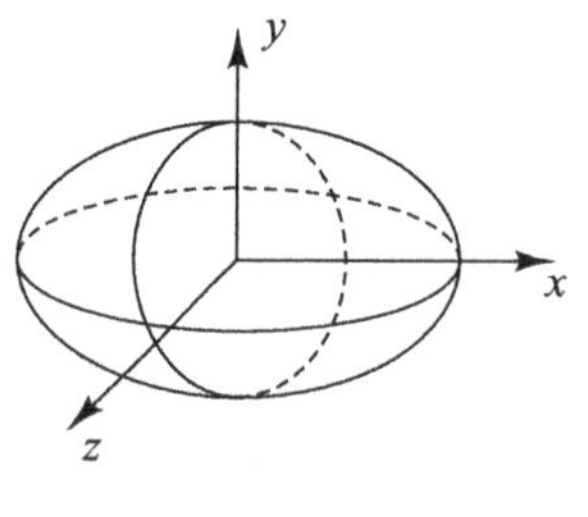

Fig. 2.8 *Lame's Stress Ellipsoid*

$$\left(\frac{X}{\sigma_I}\right)^2 + \left(\frac{Y}{\sigma_{II}}\right) + \left(\frac{Z}{\sigma_{III}}\right)^2 = 1 \quad (2.19\text{d})$$

This is the equation for the surface of an ellipsoid as shown in Fig. (2.8).

2.4.3 Principal Shear Stress

The planes of principal shear stress bisect the planes of principal normal stress and it can be seen from the Mohr's circle that the maximum shear stresses are related to the principal normal stresses σ_I, σ_{II} and σ_{III} as follows.

$$\tau_I = \frac{\sigma_{II} - \sigma_{III}}{2} \text{ and } \sigma_n = \frac{\sigma_{II} + \sigma_{III}}{2} \quad (2.20\text{a})$$

$$\tau_{II} = \frac{\sigma_{III} - \sigma_I}{2} \text{ and } \sigma_n = \frac{\sigma_{III} + \sigma_I}{2} \quad (2.20\text{b})$$

$$\tau_{III} = \frac{\sigma_I - \sigma_{II}}{2} \text{ and } \sigma_n = \frac{\sigma_I + \sigma_{II}}{2} \quad (2.20\text{c})$$

2.4.4 Mohr's Circle for 3-Dimensional State of Stress

In general state of 3D stress, it is possible at every point to reduce the stress into three mutually perpendicular normal principal stresses σ_I, σ_{II} and σ_{III} as shown in Fig. 2.9. Mohr's circles are drawn for three pairs of principal stress i.e., $\sigma_I - \sigma_{II}$, $\sigma_{II} - \sigma_{III}$ and $\sigma_{III} - \sigma_I$. Any plane other than these principal planes will lie within the shaded area.

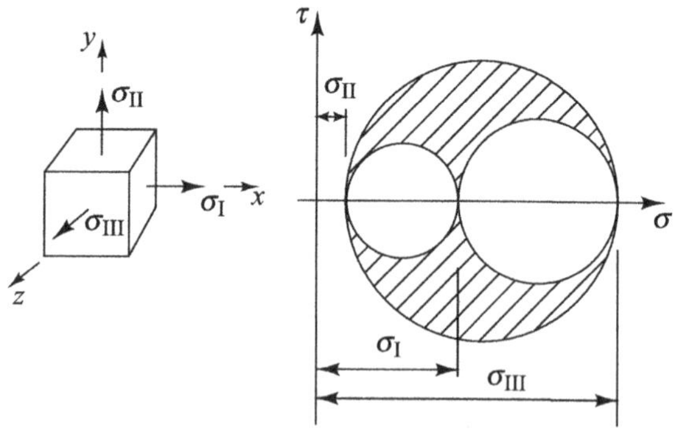

Fig. 2.9 *Mohr's Circle for 3-Dimensional Stress*

2.4.5 Decomposition into Hydrostatic and Pure Shear State

A body under normal stress σ in all three directions is called to be under hydrostatic stress σ and its stress tensor is as follows.

$$[\tau_{ij}] = \begin{vmatrix} \sigma & 0 & 0 \\ 0 & \sigma & 0 \\ 0 & 0 & \sigma \end{vmatrix} \qquad (2.21)$$

Similarly, a body can be under a state of pure shear if the normal stress is zero in all three directions and its stress tensor is as follows.

$$[\tau_{ij}] = \begin{vmatrix} 0 & \tau_{yx} & \tau_{zx} \\ \tau_{xy} & 0 & \tau_{zy} \\ \tau_{xz} & \tau_{yz} & 0 \end{vmatrix} \qquad (2.22)$$

If, $\sigma = (\sigma_x + \sigma_y + \sigma_z)$

the state of stress at any point can be decomposed into these two states as follows.

$$[\tau_{ij}] = \begin{vmatrix} \tau_{xx} & \tau_{yx} & \tau_{zx} \\ \tau_{xy} & \tau_{yy} & \tau_{zy} \\ \tau_{xz} & \tau_{yz} & \tau_{zz} \end{vmatrix} = \begin{vmatrix} \sigma & 0 & 0 \\ 0 & \sigma & 0 \\ 0 & 0 & r \end{vmatrix} + \begin{vmatrix} \sigma_x - \sigma & \tau_{yx} & \tau_{zx} \\ \tau_{xy} & \sigma_y - \sigma & \tau_{zy} \\ \tau_{xz} & \tau_{yz} & \sigma_z - \sigma \end{vmatrix} \qquad (2.23)$$

Since first stress invariant, $I_1 = (\sigma_x - \sigma) + (\sigma_y - \sigma) + (\sigma_z - \sigma) = 0$, in second matrix, it satisfies the necessary condition of pure shear.

EXAMPLES

Example 2.1 A bar is to carry an axial load of 70 MPa. Determine the stresses on a plane inclined at 30° to the direction of loading as shown in Fig. 2.10. Also locate the plane of maximum shear stress and determine the stresses on it.

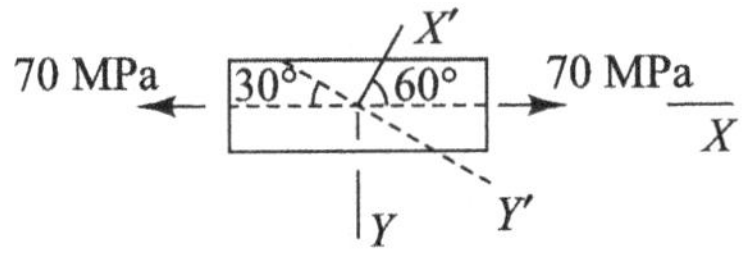

Fig. 2.10

Solution Using Eqs (2.1b) and (2.2) for normal stress and shear stress on inclined plane as shown in Fig. 2.11, where,

$$\sigma_x = 70 \text{ MPa}, \quad \sigma_y = \tau_{xy} = 0, \quad \theta = 60^\circ$$

$$\sigma_{x'} = \frac{\sigma_x + \sigma_y}{2} + \frac{\sigma_x - \sigma_y}{2} \cos 2\theta - \tau_{xy} \sin 2\theta$$

$$= \frac{70}{2} + \frac{70}{2} \cos 120^\circ = 17.5 \text{ MPa}$$

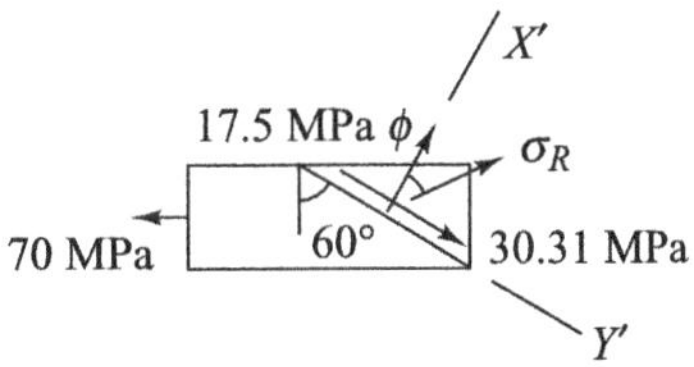

Fig. 2.11

$$\tau_{x'y'} = \frac{\sigma_x - \sigma_y}{2} \cos 2\theta - \tau_{xy} \sin 2\theta$$

$$= \frac{70}{2} \sin 120^\circ = 30.31 \text{ MPa}$$

Resultant stress,

$$\sigma_r = \sqrt{\sigma_{x'}^2 + \tau_{x'y'}^2} = \sqrt{17.5^2 + 30.31^2}$$

$$= 35 \text{ MPa}$$

Inclination of resultant stress σ_r w.r.t. normal stress $\sigma_{x'}$,

$$\theta = \tan^{-1} \frac{\tau_{x'y'}}{\sigma_{x'}} = \tan^{-1} \frac{30.31}{17.5} = 60^\circ$$

Maximum shear stress: Plane of maximum shear stress is given by Eq. (2.5),

$$\tan 2\theta_s = \frac{(\sigma_x - \sigma_y)/2}{\tau_{xy}} = \frac{70}{0} = 0$$

or, $2\theta_s = 90^\circ$

Maximum shear stress by Eq. (2.6),

$$\tau_{x'y'\max} = \pm\sqrt{\left(\frac{\sigma_x - \sigma_y}{2}\right)^2 + \tau_{xy}^2} = \pm 35 \text{ MPa}$$

Normal stress at the plane of maximum shear stress (Eq. (2.7),

$$\sigma_{x'} = \frac{\sigma_x + \sigma_y}{2} = 35 \text{ MPa}$$

Example 2.2 Solve Example 2.1 by Mohr's circle method.

Solution In Fig. 2.12, diameter OA represents the state of stress in $x - y$ plane. Draw a circle through OA. Perpendicular diameter BC represents plane of maximum shear stress. Draw DC at $2\theta = 120^\circ$ to OA. Point D represents the stress on the plane inclined at 60° to the vertical plane.

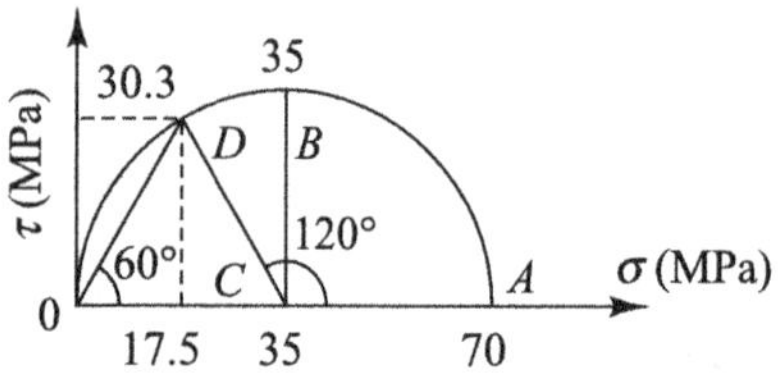

Fig. 2.12

Resultant stress, $\sigma_r = OD = 35$ MPa

Inclination of θ_r w.r.t. its normal = $\angle DOC = 60^\circ$

Example 2.3 A plane element is subjected to stress in $x - y$ plane as follows,

$\sigma_x = 40$ MPa (Compression),

$\sigma_y = 80$ MPa (Tension),

and, $\tau_{xy} = 60$ MPa (Clockwise).

Determine the stresses at a plane having its normal at 45° anticlockwise to the direction of the tensile stress 80 MPa. Also determine principal stresses and maximum shear stress.

Solution Rotation angle of $x - y$ plane (Fig. 2.13),

$2\theta = 2 \times 135 = 270^\circ$ (anticlockwise)

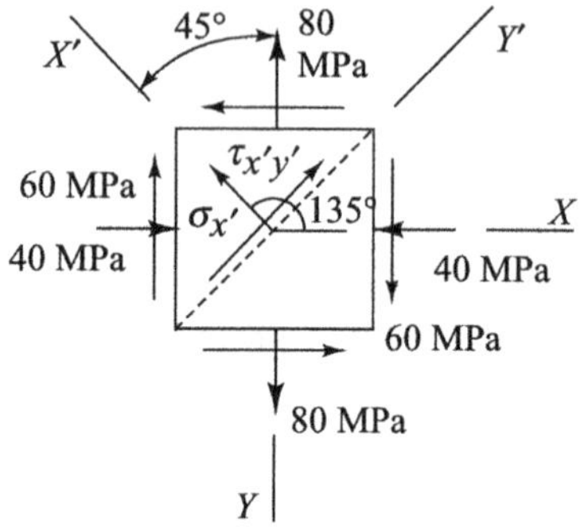

Fig. 2.13

Normal stress and shear stress on $x' - y'$ plane are given by Eqs. (2.1b) and (2.2),

$$\sigma_{x'} = \frac{\sigma_x + \sigma_y}{2} + \frac{\sigma_x - \sigma_y}{2}\cos 2\theta - \tau_{xy}\sin 2\theta$$

$$= \frac{-40 + 80}{2} + \frac{-40 - 80}{2}\cos 270^\circ - 60$$

$$\sin 270^\circ = 80 \text{ MPa}$$

$$\tau_{x'y'} = \frac{\sigma_x - \sigma_y}{2}\sin 2\theta + \tau_{xy}\cos 2\theta$$

$$= \frac{-40 - 80}{2}\sin 270^\circ + 60\cos 270^\circ$$

$$= 60 \text{ MPa}$$

Principal plane angle (Eq. (2.3),

$$\tan 2\theta_p = -\frac{\tau_{xy}}{(\sigma_x - \sigma_y)/2} = \frac{60}{60} = 1$$

or, $2\theta_p = 45^\circ, 225^\circ$ (anticlockwise)

Principal stresses are given by Eq. (2.4),

$$\sigma_{x'\max/\min} = \frac{\sigma_x + \sigma_y}{2} \pm \sqrt{\left(\frac{\sigma_x - \sigma_y}{2}\right)^2 + \tau_{xy}^2}$$

$$= \frac{-40 + 80}{2} \pm \sqrt{\left(\frac{-40 - 80}{2}\right)^2 + 60^2}$$

= 104.85 MPa (Tension),

– 64.85 MPa (Compression)

Maximum shear stress by Eq. (2.6),

$$\tau_{x'y',\max} = \pm\sqrt{\left(\frac{\sigma_x - \sigma_y}{2}\right)^2 + \tau_{xy}^2}$$

$$= \pm\sqrt{\left(\frac{-40-80}{2}\right)^2 + 60^2}$$

= + 84.85 MPa

Normal stress at the plane of maximum shear stress (Eq. (2.7)),

$$\sigma_{x'} = \frac{\sigma_x + \sigma_y}{2} = \frac{-40+80}{2} = 20 \text{ MPa}$$

Maximum shear plane angle,

$2\theta_s = 45° + 90° = 135°$

Example 2.4 Solve Example 2.3 using Mohr's circle.

Solution Plot points A and B of coordinates (– 40, 60) and (80, – 60), respectively as in Fig. 2.14. Draw Mohr's circle through diameter AB. Point C on diameter CD at anticlockwise 270° to diameter, AB, gives stress in $x'-y'$ plane.

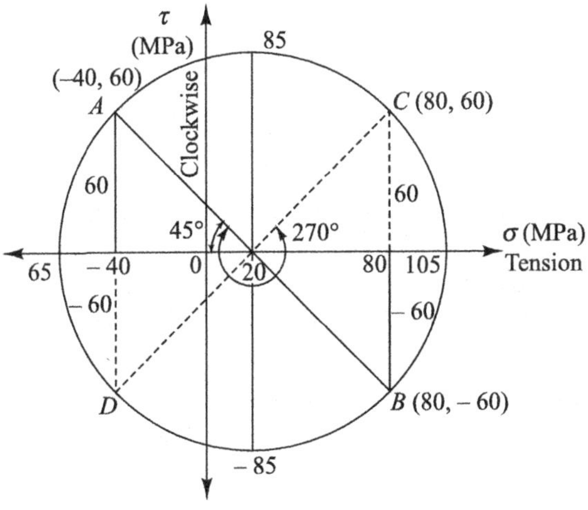

Fig. 2.14

Example 2.5 A plane element in a body is subjected to a normal stress of 25 MPa in +ve x-direction and counterclockwise shear of 100 MPa as shown in Fig. 2.15. Calculate the stresses at two planes, (i) 20° clockwise and (ii) 20° anticlockwise to x-axis.

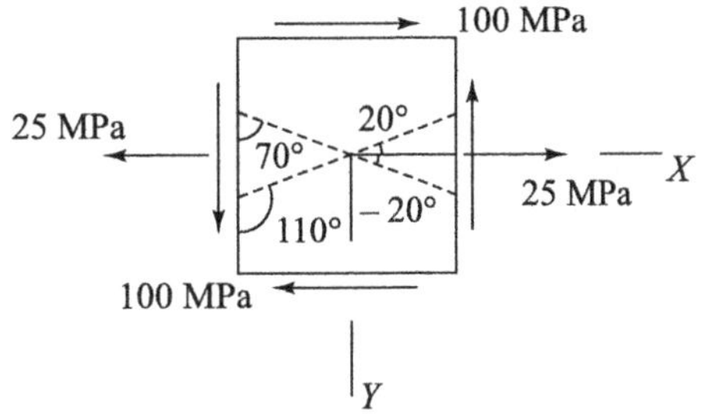

Fig. 2.15

Solution (i) Plane 20° clockwise to x-axis. Inclination with vertical (Fig. 2.15), $\theta = 70°$

Normal stress,

$$\sigma_{x'} = \frac{\sigma_x + \sigma_y}{2} + \frac{\sigma_x - \sigma_y}{2}\cos 2\theta - \tau_{xy}\sin 2\theta$$

$$= \frac{25}{2} + \frac{25}{2}\cos 140° - (-100)\sin 140°$$

= 67.2 MPa (Tension)

Shear stress,

$$\tau_{x'y'} = \frac{\sigma_x - \sigma_y}{2}\sin 2\theta + \tau_{xy}\cos 2\theta$$

$$= \frac{25}{2}\sin 140° - 100\cos 140°$$

= 84.64 MPa (Clockwise)

(ii) Plane 20° anticlockwise to x-axis. Inclination with vertical (Fig. 2.15), $\theta = 110°$

Normal stress,

$$\sigma_{x'} = \frac{25}{2} + \frac{25}{2}\cos 220° - (-100)\sin 220°$$

= – 61.35 MPa (Compression)

Shear Stress,

$$\tau_{x'y'} = \frac{25}{2}\sin 220° - 100\cos 220°$$

= 68.57 MPa (Clockwise)

Example 2.6 Solve Example 2.5 using Mohr's circle.

Solution Plot points A and B of coordinates (0, 100) and (25, – 100), respectively as in Fig. 2.16. Draw Mohr's circle through diameter AB. Point B represents stresses on vertical face. Point C on diameter CD at $2\theta = 140°$ anticlockwise to diameter AB, represents stress on plane at 20° clockwise to x-axis and point E on diameter EF at $2\theta = 220°$ anticlockwise to diameter AB, represents stress on plane at 20° anticlockwise to x-axis.

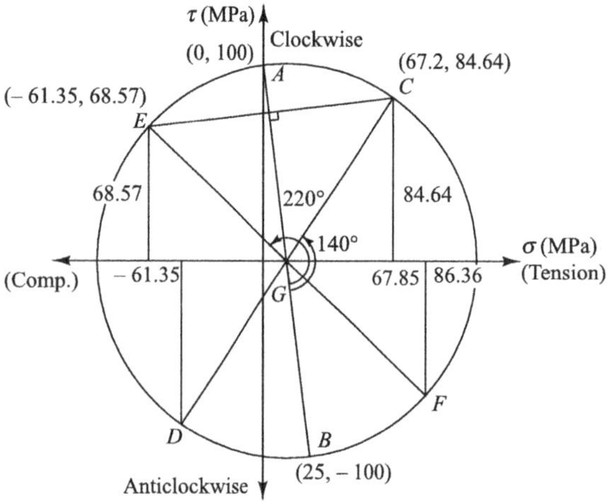

Fig. 2.16

Example 2.7 Two-dimensional stresses at a point on a plane P are 67.2 MPa (tension) and 84.64 MPa (shear, clockwise) and on another plane Q the stresses are 61.35 MPa (compression) and 68.57 MPa (shear, clockwise). Find the stresses on a plane R inclined at 20° clockwise to Q.

Solution Express the stresses on plane P in terms of Eqs (2.1) and (2.2).

$$67.2 = \frac{-61.35 + \sigma_y}{2} + \frac{-61.35 - \sigma_y}{2} \cos 2\theta - 68.57 \sin 2\theta \qquad \text{(i)}$$

$$84.64 = \frac{-61.35 - \sigma_y}{2} \sin 2\theta + 68.57 \cos 2\theta \qquad \text{(ii)}$$

where

σ_y = normal stress on the plane perpendicular to Q,

θ = angle between the planes P and Q.

From Eq. (i) and (ii),

$$\tan \theta = -\frac{128.55}{153.21}$$

or, $\theta \approx -40°$, i.e. plane P is at 40° clockwise to plane Q.

Substituting in Eq. (iii), σ_y = 86.36 MPa.

Angle of inclination of plane R w.r.t. Q,

$$2\theta = 2 \times (-20°) = -40°$$

Normal stress on plane R,

$$\sigma_{x'} = \frac{-61.35 + 86.36}{2} - \frac{61.35 + 86.36}{2} \times \cos(-40°) - 68.57 \sin(-40°)$$

$$\simeq 0$$

$$\tau_{x'y'} = \frac{-61.35 - 86.36}{2} \sin(-40°) + 68.57 \cos(-40°) = 100 \text{ MPa}$$

Example 2.8 Solve Example 2.7 using Mohr's circle.

Solution Plot two points C and E of coordinates (67.2, 84.64) and (–61.35, 68.57), respectively as in Fig. 2.16. Join C and E. Draw perpendicular bisector AB of CE intersecting normal stress axis at point G. With point G as center, draw Mohr's circle passing through C and E. Point A on diameter AB at an angle $2\theta = -40°$, clockwise to point E gives stresses on plane R.

Example 2.9 Tensile stresses of 160 MPa and 40 MPa are acting on two perpendicular planes in a body. Determine the location of a plane on which the resultant stress is most inclined to its normal.

Solution Let the stresses be $\sigma_{x'}$ and $\tau_{x'y'}$ on the plane at which the resultant stress σ_r is most inclined to its normal. If the inclination of $\sigma_{x'}$ is θ w.r.t. the stress 160 MPa, then,

$$\sigma_{x'} = \frac{160 + 40}{2} + \frac{160 - 40}{2} \cos 2\theta$$

$$= 100 + 60 \cos 2\theta \qquad \text{(i)}$$

$$\tau_{x'y'} = \frac{160-40}{2}\sin 2\theta = 60\sin 2\theta \qquad \text{(ii)}$$

$$\sigma_r = \sqrt{\sigma_{x'}^2 + \tau_{x'y'}^2}$$

$$= \sqrt{(100+60\cos 2\theta)^2 + (60\sin 2\theta)^2} \qquad \text{(iii)}$$

Inclination of σ_r with respect to normal $\sigma_{x'}$,

$$\phi = \tan^{-1}\frac{\tau_{x'y'}}{\sigma_{x'}} = \tan^{-1}\frac{60\sin 2\theta}{100+60\cos 2\theta}$$

For maximum inclination ϕ, differentiate with respect to θ,

$$\frac{d\phi}{d\theta} = 0 = 2\times 3600 + 12000\cos 2\theta$$

or, $\theta = 63.435^\circ$, and corresponding $\phi = 36.87^\circ$

Substituting θ in Eq. (iii), $\sigma_r = 80$ MPa

Example 2.10 Solve Example 2.9 using Mohr's circle.

Solution Plot principal stress points (160, 0) and (40, 0) and draw a circle through its center C as shown in Fig. 2.17. The resultant is given by, tangent OD = 80 MPa. Its inclination with its normal = ϕ and with major principal stress 160 MPa = θ, can be read from Fig. 2.17.

Angle of inclination of resultant stress σ_r with normal stress σ_x

$$\sin\phi = \frac{DC}{OC} = \frac{60}{100} = 0.6$$

$$\phi = 36.874$$

Inclination with major principal plane

$$\theta = \frac{1}{2}(90^\circ + \phi) = 63.435^\circ$$

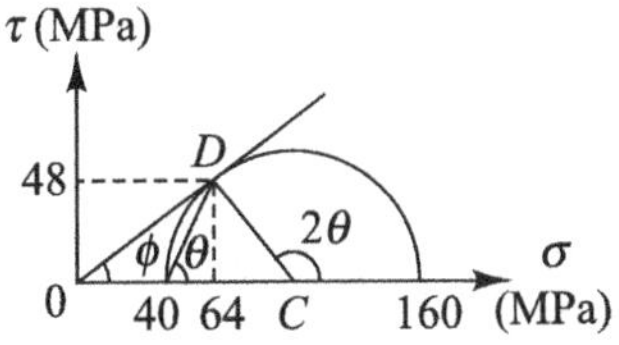

Fig. 2.17

Example 2.11 The stresses on a plane surface in A, B and C directions at a point O are as shown in Fig. 2.18. Determine the principal stresses and their location.

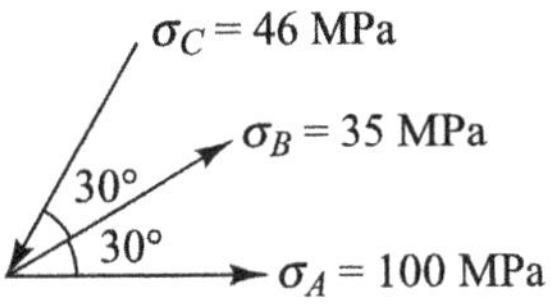

Fig. 2.18

Solution Using Eqs (2.1) and (2.2), the stresses on A, B and C are related as follows.

$$\sigma_B = \frac{\sigma_A+\sigma_y}{2} + \frac{\sigma_A-\sigma_y}{2}\cos 2\theta - \tau_{xy}\sin\theta$$

or, $$35 = \frac{100+\sigma_y}{2} + \frac{100-\sigma_y}{2}\cos 60^\circ - \tau_{xy}\sin 60^\circ \qquad \text{(i)}$$

and, $$-46 = \frac{100+\sigma_y}{2} + \frac{100-\sigma_y}{2}\cos 120^\circ - \tau_{xy}\sin 120^\circ \qquad \text{(ii)}$$

where σ_y = stress in the perpendicular direction of A,

and, τ_{xy} = shear stress across A

Solving Eqs (i) and (ii),

$\sigma_y = -62$ MPa and

$\tau_{xy} = 28.29$ MPa

Principal stresses are given by Eq. (2.4),

$$\sigma_{x'\max/\min} = \frac{\sigma_x+\sigma_y}{2} \pm \sqrt{\left(\frac{\sigma_x-\sigma_y}{2}\right)^2 + \tau_{xy}^2}$$

$$= \frac{100-62}{2} \pm \sqrt{\left(\frac{100+62}{2}\right)^2 + 28.29^2}$$

= 104.8 MPa (Tension),

– 66.8 MPa (Compression)

Location of principal stresses w.r.t. A is given by Eq. (2.3),

$$2\theta_p = \tan^{-1}\frac{-\tau_{xy}}{\left(\frac{\sigma_x - \sigma_y}{2}\right)} = \tan^{-1}\frac{-28.29}{\left(\frac{100+62}{2}\right)}$$

$$= -19.25° \quad \text{(clockwise)}$$

Example 2.12 Solve example 2.11 using Mohr's circle.

Solution Plot the three stresses on a straight line $C'O'B'A$ as shown in Fig. 2.19. Draw perpendicular lines $C'C$, $B'B$, $A'A$. Take any point B'' on $B'B$, and draw $B''A$ at 30° clockwise to $B'B$ since plane A is at 30° clockwise to plane B in Fig. 2.19. Similarly draw B'' C at 30° anticlockwise to $B'B$ as plane C is at 30° to plane B. The perpendicular bisector of $B''C$ and $B''A$ intersects at D. With D as center, draw the required Mohr's circle through points A and C and intersecting $B'B$ at B. The principal stresses can be read at the diameter parallel to the line $C'O'B'A'$.

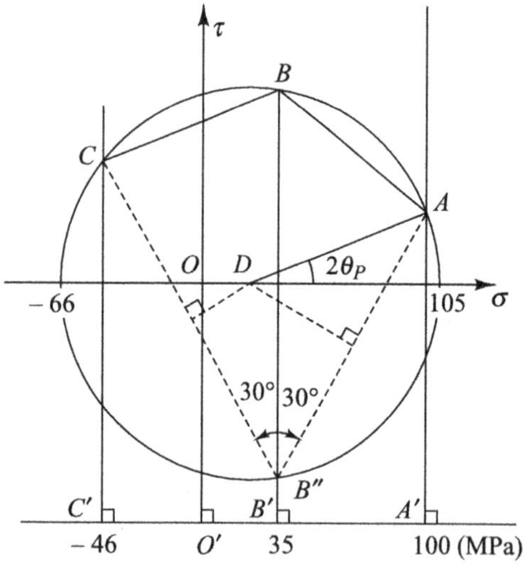

Fig. 2.19

Example 2.13 A cylindrical pressure tank of mean diameter 1.6 m, is fabricated by butt welding a 8 mm thick plate along a helix which forms an angle of 20° with the transverse plane. The end caps are spherical having wall thickness 10 mm. If the internal gauge pressure is 0.8 MPa, determine,

(a) stresses in the direction and perpendicular to the weld,

(b) normal stress and maximum shear stress in the spherical cap.

Solution **(a) Cylindrical tank:**

First consider the equilibrium of forces at a transverse cut for evaluating longitudinal stress σ_1 as shown in Fig. 2.20.

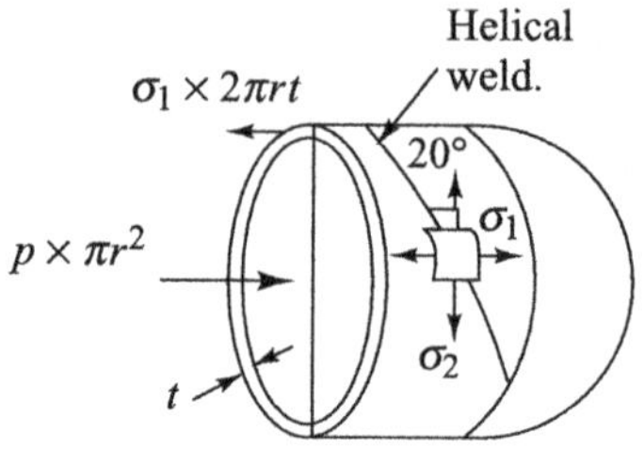

Fig. 2.20 *Transverse Section*

Inward force, $p \times \pi r^2$ = Outward force, $\sigma_1 \times \pi r \times t$

$$\text{or,} \quad \sigma_1 = \frac{pr}{2t} = \frac{0.8 \times 800}{2 \times 8} = 40 \text{ MPa}$$

Similarly consider the section of a unit length of tank as shown in Fig. 2.20 for circumferential stress σ_2,

$$2 \times \sigma_2 \times 1 \times t = p \times 1 \times 2r$$

$$\text{or,} \quad \sigma_2 = \frac{pr}{t} = \frac{0.8 \times 800}{8} = 80 \text{ MPa}$$

Let stresses across and along the weld be $\sigma_{x'}$ and $\tau_{x'y'}$. Angle between $\sigma_{x'}$ and σ_1 = 20°. Using Eqs (2.1) and (2.2),

$$\sigma_{x'} = \frac{40+80}{2} + \frac{40-80}{2}\cos 40°$$

$$= 44.68 \text{ MPa} \quad \text{(tension)}$$

$$\tau_{x'y'} = \frac{40-80}{2}\sin 40°$$

$$= -12.86 \text{ MPa} \quad \text{(Anticlockwise)}$$

(b) Spherical cap:

The stress in spherical cap is same in all directions due to symmetry. A section of a sphere along any diameter will be similar to the transverse section of the cylinder in Fig. 2.20.

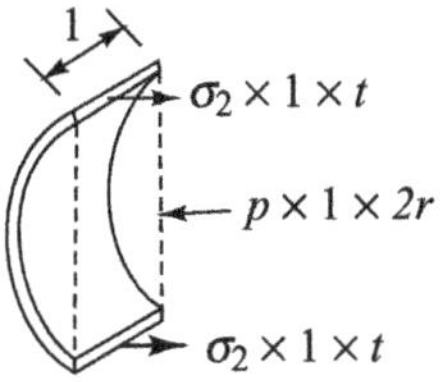

Fig. 2.21 *Longitudinal Section*

$$\therefore \quad \sigma_1 = \sigma_2 = \frac{pr}{2t} = \frac{0.8 \times 800}{2 \times 10} = 32 \text{ MPa}$$

Shear stress in the plane of the sphere is zero. The shear stress on the walls of the vessel is greater if third dimension is also considered.

Example 2.14 A horizontal force 1 kN is applied at the end of a cantilever bent of diameter 40 mm as shown in Fig. 2.22. Determine maximum principal normal stresses 200 mm above the base.

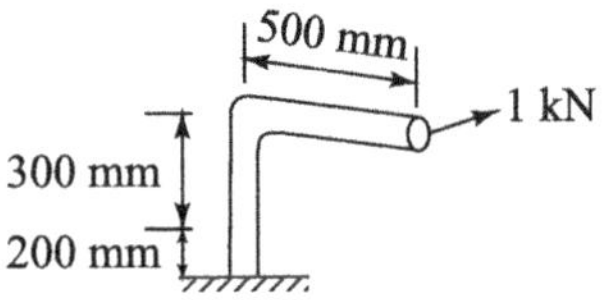

Fig. 2.22 *Cantilever Bent*

Solution The maximum stress will occur at a point P on the circumference as shown in Fig. 2.23. Consider free body diagram of a small element at point P,

Normal stress due to bending,

$$\sigma_y = \frac{M \times y}{I} = \frac{(1000 \times 300) \times 20}{\pi \times 40^4 / 64}$$
$$= 47.75 \text{ MPa}$$

Shear stress due to torsion,

$$\tau_{yx} = \frac{T \times y}{J} = \frac{5 \times 10^5 \times 20}{\pi \times 40^4 / 32} = 39.79 \text{ MPa}$$

Principal stresses,

$$\sigma_{x',\text{max/min}} = \frac{47.75}{2} + \sqrt{(47.75/2)^2 + 39.79^2}$$
$$= 70.28 \text{ MPa}, -22.53 \text{ MPa}$$

$$2\theta_p = \tan^{-1}\frac{-\tau_{xy}}{(\sigma_x - \sigma_y)/2} = \tan^{-1}\frac{-39.79}{47.75/2} = -59^\circ$$

The orientation of principal stresses is shown in Fig. 2.23, which can be verified by Eq. (2.1) or by Mohr's circle.

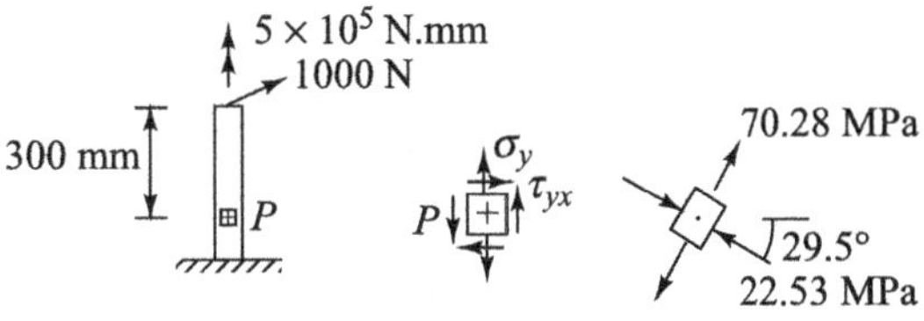

Fig. 2.23 *Free Body Diagrams*

Example 2.15 A cantilever 50 mm diameter is to carry an axial load 50 kN at an eccentricity of 5 mm below horizontal diameter in the vertical plane of symmetry along with a torque of 1.5×10^6 N mm as shown in Fig. 2.24. Calculate the stresses on a plane, normal of which is inclined at 30° clockwise to axis of the cantilever at a point on the lower edge of the vertical plane of symmetry.

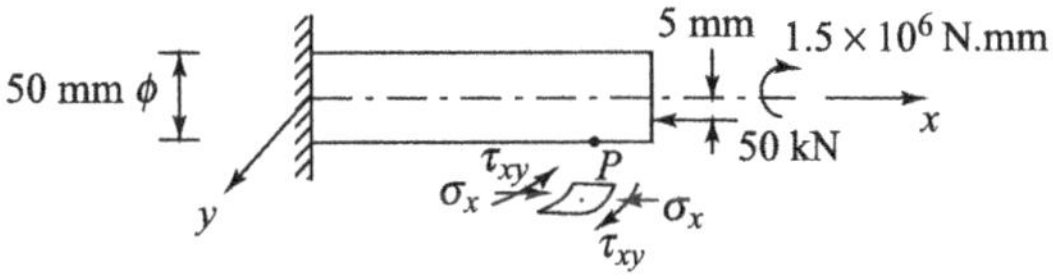

Fig. 2.24

Solution Consider an element at a lowermost point P.

Direct axial stress,

$$\sigma_x = \frac{-50000}{\pi \times 50^2 / 4}$$
$$= -25.46 \text{ MPa} \quad \text{(Compression)}$$

Bending compressive stress at P due to eccentric load,

$$\sigma_x = \frac{-M \times y}{I} = \frac{-(50000 \times 5) \times 25}{\pi \times 50^4 / 64}$$
$$= -20.37 \text{ MPa} \quad \text{(Compression)}$$

Total axial load,

$$\sigma_x = -25.46 - 20.37$$
$$= -45.83 \text{ MPa} \quad \text{(Compression)}$$

Shear stress due to torsion,

$$\tau_{xy} = \frac{T \times y}{J} = \frac{1.5 \times 10^6 \times 25}{\pi \times 50^4 / 32}$$

$$= 61.11 \text{ MPa} \quad \text{(Clockwise)}$$

(Complementary shear τ_{yx} is not shown in Fig. 2.24 for simplicity)

Plane AB shown in Fig. 2.25 has its normal inclined at 30° clockwise to the axial direction σ_x. Stresses on plane AB,

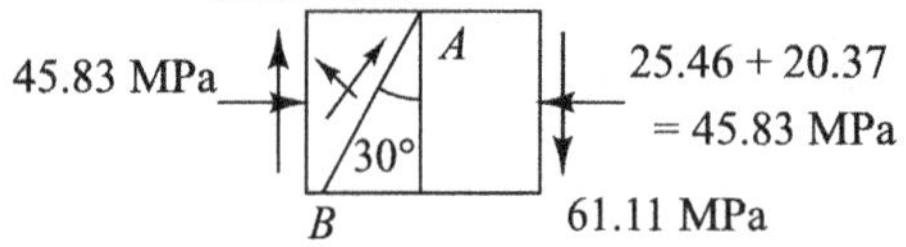

Fig. 2.25 *Element at P*

$$\sigma_{x'} = \frac{-45.83}{2} + \frac{-45.83}{2} \cos -60° - 61.11$$

$$\sin -60° = 18.55 \text{ MPa} \quad \text{(Tension)}$$

$$\tau_{x'y} = \frac{-45.83}{2} \sin -60° + 61.11 \cos -60°$$

$$= 50.4 \text{ MPa} \quad \text{(Clockwise)}$$

Example 2.16 The state of stress at a point in xyz reference is as follows.

$$\tau_{ij} = \begin{vmatrix} 80 & -60 & 0 \\ -60 & -40 & 0 \\ 0 & 0 & 40 \end{vmatrix} \text{MPa}$$

Determine the stress tensor in the direction $x'y'z'$ by rotating xyz through 30° anticlockwise about z-direction as shown in Fig. 2.26.

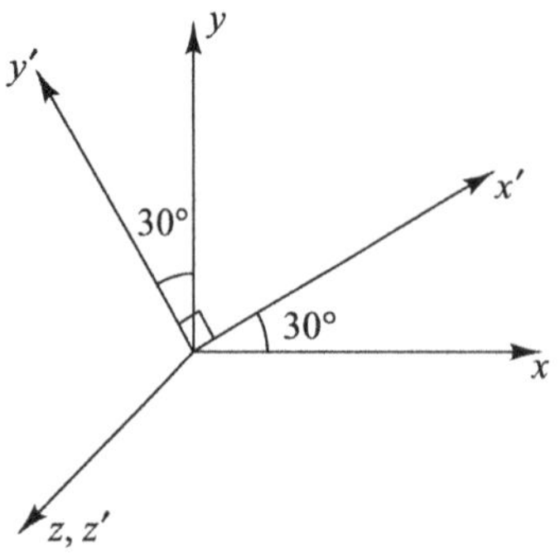

Fig. 2.26

Solution Direction cosines,

$c_{x'x} = \cos 30° = 0.866$, $c_{y'x} = \cos 120° = -0.5$,
$c_{x'y} = \cos 60° = 0.5$, $c_{y'y} = \cos 30° = 0.866$,
$c_{x'z} = \cos 90° = 0$, $c_{y'z} = \cos 90° = 0$,
$c_{z'x} = \cos 90° = 0$, $c_{z'y} = \cos 90° = 0$,
$c_{z'z} = \cos 0° = 1$

Using Eq. (2.10),

$$\tau_{ns} = \sum_{ij} \tau_{ij} \times c_{ni} \times c_{sj}$$

where all combinations of $n, s = x', y', z'$ and $i, j = x, y, z$ are exhausted.

(1) $n = x', s = x'$

$$\tau_{x'x'} = c_{x'x} c_{x'x} \tau_{xx} + c_{x'x} c_{x'y} \tau_{xy} + c_{x'x} c_{x'z} \tau_{xz}$$
$$+ c_{x'y} c_{x'x} \tau_{yx} + c_{x'y} c_{x'y} \tau_{yy} + c_{x'y} c_{x'z} \tau_{yz}$$
$$+ c_{x'z} c_{x'x} \tau_{zx} + c_{x'z} c_{x'y} \tau_{yz} + c_{x'z} c_{x'z} \tau_{zz}$$
$$= 0.866^2 \times 80 - 0.866 \times 0.5 \times 60 + 0 - 0.5 \times 0.866 \times 60 - 0.5^2 \times 40 + 0$$
$$= -1.96 \text{ MPa}$$

(2) $n = x', s = y'$

$$\tau_{x'y'} = c_{x'x} c_{y'x} \tau_{xx} + c_{x'x} c_{y'y} \tau_{xy} + c_{x'x} c_{y'z} \tau_{xz}$$
$$+ c_{x'y} c_{y'x} \tau_{yx} + c_{x'y} c_{y'y} \tau_{yy} + c_{x'y} c_{y'z} \tau_{yz}$$
$$+ c_{x'z} c_{y'x} \tau_{zx} + c_{x'z} c_{y'y} \tau_{yz} + c_{x'z} c_{y'z} \tau_{zz}$$
$$= -0.866 \times 0.5 \times 80 - 0.866^2 \times 60 + 0.5 \times 0.5 \times 60 - 0.5 \times 0.866 \times 40$$
$$= -81.96 \text{ MPa}$$

(3) $n = x', s = z'$

$$\tau_{x'z'} = c_{x'x} c_{z'x} \tau_{xx} + c_{x'x} c_{z'y} \tau_{xy} + c_{x'x} c_{z'z} \tau_{xz}$$
$$+ c_{x'y} c_{z'x} \tau_{yx} + c_{x'y} c_{z'y} \tau_{yy} + c_{x'y} c_{z'z} \tau_{yz}$$
$$+ c_{x'z} c_{z'x} \tau_{zx} + c_{x'z} c_{z'y} \tau_{yz} + c_{x'z} c_{z'z} \tau_{zz} = 0$$

(4) $n = y', s = y'$

$$\tau_{y'y'} = c_{y'x} c_{y'x} \tau_{xx} + c_{y'x} c_{y'y} \tau_{xy} + c_{y'x} c_{y'z} \tau_{xz}$$
$$+ c_{y'y} c_{y'x} \tau_{yx} + c_{y'y} c_{y'y} \tau_{yy} + c_{y'y} c_{y'z} \tau_{yz}$$
$$+ c_{y'z} c_{y'x} \tau_{zx} + c_{y'z} c_{y'y} \tau_{yz} + c_{y'z} c_{y'z} \tau_{zz}$$
$$= 0.5^2 \times 80 + 0.866 \times 0.5 \times 60 + 0 + 0.5 \times 0.866 \times 60 - 0.866^2 \times 40 + 0$$
$$= 41.96 \text{ MPa}$$

(5) $n = y', s = z'$

$$\tau_{y'z'} = c_{y'x} c_{z'x} \tau_{xx} + c_{y'x} c_{z'y} \tau_{xy} + c_{y'x} c_{z'z} \tau_{xz}$$
$$+ c_{y'y} c_{z'x} \tau_{yx} + c_{y'y} c_{z'y} \tau_{yy} + c_{y'y} c_{z'z} \tau_{yz}$$
$$+ c_{y'z} c_{z'x} \tau_{zx} + c_{y'z} c_{z'y} \tau_{yz} + c_{y'z} c_{z'z} \tau_{zz} = 0$$

(6) $n = z', s = z'$

$$\tau_{z'z'} = c_{z'x} c_{z'x} \tau_{xx} + c_{z'x} c_{z'y} \tau_{xy} + c_{z'x} c_{z'z} \tau_{xz}$$
$$+ c_{z'y} c_{z'x} \tau_{yx} + c_{z'y} c_{z'y} \tau_{yy} + c_{z'y} c_{z'z} \tau_{yz}$$
$$+ c_{z'z} c_{z'x} \tau_{zx} + c_{z'z} c_{z'y} \tau_{yz} + c_{z'z} c_{z'z} \tau_{zz}$$
$$= 1 \times 1 \times 40 = 40 \text{ MPa}$$

Since the stress tensor is symmetric, all the terms can now be written as follows.

$$\tau_{ns} = \begin{vmatrix} -1.96 & -81.96 & 0 \\ -81.96 & 41.96 & 0 \\ 0 & 0 & 40 \end{vmatrix}$$

Alternately these calculations can be done by matrix multiplication as follws.

$$\text{Stress tensor, } [\tau_{ij}] = \begin{vmatrix} 80 & -60 & 0 \\ -60 & -40 & 0 \\ 0 & 0 & 40 \end{vmatrix} \text{MPa}$$

Direction cosine matrix for transformation,

$$[C] = \begin{vmatrix} 0.866 & 0.5 & 0 \\ -0.5 & 0.866 & 0 \\ 0 & 0 & 1 \end{vmatrix}$$

Using Eq. (2.10d) in matrix notation,

$$[\tau_{ns}] = [C] \times [\tau_{ij}] \times [C]^T$$

$$\text{or, } [\tau_{ns}] = \begin{vmatrix} 0.866 & 0.5 & 0 \\ -0.5 & 0.866 & 0 \\ 0 & 0 & 1 \end{vmatrix} \begin{vmatrix} 80 & -60 & 0 \\ -60 & -40 & 0 \\ 0 & 0 & 40 \end{vmatrix}$$

$$\begin{vmatrix} 0.866 & -0.5 & 0 \\ -0.5 & 0.866 & 0 \\ 0 & 0 & 1 \end{vmatrix}$$

$$\text{or } [\tau_{ns}] = \begin{vmatrix} -1.96 & -81.96 & 0 \\ -81.96 & 41.96 & 0 \\ 0 & 0 & 40 \end{vmatrix} \text{MPa}$$

Example 2.17 Solve Example 2.16 if the rotation of *xyz* reference is 30° clockwise.

Solution Direction cosine matrix for transformation (Fig. 2.27)

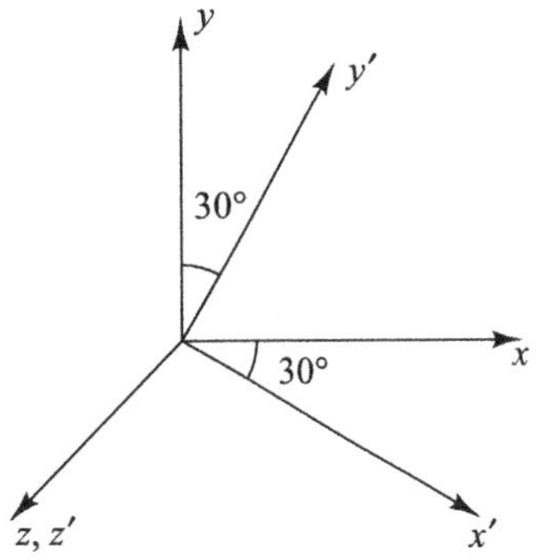

Fig. 2.27

$$[C] = \begin{vmatrix} 0.866 & -0.5 & 0 \\ 0.5 & 0.866 & 0 \\ 0 & 0 & 1 \end{vmatrix}$$

Using Eq. (2.10d)

$$[\tau_{ns}] = [C] \times [\tau_{ij}] \times [C]^T$$

$$\text{or, } [\tau_{ns}] = \begin{vmatrix} 0.866 & -0.5 & 0 \\ 0.5 & 0.866 & 0 \\ 0 & 0 & 1 \end{vmatrix} \begin{vmatrix} 80 & -60 & 0 \\ -60 & -40 & 0 \\ 0 & 0 & 40 \end{vmatrix}$$

$$\begin{vmatrix} 0.866 & 0.5 & 0 \\ -0.5 & 0.866 & 0 \\ 0 & 0 & 1 \end{vmatrix}$$

$$= \begin{vmatrix} 101.96 & 21.96 & 0 \\ 21.96 & -61.96 & 0 \\ 0 & 0 & 40 \end{vmatrix} \text{MPa}$$

Example 2.18 Figure 2.28 shows the state of stress at a point. Determine the principal normal and shear stresses and the orientation at which they act. Also determine the octahedral normal and shear stresses.

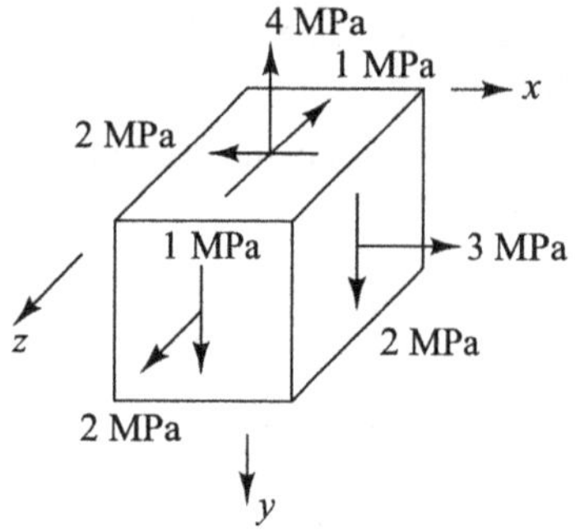

Fig. 2.28

Solution

$$\text{Stress tensor, } [\tau_{ij}] = \begin{vmatrix} 3 & 2 & 0 \\ 2 & 4 & 1 \\ 0 & 1 & 2 \end{vmatrix} \text{MPa}$$

Stress invariants,

$$I_1 = \tau_{xx} + \tau_{yy} + \tau_{zz} = 3 + 4 + 2 = 9$$

$$I_2 = \tau_{xx}\,\tau_{yy} + \tau_{yy}\,\tau_{zz} + \tau_{zz}\,\tau_{xx} - \tau_{xy}^2 - \tau_{yz}^2 - \tau_{zx}^2$$
$$= 3 \times 4 + 4 \times 2 + 2 \times 3 - 2^2 - 1^2 - 0$$
$$= 21$$

$$I_3 = \begin{vmatrix} \tau_{xx} & \tau_{xy} & \tau_{zx} \\ \tau_{xy} & \tau_{xy} & \tau_{zy} \\ \tau_{xz} & \tau_{yz} & \tau_{zz} \end{vmatrix} = \begin{vmatrix} 3 & 2 & 0 \\ 2 & 4 & 1 \\ 0 & 1 & 2 \end{vmatrix} = 13$$

Substituting in Eq. (2.13b),

$$\tau_{nn}^3 - I_1\,\tau_{nn}^2 + I_2\,\tau_{nn} - I_3 = 0$$

or, $$\tau_{nn}^3 - 9\,\tau_{nn}^2 + 21\,\tau_{nn} - 13 = 0$$

Solving by hit and trial, $\tau_{nn} = 1, 2.27, 5.73$

Using Eqs (2.11) and (2.12), for direction cosines of the plane,

when, principal normal stress $\sigma_I = 1$ MPa,

$$(c_{nx})^2 + (c_{ny})^2 + (c_{nz})^2 = 1$$
$$(3 - 1) \times c_{nx} + 2 \times c_{ny} = 0$$
$$2 \times c_{nx} + (4 - 1) \times c_{ny} + 1 \times c_{nz} = 0$$
$$1 \times c_{ny} + (2 - 1) \times c_{nz} = 0$$

Solution of these equations gives,

$$c_{nx} = c_{ny} = -\,c_{nz} = \pm 1/\sqrt{3}$$

Again using Eqs (2.11) and (2.12), for direction cosines of the plane, when, principal normal stress $\sigma_{II} = 2.27$ MPa,

$$(c_{nx})^2 + (c_{ny})^2 + (c_{nz})^2 = 1$$
$$(3 - 2.27) \times c_{nx} + 2 \times c_{ny} = 0$$
$$2 \times c_{nx} + (4 - 2.27) \times c_{ny} + 1 \times c_{nz} = 0$$
$$1 \times c_{ny} + (2 - 2.27) \times c_{nz} = 0$$

Solution of these equations gives,

$$-c_{nx} = \pm\, 0.5811,$$
$$c_{ny} = \pm\, 0.2121,$$

and $$c_{nz} = \pm\, 0.7855$$

Similarly, when principal normal stress σ_{III} = 5.73 MPa, the direction cosines are,

$$c_{nx} = \pm\, 0.5776,$$
$$c_{ny} = \pm\, 0.7884,$$

and, $$c_{nz} = \pm\, 0.2113$$

Normal octahedral stress, Eq. (2.14),

$$\tau_{nn,\text{oct}} = \frac{\sigma_I + \sigma_{II} + \sigma_{III}}{3} = \frac{1 + 2.27 + 5.73}{3}$$
$$= 3 \text{ MPa}$$

Octahedral shear stress is given by Eq. (2.18),

$$\tau_{ns,\text{oct}} = \frac{1}{3}\sqrt{\begin{matrix}(\sigma_I - \sigma_{II})^2 + (\sigma_{II} - \sigma_{III})^2 \\ +(\sigma_{III} - \sigma_I)^2\end{matrix}}$$

$$= \frac{1}{3}\sqrt{\begin{matrix}(1 - 2.27)^2 + (2.27 - 5.73)^2 \\ +(5.73 - 1)^2\end{matrix}} = 2 \text{ MPa}$$

Principal shear stress and normal stress on the planes of principal shear stress are given by Eq. (2.20).

$$\tau_I = \frac{2.27 - 1}{2} = \pm\, 0.635 \text{ MPa},$$

$$\sigma_n = \frac{1 + 2.27}{2} = 1.635 \text{ MPa}$$

$$\tau_{II} = \frac{5.73 - 2.27}{2} = \pm\, 1.73 \text{ MPa},$$

$$\sigma_n = \frac{2.27 + 5.73}{2} = 4 \text{ MPa}$$

$$\tau_{III} = \frac{5.73 - 1}{2} = \pm 2.365 \text{ Mpa},$$

$$\sigma_n = \frac{5.73 + 1}{2} = 3.365 \text{ MPa}$$

The planes of principal shear stress bisect the planes of principal normal stress.

Example 2.19 Figure 2.28 shows the state of stress at a point. Determine the stress tensor on $x'y'z'$ reference frame rotated such that x'-axis makes equal angle with x, y, z directions in first quadrant and y'-axis makes 60° with x-axis.

Solution Using simpler notations for direction cosines,

let l_1, m_1, n_1 = direction cosines of x' w.r.t. x, y, z axis,

l_2, m_2, n_2 = direction cosines of y' w.r.t. x, y, z axis,

l_3, m_3, n_3 = direction cosines of z' w.r.t. x, y, z axis

Since x', y' and z' are mutually perpendicular to each other, their direction cosines will be related as follows.

$$l_1 \times l_2 + m_1 \times m_2 + n_1 \times n_2 = 0 \quad \text{(i)}$$

$$l_2 \times l_3 + m_2 \times m_3 + n_2 \times n_3 = 0 \quad \text{(ii)}$$

$$l_3 \times l_1 + m_3 \times m_1 + n_3 \times n_1 = 0 \quad \text{(iii)}$$

Using Eq. (2.12) for relating direction cosines of x'-axis,

$$l_1^2 + m_1^2 + n_1^2 = 1 \quad \text{(iv)}$$

Since x'-axis is inclined equally w.r.t. x, y and z directions,

$$l_1 = m_1 = n_1 = 1/\sqrt{3} = \pm 0.5773$$

Using only +ve value for further calculations, substitute these values in Eq. (i),

$$l_2 + m_2 + n_2 = 0 \quad \text{(v)}$$

But, $l_2 = \cos 60° = 0.5$

$\therefore \quad m_2 = -0.5 - n_2$

Substituting in Eq. (2.12),

$$0.5^2 + (-0.5 - n_2)^2 + n_2^2 = 1$$

or, $n_2 = 0.309$ and $m_2 = -0.809$

Similarly using Eqs (ii), (iii) and (2.12),

$$l_3 = -0.6455, \quad m_3 = -0.1103, \quad n_3 = 0.7557$$

Stress tensor in rotated $x'y'z'$ reference direction is obtained by using Eq. (2.10d) in matrix notation,

$$[\tau_{ns}] = [C] \times [\tau_{ij}] \times [C]^T$$

$$[\tau_{ns}] = \begin{bmatrix} 0.5773 & 0.5773 & 0.5773 \\ 0.5 & -0.809 & 0.309 \\ -0.6455 & -0.1103 & 0.7557 \end{bmatrix} \begin{bmatrix} 3 & 2 & 0 \\ 2 & 4 & 1 \\ 0 & 1 & 2 \end{bmatrix}$$

$$\begin{bmatrix} 0.5573 & 0.5 & -0.6455 \\ 0.5773 & -0.809 & -0.1103 \\ 0.5773 & 0.309 & 0.7557 \end{bmatrix}$$

$$= \begin{vmatrix} 5.0 & -1.29 & -1.0 \\ -1.29 & 1.44 & 0.144 \\ -1.0 & 0.144 & 2.56 \end{vmatrix} \text{MPa}$$

Example 2.20 Tensile stresses of 100 MPa and 50 MPa are acting on two perpendicular planes in a body. Determine the resultant stress inclined at 35° to the major principal plane using stress ellipsoid.

Solution Draw two circles corresponding to the principal stress 100 MPa and 50 MPa as shown in Fig. 2.29. Draw a line *ABO* at 35° to the major principal stress. Intersection point *C* of a horizontal line from *B* and a vertical line from *A*, is the location of the resultant stress $\sigma_r = OC = 86.8$ MPa. Similarly, the resultant stress at any plane can be obtained and the locus of point *C* will be an ellipse. Mathematically,

$$OD = \sigma_1 \cos 35° = 100 \cos 35°$$

$$CD = \sigma_2 \sin 35° = 50 \sin 35°$$

From Eqs (2.1a) and (2.2),

$$\sigma_{x'} = \sigma_1 \cos^2 \theta + \sigma_2 \sin^2 \theta \quad \text{(i)}$$

and $\tau_{x'y'} = (\sigma_1 - \sigma_2) \times \sin \theta \times \cos \theta$ (ii)

Resultant stress of (i) and (ii),

$$\sigma_r = \sqrt{(\sigma_{x'})^2 + (\tau_{x'y'})^2}$$

$$= \sqrt{(\sigma_1 \cos^2 \theta + \sigma_2 \sin^2 \theta)^2 + ((\sigma_1 - \sigma_2) \sin \theta \cos \theta)^2}$$

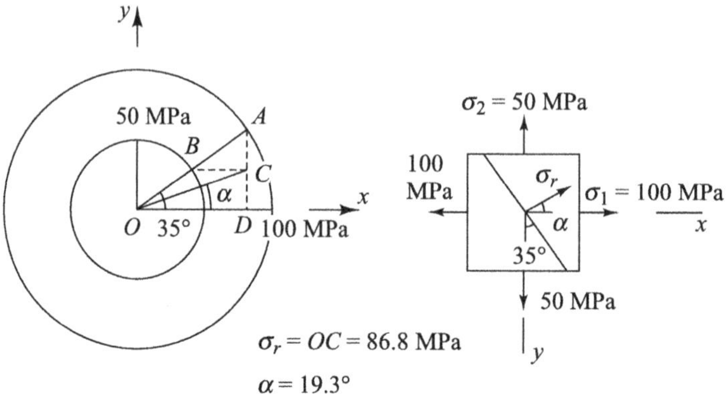

Fig. 2.29

$$= \sqrt{(100\cos 35^\circ)^2 + (50\sin 35^\circ)^2}$$
$$= 86.8 \text{ MPa}$$

Example 2.21 Write a computer program to calculate normal and shear stress on any inclined plane. It should calculate principal stresses and their location. Input data for the program will be the direct and shear stresses on two mutually perpendicular planes (σ_x, σ_y, τ_{xy}) and the angle of the inclined plane.

Solution The following computer program in Qbasic will solve for $\sigma_{x'}$ = FX1, $\tau_{x'y'}$ = SXY1, principal direct and shear stresses = FMAX, FMIN, SMAX. Location of principal direct stress = TP.

```
REM CALCULATE - DIRECT AND SHEAR
'               FORCE ON PLANE
'               INCLINED AT ANGLE
'               THETA DEG. TO X-AXIS
'     CALCULATE - PRINCIPAL STRESSE
'               ANI THEIR LOCATION
' SIGN CONVENTION - TENSION +VE, COMP.-ve,
' SHEAR CLOCKWISE +VE, THETA
' ANTICLOCKWISE +VE
' FX, FY = DIRECT FORCE IN X, Y
' DIRECTIONS
' SXY = SHEAR FORCE IN XY DIRECTION
CLS
INPUT FX, FY, SXY, THETA
FP = (FX + FY)/2
FM = (FX - FY)/2
T2 = 2 * THETA * 3.1415927/180
FX1 = FP + FM * COS (T2) - SXY * SIN (T2)
SXY1 = FM * SIN (T2) + SXY * COS (T2)
SMAX = SQR (FM ^ 2 + SXY ^ 2)
FMAX = FP + SMAX
FMIN = FP - SMAX
TP = .5 * ATN (-SXY/FM)* 180/3.1415927
PRINT "FX1 = "; FX1, "SXY1 = "; SXY1
PRINT "FMAX = ";FMAX; "FMIN = "; FMIN;;
PRINT " ANGLE TP = ";  TP
PRINT "SMAX = ";SMAX
END
```

If screen input for Ex. 2.3 is : – 40, 80, 60, 135 then screen output will be as follows,

FX1 = 80 SXY1 = 60

FMAX = 104.8528 FMIN = –64.85281

ANGLE TP = 22.5
SMAX = 84.85281

Example 2.22 Figure 2.30(a) shows an element under axial stress 20 MPa and 50 MPa along with shear stress 40 MPa in x–y plane. Determine the state of stress in x' – y' plane at 30° counter-clockwise to x–y plane.

Solution Equation (2.11b) can be used to calculate stresses in x'– y'plane and Eq. (2.13a) will give principal stresses. The directive cosines in transfer matrix are drived from Fig. 2.30(b) as follows.

$$c_{x'x} = \cos\theta$$
$$c_{x'y} = \cos(90° + \theta) = -\sin\theta$$
$$c_{y'x} = \cos(90° - \theta) = \sin\theta$$
$$c_{y'y} = \cos\theta$$

Substitute these values in Eq. (2.10b) for stresses in x'– y'reference axes as follows.

$$\begin{bmatrix} \sigma_{x'x'} & \tau_{x'y'} \\ \tau_{y'x'} & \sigma_{y'y'} \end{bmatrix} = \begin{bmatrix} c_{x'x} & c_{x'y} \\ c_{y'x} & c_{y'y} \end{bmatrix} \begin{bmatrix} \sigma_{xx} & \tau_{xy} \\ \tau_{yx} & \sigma_{yy} \end{bmatrix} \begin{bmatrix} c_{x'x} & c_{y'x} \\ c_{x'y} & c_{y'y} \end{bmatrix}$$

$$= \begin{bmatrix} 0.866 & -0.5 \\ 0.5 & 0.866 \end{bmatrix} \begin{bmatrix} 20 & 40 \\ 40 & -50 \end{bmatrix} \begin{bmatrix} 0.866 & 0.5 \\ -0.5 & 0.866 \end{bmatrix}$$

$$= \begin{bmatrix} -32.14 & 50.3 \\ 50.3 & 2.14 \end{bmatrix}$$

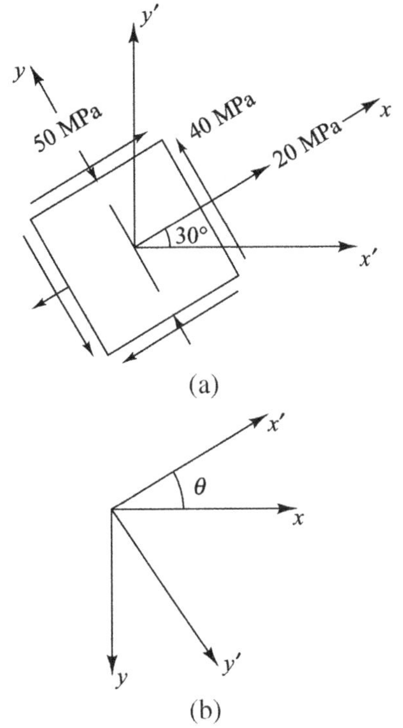

Fig. 2.30

Principal stresses are given by Eq.(2.13a),

$$\begin{bmatrix} -32.14 - \sigma & 50.3 \\ 50.3 & 2.14 - \sigma \end{bmatrix} = 0$$

which gives principal stresses

$$\sigma = 38.14 \text{ MPa and } -68.14 \text{ MPa}$$

EXERCISES

PROBLEMS

2.1 For the state of plane stress on the square element *ABCD* shown in Fig. 2.31, determine the stresses on plane *AA'* by considering the equilibrium of the triangular portion *AA'B*. Verify using Eqs (2.1) and (2.2).

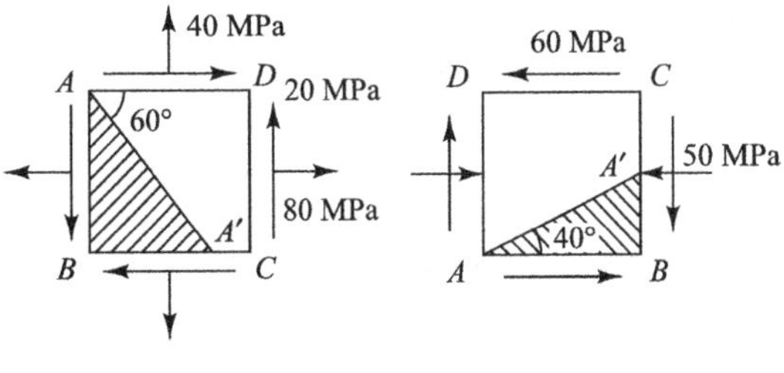

Fig. 2.31

2.2 Direct stresses of 20 MPa (tension) and 80 MPa are acting on two perpendicular planes. What shearing stress can be applied if the direct stress cannot exceed 150 MPa.

2.3 In a plane stressed body, the resultant stress at a plane *P* is 70 MPa inclined at 15° to its normal and at another plane *Q*, the resultant stress is 50 MPa inclined at 30° to the normal stress. Determine the angle between the planes *P* and *Q*. Also find the principal stress in the body.

2.4 Determine the maximum load *P* that can be applied to the shaft shown in Fig. 2.32, if the maximum tensile stress is not to exceed 150 MPa.

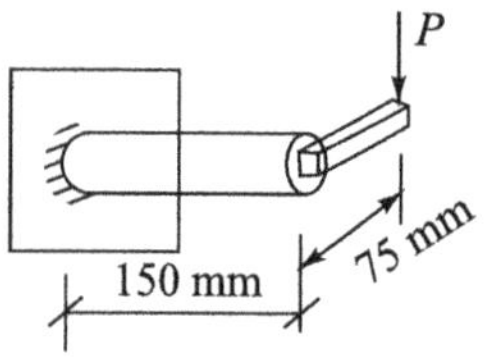

Fig. 2.32

2.5 A cantilever T-beam of cross-section shown in Fig. 2.33, is to carry an u.d.l. of 5 kN/m. Determine the maximum stresses at 3 m from the free end, at a point 25 mm above the bottom fiber.

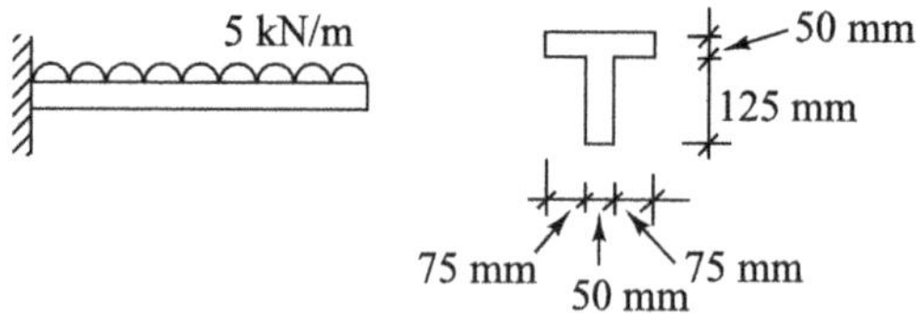

Fig. 2.33

2.6 Two vertical forces are applied at the free end of a cantilever bar of diameter 60 mm as shown in Fig. 2.34. Determine the maximum stresses at the top fiber 20 cm away from the free end.

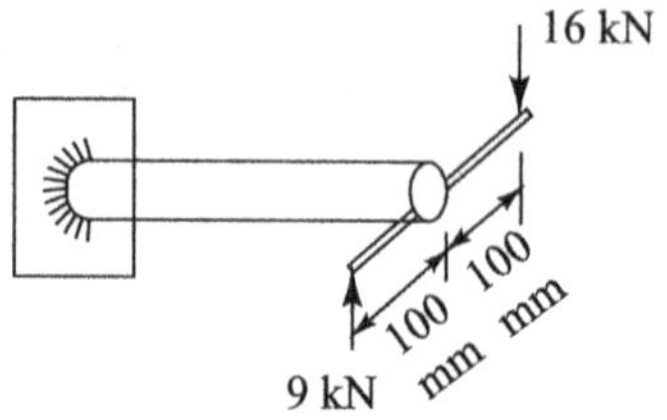

Fig. 2.34

2.7 A steel pipe having inside diameter 200 mm is fabricated by welding 6 mm thick plate along a helix which forms an angle 15° with the plane perpendicular to the axis of the pipe. Determine the normal and tangential stresses along the weld if the pipe is under torque of 5 kN.m and internal pressure 3 MPa (Fig. 2.35).

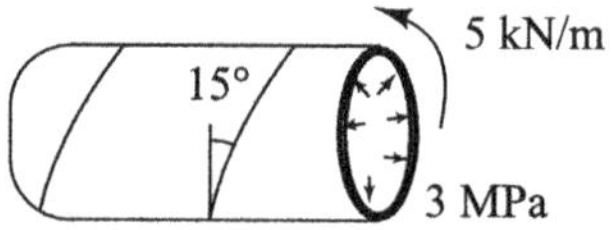

Fig. 2.35

2.8 Determine the magnitudes of principal normal and tangential stresses for the state of stress shown in Fig 2.36. Also determine octahedral stresses and normal stresses on a plane inclined equally with the three +ve *xyz* directions.

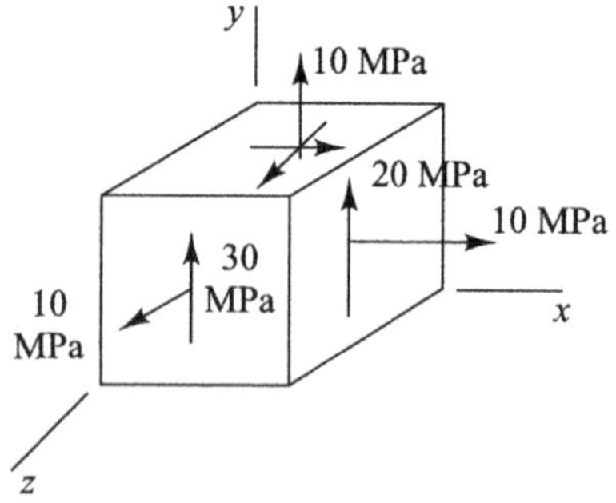

Fig. 2.36

2.9 Derive a relation for octahedral shear stress in terms of stress invariants.

2.10 The state of stress at a point is given by the stress tensor as follows.

$$[\tau_{ij}] = \begin{vmatrix} 3 & 2 & 0 \\ 2 & 4 & 1 \\ 0 & 1 & 2 \end{vmatrix} \text{MPa}$$

Determine the stress tensor in the rotated $x'y'z'$directions if x' lies in first quadrant such that the direction cosine w.r.t. x-axis is half than the direction cosines w.r.t. y and z axes. One direction for other two $y'$and z'axes are also numerically half of the other direction cosines.

2.11 Determine octahedral shear and normal stress for the state of stress at a point as follows.

$$[\tau_{ij}] = \begin{vmatrix} 10 & 5 & 6 \\ 5 & 3 & 8 \\ 6 & 8 & 5 \end{vmatrix}$$

OBJECTIVE QUESTIONS

2.12 A plate is under bi-axial stress σ_1 and σ_2 in two mutually perpendicular directions alongwith stress τ then the maximum normal stress in the plate is,

(a) $\dfrac{\sigma_1+\sigma_2}{2}+\dfrac{1}{2}\sqrt{(\sigma_1+\sigma_2)^2+4\tau^2}$

(b) $\dfrac{\sigma_1-\sigma_2}{2}+\dfrac{1}{2}\sqrt{(\sigma_1+\sigma_2)^2+4\tau^2}$

(c) $\dfrac{\sigma_1+\sigma_2}{2}+\dfrac{1}{2}\sqrt{(\sigma_1-\sigma_2)^2+4\tau^2}$

(d) $\dfrac{\sigma_1+\sigma_2}{2}-\dfrac{1}{2}\sqrt{(\sigma_1+\sigma_2)^2+4\tau^2}$

2.13 A machine shaft is subjected to axial tensile stress σ and a shear stress τ due to torsion. The maximum shear induced will be,

(a) $\sqrt{\sigma^2+\tau^2}$ (b) $\sqrt{\sigma^2/2+\tau^2}$

(c) $\dfrac{1}{2}\sqrt{\sigma^2+4\tau^2}$ (d) $\dfrac{1}{4}\sqrt{\sigma^2+4\tau^2}$

2.14 A bar is under axial stress of 40 MPa. The maximum shear stress in the bar is,

(a) 0 MPa (b) 5 MPa
(c) 10 MPa (d) 20 MPa

2.15 The diameter of Mohr's circle for a thin plate under tensile stress of 20 MPa in both x and y directions, is equal to,

(a) zero (b) 10 MPa
(c) 20 MPa (d) 40 MPa

2.16 A thin plate is under tensile stress of 80 MPa in x-direction and shear stress τ_{xy} = 15 MPa (clockwise). If tensile stress of 20 MPa is acting in y-direction, then the normal stress in the direction of 45° anticlockwise to x-direction will be,

(a) 35 MPa (b) 65 MPa
(c) 95 MPa, (d) 115 MPa

2.17 A thin plate is under tensile stress of 80 MPa in x-direction and shear stress τ_{xy} = 15 MPa (clockwise). If tensile stress of 20 MPa is acting in y-direction, then the normal stress in the direction of 45° clockwise to x-direction will be,

(a) 35 MPa (b) 65 MPa
(c) 95 MPa (d) 115 MPa

2.18 A plate is under tensile and compressive stress of 40 MPa in two mutually perpendicular directions. If shear stress of 30 MPa is also acting in these two directions, then the maximum stress in the plate is,

(a) 50 MPa (b) 70 MPa
(c) 80 MPa (d) 90 MPa

2.19 The normal and shear stresses at a point on plane A are 40 MPa (tension) and 30 MPa (clockwise), respectively. The normal and shear stresses at the same point on another plane B are 40 MPa (compressive) and 30 MPa (anticlockwise), respectively. The angle between plane A and B is,

(a) 45° (b) 90°
(c) 135° (d) 180°

2.20 The normal stress on two perpendicular planes A and B are 60 MPa and 20 MPa, respectively. The shear stress on these planes is 20 MPa. The plane having zero shear stress is inclined to plane B at,

(a) 7.5° (b) 15°
(c) 22.5° (d) 30°

2.21 The first stress invariant for the stress tensor,

$$[\tau_{ij}] = \begin{vmatrix} 10 & -20 & 30 \\ -20 & 50 & 60 \\ 30 & 60 & -60 \end{vmatrix}$$

is equal to,

(a) 0 (b) 10 MPa
(c) – 10 MPa (d) 100 MPa

2.22 The state of stress given by the stress tensor is of pure shear as follows.

(a) $\begin{vmatrix} 5 & 37 & -65 \\ 37 & 5 & 15 \\ -65 & 15 & -10 \end{vmatrix}$ (b) $\begin{vmatrix} 0 & 0 & 25 \\ 0 & 0 & -15 \\ 25 & -15 & -10 \end{vmatrix}$

(c) $\begin{vmatrix} 12 & -2 & 25 \\ -2 & 0 & 15 \\ 25 & 15 & -10 \end{vmatrix}$ (d) $\begin{vmatrix} 0 & 0 & 25 \\ 0 & 0 & 15 \\ 25 & 15 & -10 \end{vmatrix}$

3

Analysis of Strains

CONCEPT REVIEW

3.1 INTRODUCTION

It may not be possible to measure stress at every location but it can be calculated if the corresponding strains are known by measurements using strain gauges. A strain gauge can measure direct or linear strain. Shear strains are calculated by measuring direct strains in 3 or 4 directions using a set of strain gauges called strain rosette. It is therefore required to derive a relation between strains in various directions. Since strains are proportional to stresses, the relations are similar to those in Chapter 2.

3.2 TRANSFORMATION OF PLANE STRAINS

Consider a thin plate under two-dimensional stresses σ_x, σ_y and τ_{xy} as shown in Fig. 3.1, which causes strains ε_x, ε_y and γ_{xy} in their respective directions. $\varepsilon_{x'}$, $\varepsilon_{y'}$ and $\gamma_{x'y'}$ are the strains in rotated $x' - y'$ coordinates. A triangular element ABC is distorted to $A'B'C'$ such that,

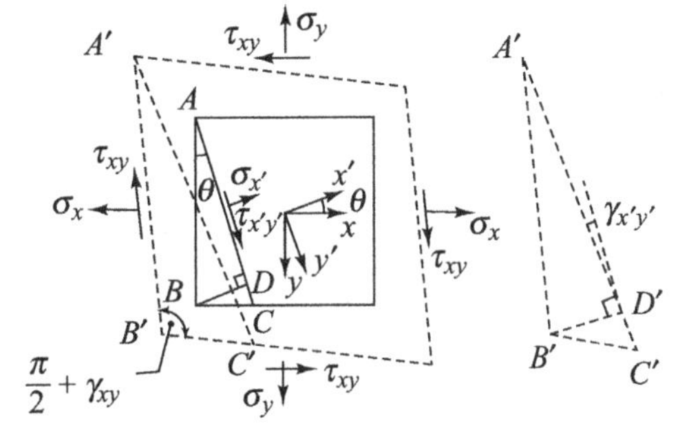

Fig. 3.1

$$\angle A'B'C' = 90^\circ + \gamma_{xy}$$

$$B'C' = BC\,(1 + \varepsilon_x)$$

$$A'B' = AB\,(1 + \varepsilon_y)$$

$$A'C' = AC\,(1 + \varepsilon_{y'})$$

$$(A'C')^2 = (B'C')^2 + (A'B')^2 - 2 \times B'C' \times A'B' \times \cos(90^\circ + \gamma_{xy})$$

or, $$AC^2\,(1 + \varepsilon_{y'})^2 = BC^2\,(1 + \varepsilon_x)^2 + AB^2\,(1 + \varepsilon_y)^2 + 2 \times BC \times AB\,(1 + \varepsilon_x)(1 + \varepsilon_y)\sin\gamma_{xy}$$

Neglecting higher order terms of strains and substituting $\sin \gamma_{xy} \approx \gamma_{xy}$,

$$\text{or,}\quad \varepsilon_{y'} = \frac{BC^2}{AC^2}\varepsilon_x + \frac{AB^2}{AC^2}\varepsilon_y + \frac{BC}{AC}\times\frac{AB}{AC}\times\gamma_{xy}$$

$$\text{or,}\quad \varepsilon_{y'} = \varepsilon_x \sin^2\theta + \varepsilon_y \cos^2\theta + \frac{\gamma_{xy}}{2}\sin 2\theta \tag{3.1a}$$

For strains in x'-directions, substitute $\theta + 90°$ for θ,

$$\text{or,}\quad \varepsilon_{x'} = \varepsilon_x \sin^2(90° + \theta) + \varepsilon_y \cos^2(90° + \theta) + \frac{\gamma_{xy}}{2}\sin 2(90° + \theta)$$

$$\text{or,}\quad \varepsilon_{x'} = \varepsilon_x \cos^2\theta + \varepsilon_y \sin^2\theta - \frac{\gamma_{xy}}{2}\sin 2\theta \tag{3.1b}$$

$$\text{or,}\quad \varepsilon_{x'} = \frac{\varepsilon_x + \varepsilon_y}{2} + \frac{\varepsilon_x - \varepsilon_y}{2}\cos 2\theta - \frac{\gamma_{xy}}{2}\sin 2\theta \tag{3.1c}$$

which is similar to Eq. (2.1).

For shear strain $\gamma_{x'y'}$, consider $\Delta A'B'D'$, where $B'D'$ is parallel to BD of undistorted ΔABC.

$$\angle A'D'B' = 90° - \gamma_{x'y'}$$

$$(A'B')^2 = (B'D')^2 + (A'D')^2 - 2\times B'D' \times A'D' \times \cos(90° - \gamma_{x'y'})$$

$$\text{or,}\quad AB^2(1+\varepsilon_y)^2 = BD^2(1+\varepsilon_{x'})^2 + AD^2(1+\varepsilon_{y'})^2 - 2\times BD\times AD\times(1+\varepsilon_{x'})(1+\varepsilon_{y'})\sin\gamma_{x'y'}$$

Neglecting higher order terms of strains and substituting $\sin\gamma_{x'y'} \approx \gamma_{xy}$,

$$\text{or,}\quad \varepsilon_y = \frac{BD^2}{AB^2}\varepsilon_{x'} + \frac{AD^2}{AB^2}\varepsilon_{y'} - \frac{BD}{AB}\times\frac{AD}{AB}\times\gamma_{x'y'}$$

$$\text{or,}\quad \gamma_{x'y'} = \frac{\varepsilon_{x'}\sin^2\theta + \varepsilon_{y'}\cos^2\theta - \varepsilon_y}{\sin\theta\times\cos\theta}$$

Substituting values of $\varepsilon_{x'}$ $\varepsilon_{y'}$ from Eq. (3.1),

$$\frac{\gamma_{x'y'}}{2} = \frac{\varepsilon_x - \varepsilon_y}{2}\sin 2\theta + \frac{\gamma_{xy}}{2}\cos 2\theta \tag{3.2}$$

which is similar to Eq. (2.2).

3.2.1 Sign Conventions

The sign of strains will be positive corresponding to positive state of stresses. Thus the positive state of strains for the square plate in Fig. 3.1 is shown by the rhombus in dotted line corresponding to the positive state of stress in the $x - y$ reference plane.

3.2.2 Principal Strains

As Eqs (2.1), (2.2) and (3.1), (3.2) are similar and the derivation of equations for principal strains is based on these two equations, they can be written by replacing σ_x, σ_y with $\varepsilon_{x'}$ ε_y and τ_{xy} with $\gamma_{xy}/2$ in Eq. (2.3) through (2.7) as follows.

Principal strains,

$$\varepsilon_{\text{max/min}} = \frac{\varepsilon_x + \varepsilon_y}{2} \pm \frac{1}{2}\sqrt{(\varepsilon_x - \varepsilon_y)^2 + (\gamma_{xy})^2} \tag{3.3}$$

Shear strain on these planes is zero and therefore shear stress on these planes is zero. This shows that the principal strains and the principal stresses act on same plane.

$$\text{Principal strain angle, } \tan 2\theta_p = -\frac{\gamma_{xy}}{\varepsilon_x - \varepsilon_y} \tag{3.4}$$

Similarly, maximum shear strain is given by,

$$\frac{\gamma_{\max}}{2} = \pm\frac{1}{2}\sqrt{(\varepsilon_x - \varepsilon_y)^2 + (\gamma_{xy})^2} = \frac{\varepsilon_{\max} - \varepsilon_{\min}}{2} \tag{3.5}$$

Normal strain on the plane of maximum shear strain,

$$\varepsilon = \frac{\varepsilon_x + \varepsilon_y}{2} \tag{3.6}$$

and principal shear strain angle,

$$\tan 2\theta_s = \frac{\varepsilon_x - \varepsilon_y}{\gamma_{xy}} \tag{3.7}$$

3.2.3 Mohr's Circle for Strains

The similarity argument of Art. 3.3 is also applicable for Mohr's circle of stress and Mohr's circle of strain is drawn by replacing σ_x, σ_y with ε_x, ε_y and τ_{xy} with $\gamma_{xy}/2$.

3.3 GENERAL THREE-DIMENSIONAL STATE OF STRAIN

Equation (2.10d) for general three dimensional state of stress can similarly be written for strains as follows.

$$\varepsilon_{ns} = \sum_i \sum_j \varepsilon_{ij} \times c_{ni} \times c_{sj} \quad (3.8a)$$

where $\quad i, j = x, y, z$
$\quad n, s = x', y', z'$

Equation (3.8a), can be written in matrix form as follows,

$$\begin{bmatrix} \varepsilon_{x'x'} & \varepsilon_{x'y'} & \varepsilon_{x'z'} \\ \varepsilon_{y'x'} & \varepsilon_{y'y'} & \varepsilon_{y'z'} \\ \varepsilon_{z'x'} & \varepsilon_{z'y'} & \varepsilon_{z'z'} \end{bmatrix} = \begin{bmatrix} c_{x'x} & c_{x'y} & c_{x'z} \\ c_{y'x} & c_{y'y} & c_{y'z} \\ c_{z'x} & c_{z'y} & c_{z'z} \end{bmatrix} \times \begin{bmatrix} \varepsilon_{xx} & \varepsilon_{xy} & \varepsilon_{xz} \\ \varepsilon_{yx} & \varepsilon_{yy} & \varepsilon_{yz} \\ \varepsilon_{zx} & \varepsilon_{zy} & \varepsilon_{zz} \end{bmatrix} \times \begin{bmatrix} c_{x'x} & c_{y'x} & c_{z'x} \\ c_{x'y} & c_{y'y} & c_{z'y} \\ c_{x'z} & c_{y'z} & c_{z'z} \end{bmatrix} \quad (3.8b)$$

All diagonal terms of strain tensors $\varepsilon_{ns} = \varepsilon_{x'x'}, \varepsilon_{y'y'}, \varepsilon_{z'z'}$ and $\varepsilon_{ij} = \varepsilon_{xx}, \varepsilon_{yy}, \varepsilon_{zz}$ are direct strains in their respective directions and all other non-diagonal terms are half of the respective shear strains, e.g. $\varepsilon_{x'x'} = \gamma_{x'y'}/2$.

Principal strains are given by the roots of Eq. (3.9) similar to Eq. (2.13b),

$$\varepsilon_{nn}^{\ 3} - J_1 \varepsilon_{nn}^{\ 2} + J_2 \varepsilon_{nn} - J_3 = 0 \quad (3.9)$$

where strain invariants,

$$J_1 = \varepsilon_{xx} + \varepsilon_{yy} + \varepsilon_{zz}$$

$$J_2 = \varepsilon_{xx}\varepsilon_{yy} + \varepsilon_{yy}\varepsilon_{zz} + \varepsilon_{zz}\varepsilon_{xx} - \frac{1}{4}(\gamma_{xy}^2 + \gamma_{yz}^2 + \gamma_{zx}^2)$$

$$J_3 = \begin{vmatrix} \varepsilon_{xx} & \gamma_{xy}/2 & \gamma_{xz}/2 \\ \gamma_{yx}/2 & \varepsilon_{yy} & \gamma_{yz}/2 \\ \gamma_{zx}/2 & \gamma_{yz}/2 & \varepsilon_{zz} \end{vmatrix}$$

Location of principal strains is then given by Eqs. (3.10) and (2.12),

$$(\varepsilon_{xx} - \varepsilon_{nn}) \times c_{nn} + \varepsilon_{xy} \times c_{ny} + \varepsilon_{xz} \times c_{nz} = 0 \quad (3.10a)$$

$$\varepsilon_{yx} \times c_{nx} + (\varepsilon_{yy} - \varepsilon_{nn}) \times c_{ny} + \varepsilon_{yz} \times c_{nz} = 0 \quad (3.10b)$$

$$\varepsilon_{zx} \times c_{nx} + \varepsilon_{zy} \times c_{ny} + (\varepsilon_{zz} - \varepsilon_{nn}) \times c_{nz} = 0 \quad (3.10c)$$

Similar to hydrostatic stress and pure shear stress conditions of Eq. (2.23), all strain tensors can be divided into deviatorial and spherical strains as follows.

$$\varepsilon_{ij} = \begin{bmatrix} \varepsilon_{xx} & \varepsilon_{xy} & \varepsilon_{xz} \\ \varepsilon_{yx} & \varepsilon_{yy} & \varepsilon_{yz} \\ \varepsilon_{zx} & \varepsilon_{zy} & \varepsilon_{zz} \end{bmatrix} \begin{bmatrix} \varepsilon_{xx} - \varepsilon & \varepsilon_{xy} & \varepsilon_{xz} \\ \varepsilon_{yx} & \varepsilon_{yy} - \varepsilon & \varepsilon_{yz} \\ \varepsilon_{zx} & \varepsilon_{zy} & \varepsilon_{zz} - \varepsilon \end{bmatrix} + \begin{bmatrix} \varepsilon & 0 & 0 \\ 0 & \varepsilon & 0 \\ 0 & 0 & \varepsilon \end{bmatrix} \quad (3.11)$$

where $\quad \varepsilon = \dfrac{1}{3}\,(\varepsilon_{xx} + \varepsilon_{xy} + \varepsilon_{xz}) \quad (3.12)$

3.4 RELATION BETWEEN *E*, *G* AND μ

Consider shear stress τ acting on a plane element *ABCD*. The corresponding Mohr's circles for the stress and strain show on that the principal stress σ and principal strain ε act at 45° as shown in Figs 3.2 (a – c). It is also seen from the Mohr's circles that the magnitude of principal normal stress $\sigma = \tau$ and the magnitude of principal strain $\varepsilon = \gamma/2$.

Substituting in Eq. (1.4),

$$\gamma = \frac{\tau}{G}$$

or, $\quad 2\varepsilon = \dfrac{\sigma}{G} \quad$ (i)

The principal strain and principal stress are related by Eq. (1.23a),

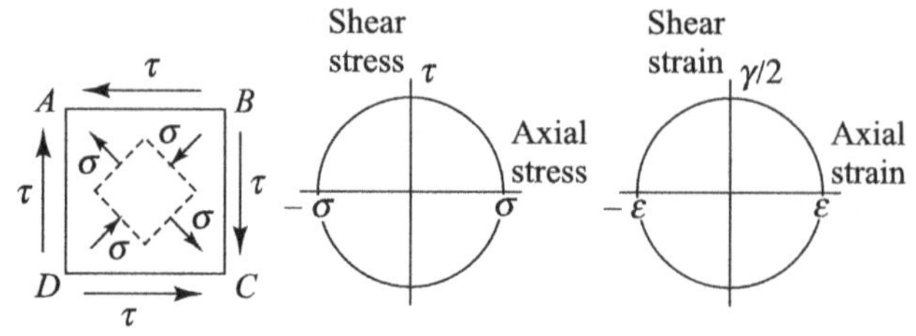

(a) Element $ABCD$ under shear (b) Mohr's circle for stress (c) Mohr's circle for strain

Fig. 3.2

$$\varepsilon = \frac{1}{E}\,[\sigma - \mu \times (-\sigma)] = \frac{\sigma(1+\mu)}{E} \qquad \text{(ii)}$$

From Eqs (i) and (ii), $G = \dfrac{E}{2(1+\mu)}$ (3.13)

The generalized Hook'e law, Eq. (1.23), can thus be written as,

$$\begin{Bmatrix} \varepsilon_x \\ \varepsilon_y \\ \varepsilon_z \\ \gamma_{xy} \\ \gamma_{yz} \\ \gamma_{xz} \end{Bmatrix} = \frac{1}{E}\begin{bmatrix} 1 & -\mu & -\mu & 0 & 0 & 0 \\ -\mu & 1 & -\mu & 0 & 0 & 0 \\ -\mu & -\mu & 1 & 0 & 0 & 0 \\ 0 & 0 & 0 & 2(1+\mu) & 0 & 0 \\ 0 & 0 & 0 & 0 & 2(1+\mu) & 0 \\ 0 & 0 & 0 & 0 & 0 & 2(1+\mu) \end{bmatrix} \begin{Bmatrix} \sigma_x \\ \sigma_y \\ \sigma_z \\ \tau_{xy} \\ \tau_{yz} \\ \tau_{xz} \end{Bmatrix} \qquad (3.14a)$$

or, $\{\varepsilon\} = [N] \times \{\sigma\}$ (3.14b)

$[E]$ matrix in Eq. (1.23g) is the inverse of $[N]$ matrix of Eq. (3.14b).

3.4.1 Plane Stress Condition

It is a two-dimensional state of stress condition in which $\sigma_z = \tau_{xz} = \tau_{yz} = 0$. Equation (3.14) is then reduced to Eq. (3.15) as follows.

$$\begin{Bmatrix} \varepsilon_x \\ \varepsilon_y \\ \gamma_{xy} \end{Bmatrix} = \frac{1}{E}\begin{bmatrix} 1 & -\mu & 0 \\ -\mu & 1 & 0 \\ 0 & 0 & 2(1+\mu) \end{bmatrix} \begin{Bmatrix} \sigma_x \\ \sigma_y \\ \tau_{xy} \end{Bmatrix} \qquad (3.15)$$

Equation 3.15 is written as Eq. (1.23) to express stresses in terms of strains. ε_z is non-zero in case of plane stress condition and it is given by equating $\sigma_z = 0$ in Eq. (1.23c).

3.4.2 Plane Strain Condition

Equation (1.23o) for plane strain case is written as Eq. (3.16) to express strains in terms of stresses as follows.

$$\begin{Bmatrix} \varepsilon_x \\ \varepsilon_y \\ \gamma_{xy} \end{Bmatrix} = \frac{1+\mu}{E}\begin{bmatrix} 1-\mu & -\mu & 0 \\ -\mu & 1-\mu & 0 \\ 0 & 0 & 2 \end{bmatrix} \begin{Bmatrix} \sigma_x \\ \sigma_y \\ \tau_{xy} \end{Bmatrix} \qquad (3.16)$$

σ_z is not zero in case of plane strain condition and it is given by equating $\varepsilon_z = 0$, in Eq. (1.23c).

EXAMPLES

Example 3.1 A 10 mm × 10 mm square element in a state of plane strain deforms such that the horizontal side reduces by 6.4×10^{-6} m, the vertical side increases by 3.6×10^{-3} m and the angle at the lower left corner reduces by 0.16×10^{-3} rad in $x-y$ coordinates as shown in Fig. 3.3. Determine the strains associated with axes rotated through 25° clockwise.

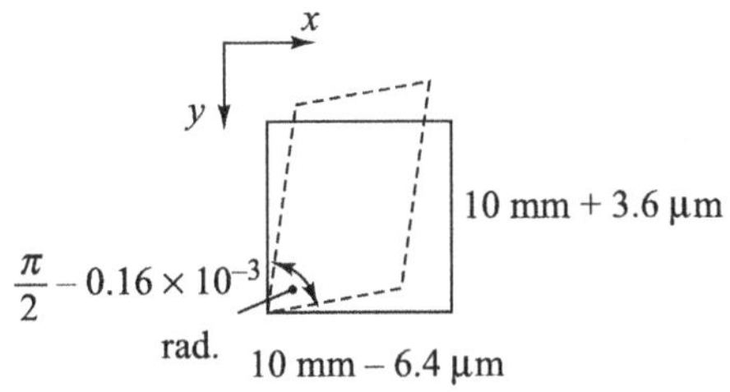

Fig. 3.3

Solution Strain components,

$$\varepsilon_x = \frac{-6.4 \times 10^{-6}\text{ m}}{10 \times 10^{-3}\text{ m}} = -640 \times 10^{-6}$$

$$\varepsilon_y = \frac{3.6 \times 10^{-6}\text{ m}}{10 \times 10^{-3}\text{ m}} = 360 \times 10^{-6}$$

$$\gamma_{xy} = -160 \times 10^{-6}$$

Using Eq. (3.1c), normal strain in x' direction at orientation angle, $\theta = -25°$ (anticlockwise rotation is taken +ve in derivation of Eq. (3.1)).

$$\varepsilon_{x'} = \frac{\varepsilon_x + \varepsilon_y}{2} + \frac{\varepsilon_x - \varepsilon_y}{2}\cos 2\theta - \frac{\gamma_{xy}}{2}\sin 2\theta$$

$$= \frac{-640 + 360}{2 \times 10^6} + \frac{-640 - 360}{2 \times 10^6}\cos(-50°) - \frac{-160}{2 \times 10^6}\sin(-50°)$$

$$= -522.677 \times 10^{-6}$$

Similarly Eq. (3.1a), gives normal strain in y' direction,

$$\varepsilon_{y'} = \frac{\varepsilon_x + \varepsilon_y}{2} - \frac{\varepsilon_x - \varepsilon_y}{2}\cos 2\theta + \frac{\gamma_{xy}}{2}\sin 2\theta$$

$$= \frac{-640 + 360}{2 \times 10^6} - \frac{-640 - 360}{2 \times 10^6}\cos(-50°) + \frac{-160}{2 \times 10^6}\sin(-50°)$$

$$= 242.677 \times 10^{-6}$$

Shear strain in $x'y'$ direction is given by Eq. (3.2) as follows,

$$\frac{\gamma_{x'y'}}{2} = \frac{\varepsilon_x - \varepsilon_y}{2}\sin 2\theta + \frac{\gamma_{xy}}{2}\cos 2\theta$$

$$= \frac{-640 - 360}{2 \times 10^6}\sin(-50°) + \frac{-160}{2 \times 10^6}\cos(-50°)$$

$$= 663.2 \times 10^{-6}$$

Example 3.2 Determine axial and shear strains in x, y directions for the element of Example 2.3. Also calculate the principal strains and verify the results using Mohr's circle (Taking E = 200 GPa and $\mu = 0.3$).

Solution From Eq. (3.13),

$$G = \frac{E}{2(1 + \mu)} = \frac{200000}{2 \times 1.3} = 76923 \text{ MPa}$$

From Example 2.3, $\sigma_x = -40$ MPa, $\sigma_y = 80$ MPa, and $\tau_{xy} = 60$ MPa,

Substituting in Eq. (1.23),

$$\varepsilon_x = \frac{1}{E}[\sigma_x - \mu \times \sigma_y] = \frac{-40 - 0.3 \times 80}{200000}$$

$$= -3.2 \times 10^{-4}$$

$$\varepsilon_y = \frac{1}{E}[\sigma_y - \mu \times \sigma_x] = \frac{80 + 0.3 \times 40}{200000}$$

$$= 4.6 \times 10^{-4}$$

$$\gamma_{xy} = \frac{1}{G}\ \tau_{xy} = \frac{60}{76923} = 7.8 \times 10^{-4}$$

Principal strains from Eq. (3.6),

$$\varepsilon_{\max/\min} = \frac{\varepsilon_x + \varepsilon_y}{2} \pm \frac{1}{2}\sqrt{(\varepsilon_x - \varepsilon_y)^2 + (\gamma_{xy})^2}$$

$$= \frac{(-3.2 + 4.6)\times 10^{-4}}{2} \pm \frac{\sqrt{[(-3.2+4.6)\times 10^{-4}]^2 + [7.8\times 10^{-4}]^2}}{2}$$

$$= 7 \times 10^{-5} \pm 5.5 \times 10^{-4}$$

$$= 6.21 \times 10^{-4},\ -4.8 \times 10^{-4}$$

Maximum shear strain from Eq. (3.5),

$$\frac{\gamma_{\max}}{2} = \pm\frac{1}{2}\sqrt{(\varepsilon_x - \varepsilon_y)^2 + (\gamma_{xy})^2} = 5.5 \times 10^{-4}$$

Figure 3.4 shows Mohr's circle for strains with these values.

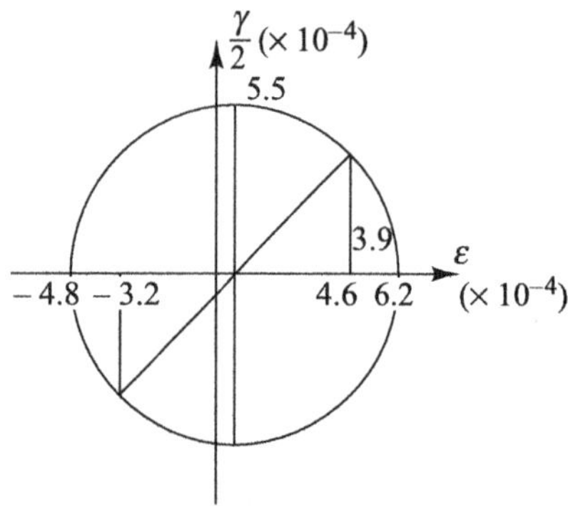

Fig. 3.4 *Mohr's Circle for Strains*

Example 3.3 A shaft of diameter 50 mm is subjected to torque and axial load. A rectangular strain gauge rosette on the surface of the shaft gives micro-strain readings; $\varepsilon_A = 1000 \times 10^{-6}$, $\varepsilon_B = -200 \times 10^{-6}$, $\varepsilon_C = -300 \times 10^{-6}$. Gauges A and C are in-line with and perpendicular to the axis of the shaft as shown in Fig. 3.5. If the modulus of elasticity, $E = 200$ GPa and Poisson's ratio, $\mu = 0.3$, calculate the values of axial load and torque.

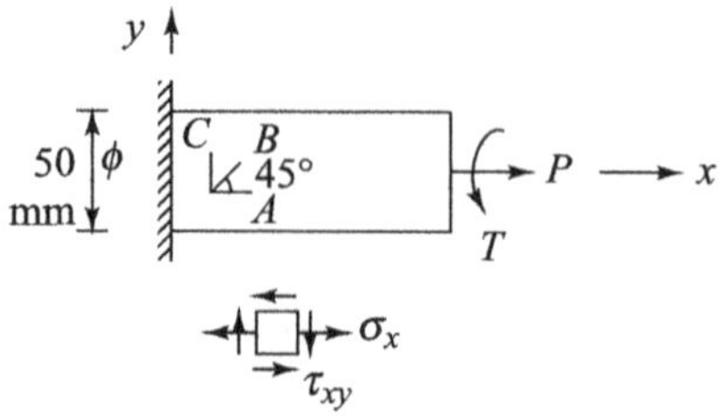

Fig. 3.5

Solution Using Eq. (3.13),

$$G = \frac{E}{2(1+\mu)} = \frac{200000}{2\times 1.3} = 76923.077 \text{ MPa}$$

Consider mutually perpendicular direction A, C as x, y directions in Eq. (3.4b)

$$\varepsilon_B = \frac{\varepsilon_A + \varepsilon_C}{2} + \frac{\varepsilon_A - \varepsilon_C}{2}\cos 2\times 45^\circ - \frac{\gamma_{AC}}{2}\sin 2\times 45^\circ$$

or, $$-200\times 10^{-6} = \frac{1000\times 10^{-6} - 300\times 10^{-6}}{2} - \frac{\gamma_{AC}}{2}$$

or, $\gamma_{AC} = 11 \times 10^{-6}$

Shear stress, due to torque,

$$\tau_{AC} = G \times \gamma_{AC}$$

$$= 76923.077 \times 11 \times 10^{-6}$$

$$= 84.61 \text{ MPa}$$

Torque, $$T = \frac{\tau_{AC}}{r} \times J$$

$$= \frac{84.61}{25} \times \frac{\pi \times 50^4}{32}$$

$$= 2076641.8 \text{ N. mm}$$

$$= 2.077 \text{ kN. m}$$

Axial stress,

$$\sigma = E \times \varepsilon_A$$

$$= 200000 \times 1000 \times 10^{-6}$$

$$= 200 \text{ MPa}$$

Axial load, $P = \sigma \times A = 200 \times \frac{\pi}{4} \times 50^2$

$= 392699$ N $= 392.7$ kN

Example 3.4 A delta rosette shown in Fig. 3.6, measures micro-strains as follows.

$\varepsilon_A = -100 \times 10^{-6}, \quad \varepsilon_B = 700 \times 10^{-6},$

$\varepsilon_C = 600 \times 10^{-6}$

Determine principal stresses and their location if modulus of elasticity, $E = 200$ GPa, and Poisson's ratio, $\mu = 0.3$.

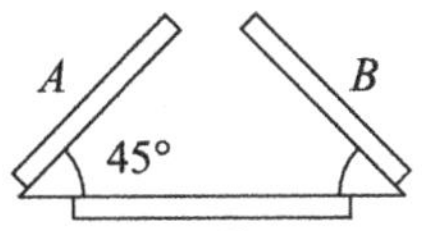

Fig. 3.6

Solution Substituting

$\varepsilon_x = \varepsilon_B = 700 \times 10^{-6}$

$\varepsilon_y = \varepsilon_A = -100 \times 10^{-6}$

$\varepsilon_{x'} = \varepsilon_C = 600 \times 10^{-6}$

and, $\theta = 45°$ in Eq. (3.1b),

$$600 = 700 \times \frac{1}{2} - 100 \times \frac{1}{2} - \frac{\gamma_{xy}}{2}$$

or, $\gamma_{xy} = 600 \times 10^{-6}$

Principal strains,

$$\varepsilon_{max/min} = \frac{\varepsilon_x + \varepsilon_y}{2} \pm \frac{1}{2}\sqrt{(\varepsilon_x - \varepsilon_y)^2 + (\gamma_{xy})^2}$$

or, $$\varepsilon_{max/min} = \frac{700 - 100}{2 \times 10^6} \pm \frac{\sqrt{800^2 + 600^2}}{2 \times 10^6}$$

$$= 800 \times 10^{-6}, -200 \times 10^{-6}$$

Principal strain angle,

$$2\theta_P = \tan^{-1}\frac{-\gamma_{xy}}{\varepsilon_x - \varepsilon_y}$$

$$= \tan^{-1}\frac{-600}{700 + 100}$$

$= -36.87°$ (clockwise w.r.t. B)

Corresponding principal stresses are given by Eq. (1.23),

$$\varepsilon_{max} = \frac{1}{E}[\sigma_{max} - \mu \times \sigma_{min}]$$

$$\varepsilon_{min} = \frac{1}{E}[\sigma_{min} - \mu \times \sigma_{max}]$$

or, $\sigma_{max} - 0.3\ \sigma_{min} = 800 \times 10^{-6} \times 200000$

and, $-0.3 \times \sigma_{max} + \sigma_{min} = -200 \times 10^{-6} \times 200000$

which gives, $\sigma_{max} = 162.64$ MPa

and $\sigma_{min} = 8.79$ MPa

Example 3.5 A strain rosette measures micro-strain as shown in Fig. 3.7. Determine principal strains and their locations.

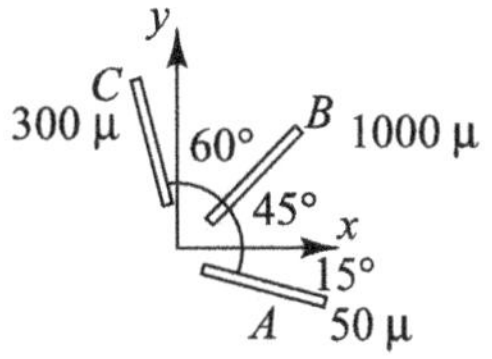

Fig. 3.7

Solution Substituting

$\varepsilon_{x'} = \varepsilon_A$, ε_B and ε_C in Eq. (3.1b),

$$50 \times 10^{-6} = \varepsilon_x \cos^2(-15°) + \varepsilon_y \sin^2(-15°) - \frac{\gamma_{xy}}{2}\sin(-2 \times 15°) \quad \text{(i)}$$

$$1000 \times 10^{-6} = \varepsilon_x \cos^2 45° + \varepsilon_y \sin^2 45° - \frac{\gamma_{xy}}{2}\sin(-2 \times 45°) \quad \text{(ii)}$$

$$300 \times 10^{-6} = \varepsilon_x \cos^2 105° + \varepsilon_y \sin^2 105° - \frac{\gamma_{xy}}{2}\sin(-2 \times 105°) \quad \text{(iii)}$$

Equations (i), (ii) and (iii) give,

$\varepsilon_x = 305.66 \times 10^{-6}, \quad \varepsilon_y = 594.34 \times 10^{-6}$

and $\gamma_{xy} = -1100 \times 10^{-6}$

Substituting in Eq. (3.6), principal strains,

$$\varepsilon_{max/min} = \frac{305.66 + 594.34}{2 \times 10^6} \pm$$

$$\frac{1}{2\times 10^6}\sqrt{(305.66-594.34)^2+1100^2}$$

$$= 1018.62\times 10^{-6}, -118.62\times 10^{-6}$$

Example 3.6 Solve Ex. 3.5 graphically.

Solution Take a horizontal reference line and mark 0,50,300 and 1000 micro-strain to some scale and draw parallel vertical lines from these points as shown in Fig. 3.8. Take a point B' on line 1000×10^{-6} and draw AB' at 60° clockwise since strain gauge A is at 60° clockwise to B. Similarly, strain gauge C is at 60° anticlockwise to B, therefore draw $B'C$ at 60° anticlockwise direction as shown in Fig. 3.8. Draw DF and EF, the perpendicular bisector of $B'C$ and $B'A$, respectively, intersecting at F. With F as center, draw Mohr's circle through A, C and B', which intersects line from 1000×10^{-6} at B. Points A, B and C represent the state of strain in strain gauges A, B and C. Horizontal diameter parallel to initial horizontal reference line gives the value of principal strains as 1020×10^{-6} and -120×10^{-6}.

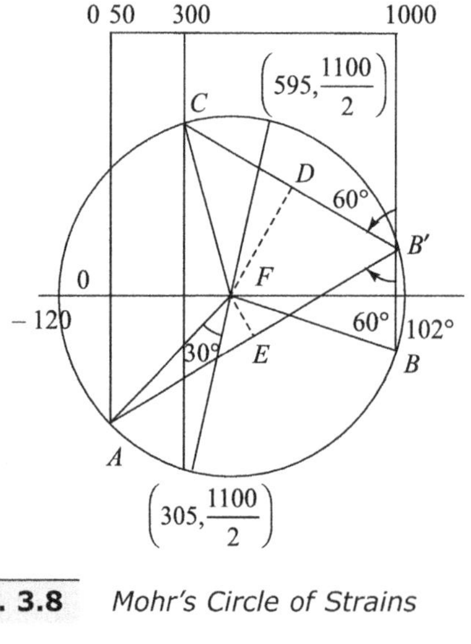

Fig. 3.8 *Mohr's Circle of Strains*

Example 3.7 A circle of diameter 100 mm is marked on a steel plate. The circle deforms in an ellipse when the plate is stressed as shown in Fig. 3.9. Determine the major and minor axes of the ellipse and their location, if $E = 200$ GPa and $\mu = 0.3$.

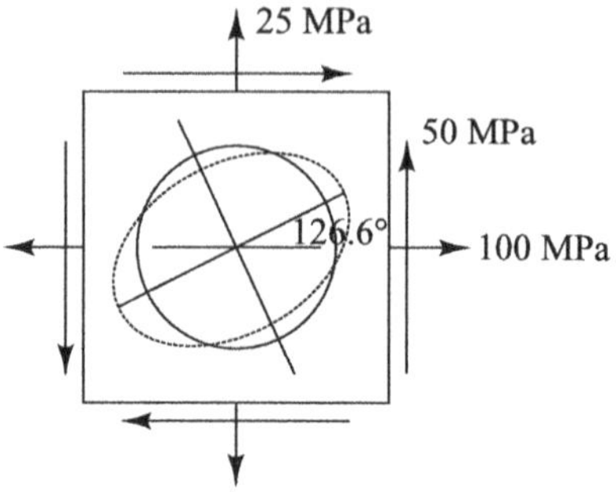

Fig. 3.9

Solution Principal stresses are given by,

$$\sigma_{\max/\min} = \frac{\sigma_x+\sigma_y}{2} \pm \sqrt{\left(\frac{\sigma_x-\sigma_y}{2}\right)^2 + \tau_{xy}^2}$$

$$\sigma_{\max/\min} = \frac{100+25}{2} \pm \sqrt{\left(\frac{100-25}{2}\right)^2 + 50^2}$$

$$= 125 \text{ MPa, } 0 \text{ MPa}$$

Corresponding principal strains using Eq. (1.23),

$$\varepsilon_{\max} = \frac{1}{200000}(125 - 0.3\times 0) = 625\times 10^{-6}$$

$$\varepsilon_{\min} = \frac{1}{200000}(0 - 0.3\times 125)$$

$$= -187.5\times 10^{-6}$$

Location of major principal axis,

$$\theta_p = \frac{1}{2}\tan^{-1}\frac{-\tau_{xy}}{(\sigma_x-\sigma_y)/2}$$

$$= \frac{1}{2}\tan^{-1}\frac{50}{(100-25)/2} = 26.56^\circ$$

(anticlockwise)

Longer, diameter $= D + 625 \times 10^{-6} \times D$
$= 100 + 0.0625$
$= 100.0625$ mm

Smaller diameter $= D - 187.5 \times 10^{-6} \times D$
$= 100 - 0.01875$
$= 99.98125$ mm

Example 3.8 Determine the principal strains for a body that has undergone strains in xyz directions as follows.

$$\varepsilon_{ij} = \begin{vmatrix} 2 & 0 & 1 \\ 0 & 3 & 2 \\ 1 & 2 & 4 \end{vmatrix} \times 10^{-4}$$

Solution Strain invariants,

$$J_1 = \varepsilon_x + \varepsilon_y + \varepsilon_z = (2+3+4)\times 10^{-4} = 9\times 10^{-4}$$

$$J_2 = \varepsilon_{xx}\varepsilon_{yy} + \varepsilon_{yy}\varepsilon_{zz} + \varepsilon_{zz}\varepsilon_{xx} - (\varepsilon_{xy}^2 + \varepsilon_{yz}^2 + \varepsilon_{zx}^2)$$
$$= [2\times 3 + 3\times 4 + 4\times 2 - (0 + 1\times 1 + 2\times 2)]\times 10^{-8} = 21\times 10^{-8}$$

$$J_3 = \begin{vmatrix} 2 & 0 & 1 \\ 0 & 3 & 2 \\ 1 & 2 & 4 \end{vmatrix} \times 10^{-4} = 13\times 10^{-12}$$

Substituting in Eq. (3.12),

$$\varepsilon_{nn}^3 - J_1\varepsilon_{nn}^2 + J_2\varepsilon_{nn} - J_3 = 0$$

Roots of this equations by hit and trial give the principal strains,

$$\varepsilon_{nn}\times 10^4 = 1,\ 2.268,\ 5.732.$$

Direction cosines are given by Eqs (3.13) and (2.12).

Substituting $\varepsilon_{nn} = 1\times 10^{-4}$ in Eqs. (3.13a) and (3.10)

$$(2-1)\,c_{nx} + c_{nz} = 0 \qquad \text{(i)}$$

$$(3-1)\,c_{ny} + 2c_{nz} = 0 \qquad \text{(ii)}$$

Equation 2.12; $(c_{nx})^2 + (c_{ny})^2 + (c_{nz})^2 = 1$ (iii)

Equation (i) through (iii) give,

$$c_{nx} = c_{ny} = -c_{nz} = \pm 0.5773$$

Similarly for $\varepsilon_{nn} = 2.268\times 10^{-4}$

$$c_{nx} = \pm 0.7886,\ -c_{ny} = \pm 0.5773$$

and $c_{nz} = \pm 0.2113$

For $\varepsilon_{nn} = 5.732\times 10^{-4}$

$$c_{nx} = \pm 0.2113,\quad c_{ny} = \pm 0.5773$$

and $c_{nz} = \pm 0.7887$

Example 3.9 The state of strain in at a point in xyz reference frame is given by the strain tensor as follows.

$$\varepsilon_{ij} = \begin{vmatrix} 100 & -200 & 100 \\ -200 & 300 & -100 \\ 100 & -100 & 200 \end{vmatrix} \times 10^{-6}$$

Determine the strain tensor if the reference frame is rotated about y-axis so that $c_{x'x} = 0.866$

Solution Figure Eq. 3.10 shows the rotated reference frame. The corresponding direction cosines are given by,

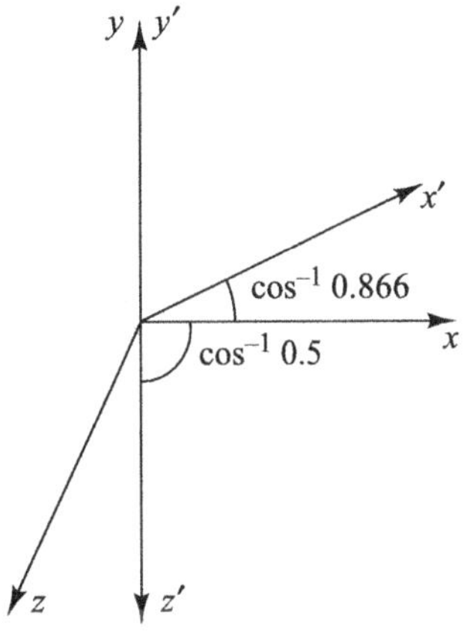

Fig. 3.10

$$\begin{vmatrix} c_{x'x} & c_{x'y} & c_{x'z} \\ c_{y'x} & c_{y'y} & c_{y'z} \\ c_{z'x} & c_{z'y} & c_{z'z} \end{vmatrix} = \begin{vmatrix} 0.866 & 0 & -0.5 \\ 0 & 1 & 0 \\ +0.5 & 0 & 0.866 \end{vmatrix}$$

Strain tensor in rotated reference frame is given by Eq. (3.8b),

$$\varepsilon_{ns} = \begin{vmatrix} 0.866 & 0 & -0.5 \\ 0 & 1 & 0 \\ 0.5 & 0 & 0.866 \end{vmatrix} \times \begin{vmatrix} 100 & -200 & 100 \\ -200 & 300 & -100 \\ 100 & -100 & 200 \end{vmatrix}$$

$$\times \begin{vmatrix} 0.866 & 0 & 0.5 \\ 0 & 1 & 0 \\ -0.5 & 0 & 0.866 \end{vmatrix} \times 10^{-6}$$

$$= \begin{vmatrix} 38.4 & -123.2 & 6.7 \\ -123.2 & 300.0 & -186.6 \\ 6.7 & -186.6 & 261.6 \end{vmatrix} \times 10^{-6}$$

Example 3.10 Write a computer program to calculate strains in $x-y$ directions from the data of a strain rosette as shown in Fig. 3.11.

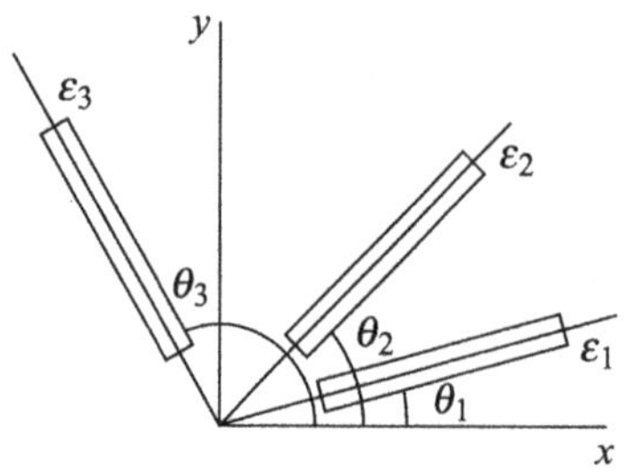

Fig. 3.11

Solution Strains in $x-y$ direction: ε_x, ε_y, τ_{xy} are related to ε_1, ε_2 and ε_3 as follows.

$$\begin{vmatrix} \varepsilon_1 \\ \varepsilon_2 \\ \varepsilon_3 \end{vmatrix} = \begin{vmatrix} \cos^2\theta_1 & \sin^2\theta_1 & -\sin\theta_1\times\cos\theta_1 \\ \cos^2\theta_2 & \sin^2\theta_2 & -\sin\theta_2\times\cos\theta_2 \\ \cos^2\theta_3 & \sin^2\theta_3 & -\sin\theta_3\times\cos\theta_3 \end{vmatrix} \times \begin{vmatrix} \varepsilon_x \\ \varepsilon_y \\ \tau_{xy} \end{vmatrix}$$

or, $(SR) = (TR) \times (SXY)$

or, $(SXY) = (TR)^{-1} \times (SR)$

This program algorithm is used in the computer program in Fortran as follows.

```
   DIMENSION SXY (3,1),
   SR (3,1), TR (3,3), TRI (3,3)
   WRITE (*,*) 'STRAIN, THEETA'
   DO 10 I = 1,3
   READ (*, *) SR (I,1), THEETA
   T = THEETA *3.141593/180.
   TR (I,1) = COS (T) *COS (T)
   TR (I,2) = SIN (T) *SIN (T)
   TR (I,3) = -SIN (T) *COS (T)
10 CONTINUE
   CALL INV (TR, TRI, 3)
   CALL MUL (TRI, SR, SXY, 3, 3, 1)
   DO 20 I = 1, 3
   Write (*, *) SXY (I, 1)
20 CONTINUE
   STOP
   END
```

Subroutines for matrix inversion and multiplication: INV and MUL as given in Appendix B have to be added in this program.

EXERCISES

PROBLEMS

3.1 Determine the longitudinal strain at 15° anticlockwise to x-axis and shear strain associated with this direction when, $\varepsilon_x = 200 \times 10^{-6}$, $\varepsilon_y = 150 \times 10^{-6}$ and $\gamma_{xy} = -250 \times 10^{-6}$. Also determine the principal strains and their direction.
(**Note**: The positive shear strain γ_{xy} associated with the reference frame taken in this chapter in Fig. 3.1, means that the bottom left corner angle increases from $\pi/2$ to $\pi/2 + \gamma_{xy}$. However it is written negative if y-axis is assumed upwards)

3.2 A rectangular strain gauge measures micro-strains as follows.
$\varepsilon_1 = 50 \times 10^{-6}$ at $\theta = 0°$
$\varepsilon_2 = 400 \times 10^{-6}$ at $\theta = 45°$
$\varepsilon_3 = -350 \times 10^{-6}$ at $\theta = 90°$
Determine principal strains and stresses and their location. Take E = 200 GPa and $\mu = 0.3$.

3.3 A delta rosette measures micro-strains, $\varepsilon_1 = -300 \times 10^{-6}$ at $\theta = 0°$, $\varepsilon_2 = 600 \times 10^{-6}$ at $\theta = 60°$ and $\varepsilon_3 = -200 \times 10^{-6}$ at $\theta = 120°$. Determine principal stresses and their location. Take E = 200 GPa and $\mu = 0.3$.

3.4 A strain rosette measures micro-strains as shown in Fig. 3.12. Determine the reading in the direction of strain gauge 3.

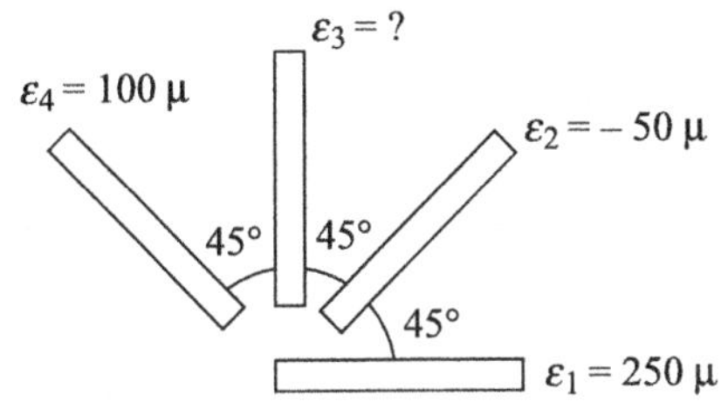

Fig. 3.12

3.5 A beam 100 mm wide, 200 mm deep is to carry a uniformly distributed load over a simply supported span of 5 m. A rectangular strain rosette measures micro-strains near end support at neutral axis, $\varepsilon_1 = -150 \times 10^{-6}$, $\varepsilon_2 = 100 \times 10^{-6}$ and $\varepsilon_3 = 150 \times 10^{-6}$. If ε_1 is along longitudinal axis of the beam and ε_3 is in its perpendicular direction, determine the udl. Take $E = 200$ GPa and $\mu = 0.3$.

3.6 The inside diamter of a pressure vessel is 500 mm and its wall sare 10 mm thick. Two strain gauges measure micro-strains, 50×10^{-6} and 250×10^{-6} along longitudinal and transverse directions, respectively. If the shear modulus, $G = 80$ GPa, determine the gauge pressure inside the vessel.

3.7 Solve Question 3.3 graphically.

3.8 Determine the maximum positive principal strain and its location for the state of strain in a body as follows.

$$\varepsilon_{ij} = \begin{vmatrix} 100 & -200 & 0 \\ -200 & -300 & 100 \\ 0 & 100 & 0 \end{vmatrix} \times 10^{-6}$$

3.9 If the stress in x direction σ_x is written in terms of the strain in x direction ε_x, strain invariant J_1 and Lame's constants u and n such that,

$$\sigma_x = 2u\varepsilon_x + nJ_1$$

Derive the expressions for u and n in terms of modulus of elasticity E and Poisson's ratio μ.

3.10 Derive the stress tensor corresponding to the strain tensor given in Question 3.8. Take $E = 200$ GPa and $\mu = 0.25$.

3.11 Determine the strain invariant for the strain tensor,

$$\varepsilon_{ij} = \begin{vmatrix} 600 & 125 & 0 \\ 125 & -300 & 300 \\ 0 & 300 & -100 \end{vmatrix} \times 10^{-6}$$

Also determine the spherical and deviatorial components of the strain tensor.

3.12 Extend the computer program given in Ex. 3.10 to compute the principal stresses and strains and the stresses in $x - y$ and any other $x' - y'$ directions.

OBJECTIVE QUESTIONS

3.13 Which quantity will not be zero for a plane strain problem in $x - y$ plane.
(a) shear strain in $x - z$ plane
(b) normal strain in z direction
(c) normal stress in z direction
(d) shear stress in $y - z$ plane

3.14 In the Mohr's circle for strains, radius of Mohr's circle gives the,
(a) Minimum value of normal strain
(b) Maximum value of normal strain
(c) Maximum value of shear strain
(d) Half of maximum value of shear strain

3.15 The principal strains at a point are + 800 μ, + 400 μ and –1200 μ. The volumetric strain in equal to, (μ denotes 10^{-6})
(a) + 1200 μ (b) + 800 μ
(c) – 1200 μ (d) zero

3.16 In a plane strain, the shear stress = 12 μ in x-y plane, the normal strain = 0 in x and y direction. The diameter of Mohr's circle will be, (μ = 10^{-6})
(a) 6 μ (b) 8 μ
(c) 12 μ (c) 24 μ

3.17 The strain components in a plane strain problem are $\varepsilon_x = 800\ \mu$, $\varepsilon_y = 100\ \mu$ and $\tau_{xy} = -800\ \mu$. What are the principal strain for these strains.

(a) 981 μ, –81 μ (b) 891 μ, –18 μ
(c) 881 μ, 71 μ (c) 839 μ, –81 μ

3.18 A plane rectangular element in subjected to two normal strains $\varepsilon_x = \varepsilon_y = \varepsilon$ as shown in Fig. 3.13. Which of the following statements are not true in this regard.

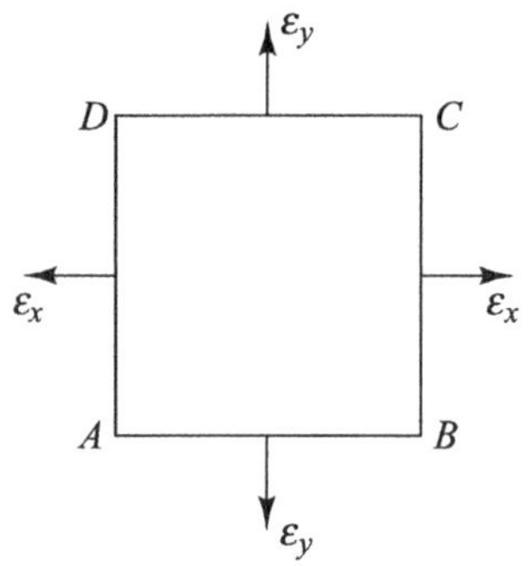

Fig. 3.13

(a) The planes *AB* and *BC* are principal planes.
(b) The diameter of Mohr's circle will be zero.
(c) Shear strain will act on planes inclined to planes *AB* and *BC*.
(d) Normal strain will be equal an all inclined planes.

3.19 Match the items of List I and List II for the Mohr's circle shown in Fig. 3.14 using the codes given below the list.

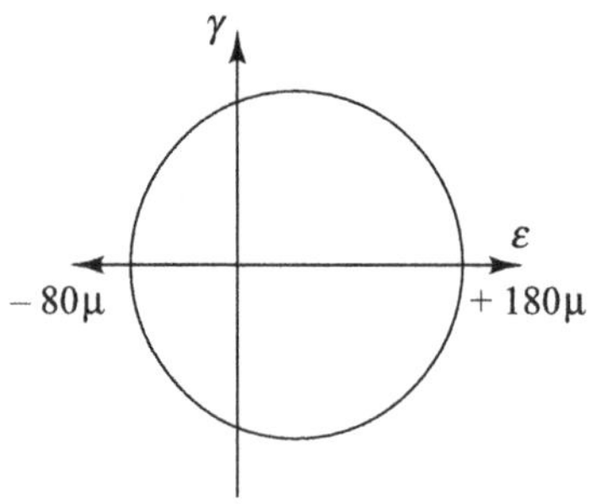

Fig. 3.14

List I	List II
A. Maximum principal strain	1. 260 μ
B. Minimum principal strain	2. – 80 μ
C. Maximum shear strain	3. 130 μ
D. Normal strain on the plane of maximum shear strain	4. 180 μ
	5. 50 μ

Answer code:

	A	B	C	D
(a)	1	5	2	4
(b)	4	2	3	5
(c)	4	2	1	5
(d)	1	5	3	4

3.20 The normal and shear strains at a point on plane *A* are 400 μ and 600 μ (clockwise). The normal and shear strains at the same point an another plane *B* are –400 μ and 600 μ (anticlockwise). The angle between planes *A* and *B* is,

(a) 45° (b) 15° (c) 90° (d) 135°

3.21 The state of strain is spherical strain as follows.

(a) $\begin{vmatrix} 100 & -200 & 300 \\ -200 & 500 & 600 \\ 300 & 600 & -700 \end{vmatrix} \times 10^{-6}$

(b) $\begin{vmatrix} 100 & -200 & 300 \\ -200 & 500 & 600 \\ 300 & 600 & -600 \end{vmatrix} \times 10^{-6}$

(c) $\begin{vmatrix} 500 & 0 & 0 \\ 0 & 900 & 0 \\ 0 & 0 & -200 \end{vmatrix} \times 10^{-6}$

(d) $\begin{vmatrix} -200 & 100 & 300 \\ 100 & 500 & 600 \\ 300 & 600 & -600 \end{vmatrix} \times 10^{-6}$

4

Torsion

CONCEPT REVIEW

4.1 ANALYSIS OF CIRCULAR SECTION—THE TORSION EQUATION

Circular shafts transmit power normally in machines through torsion. Figure 4.1 shows a cylindrical bar of length L and radius R, rigidly fixed at one end and subjected to a torque, $T = F \times d$. It is assumed that,

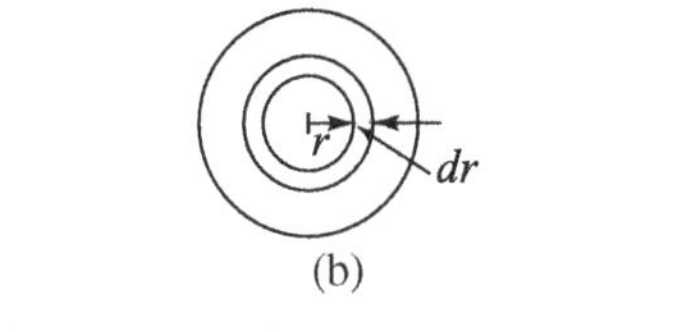

Fig. 4.1 *Circular Bar under Torsion*

(1) Plane section remains plane after torsion and there is no warping.
(2) All cross-sections normal to longitudinal axis at equal distance undergo same relative rotation.

Consider a line AB on the surface of the bar parallel to its longitudinal axis. Line AB is displaced to AC by an angle γ_s after the torsion T is applied. From Fig. 4.1,

$$\text{Shear strain } \gamma_s = \frac{BC}{L} \quad \text{(i)}$$

and,

$$\text{Angle of twist, } \theta = \frac{BC}{R} \quad \text{(ii)}$$

From Eqs (i) and (ii),

$$\gamma_s = \frac{R\theta}{L} \quad \text{(iii)}$$

Similarly, the shear strain γ at any radius r is given by,

$$\gamma = \frac{r\theta}{L}$$

Figure 4.1 shows the applied torsion $F \times d$ in alternate way by double arrow T using right hand thumb rule, i.e. if the fingers of right hand point towards the direction of the twist, then thumb points in the direction of the torsion applied T.

Shear stress τ is related to shear strain γ as follows,

$$\tau = \gamma \times G = \frac{r\theta G}{L} \tag{4.1}$$

and therefore shear stress τ varies with radius r, from O at the center to a maximum $\tau_{max} = R\theta G/L$ at the surface. Now consider shear stress τ acting on an elementary annular area of width dr, at radius r as shown in Fig. 4.1(b).

Shear force on the elementary area

$$= \tau \times \text{area} = \tau \times (2\pi r) \times dr$$

Torque due to shear force on the elementary area

$$= \tau \times (2\pi r \times dr) \times r$$

The external torque T must be equal to the torque to internal forces. Thus integrating over the entire cross-section.

$$T = \int_O^R \tau \times (2\pi r \times dr) \times r$$

$$= \int_O^R \frac{r\theta G}{L} \times (2\pi r \times dr) \times r \tag{4.2}$$

or, $$T = \frac{\pi R^4}{2} \times \frac{\theta G}{L} = \frac{JG\theta}{L} \tag{4.3}$$

where $J = \pi R^4/2$, is called polar second moment of the cross-section. Torsion equation is given by combining Eqs (4.1) and (4.3) as follows.

$$\frac{T}{J} = \frac{G\theta}{L} = \frac{\tau}{r} \tag{4.4}$$

4.2 GENERAL EQUATION FOR TORSION—ST. VENANT THEORY

Consider a prismatic bar of constant arbitrary cross-section subjected to torsion T as shown in Fig. 4.2. z-axis is taken to coincide with the geometric axis of the bar. The two assumptions used in the analysis of circular sections, do not hold good for non-circular sections due to warping. In his so called semi-inverse method, St.Venant assumed that the projections of a warped cross-section rotate as a rigid body and the warping is taken to be identical for all the cross-sections, i.e. independent of the axial location.

Figure 4.2(b) shows a partial cross-section of the bar at a distance z from the origin. It is assumed that no rotation occurs at $z = 0$, and the angle of twist θ per unit length is small. Displacements u, v and w of point A to A' in x, y and z directions, respectively are,

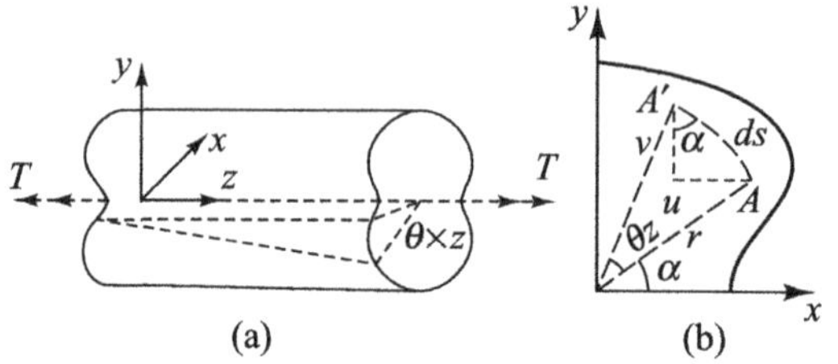

Fig. 4.2 *Bar of an Arbitary Section under Torsion*

$$u = -r\theta z \times \sin\alpha = -y\theta z \tag{4.5a}$$

$$v = r\theta z \times \cos\alpha = x\theta z \tag{4.5b}$$

$$w = f(x, y) \tag{4.5c}$$

where x and y are the coordinates or point A and w is the warping which is a function x, y coordinates only. Substituting Eq. (4.5) in strain-displacement Eq. (1.19).

$$\varepsilon_x = \frac{\partial u}{\partial x} = 0; \ \varepsilon_y = \frac{\partial v}{\partial y} = 0;$$

$$\varepsilon_z = \frac{\partial w}{\partial z} = 0; \ \gamma_{xy} = \frac{\partial u}{\partial y} + \frac{\partial v}{\partial x} = 0 \tag{4.6a}$$

$$\gamma_{xz} = \frac{\partial u}{\partial z} + \frac{\partial w}{\partial x} = -y\theta + \frac{\partial w}{\partial x} \tag{4.6b}$$

$$\gamma_{yz} = \frac{\partial v}{\partial z} + \frac{\partial w}{\partial y} = x\theta + \frac{\partial w}{\partial y} \tag{4.6c}$$

Again substituting Eq. (4.6) in three-dimensional stress–strain Eq. (1.23),

$$\sigma_x = \sigma_y = \sigma_z = \tau_{xy} = 0 \tag{4.7a}$$

$$\tau_{xz} = \tau_{zx} = G\left(-y\theta + \frac{\partial w}{\partial x}\right) \tag{4.7b}$$

$$\tau_{yz} = \tau_{zy} = G\left(x\theta + \frac{\partial w}{\partial y}\right) \tag{4.7c}$$

Thus the state of stress is pure shear for the assumed deformation. Substituting Eq. (4.7) in equilibrium Eq. (1.17) (neglecting body forces),

$$\frac{\partial \tau_{xz}}{\partial x}+\frac{\partial \tau_{yz}}{\partial y}=0$$

or
$$\frac{\partial^2 w}{\partial x^2}+\frac{\partial^2 w}{\partial y^2}=0 \qquad (4.8)$$

This is Laplace equation. The solution for torsion problem must satisfy Eq. (4.8) along with boundary condition.

4.2.1 Prandtl's Stress Function

In 1903, Prandtl proposed that the two non-zero stresses τ_{xz} and τ_{yz} be drfined in terms of a stress function $\phi(x, y)$ as follows,

$$\tau_{xz}=\frac{\partial \phi}{\partial y} \qquad (4.9a)$$

$$\tau_{yz}=-\frac{\partial \phi}{\partial x} \qquad (4.9b)$$

so that the equilibrium Eq. (1.17) is satisfied as follows,

$$\frac{\partial \tau_{xz}}{\partial x}+\frac{\partial \tau_{yz}}{\partial y}=\frac{\partial^2 \phi}{\partial x \partial y}-\frac{\partial^2 \phi}{\partial x \partial y}=0$$

Substituting Eq. (4.9) in Eq. (4.7),

$$\frac{\partial \phi}{\partial y}=G\left(-y\theta+\frac{\partial w}{\partial x}\right) \qquad (4.10a)$$

$$-\frac{\partial \phi}{\partial x}=G\left(x\theta+\frac{\partial w}{\partial y}\right) \qquad (4.10b)$$

Displacement function is eliminated by differentiating Eq. (4.10a) w.r.t. y and Eq. (4.9b) w.r.t. x and their subtraction gives the compatibility Eq. (4.11) as follows:

$$\frac{\partial^2 \phi}{\partial x^2}+\frac{\partial^2 \phi}{\partial y^2}=-2G\theta \qquad (4.11)$$

This is Poisson's equation in which stress function ϕ is still unknown.

For boundary condition, consider the longitudinal boundary of the bar which is free of the external loads as per Eq. (4.7a). Figure 4.3 shows an elementary area at the boundary of the bar. The resultant shear stress τ of the two non-zero shear stresses τ_{zx} and τ_{zy} acting on the elementary area must be tangential to the boundary. The normal component of shear η must be zero as there are no normal forces acting at the boundary. Thus for equilibrium,

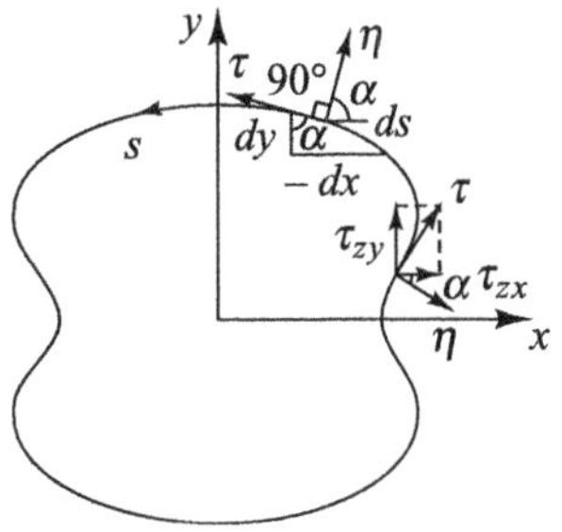

Fig. 4.3

$$\tau_{xz}\cos\alpha+\tau_{yz}\sin\alpha=0 \qquad (4.12)$$

where,
$$\cos\alpha=\frac{dy}{ds} \qquad (4.13a)$$

$$\sin\alpha=-\frac{dx}{ds} \qquad (4.13b)$$

Minus sign is used with dx/ds since ds is assumed to increase in –ve x direction. Substitute Eqs (4.13) and (4.9) in Eq. (4.12),

$$\frac{\partial \phi}{\partial y}\frac{dy}{ds}+\frac{\partial \phi}{\partial x}\frac{dx}{ds}=0$$

or,
$$\frac{d\phi}{ds}=0$$

Thus the value of ϕ is an arbitrary constant at the boundary. As the stresses depend on the derivative of ϕ rather than its value, it is convenient to assume $\phi=0$ on the boundary. In case of multiple connected cross-sections, only one of the boundary can be assigned the value zero.

Now for the applied torque T, consider the summation of moment of the stresses τ_{zx} and τ_{zy} about the geometrical axis of the bar as follows.

$$T=\iint (x\tau_{zy}-y\tau_{zx})dx\,dy$$

$$=-\iint x\frac{\partial \phi}{\partial x}dx\,dy-\iint y\frac{\partial \phi}{\partial y}dx\,dy$$

$$=-\int dy\int x\frac{\partial \phi}{\partial x}dx-\int dx\int y\frac{\partial \phi}{dy}dy$$

Integrating by parts,

$$T = -\int x\ \phi\Big|_{x_1}^{x_2} dy + \int\int \phi\, dx\, dy$$

$$-\int y\,\phi\Big|_{y_1}^{y_2} dx + \int\int \phi\, dx\, dy$$

where x_1, x_2, y_1, y_2 are the points on the lateral surface. As ϕ is constant on the surface, the values of $\phi_2 - \phi_1 = 0$, corresponding to x_1, x_2, y_1, y_2. Therefore,

$$T = 2\int\int \phi\, dx\, dy \qquad (4.14)$$

Equation (4.24) represents twice the volume under ϕ surface.

4.2.2 Prandtl's Membrane Analogy

Equation 4.13 is of the same form as the equation for a small deflection of a membrane-like soap film subjected to a small pressure. This membrane analogy provides a useful experimental technique. It is also used as a basis for obtaining approximate analytical solution for various types of sections. Consider a membrane stretched and supported on the boundary of a hole cut in a plate as shown in Fig. 4.4. The shape of the hole is the same as that of the bar under torsion while size may be different.

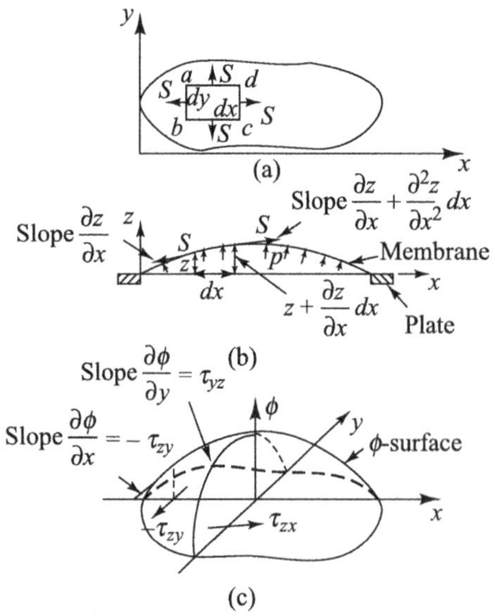

Fig. 4.4 *Membrane Analogy*

The membrane inflates under a small pressure p from the bottom. Equation for the deflected shape is derived from equilibrium considerations of a small element $abcd$ of size $dx \times dy$. The weight of the membrane is neglected and the tensile force S per unit length is considered to be constant. The angles of inclination of S on sides ab and dc are $\frac{\partial z}{\partial x}$ and $\frac{\partial z}{\partial x} + \frac{\partial}{\partial x}\left(\frac{\partial z}{\partial x}\right)dx$ and on sides ad and bc are $\frac{\partial z}{\partial y}$ and $\frac{\partial z}{\partial y} + \frac{\partial}{\partial y}\left(\frac{\partial z}{\partial y}\right)dy$.

Equating vertical components of the forces (approximately),

$$-(S\,dy)\frac{\partial z}{\partial x} + (S\,dy)\left(\frac{\partial z}{\partial x} + \frac{\partial^2 z}{\partial x^2}dx\right)$$

$$-(S\,dx)\frac{\partial z}{\partial y} + (S\,dx)\left(\frac{\partial z}{\partial y} + \frac{\partial^2 z}{\partial y^2}dy\right) + p\,dx\,dy = 0$$

or,

$$\frac{\partial^2 z}{\partial x^2} + \frac{\partial^2 z}{\partial y^2} = -\frac{p}{S} \qquad (4.15)$$

This is also Poisson's equation and the comparison of Eqs (4.11) and (4.15) shows the quantities to be analogous as follows.

Torsion	*Membrane*
ϕ	z
G	$1/S$
2θ	p
τ_{zx}, τ_{zy}	$\frac{\partial z}{\partial y}, -\frac{\partial z}{\partial x}$
T	Twice volume under membrane

Rectangular Sections The solution for the torsion problem of non-circular sections is complicated and can be obtained by numerical methods using computer. Figure 4.5 shows the membrane analogy for this shape. It can be visualized from this that the steepest slope is along the mid-axis A-A and therefore the greatest shear stress occurs at the midpoints of the longer sides. The slope of the ϕ-surface is about zero at the corners of the section and therefore the shear stress is zero at the corners. The formulae for maximum shear stress τ_{max} and angle of twist θ can be derived either by the membrane analogy or by the theory of elasticity as follows:

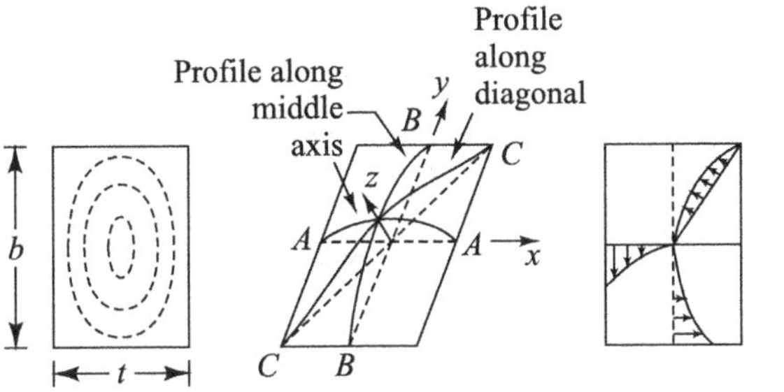

(a) Contours of membrane over square hole (b) Membrane profile in different directions (c) Shear stress distribution

Fig. 4.5

$$\tau_{max} = \frac{T}{\alpha b t^2} \tag{4.16}$$

$$\theta = \frac{TL}{\beta b t^3 G} \tag{4.17}$$

where the coefficients α and β are given in Table 4.1. (Refer books on the theory of elasticity for further details of the solution).

Thin Walled Members of Open Corss-section While applying the membrane analogy to a bar of thin rectangular cross-section, the membrane can be assumed to be of a constant cylindrical shape neglecting the slope at the thin ends, i.e. $\partial z/\partial y = 0$, as shown in Fig. 4.6. Substituting in Eq. (4.15), integrating twice and using boundary condition that $dz/dx = 0$ at $x = 0$ and $z = 0$ at $x = t/2$ gives,

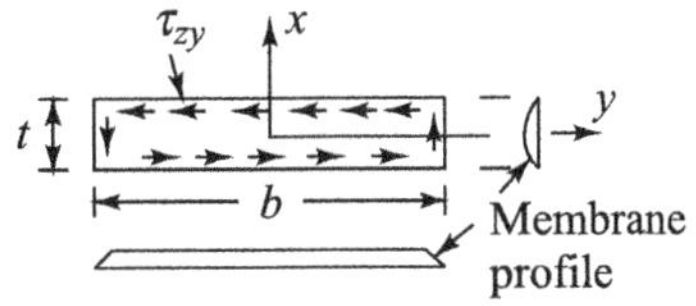

Fig. 4.6 *Membrane Analogy for Thin Rectangular Section*

$$z = \frac{1}{2}\frac{p}{S}\left[\left(\frac{t}{2}\right)^2 - x^2\right] \tag{4.18}$$

Now the volume under the parabolic cylindrical membrane,

$$V = \text{Area of the parabola} \times \text{length}$$

$$= \frac{2 \times z \times t}{3} \times b$$

where

$$z = pt^2/8S \text{ at } x = 0 \text{ in Eq. (4.18)}$$

or,
$$V = \frac{pbt^2}{12S}$$

According to membrane analogy, $1/S$ is replaced by G, p by 2θ and $2V$ by the applied torsion T which gives the expression for the angle of twist per unit length as follows.

$$\theta = \frac{3T}{Gbt^3} \tag{4.19}$$

The membrane analogy also requires that,

$$\tau_{zy} = \frac{\partial z}{\partial x} = 2G\theta x$$

The maximum shear will occur at $x = t/2$.

or,
$$\tau_{max} = G\theta t = \frac{3T}{bt^2} \tag{4.20}$$

One can easily visualize the solution of the torsion problem for other type of thin walled sections using the membrane analogy. For example the complex probzlem of the torsion of an I-section

Table 4.1 *Coefficients for Rectangular Bars*

b/t	1.0	1.5	1.75	2.0	2.5	3.0	4.0	6.0	10.0	∞
α	0.208	0.231	0.239	0.246	0.258	0.267	0.282	0.299	0.313	0.333
β	0.141	0.196	0.214	0.229	0.249	0.263	0.281	0.299	0.313	0.333

can be simplified by using the membrane analogy of an equivalent rectangular sections as shown in Fig. 4.7. Thus neglecting the stress concentration

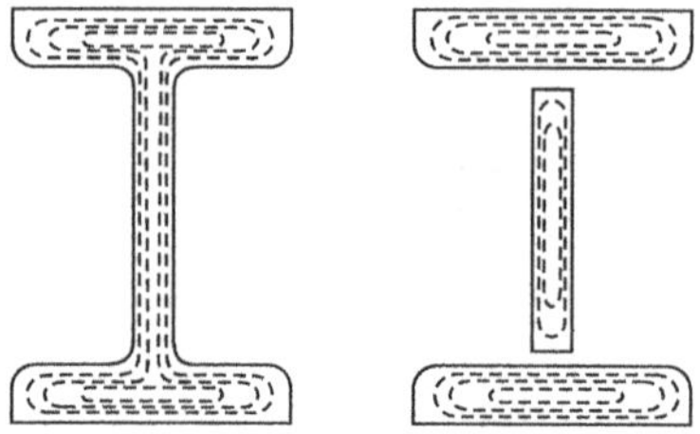

(a) Membrane analogy for I – shape (b) Equivalent section

Fig. 4.7

at the corners, Eqs (4.19) and (4.20) are also applicable to the thin walled open cross-sections. For the channel and I sections of Fig. 4.8, the angle of twist per unit length and the maximum shear stress is obtained by summing the torsional rigidity of the three rectangular com-ponents as follows.

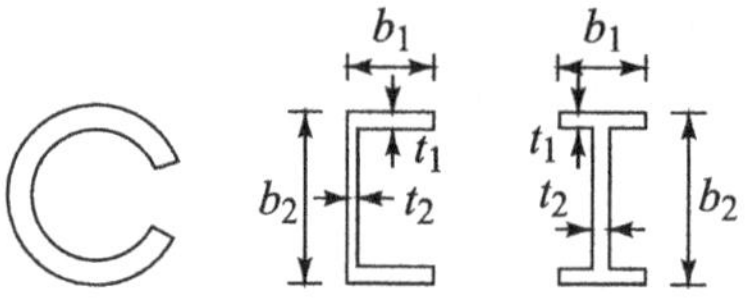

(a) (b) (c)

Fig. 4.8 *Thin walled open cross sections*

$$\theta = \frac{3T}{G(2 \times b_1 t_1^3 + b_2 t_2^3)} \tag{4.21}$$

$$\tau_{\max} = G\theta t_i = \frac{3T \times t_i}{2 \times b_1 t_1^3 + b_2 t_2^3} \tag{4.22}$$

where t_i is the larger of t_1 and t_2.

Thin Walled Hollow Sections For hollow shafts under torsion, the membrane analogy holds good if a horizontal weightless plate of the same size as the hollow, floats on the membrane as shown in Fig. 4.9. The applied torque will now be proportional to the total volume under the floating plate and the membrane. As the walls of the shaft are assumed to be thin, the curvature of the membrane is neglected, i.e. the slope is constant over a given thickness t and therefore the shearing stress is given by,

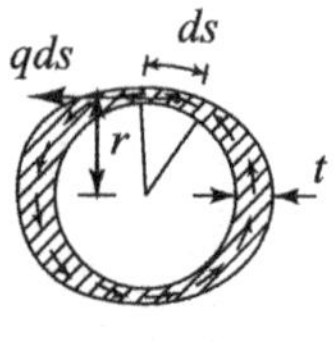

(a) Hollow shaft under torsion

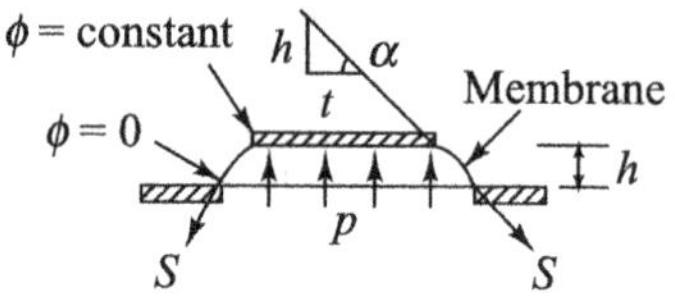

(b) Membrane analogy

Fig. 4.9

$$\tau = \frac{h}{t} \text{ or } h = \tau t \tag{4.23}$$

As the thickness of the shaft can vary along the circumference and h is constant it follows that the product of shearing stress and the thickness of the tube is constant for the shaft, i.e.;

$$\tau_1 t_1 = \tau_2 t_2 = \tau_3 t_3 = \cdots = q \tag{4.24}$$

where q is the constant shear force per unit length of the periphery and is termed as shear flow. Torque produced by the shear force $q \times ds$ acting on an infinitesimal element is $q \times ds \times r$.

Therefore the torque applied

$$T = \oint q \times r \times ds = q \oint r \times ds \tag{4.25}$$

The product $r \times ds$ is twice the area of the triangle of altitude r and base ds. Thus the integral of $r \times ds$ is equal to twice the area A enclosed by the mean perimeter of the shaft, i.e.,

$$T = 2\,Aq = 2Ah \tag{4.26}$$

For the angle of twist again consider the membrane analogy in Fig. 4.10 in which $h << t$ and therefore $\sin \alpha \approx \tan \alpha \approx h/t$. Vertical equilibrium of the membrane tension S and internal pressure p yields,

$$pA = \oint S \sin \alpha\, ds = \oint \frac{h}{t} S\, ds$$

where s is the mean perimeter of the shaft, S is taken out of the integration since it is constant along the perimeter and then from the analogy,

$$\theta = \frac{h}{2GA}\oint \frac{ds}{t} = \frac{T}{4GA^2}\oint \frac{ds}{t}$$

$$= \frac{1}{2GA}\oint \tau ds \qquad (4.27)$$

Equations (4.26) and (4.27) are known as Bredt's formulae. They are also applicable to multiply connected tubes as shown in Fig. 4.10. As the stress concentration is neglected in these derivations, the internal angles of rectangular or square hollow sections should be rounded for free flow of the shear stress.

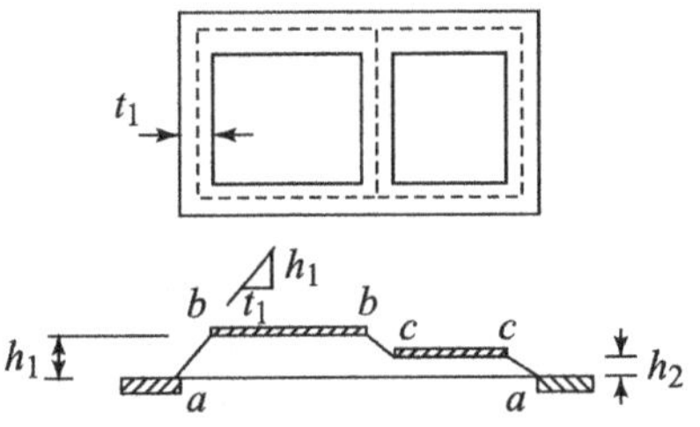

Fig. 4.10 (a) *Multiply Connected Thin Walled Section* (b) *Membrane Analogy*

4.3 TORSION IN THE INELASTIC RANGE

Figure 4.11(a) shows the variation of shear stress along the diameter of a circular shaft under torsion upto elastic limit. If the torsion is increased further then yielding will start at the outer fibers. Figure 4.11(b) show the shear stress variation if the material is assumed to behave ideally plastic beyond the elastic limit. Yielding moves progressively inward as the load increases till the yielding approach the middle of the bar as shown in Fig. 4.11(c). The corresponding torque is the ultimate torque T_u for the shaft and is given by Eq. (4.28) as follows.

$$T_u = \int_0^R \tau_y (2\pi r \times dr) \times r = \frac{2\pi R^3 \tau_y}{3} \qquad (4.28)$$

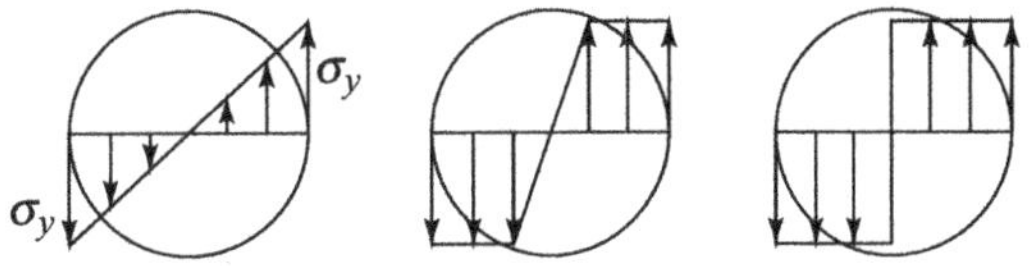

(a) Fully elastic state (b) Elastic–plastic state (c) Fully plastic state

Fig. 4.11 *Shear Stress Distribution Due to Torsion*

Further twisting of the shaft occurs without increase in torque beyond the ultimate torque T_u. The torque T_y corresponding to the elastic limit in Fig. 4.11(a) is given by Eq. (4.4) by $\tau = \tau_y$ and $r = R$.

$$T_y = \frac{\tau_y J}{R} = \frac{\pi R^3 \tau_y}{2} \qquad (4.29)$$

From Eqs (4.28) and (4.29),

$$\frac{T_u}{T_y} = \frac{4}{3} \qquad (4.30)$$

Thus if the effect of strain hardening is neglected then only one-third increase in the torque beyond the elastic limit will bring the shaft to its ultimate load taking capacity.

4.3.1 Residual Stress

When shaft is loaded beyond the elastic limit then the outer fibers are deformed with a permanent strain which remains even after the load is removed. This residual strain causes residual stress.

4.4 ELASTIC, LINEAR STRAIN HARDENING MATERIALS

Shear stress–strain relation for several materials can be represented by Fig. 4.12 where the shear modulus G, i.e. the slope of the curve reduces after onset of yield. Figure 4.13 shows the stress distribution for a circular shaft of such a material under torsion. It is assumed that the strain varies linearly on the cross-section, i.e. radii remains straight during plastic torsion of the shaft and accordingly the shear stress at any distance r is given by,

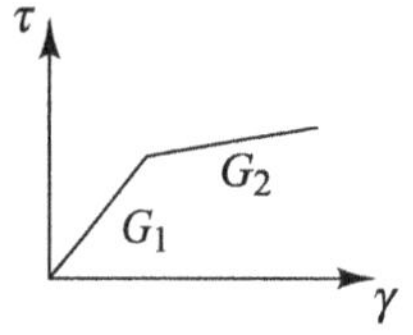

Fig. 4.12

$$\tau = \frac{r}{a}\tau_0 \quad \text{for } 0 <= r <= a,$$

$$= (1 - \alpha)\,\tau_0 + \alpha\tau_0 \frac{r}{a} \quad \text{for } a <= r <= R$$

where, $\alpha = G_2/G_1$. The applied torque is then given by

$$T = \int_0^a \left(\frac{r}{a}\tau_0\right)(2\pi r \times dr)r$$

$$+ \int_a^R \left[(1-\alpha)\tau_0 + \alpha\tau_o \frac{r}{a}\right](2\pi r \times dr)r$$

$$= 2\pi\tau_0 \left[\frac{a^3}{4} + \frac{1-\alpha}{3}(R^3 - a^3) + \frac{\alpha}{4a}(R^4 - a^4)\right]$$

The angle of twist is then given by Eq. (4.1),

$$\frac{d\theta}{dx} = \frac{\tau_0}{aG}.$$

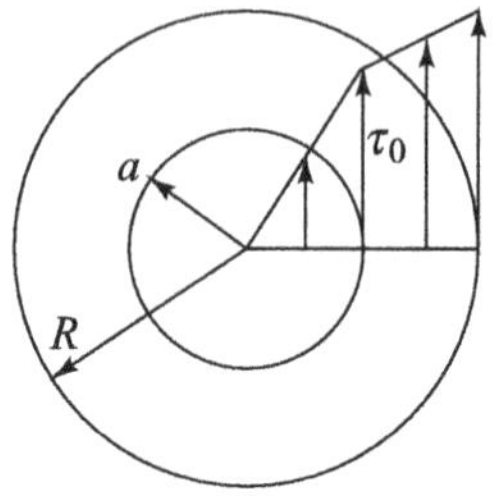

Fig. 4.13

4.4.1 Non-Circular Sections

The complex problem of the torsion of non-circular and hollow sections is simplified by the use of the sand heap analogy. Sand is poured on a horizontal plate having the same shape as the cross-section of the bar under torsion till it is incapable of carrying any more sand. The torque T_u is then given by twice the volume of the sand heap similar to the membrane analogy. A combination of membrane and sand heap analysis is used for the elastic–plastic range of torsion by building a roof

EXAMPLES

Example 4.1 A solid circular shaft of 50 mm diameter is under torsion as shown in Fig. 4.14(a). Determine the maximum shear stress and the angle of twist in the shaft. Also sketch the shear stress distribution in between B and C. Take $G = 80$ GPa.

Solution Figure 4.14(b) shows the variation of torsion along the length of the shaft. Maximum shear stress corresponding to maximum torsion in section A–B is given by Eq. (4.4),

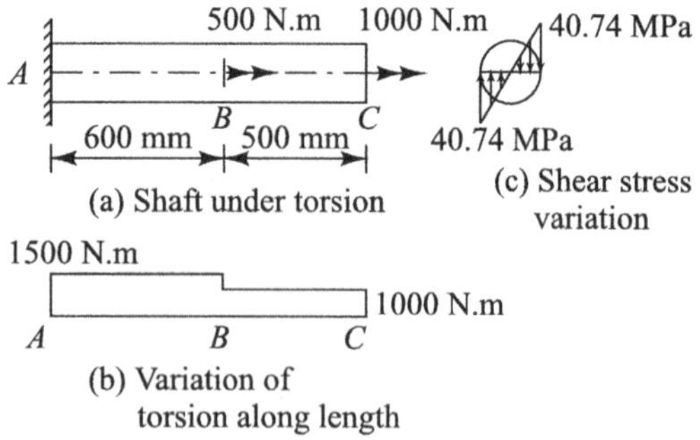

Fig. 4.14

$$\tau_{max} = \frac{rT}{J} = \frac{25 \times 1500000}{\pi \times 25^4 / 2}$$
$$= 61.1 \text{ MPa}$$

The free end will undergo maximum rotation as given by Eq. (4.3),

$$\theta_{A-C} = \left(\frac{TL}{GJ}\right)_{A-B} + \left(\frac{TL}{GJ}\right)_{B-C}$$

$$= \frac{1500000 \times 600}{80000 \times \pi \times 25^4 / 2} + \frac{1000000 \times 500}{80000 \times \pi \times 25^4 / 2}$$
$$= 0.0285 \text{ rad}$$

Maximum shear stress for a section in between *B* and *C*,

$$\tau = \frac{RT}{J} = \frac{25 \times 1000000}{\pi \times 25^4 / 2} = 40.74 \text{ MPa}$$

Figure 4.14(c) shows the shear stress variation along diameter of a section between *B* and *C*.

Example 4.2 A circular shaft is fixed at one end and free at the other as shown in Fig. 4.15. It is loaded with uniformly distributed torsion *t*/unit length along its length *L*. Determine the rotation of the free end if the torsional rigidity of the shaft is *GJ*.

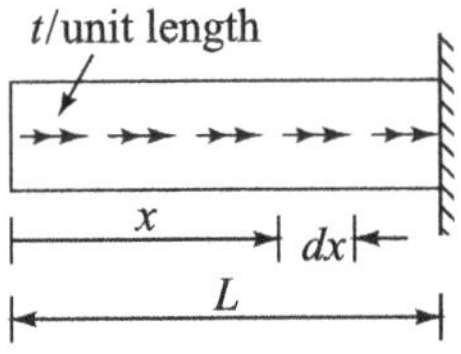

Fig. 4.15

Solution Consider an elementary section of length *dx* at a distance *x* from the free end. The torsion applied at the section is $t \times x$. The angle of twist $d\theta$ of the elementary section is given by Eq. (4.3),

$$\frac{d\theta}{dx} = \frac{T}{JG} = \frac{t \times x}{JG}$$

or,
$$d\theta = \frac{(t \times x) \times dx}{GJ}$$

Integrating from 0 to *L*, the angle of twist for the free end,

$$\theta = \frac{tL^2}{2GJ}$$

Example 4.3 Determine using a computer program the angle of twist of the free end of an uniformly tapering solid circular shaft 2 m long under a uniformly distributed torsion 500 Nm/m as shown in Fig. 4.16. Take $G = 80$ GPa.

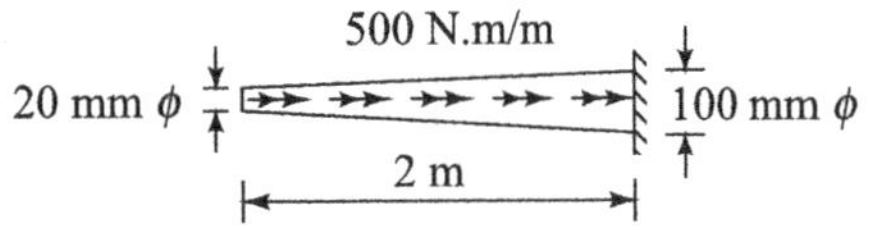

Fig. 4.16

Solution The computer program in Qbasic divides the shaft in small segments and the angle of twist is calculated and added for each segment considering it to be a uniform circular shaft of diameter equal to its mean diameter.

```
DIM DS (40)
INPUT " G = ", G
INPUT "Length of the shaft, L = ", L
INPUT " Torsion per unit length, t = ", T
INPUT " Dia of ends, D1, D2 = " D1, D2
INPUT " Number of segments, NS = ", NS
REM - Length of each segment = LS
REM - Mean Dia of each segment = DS
LS = L / NS
TS = T * LS /2
D = (D2 – D1) / NS
DS (1) = D1 + D / 2
FOR I = 2 TO NS
DS (I) = DS (I – 1) + D
NEXT I
REM – Calculate the angle of twist for
REM – each segment
FOR I = 1 TO NS
J = 3.14159 * DS (I) ^ 4 / 32
THEETA = THEETA + TS * LS / (G * J)
TS = TS + T * LS
NEXT I
```

```
PRINT "THEETA = "; THEETA
END
```

The output/input is as follows.

$G = 80000$

Length of the shaft, $L = 2000$

Torsion per unit length, $t = 500$

Dia. of ends, $D1$, $D2 = 20$, 100

Number of segments, NS = 40

THEETA = 1.489536E–02

Example 4.4 A solid shaft is to transmit 135 kW at 700 rpm. Find the suitable diameter for the shaft if the maximum shear stress should not exceed 100 MPa and the maximum torque exceeds the mean by 20%.

Solution Average power $= \dfrac{2\pi N T_{av}}{60 \times 1000}$ kW

or $$T_{av} = \frac{135 \times 60 \times 1000}{2\pi \times 700}$$

$$= 1841.65 \quad \text{N.m}$$

Maximum torque $T = 1.2 \times 1841.65 = 2210$ N.m

Substitute T_{max} is Eq. (4.4),

$$T_{max} = \frac{\tau J}{r} = \frac{\tau \pi R^4 / 2}{R}$$

or $$R = \sqrt[3]{\frac{2T_{max}}{\pi \tau}}$$

$$= \sqrt[3]{\frac{2 \times 2210000}{100\pi}}$$

$$= 24.14 \text{mm}$$

Example 4.5 Design a solid circular shaft for a 200 HP motor operating at 6 Hz. The allowable shear stress is 50 MPa and the angle of twist is limited to 3° in a length of 2.5 m. Take $G = 80$ GPa.

Solution For conversion, 1 HP = 745.7 Watt or N m/s

$$\text{Power} = 200 \times 745.7 = 149140 \text{ Nm/s}$$

As power is the rate of doing work,

$$\text{Power} = \text{Torque} \times \text{Angular velocity}$$

or, $$P = T \times \omega = T \times 2\pi f$$

where f = revolution/sec or Hertz

Substituting the values in Eq. (i),

$$149140 = T \times 2\pi \times 6$$

or Torque, $T = 3956.06$ N.m

(a) Maximum shear criteria – From Eq. (4.4),

$$\frac{J}{r} = \frac{T}{\tau}$$

or $$\frac{\pi r^4}{2 \times r} = \frac{3956.06 \times 1000}{50}$$

or $$r = 36.39 \text{ mm}$$

$$\text{Diameter} = 73.86 \text{ say } 74 \text{ mm}$$

(b) Maximum twist criteria –From Eq. (4.4),

$$J = \frac{TL}{G\theta}$$

or, $$\frac{\pi r^4}{2} = \frac{3956.06 \times 1000 \times 2500}{80000 \times 3\pi / 180}$$

or, $$r = 35.04 \text{ mm}$$

$$\text{Diameter} = 70.02 \text{ say } 70 \text{ mm}$$

Therefore adopt bigger of 74 and 70, i.e. 74 mm diameter.

Example 4.6 A hollow steel shaft 1.5 m long is to transmit a torque of 80 kNm. Design the shaft if the angle of twist should not exceed 1° and the allowable shear stress is 90 MPa. Take $G = 85$ GPa.

Solution From Eq. (4.4) ,

$$J = \frac{TL}{G\theta} = \frac{80 \times 10^6 \times 1500}{85000 \times \pi / 180}$$

$$= 8.09 \times 10^7 \text{ mm}^4$$

Equation (4.4) also gives,

$$r = \frac{\tau J}{T} = \frac{90 \times 8.09 \times 10^7}{80 \times 10^6} = 91 \text{ mm}$$

Thus if D_1 and D_2 = inside and outside diameters, respectively,

then, $$D_2 = 2 \times r = 182 \text{ mm}$$

Substitute D_2 in the value of $J = 8.09 \times 10^7$

or, $$\frac{\pi}{32}(D_2^4 - D_1^4) = 8.09 \times 10^7$$

or, $$D_1 = 128.56 \text{ say } 128 \text{ mm}$$

Example 4.7 A pump is driven by a 300 HP motor operating at 330 rpm through a solid steel shaft of diameter 75 mm and a gear arrangement as shown in Fig. 4.17(a). Determine (a) the maximum shear stress in the shaft and (b) the relative rotation of the two ends A and C if $G = 85$ GPa. Assume that the shaft is adequately supported by bearing and there is no bending in the shaft.

Solution (a) Maximum shear stress

$$\text{Power, } P = \text{Torque} \times \text{angular velocity} = T \times 2\pi f$$

where,

$$P = 300 \times 745.7 = 223710 \text{ Nm/s}$$

and, $f = 330/60 = 5.5$ rev/s

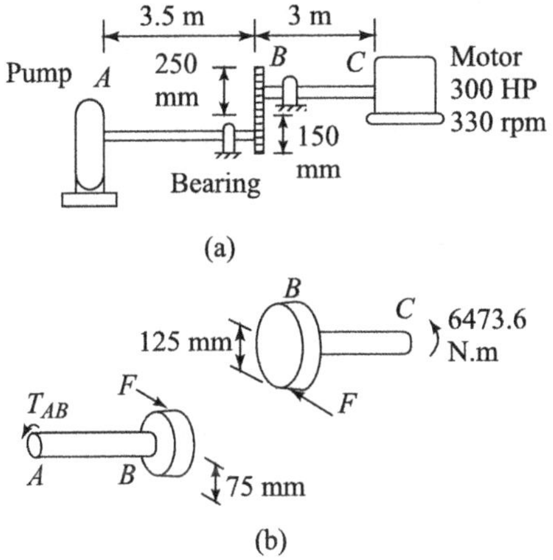

Fig. 4.17

$$\text{Torque, } T = \frac{223710}{2\pi \times 5.5} = 6473.56 \text{ Nm}$$

Maximum shear stress in shaft BC,

$$\tau = \frac{Tr}{J} = \frac{6473.56 \times 1000 \times 75/2}{\pi \times 75^4/32}$$

$$= 78.15 \text{ MPa}$$

Consider equilibrium of the free body diagram in Fig. 4.17(b) to calculate the torque in the shaft AB,

$$\Sigma M = 0, \quad F \times 125 = 6473.56 \times 1000 \text{ Nmm}$$

or,

$$F = 51788.48 \text{ N}$$

$$T_{AB} = F \times 75 = 3884136 \text{ Nmm}$$

Maximum shear stress in shaft AB,

$$\tau_{AB} = \frac{Tr}{J} = \frac{3884136 \times 75/2}{\pi \times 75^4/32}$$

$$= 46.89 \text{ MPa}$$

(b) Angle of twist: From Eq. (4.3),

$$\theta_{A/B} = \frac{T_{A/B} L_{A/B}}{GJ_{A/B}} = \frac{3884136 \times 3500}{85000 \times \pi \times 75^4/32}$$

$$= 0.051487 \text{ rad}$$

$$\theta_{B/C} = \frac{T_{B/C} L_{B/C}}{GJ_{B/C}} = \frac{6473560 \times 3000}{85000 \times \pi \times 75^4/32}$$

$$= 0.07355 \text{ rad}$$

For relative rotation of ends A and C, assume end A to be fixed while the twisting effort is at C. The angular displacement of shaft AB is transmitted through the gear ratio 75/125. Therefore the relative displacement of ends A and C,

$$\theta_{A/C} = \theta_{A/B} \times (75/125) + \theta_{B/C}$$

$$= 0.051487 \times 75/125 + 0.07355$$

$$= 0.10444 \text{ rad}$$

Example 4.8 A solid circular shaft of length 3 m and diameter 50 mm rotates at 1200 rpm by a 400 HP electric motor at its middle. It drives two machines of 150 and 250 HP at left and right ends of the shaft, respectively. Determine the maximum shear stress and the relative displacement of the two ends of the shaft. Take $G = 85$ GPa.

Solution Torque in the left half of the shaft,

$$= \frac{P}{2 \times \pi \times f} = \frac{150 \times 745.7}{2 \times 3.14159 \times 20}$$

$$= 890.115 \text{ Nm}$$

Torque in the right half of the shaft,

$$= \frac{P}{2 \times \pi \times f} = \frac{250 \times 745.7}{2 \times 3.14159 \times 20}$$

$$= 1483.524 \text{ Nm}$$

∴ Maximum shear stress due to torque,

$$\tau = \frac{Tr}{J} = \frac{1483524 \times 25}{\pi \times 50^4/32} = 60.44 \text{ MPa}$$

The angles of twist of the left and right ends relative to the mid section of the shaft are respectively,

$$\theta_1 = \frac{TL}{GJ} = \frac{890115 \times 1500}{85000 \times \pi \times 50^4/32} = 0.0256 \text{ rad}$$

$$\theta_2 = \frac{1483524 \times 1500}{85000 \times \pi \times 50^4/32} = 0.0427 \text{ rad}$$

Example 4.9 A solid steel bar of diameter 20 mm and length 3 m is fixed at both ends. It is loaded with an uniformly distributed torsion of 60 N m/m as shown in Fig. 4.18. Plot the variation of torsional load and the angle of twist along the length of the bar if the modulus and elasticity $E = 200$ GPa and the Poisson's ratio $\mu = 0.25$.

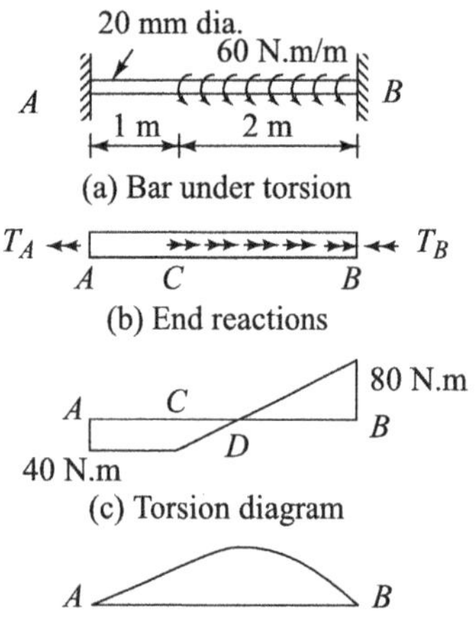

(a) Bar under torsion
(b) End reactions
(c) Torsion diagram
(d) Angle of twist variation

Fig. 4.18

Solution Using Eq. (3.16) the modulus of rigidity,

$$G = \frac{E}{2(1+\mu)} = \frac{200000}{2 \times (1 + 0.25)}$$

$$= 80000 \text{ MPa}$$

Uniformly distributed load = 60 N m/m = 60 N mm/mm. Let the fixed end reactions are T_A and T_B as shown in Fig. 4.18(b) so that,

$$T_A + T_B = 60 \times 2000$$

$$= 120000 \text{ N mm}$$

Angle or twist at end B due to T_A,

$$= \frac{-T_A \times L}{JG} = \frac{-3000\,T_A}{JG}$$

Angle of twist at end B due to uniformly distributed torque as derived in Example 3.2,

$$= \frac{tL^2}{2GJ} = \frac{60 \times 2000^2}{2JG}$$

$$= \frac{120000000}{JG}$$

Since the twist at the fixed end B is zero,

$$\frac{-3000T_A}{JG} + \frac{120000000}{JG} = 0$$

or, $\quad T_A = 40000$ N mm

Substituting in Eq. (1),

$$T_B = 8000 \text{ N mm}$$

Figure 4.18(c and d) show the variation of the torsion and the angle of twist along the length of the bar. The maximum angle of twist occurs where the torsion is zero.

Polar second moment of area

$$J = \pi r^4/2 = 15707.95 \text{ mm}^4$$

Maximum angle of twist

$$\theta_{A/D} = \frac{40000 \times 1666.67}{15707.95 \times 80000}$$

$$- \frac{60 \times 666.67 \times 666.67}{2 \times 15707.95 \times 80000}$$

$$= 0.04244 \text{ rad}$$

Example 4.10 A compound circular shaft is fixed rigidly at its both ends. It is made by fillet welding at the periphery of a 40 mm dia. bar with a hollow bar of inside and outside diameters 60 mm and 20 mm, respectively. Determine the shear stress at the weld if two torsional loads of 4 kNm and 3 kNm are applied as shown in Fig. 4.19(a).

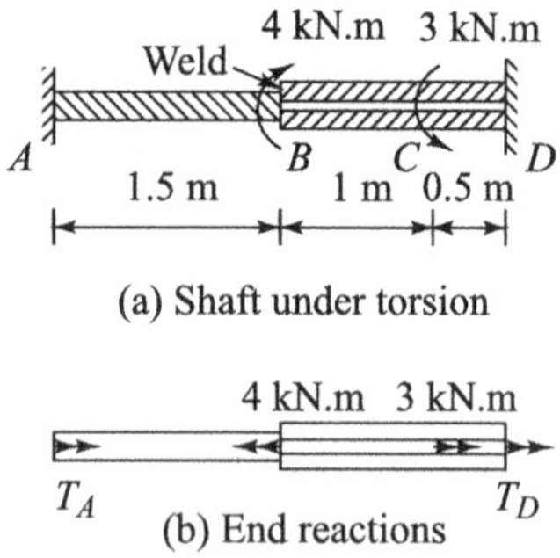

(a) Shaft under torsion
(b) End reactions

Fig. 4.19

Solution Polar second moment of area,

$$J_{AB} = \pi D^4/32 = 251327.2 \text{ mm}^4$$

$$J_{BD} = \pi(60^4 - 20^4)/32$$

$$= 1256636 \text{ mm}^4$$

Figure 4.19(b) shows the end reactions assumed at ends A and D so that,

$$T_A + T_D = 4 - 3 = 1 \text{ kN m} \quad \text{(i)}$$

Using boundary condition that $\theta_A = 0$,

$$\frac{T_A \times 1.5}{J_{AB}G} - \frac{(4 - T_A) \times 1}{J_{BD}G} + \frac{(3 + T_A - 4) \times 0.5}{J_{BD}G} = 0$$

(considering end A is free and end D is fixed)

or, $$T_A = \frac{3J_{AB}}{J_{AB} + J_{BD}} = \frac{3 \times 251327.2}{251327.2 + 1256636}$$

$$= 0.5 \text{ kN m}$$

Shear stress at the weld,

$$\tau = \frac{T_A r}{J} = \frac{0.5 \times 10^6 \times 20}{251327.2}$$

$$= 39.79 \text{ MPa}$$

Example 4.11 A circular composite shaft consists of a 50 mm diameter steel tube fitted tightly over a hollow brass tube of external and internal diameters of 30 mm and 20 mm, respectively. If the shaft is to carry a torque of 1000 Nm, calculate the maximum stresses in steel and brass. Take G = 80 GPa and 40 GPa for steel and brass, respectively.

Solution If T_1 = torque carried by steel

and, T_2 = torque carried by brass,

Total torque, $T_1 + T_2 = 1000$ N m (i)

Since the angle of twist is same in steel and brass.

$$\frac{T_1 L}{80000 \times \pi(50^4 - 30^4)/32} = \frac{T_2 L}{40000 \times \pi(30^4 - 20^4)/32}$$

or, $T_1 = 16.7385\,T_2$

Substituting in Eq. (i),

$$T_2 = 56374.55 \text{ N mm}$$

and, $$T_1 = 943625.45 \text{ N mm}$$

The maximum stress in steel,

$$\tau = \frac{Tr}{J} = \frac{943625.45 \times 25}{\pi(50^4 - 30^4)/32}$$

$$= 44.64 \text{ MPa}$$

The maximum stress in brass,

$$\tau = \frac{Tr}{J} = \frac{56374.55 \times 15}{\pi(30^4 - 20^4)/32}$$

$$= 13.25 \text{ MPa}$$

Example 4.12 Derive and expression for the torsional rigidity of an equilateral triangle and hence calculate the angle of twist due to an applied torsion of 100 Nm on an equilateral triangular prism of side 60 mm and length 2 m if the shear modulus G is equal to 80 GPa.

Solution The equation of three sides of a triangle shown in Fig. 4.20 are written as follows.

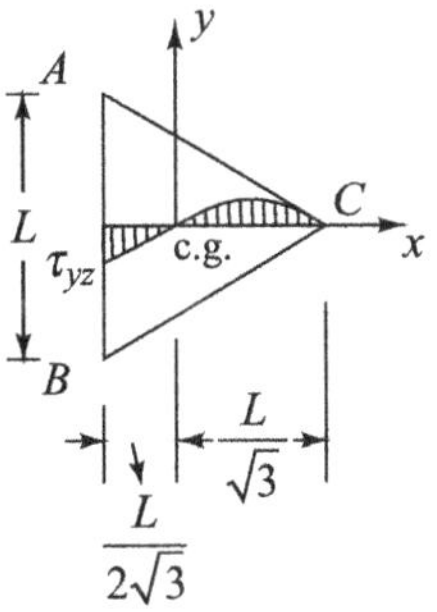

Fig. 4.20

Side AB: $x = -L/2\sqrt{3}$

Side BC: $y = x \tan 30° - L/3 = x/\sqrt{3} - L/3$

Side AC: $y = L/3 - x \tan 30° = L/3 - x/\sqrt{3}$

Choosing the stress function,

$$\phi = A[x + L/2\sqrt{3}]\,[y - x/\sqrt{3} + L/3]\,[y - L/3 + x/\sqrt{3}]$$

where A is a constant,

or, $$\phi = A[xy^2 + (y^2 + x^2)L/2\sqrt{3} - x^3/3 - L^3/18\sqrt{3}]$$

The stress function should satisfy Eq.(4.11),

$$\frac{\partial^2 \phi}{\partial x^2} + \frac{\partial^2 \phi}{\partial y^2} = -2G\theta$$

which gives, $A = \dfrac{-G\theta}{L/\sqrt{3}}$

Thus the stress function is written as follows:

$$\phi = \frac{-G\theta}{L/\sqrt{3}}[xy^2 + (y^2 + x^2)L/2\sqrt{3} - x^3/3 - L^3/18\sqrt{3}]$$

Substituting ϕ in Eq. (4.14), $T = 2\int\int \phi\, dx\, dy$ where the limits of integration for

$$y = x/\sqrt{3} - L/3 \text{ to } L/3 - x/\sqrt{3}$$

and $x = -L/2\sqrt{3}$ to $L/\sqrt{3}$

$$T = \frac{-2G\theta}{L/\sqrt{3}}\int_{-L/2\sqrt{3}}^{L/\sqrt{3}} dx \int_{x/\sqrt{3}-L/3}^{L/3-x/\sqrt{3}} [xy^2 + (y^2 + x^2)L/2\sqrt{3} - x^3/3 - L^3/18\sqrt{3}]\, dy$$

or, $T = \sqrt{3}\ G\theta L^4/80$

therefore the torsional rigidity,

$$C = \frac{T}{\theta} = \frac{\sqrt{3}\,GL^4}{80}$$

If $T = 100000$ N mm,

$G = 80$ GPa

and $L = 60$ then,

Angle of twist per mm,

$$\theta = \frac{80T}{\sqrt{3}GL^4} = \frac{80 \times 100000}{\sqrt{3} \times 80000 \times 60^4}$$

$= 4455 \times 10^{-9}$ rad

Angle of twist for 2 m length

$= 4455 \times 10^{-9} \times 2000$

$= 0.00891$ rad

The shearing stress due to torsion is given by Eq. (4.9) as follows.

$$\tau_{xz} = \frac{\partial\phi}{\partial y} = \frac{-\sqrt{3}G\theta}{L}(2xy + yL/\sqrt{3})$$

$\tau_{xz} = 0$, at $y = 0$ along x-axis.

$$\tau_{yz} = -\frac{\partial\phi}{\partial x} = \frac{\sqrt{3}G\theta}{L}(y^2 + xL/\sqrt{3} - x^2)$$

Figure 4.21 shows the variation of τ_{yz} at $y = 0$ along x-axis. τ_{yz} is maximum at the middle of a side and is zero at the corners and c.g. of the triangle.

$$\tau_{yz,\max} = \frac{\sqrt{3}G\theta L}{4}$$

$$= \frac{\sqrt{3} \times 80000 \times 4455 \times 10^{-9} \times 60}{4}$$

$= 9.26$ MPa

Example 4.13 Calculate the angle of twist and the maximum shearing stress due to a torsion of 100 Nm applied on an elliptical cylinder having major and minor axes 50 mm and 25 mm, respectively. The shear modulus G is equal to 80 GPa. Also plot the warping function $w(x, y)$.

Solution Equation (4.11) and the boundary condition $d\phi/ds = 0$, are satisfied by the stress function ϕ as follows.

$$\phi = k\left(\frac{x^2}{a^2} + \frac{y^2}{b^2} - 1\right) \quad \text{(i)}$$

where, k = constant

and, $2a$ and $2b$ = major and minor axes, respectively

Substituting Eq. (i) in Eq. (4.11) gives,

$$k = -G\theta a^2b^2/(a^2 + b^2)$$

Therefore,

$$\phi = -\frac{G\theta a^2b^2}{(a^2 + b^2)}\left(\frac{x^2}{a^2} + \frac{y^2}{b^2} - 1\right) \quad \text{(i)}$$

Substituting ϕ in Eq. (4.14),

$$T = 2\int\int \phi\, dx\, dy$$

or

$$T = -\frac{2G\theta a^2b^2}{a^2 + b^2}\left(\frac{1}{a^2}\int\int x^2 dx\, dy + \frac{1}{b^2}\int\int y^2 dx\, dy - \int\int dx\, dy\right)$$

$$= \frac{2G\theta a^2b^2}{a^2 + b^2}\left(\frac{1}{a^2}I_y + \frac{1}{b^2}I_x - A\right)$$

where I_x, I_y = second moment of area of the ellipse

and, A = area of the ellipse

$$\text{or,}\quad T = \frac{-2G\theta a^2 b^2}{a^2 + b^2}\left(\frac{1}{a^2}\frac{\pi b a^3}{4} + \frac{1}{b^2}\frac{\pi a b^3}{4} - \pi ab\right)$$

$$\text{or,}\quad T = \frac{\pi a^3 b^3 G\theta}{a^2 + b^2} \qquad \text{(ii)}$$

Thus for $T = 100$ Nm = 100000 N mm, $a = 25$ mm, $b = 12.5$ mm and $G = 80000$ N/mm^2,

Angle of twist per mm,

$$\theta = \frac{100000(25^2 + 12.5^2)}{\pi \times 25^3 \times 12.5^3 \times 80000}$$
$$= 10.186 \times 10^{-6} \text{ rad}$$

Equations (i) and (ii) give the stress function

$$\phi = -\frac{T}{\pi ab}\left(\frac{x^2}{a^2} + \frac{y^2}{b^2} - 1\right)$$

The shearing stress due to torsion is given by Eq. (4.9) as follows.

$$\tau_{xz} = \frac{\partial \phi}{\partial y} = \frac{-2Ty}{\pi ab^3} \text{ and}$$

$$\tau_{yz} = -\frac{\partial \phi}{\partial x} = \frac{2Tx}{\pi a^3 b}$$

The maximum τ_{xz} will be at $y = \pm\, b$ and maximum τ_{yz} will be at $x = \pm\, a$

$$\tau_{xz,\,\max} = \frac{2 \times 100000}{\pi \times 25 \times 12.5^2}$$
$$= 16.3 \text{ MPa}$$

$$\tau_{yz,\,\max} = \frac{2 \times 100000}{\pi \times 12.5 \times 25^2}$$
$$= 8.15 \text{ MPa}$$

The warping function $w(x, y)$ is given by the integration of Eq. (4.7)

$$\tau_{xz} = G\left(\frac{\partial w}{\partial x} - y\theta\right)$$

$$\text{or,}\quad \frac{-2Ty}{\pi ab^3} = G\left(\frac{\partial w}{\partial x} - \frac{y(a^2 + b^2)}{\pi a^3 b^3 G}\right)$$

$$\text{or,}\quad w(x, y) = \frac{T(b^2 - a^2)xy}{\pi a^3 b^3 G}$$

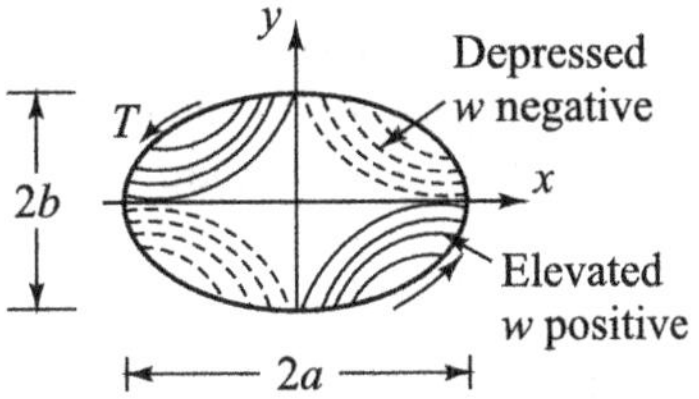

Fig. 4.21

Figure 4.21 shows the hyperbolic variation of $w(x, y)$ across the ellipse.

Example 4.14 Calculate the torsion constant J for the angle section and the channel section shown in Figs. 4.22(a and b).

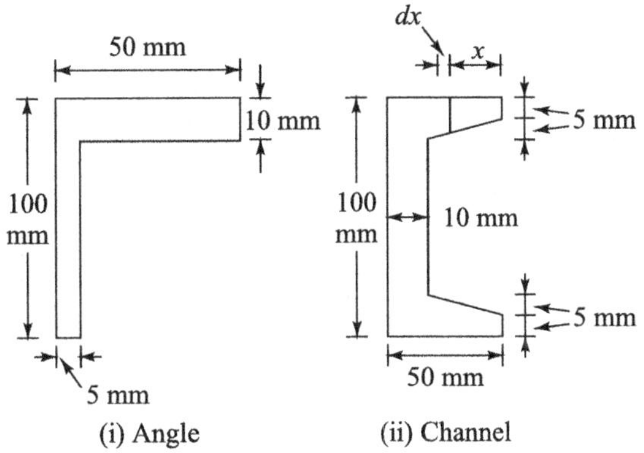

Fig. 4.22

Solution (a) Angle section: Comparing Eq. (4.4) and (4.20), J for rectangular thin walled open cross-section is given by,

$$J = \sum \frac{bt^3}{3} = \frac{50 \times 10^3}{3} + \frac{90 \times 5^3}{3}$$
$$= 20416.67 \text{ mm}^4$$

(b) Channel section: Evaluating $\Sigma bt^3/3$ by integration for the sloping flanges as follows.

$$J = \frac{100 \times 10^3}{3} + 2\int_0^{40}\left(5 + \frac{5}{40}x\right)^3 \frac{dx}{3}$$
$$= 45833.33 \text{ mm}^4$$

Example 4.15 A hollow rectangular tube of cross-section shown in Fig. 4.23(a) is subjected to a torsion of 60 kNm along its longitudinal axis. Determine the shearing stress and the angle of twist per unit length if the shear modulus G is equal to 28 GPa.

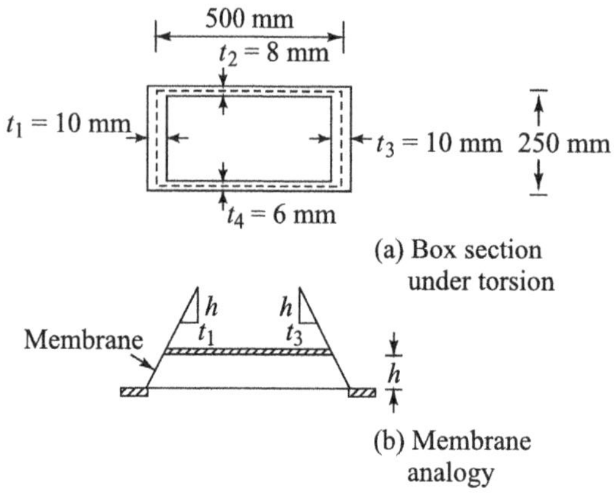

Fig. 4.23

Solution Figure 4.23(b) shows the membrane analogy for the tube. Using Eq. (4.26),

$$T = 2Ah,$$

$$h = \frac{60 \times 10^6}{2 \times 250 \times 500}$$

$$= 240 \text{ N/mm}$$

Shear stress from Eq. (4.23),

on side 1 and 3, $\tau = \frac{h}{t} = \frac{240}{10} = 24$ MPa

on side 2 $= \frac{240}{8} = 30$ MPa

on side 4 $= \frac{240}{6} = 40$ MPa

Angle of twist per unit length is given by Eq. (4.27),

$$\theta = \frac{1}{2GA}\int \tau\, ds = \frac{h}{2GA}\int \frac{ds}{t}$$

or, $$\theta = \frac{240}{2 \times 28000 \times 250 \times 500} \times \left(\frac{2 \times 250}{10} + \frac{500}{8} + \frac{500}{6}\right)$$

$$= 6.71 \times 10^{-6} \text{ rad/mm}$$

Example 4.16 Calculate the maximum torque which the concrete box girder shown in Fig. 4.24 can carry if the maximum shear stress is limited to 0.2 MPa.

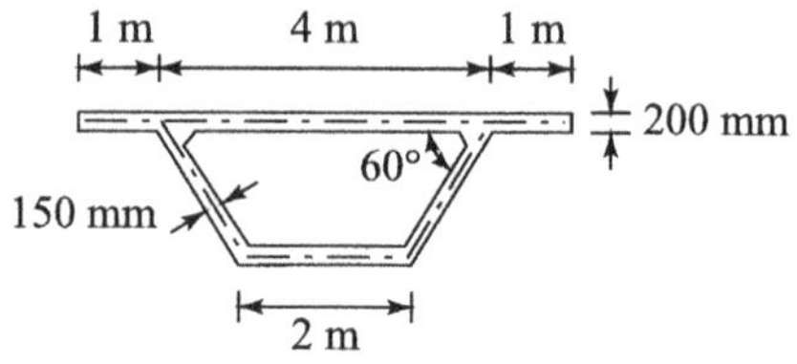

Fig. 4.24

Solution Maximum shearing stress $\tau = 0.2$ MPa. Since the shear flow is constant for a thin box section and the thickness of the bottom and side walls is less, they will be subjected to maximum shearing stress.

$$\text{shear flow, } q = 0.2 \times 150 = 30 \text{ N/mm}$$

The torsional rigidity of the box section is very large compared to the outstanding flanges and hence they can be neglected. Torque resisted by the box section using Eq. (4.26).

$$T_1 = 2Aq = 2 \times \left(\frac{4000 + 2000}{2} \times 1732\right) \times 30$$

$$= 2 \times 5196000 \times 30 = 311760000 \text{ N mm}$$

Angle of twist from Eq. (4.27),

$$\theta = \frac{1}{2GA}\oint \tau\, ds = \frac{q}{2GA}\oint \frac{ds}{t}$$

or, $$\theta = \frac{30}{2G \times 5196000} \times \left(\frac{4000}{200} + \frac{2 \times 1732}{150} + \frac{2000}{150}\right)$$

or, $\theta = 0.000163/G$ rad/mm

Since the box section and the outstanding flanges rotate together, then from Eq. (4.19) the torsion resisted by the flanges is given by,

$$T_2 = \frac{bt^3}{3} \times G\theta$$

$$= \frac{2 \times 1000 \times 200^3}{3} \times 0.000163$$

$$= 868770.85 \text{ N mm}$$

Therefore total torque resisted

$$= T_1 + T_2 = 312628770.85 \text{ N mm}.$$

Example 4.17 A thin walled section as shown in Fig. 4.25 is subjected to a torsion T. Calculate the torsional stiffness and shear stress in the section.

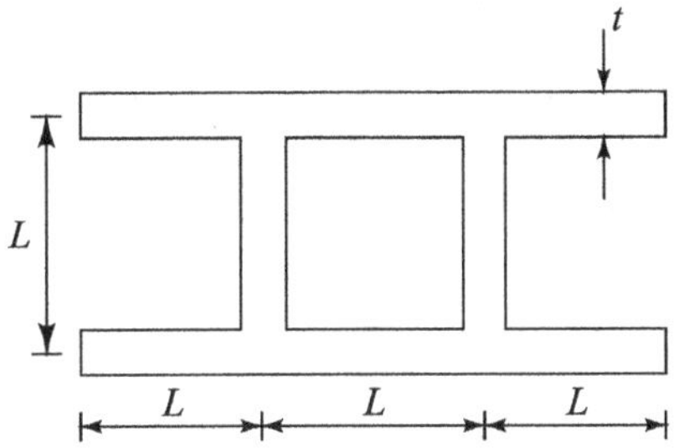

Fig. 4.25

Solution Torsional stiffness, i.e. torsion/angle or twist per unit length for rectangular outstand using Eq. (4.17),

$$\frac{T}{\theta} = \sum \beta b t^3 G = 4Lt^3G/3$$

where $\beta = 1/3$ for $b/t >> 10$ and $\sum b = L + L + L + L$ $= 4L$

Torsional stiffness of box using Eq. (4.27),

$$\frac{T}{\theta} = \frac{4GA^2}{\oint \frac{ds}{t}} = \frac{4GL^4t}{4L} = GL^3t$$

where, $A = L \times L = L^2$

and $\oint \frac{ds}{t} = \frac{L}{t} + \frac{L}{t} + \frac{L}{t} + \frac{L}{t} = 4L/t$

The rectangular outstands and the box will share the applied torsion T in proportion to their torsional stiffness.

Torsion taken by the rectangular outstands

$$= \frac{T \times 4Lt^3G/3}{4Lt^3G/3 + GL^3t} = \frac{4t^2T}{4t^2 + 3L^2}$$

Torsion taken by the box section

$$= \frac{T \times GL^3t}{4Lt^3G/3 + GL^3t} = \frac{3L^2T}{4t^2 + 3L^2}$$

Maximum shear stress in the rectangular outstand using Eq. (4.16),

$$\tau_{max} = \frac{4t^2T}{4t^2 + 3L^2} \times \frac{1}{\Sigma \alpha b t^2} = \frac{3T}{L(4t^2 + 3L^2)}$$

Maximum shear stress in the box section using Eq. (4.26)

$$\tau = \frac{T}{2At} = \frac{3L^2T}{4t^2 + 3L^2} \times \frac{1}{2L^2t}$$

$$= \frac{3T}{2t(4t^2 + 3L^2)}$$

Example 4.18 A tubular steel section shown in Fig. 4.26(a) is subjected to a torque of 110 kN m. Determine the stresses and the angle of twist per unit length of the section if G = 80 GPa.

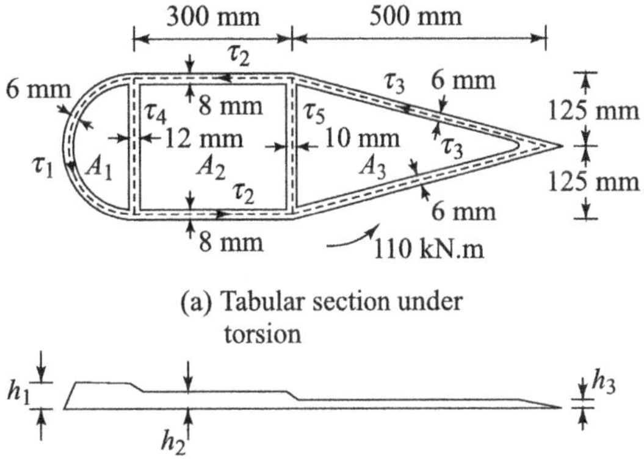

(a) Tabular section under torsion

(b) Membrane analogy

Fig. 4.26

Solution Figure 4.26 shows the membrane analogy for the multiple connected hollow section. The shear stresses τ_1, τ_2, etc. for the walls are given by Eq. (4.23),

$$\tau_1 = \frac{h_1}{t_1}, \quad \tau_2 = \frac{h_2}{t_2}, \quad \tau_3 = \frac{h_3}{t_3},$$

$$\tau_4 = \frac{h_1 - h_2}{t_4}, \quad \tau_5 = \frac{h_2 - h_3}{t_5}$$

Assuming the direction of the shearing stress as shown in Fig. 4.26, the constant shear flow condition yields,

$$\tau_1 t_1 = \tau_2 t_2 + \tau_4 t_4 \quad \text{(i)}$$

and

$$\tau_2 t_2 = \tau_3 t_3 + \tau_5 t_5 \quad \text{(ii)}$$

Using Eq. (4.26), the torque applied

$$T = 2Ah = 2A_1h_1 + 2A_2h_2 + 2A_3h_3$$
$$= 2A_1\tau_1 t_1 + 2A_2\tau_2 t_2 + 2A_3\tau_3 t_3 \quad \text{(iii)}$$

Applying Eq. (4.27), over areas A_1, A_2 and A_3 gives

$$\tau_1 s_1 + \tau_4 s_4 = 2G\theta A_1 \quad \text{(iv)}$$
$$\tau_4 s_4 + 2\tau_2 s_2 + \tau_5 s_5 = 2G\theta A_2 \quad \text{(v)}$$
$$\tau_5 s_5 + 2\tau_3 s_3 = 2G\theta A_3 \quad \text{(vi)}$$

where

$t_1 = t_3 = 6$ mm, $t_2 = 8$ mm,

$t_4 = 12$ mm, $t_5 = 10$ mm,

$A_1 = \pi r^2/2 = \pi \times 125^2/2 = 24543.67\text{ mm}^2$,

$A_2 = 300 \times 250 = 75000\text{ mm}^2$,

$A_3 = 250 \times 500/2 = 62500\text{ mm}^2$,

$s_1 = \pi r = \pi \times 125 = 392.7$ mm,

$s_2 = 300$ mm,

$s_3 = \sqrt{500^2 + 125^2} = 515.39$ mm,

$s_4 = s_5 = 250$ mm

and $T = 110 \times 10^6$ N mm

Substituting these values in Eqs (i) through (vi) yield,

$\tau_1 = 44.91$ MPa, $\tau_2 = 55.95$ MPa,

$\tau_3 = 39.5$ MPa,

$\tau_4 = -14.85$ MPa, $\tau_5 = 21.06$ MPa,

$\theta = 3.546 \times 10^{-6}$ rad/mm

Example 4.19 A circular steel bar of 60 mm diameter is twisted beyond the elastic limit such that a central core of 30 mm diameter remains elastic. Determine the torque applied if $\tau_y = 230$ MPa. Also find the residual stresses and the residual angle of twist if the shear modulus $G = 80$ GPa.

Solution Equation (4.4) applicable for the elastic zone and Eq. (4.28) is applicable for the plastic zone. The applied torque is therefore given by,

$$T = \left.\frac{\tau_y J}{R}\right|_0^{15} + \left.\frac{2\pi R^3 \tau_y}{3}\right|_{15}^{30}$$

or, $$T = \frac{\pi \times 15^3 \times 230}{2} + 2\frac{\pi \times 230}{3} \times (30^3 - 15^3)$$

$= 12599739.4$ Nmm

The residual stress and deformation conditions after removal of load, can be visualized easily by superposition of an equal and opposite torque.

If the bar remains elastic under a torque of 12599739.4 N mm, then the maximum stress,

$$\tau = \frac{T \times r}{J} = \frac{12599739.4 \times 30}{\pi \times 30^4/2}$$

$= 297.08$ MPa

Figure 4.27(a) shows the elastic–plastic stress diagram due to the applied torque of 12599739.4 N mm and Fig. 4.27(b) shows the stress diagram due to a torque 12599739.4 N mm applied in the opposite direction. Stress diagram for the residual stresses is obtained by superposition of the two figures as in Fig. 4.27(c).

Residual angle of twist: Consider the elastic core for determination of the angle of twist in the elastic plastic case,

Shear strain in the elastic core,

$$\gamma = \frac{\sigma}{G} = \frac{230}{80000} = 0.002875$$

The angle of twist per unit length (Eq. 4.1),

$$\frac{\theta}{L} = \frac{\gamma}{r} = \frac{0.002875}{15}$$

$= 0.0001917$ rad/mm

If the bar remains elastic under the torque applied in opposite directions, then the angle of twist per unit length is given by Eq. (4.3),

$$\frac{\theta}{L} = \frac{T}{JG} = \frac{12599739.4}{80000 \times \pi \times 30^4/2}$$

$= 0.0001238$ rad/mm

The residual angle of twist is given by superposition of the two cases as done for the residual shear stress.

$\therefore$ Residual angle of twist $= 0.0000679$ rad/mm.

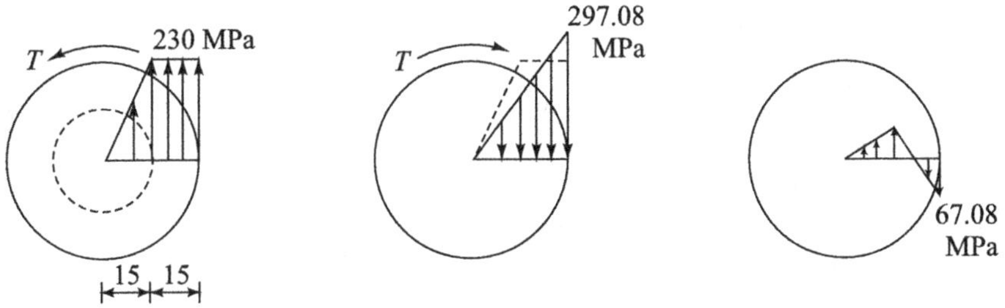

(a) Stress distribution due to applied torque

(b) Stress distribution due to torque applied in opposite direction

(c) Residual stress distribution

Fig. 4.27

EXERCISES

PROBLEMS

4.1 The shaft coupling shown in Fig. 4.28 transmits torque through 8 bolts of 20 mm diameter. Determine the torque which can be transmitted if the allowable shear stress in the bolts is 50 MPa. Neglect the plate friction.

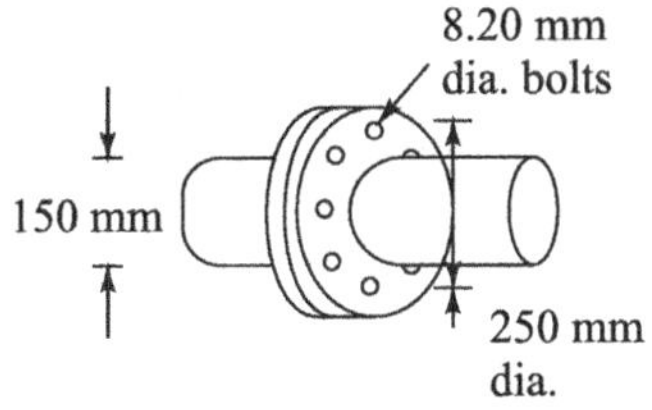

Fig. 4.28

4.2 A flange coupling is connected by 10 mm diameter bolts in three circles as shown in Fig. 4.29. It is required to transmit 1000 HP at 1200 rpm. Determine the maximum shear stress in the bolts.

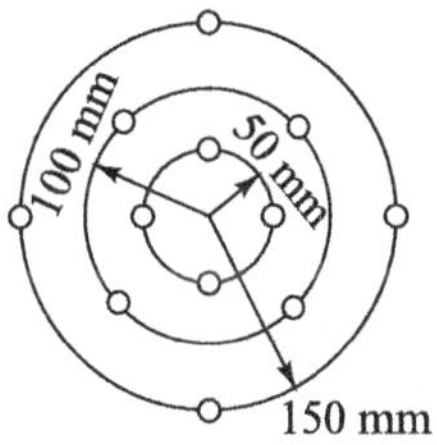

Fig. 4.29

4.3 A solid circular bar of varying cross-section is subjected to torque as shown in Fig. 4.30. Determine the maximum shear stress in the bar and the angle of twist of the free end if the shear modulus $G = 80$ GPa.

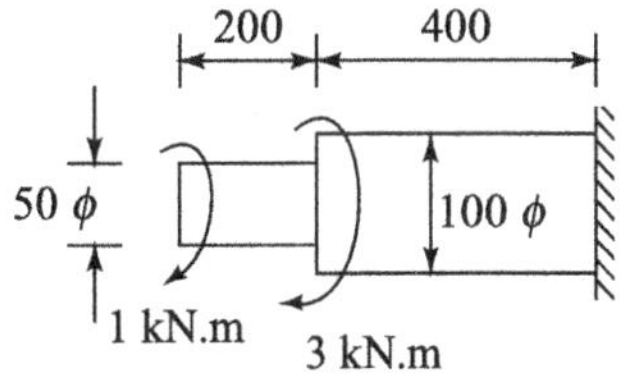

Fig. 4.30

4.4 A solid circular steel shaft of 50 mm diameter is used for power transmission through pulleys at 1440 rpm (clockwise) as shown in Fig. 4.31. Plot the torque diagram and determine the maximum stress in the shaft. Also calculate the angle of twist in each segment if the modulus of shear $G = 80$ GPa.

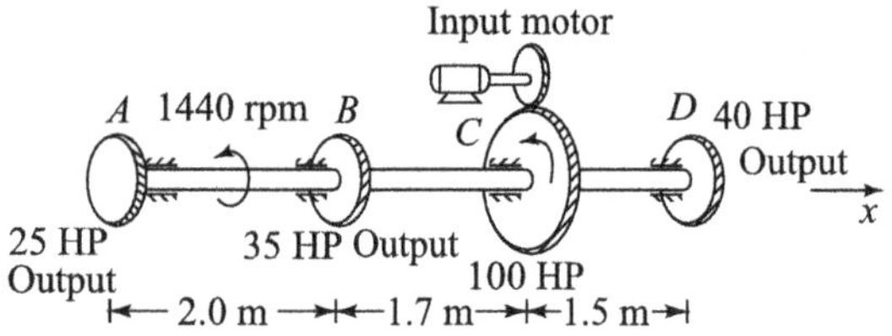

Fig. 4.31

4.5 An electric motor supplies power at 24 Hz to two machines through solid steel shafts ABC and DE as shown in Fig. 4.32. The gear at B has 60 teeth and the gear at D has 150 teeth. The machine at C consumes 90 HP and the machine at E consumes 110 HP. Determine the diameter of the shafts required if the permissible shear stress is 50 MPa. Also determine the angle of twist A with respect to E. Neglect the energy losses in the gears and the bearings. Take $G = 80$ GPa.

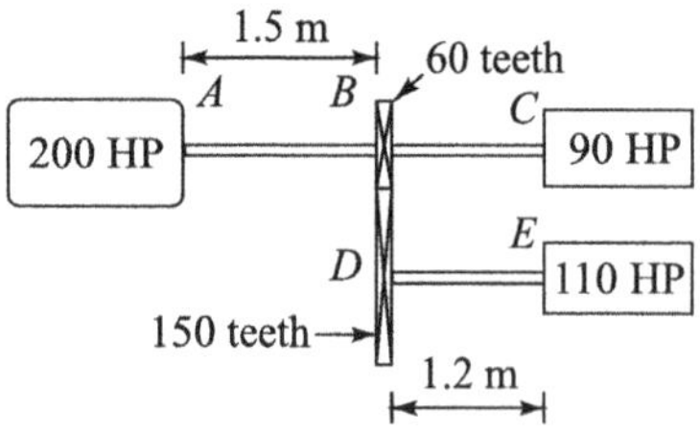

Fig. 4.32

4.6 A solid circular bar of torsional rigidity GJ, is fixed at both ends. It is subjected to a varying torque $x\, t_o/L$ as shown in Fig. 4.33. Determine the end reactions and the maximum angle of twist.

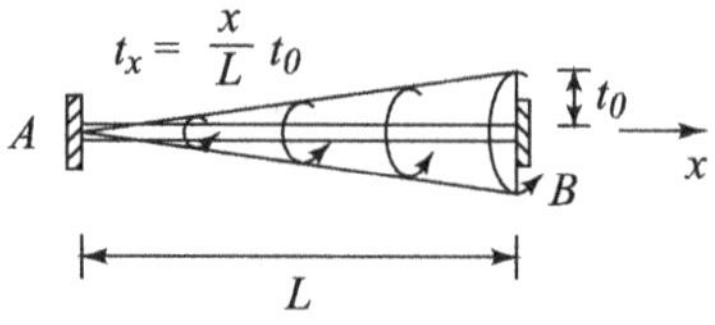

Fig. 4.33

4.7 A solid stepped circular shaft is subjected to torque as shown in Fig. 4.34. Determine the end reactions and plot the angle of twist diagram along the length of the shaft. Take $G = 80$ GPa.

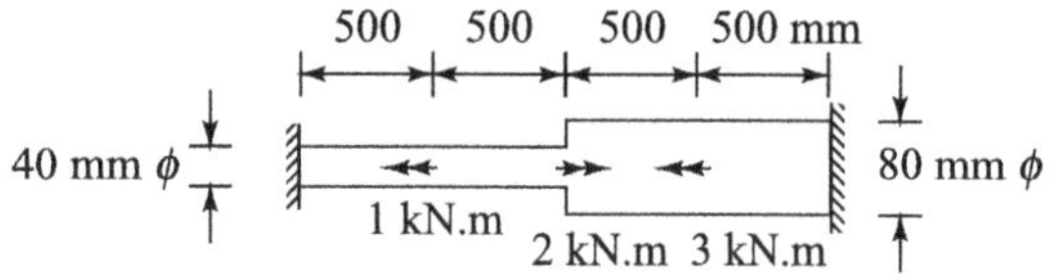

Fig. 4.34

4.8 A hollow circular shaft AB of outside and inside diameters 120 mm and 90 mm, respectively, is connected through gears to a solid circular shaft CD of diameter 75 mm as shown in Fig. 4.35. The gear at B had 150 teeth and the gear at C has 50 teeth. If end A is fixed while a torque 2500 Nm is applied at end C, find the shear stress in each shaft and the rotation of end D if the shear modulus $G = 80$ GPa.

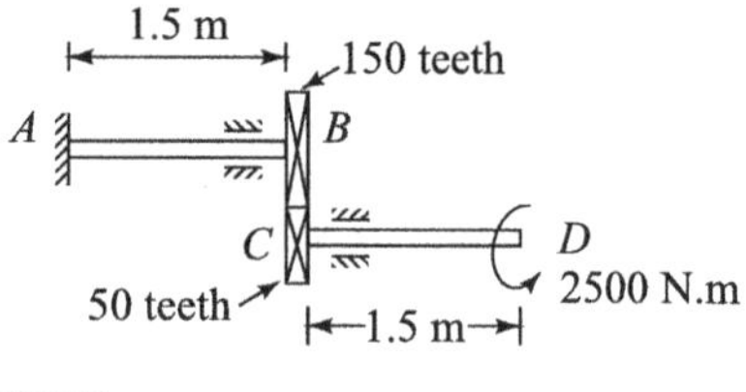

Fig. 4.35

4.9 A 50 mm square cantilever bar carries a uniformly distributed torque of 1 kN m/m as shown in Fig. 4.36. Calculate the maximum shear stress and sketch the distribution of the angle of twist along the length of the bar. Take $G = 80$ GPa.

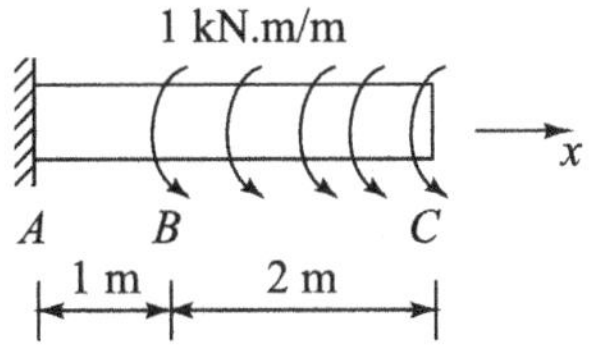

Fig. 4.36

4.10 Compare the torsional stiffness and the maximum twisting moment for the box and the rectangular open section shown in Fig. 4.37 if the material is same for the both.

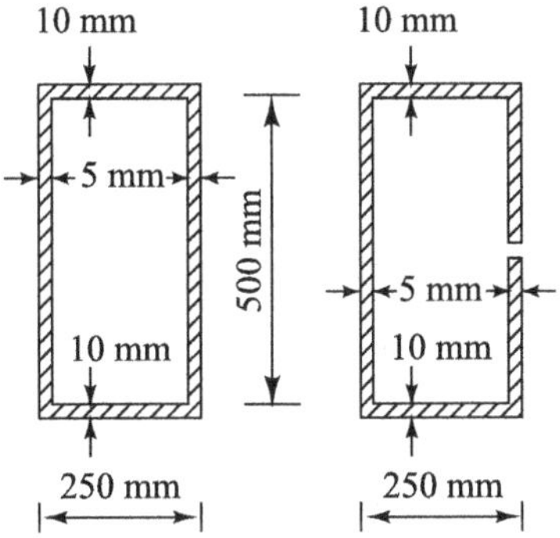

Fig. 4.37

4.11 A thin walled section as in Fig. 4.38 is under torsion T. Calculate the torsional stiffness and the shear stress in the section.

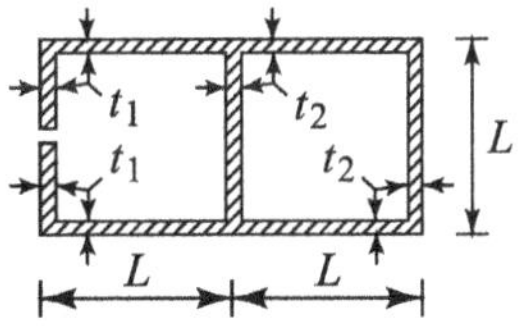

Fig. 4.38

4.12 Derive an approximate expression for the torsional rigidity of a narrow triangular section shown in Fig. 4.39 assuming that the expression for the stress function ϕ corresponding to a parabolic membrane is represented by, $\phi = G\theta[(b/2)^2 - x^2]$ where b is the width of the triangle at any distance y from the apex.

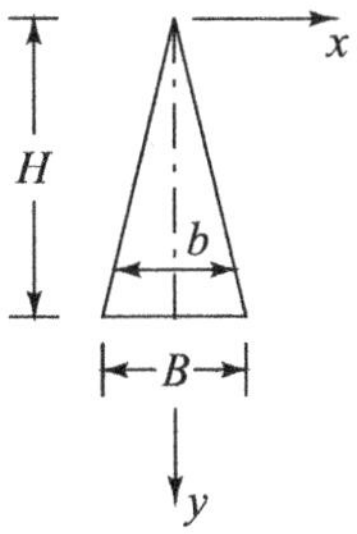

Fig. 4.39

4.13 A thin walled regular hexagonal section of sides 100 mm and thickness 5 mm, is under torsion 4.5 kNm. Determine the shearing stress and angle of twist per unit length if the shear modulus G = 27 GPa.

4.14 A hollow shaft having inside and outside diameters 50 mm and 150 mm, respectively is under torsion such that the plastic deformation has penetrated to a depth of 25 mm. Calculate the torque applied if the shaft material is assumed to be elastic upto yield stress in shear 145 MPa and perfectly plastic thereafter.

4.15 A circular steel bar of 150 mm diameter is twisted beyond elastic limit so that a central core of 50 mm diameter remains elastic. Determine the torque applied, the residual stresses and the residual angle of twist if the stress–strain relation for the material of the shaft is shown in Fig. 4.40.

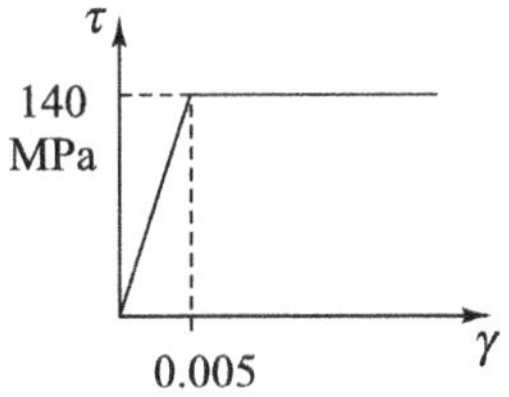

Fig. 4.40

4.16 A solid circular shaft of diameter 100 mm is under torque 10 kN m. Calculate the maximum shear stress and the angle of twist per unit length if the stress–strain relation for the material of the shaft is shown in Fig. 4.41.

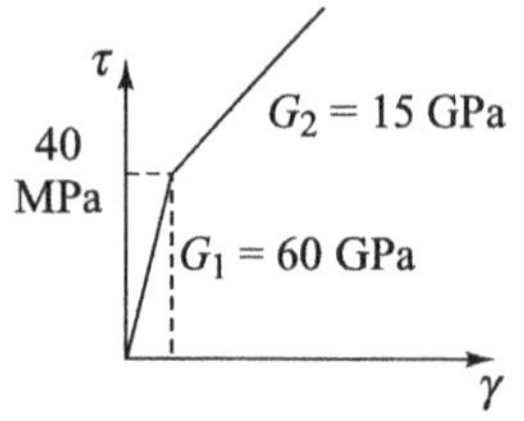

Fig. 4.41

OBJECTIVE QUESTIONS

4.17 The torsional stiffness per unit length of a circular shaft of radius r and shear modulus G is given by,

(a) $\pi r^4/2G$ (b) $\pi r^4G/2$
(c) $\pi r^4/32G$ (d) $\pi r^4G/64$

4.18 An electric motor delivers 1000 W at 1440 rpm. It will generate torque equal to,

(a) $125/6\pi$ (b) 144π
(c) $3\pi/125$ (d) $125/3\pi$

4.19 A stepped circular shaft is fixed at one end. Its torsional rigidity is $2JG$ over half of its length and JG over the rest as shown in Fig. 4.42. The angle of twist due to a torsion T at the free end is given by,

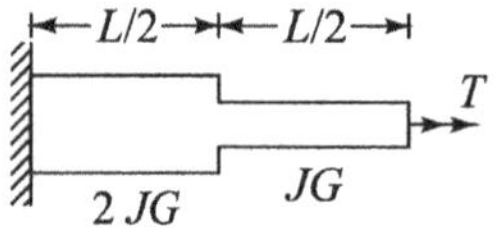

Fig. 4.42

(a) $3TL/4GJ$ (b) $TJ/4GL$
(c) $2TL/3GJ$ (d) $3TL/2GJ$

4.20 A cantilever circular shaft of diameter D has a central hole of diameter $D/2$ as shown in Fig. 4.43. If JG is the torsional rigidity of the shaft then the angle of twist due to a torsional moment T at the free end is given by,

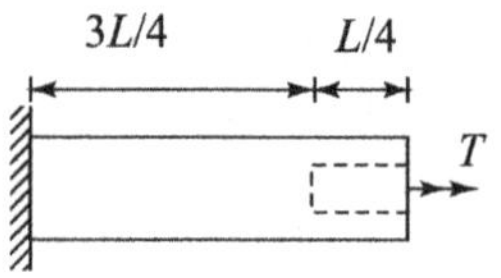

Fig. 4.43

(a) $8TJ/9GL$ (b) $8TL/9GJ$
(c) $49TL/48GJ$ (d) $9TL/64GJ$

4.21 Two identical shafts of length L and torsional rigidity GJ are connected by two gears having 250 and 50 teeth as shown in Fig. 4.44. If a torque T is applied at end B, the angle of twist of end B relative to end A is given by,

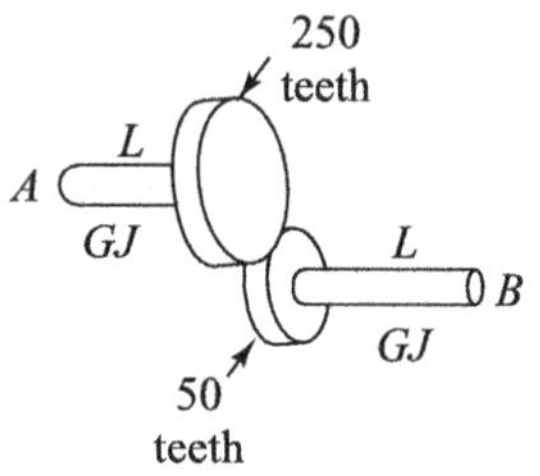

Fig. 4.44

(a) $26TL/GL$ (b) $26TL/25GJ$
(c) $25TL/26GJ$ (d) $TL/6GJ$

4.22 A circular shaft fixed at both ends carries a torsion T as shown in Fig. 4.45. The reaction at end A is given by,

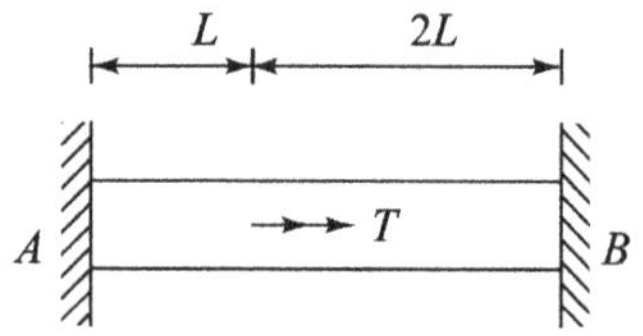

Fig. 4.45

(a) $T/4$ (b) $2T/3$
(c) $T/2$ (d) $T/3$

4.23 A stepped circular shaft fixed at both ends carries a torsion T at its mid-point as shown in Fig. 4.46. If its torsional rigidity GJ is twice over half of its length L then the reaction at end A is given by,

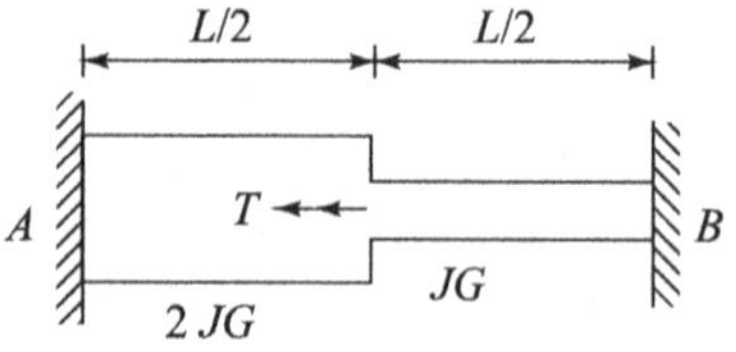

Fig. 4.46

(a) $T/4$ (b) $T/3$
(c) $T/2$ (d) $2T/3$

4.24 The maximum shear stress in a thin walled H section of size 250 mm × 250 mm and thickness 10 mm as shown in Fig. 4.47 when subjected to a torque of 2 kN m, will be,

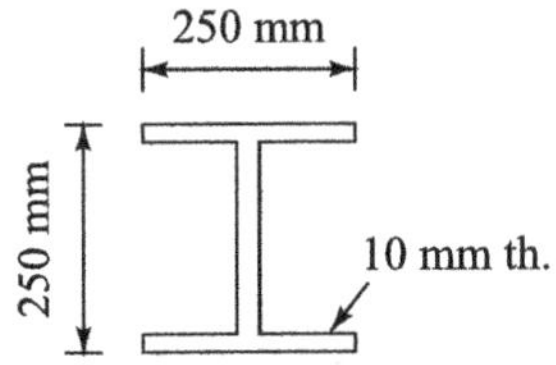

Fig. 4.47

(a) 2.4 MPa (b) 24 MPa
(c) 8 MPa (d) 80 MPa

4.25 If the shear modulus $G = 80$ GPa, the angle of twist for the H section of 4.24 will be ,
(a) 5.73°/m (b) 2.36°/m
(c) 0.28°/m (d) $4.13° \times 10^{-3}$/m

4.26 A hollow aluminium square tube of size 250 mm 250 mm has thickness 6 mm. It is subject to a torque of 21 kN m along its longitudinal axis. The shear stress will be,
(a) 24 MPa (b) 24.8 MPa
(c) 28 MPa (d) 32 MPa

4.27 If the shear modulus $G = 80$ GPa, the angle of twist for the hollow aluminium section of 4.26 will be,
(a) 0.008 rad/m (b) 0.0086 rad/m
(c) 0.006 rad/m (d) 0.0068 rad/m

5

Shearing Force and Bending Moment

CONCEPT REVIEW

5.1 INTRODUCTION

A beam is a structural member which is primarily required to withstand bending and shearing forces in addition to axial force if any. Normally axial forces are insignificant and therefore neglected. If axial force is large then the beams are further analysed as beam-columns in a separate chapter.

A beam is statically determinate if the number of unknown support reactions are equal to the number of equations of static equilibrium, i.e. $\Sigma F_x = 0$, $\Sigma F_y = 0$, $\Sigma M = 0$. A hinge in the beam can transfer shear but moment is not transferred to its either sides. A hinge gives an additional condition of static equilibrium, i.e. $\Sigma M = 0$, at the hinge. Figure 5.1 shows the number of support reactions for statically determinate beams. An additional support will make the beam statically indeterminate and a lesser number of the support reactions will make the beam an unstable mechanism. A roller support does not support the beam in its axial direction but gives support in transverse direction—upward or downward both. Except in case of long span bridges a proper hinge and roller support is rarely provided in actual structures. In general the beams are designed to withstand maximum bending moment and since a statically determinate beam will generally have a higher bending moment—an assumed hinge and roller support is on safer side.

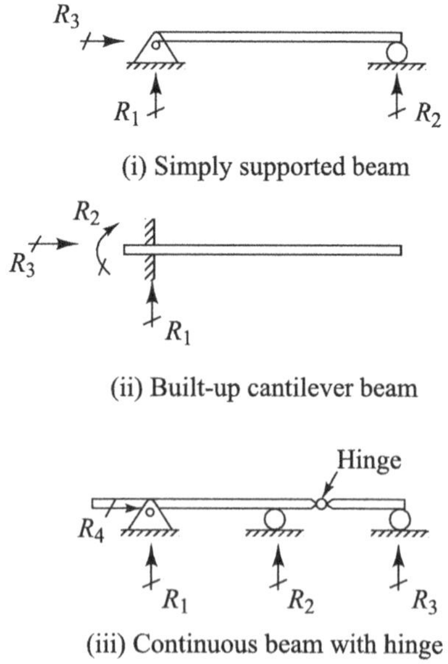

Fig. 5.1 *Statically Determinate Beams*

5.2 BENDING MOMENT AND SHEAR FORCE DIAGRAMS

Bending moment and shear force diagrams show the variation of shear force and bending moment in a structure. For cantilever beams it is simple to calculate shear force and bending moment at any section starting from the free end. In case of a simply supported beam under loads 6 kN/m. 8 kN and 20 kN.m as shown in Fig. 5.2, the support reactions R_1 and R_2 are first calculated using equilibrium equations $\Sigma F = 0$ and $\Sigma M = 0$. The shear force and bending moment at any section $A-A$ can now be calculated by considering an imaginary cut at the section and the static equilibrium of the free body diagram of either left or right of the section $A-A$ will give shear force V and bending moment M.

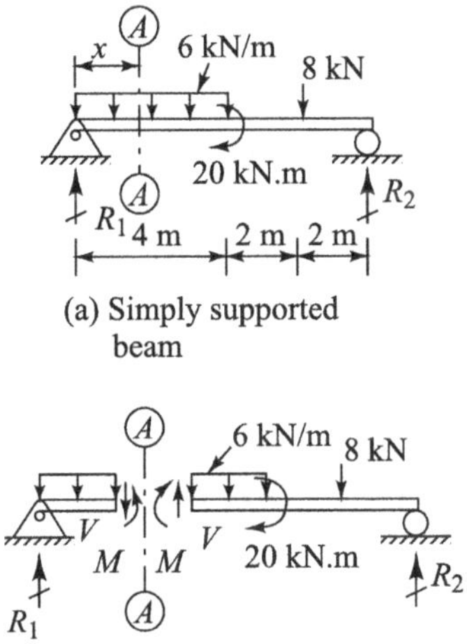

(a) Simply supported beam

(b) Free body diagram

Fig. 5.2

5.2.1 Sign Conventions

As earlier in Art. 2.2, +ve direction forces on +ve faces and –ve direction forces on –ve faces are usually taken +ve. The shear force and the bending moment applied at the ends of a beam as shown in Fig. 5.3, are taken as positive. Positive bending moment diagram is sometimes drawn on tension face of the member, otherwise +ve or –ve sign is to be marked in the diagram.

Fig. 5.3 *Positive Shear and Moment*

Shear force and bending moment diagram by parts: S.F. and B.M. diagrams for a beam under several loads can be obtained by the principle of superposition by adding S.F. and B.M. diagrams for the beam under individual force.

5.3 RELATION BETWEEN LOAD *w*, SHEAR *V* AND MOMENT *M*

Consider an elementary length dx in a simply supported beam with an arbitrary load as shown in Fig. 5.4.

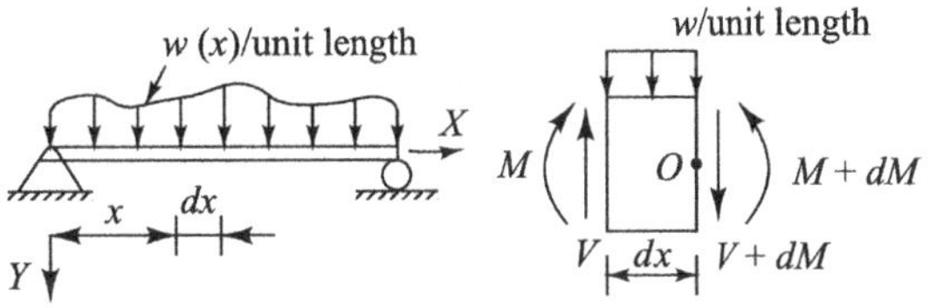

Fig. 5.4

$$\Sigma F_y = 0: \qquad V + dV - V + wdx = 0$$

or, $$\frac{dV}{dx} = -w \tag{5.1}$$

or, Slope of S.F. diagram = load intensity at that section

If there is no load at a section then the shear force will be constant and the slope of the shear force diagram will be zero. In case of a uniformly distributed load, the slope of S.F. diagram will be –ve, i.e. downward.

If, $$\frac{d}{dx}\left(\frac{dV}{dx}\right) = -\frac{dw}{dx} > 0, \tag{5.2}$$

i.e. if w decreases in x-direction (dw/dx is –ve), then the shear force diagram will concave upwards.

Taking moment about point O in the right face in Fig. 5.4,

$$\Sigma M_z = 0: \quad M + dM - M - Vdx + w\,\frac{dx^2}{2} = 0$$

Neglecting square of a small term, $dx^2 = 0$

$$\therefore \qquad \frac{dM}{dx} = V \qquad (5.3)$$

or, Slope of B.M. diagram = S.F. at the section

If the shear force is +ve, then the bending moment diagram will slope upwards. Equation (5.3) shows that the maximum or minimum bending moment occurs when either the shear force changes the sign or it is zero.

$$\text{If,} \qquad \frac{d}{dx}\left(\frac{dM}{dx}\right) = \frac{dV}{dx} > 0, \qquad (5.4)$$

i.e. if V increases in x-direction (dV/dx is +ve), then the bending moment diagram will concave upwards. Equation (5.1) is not valid at a concentrated load and Eq. (5.3) is not valid at a couple applied on the beam. They are however valid between concentrated loads and couples, respectively.

Since $V = -\int w dx$ and $M = \int V dx$, if w is n degree curve, V will be $n + 1$ degree curve and M will be $n + 2$ degree curve. In case of uniformly distributed load, w is uniform, V is linear and M is of second order parabola as shown in Fig. 5.11.

These concepts are explained by Fig. 5.5 in case of a cantilever. The shear force diagram is +ve in case (i) and –ve in case (ii). The shear force diagram slope downward in both cases as $dV/dx = -w$, is –ve. Bending moment diagrams are –ve in both cases as the deflected shape is hogging. The bending moment will slope upward in case (i) as $dM/dx = V$ is positive and will slope down in case (ii) as shear force, $V = dM/dx$ is –ve. The bending moment diagram is concave downward in both cases as $dV/dx = d^2M/dx^2$ is negative.

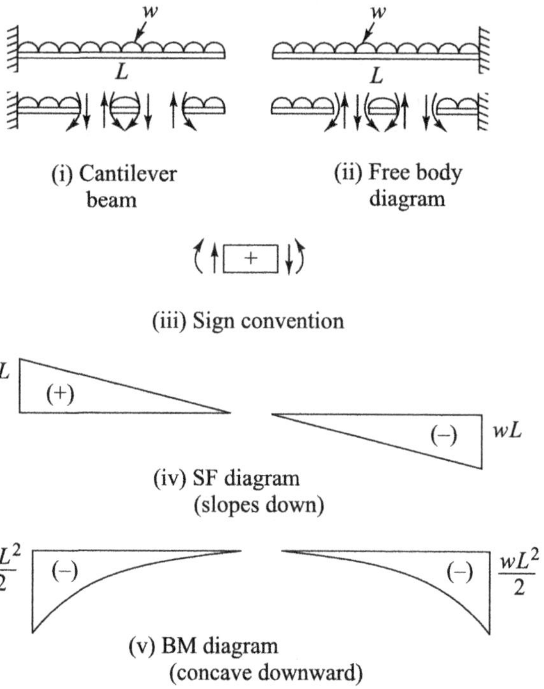

Fig. 5.5

EXAMPLES

Example 5.1 Draw shearing force and bending moment diagrams for the cantilever beam shown in Fig. 5.6.

Solution **Free body diagram:** Consider an imaginary cut at a distance x from end A. Shearing force V and bending moment M (assumed in +ve directions) are required to act for the equilibrium of the section.

S.F. Diagram Using static equilibrium equation for the free body diagram, the shear force is given by,

$$\Sigma F = 0, \quad 5 + V = 0$$

or, $\qquad V = -5$ kN

which is constant over $0 < x < 2$ m

–ve sign means V is a actually acting opposite to the direction assumed earlier in the free body diagram.

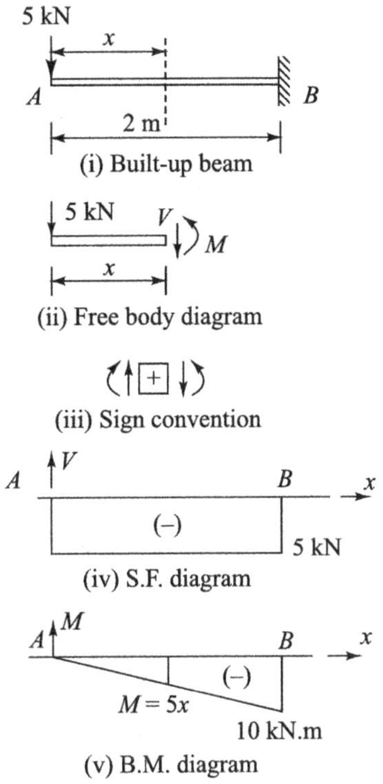

Fig. 5.6

B.M. Diagram Taking moment of the forces in the free body diagram about end A,

$\Sigma M = 0, \quad M - V \times x = 0$

or, $\quad M = -5x$ (linear variation with x)

At end B, $\quad M = -10$ kN/m

Figure 5.6 (iv) and (v) respectively show the variation of shearing force and bending moment along length of the beam.

Example 5.2 Draw shearing force and bending moment diagrams for the cantilever beam shown in Fig. 5.7.

Solution ***S.F. Diagram*** Make an imaginary cut at a distance x from end B. Forces V and M are assumed in +ve directions for the equilibrium of the section.

$\Sigma F = 0, \quad V = 0$

Shear force is zero throughout:

B.M. Diagram Using equilibrium equation,

$$\Sigma M = 0, \quad M = 10 \text{ kN m}$$

Bending moment is constant and +ve as shown in Fig. 5.7(iii).

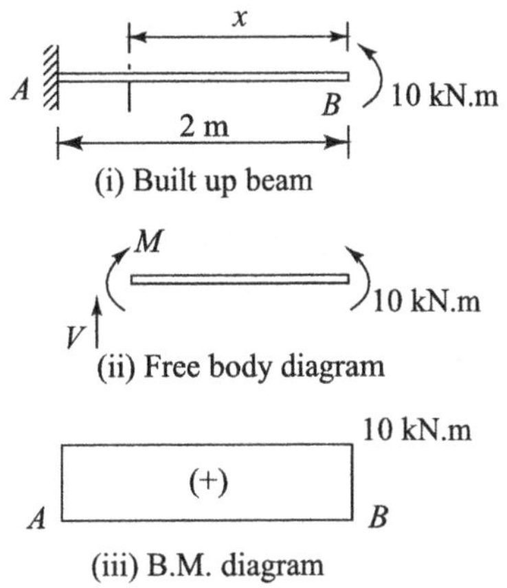

Fig. 5.7

Example 5.3 Draw shear force and bending moment diagrams for the cantilever beam shown in Fig. 5.8.

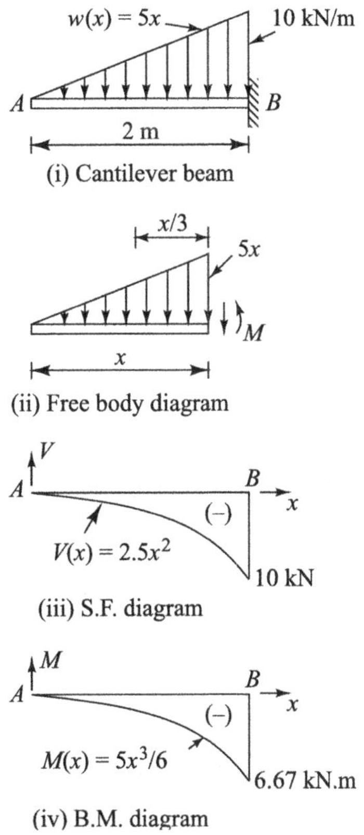

Fig. 5.8

Solution Consider the free body diagram of a section of the cantilever. Forces V and M are assumed in +ve directions for the equilibrium of the section.

S.F. Diagram Using equilibrium equation for the free body diagram,

$$\Sigma F = 0, \quad V + \frac{5x \times x}{2} = 0$$

or, $\quad V = -2.5x^2 = -10$ kN when $x = 2$ m

–ve sign means V is actually acting opposite to the direction assumed earliar in the free body diagram. Figure 5.8(iii) shows second degree curve for shear force V.

B.M. Diagram Take moment of the forces in the free body diagram about right end, considering the triangular distributed load $5x \times x/2$ to act through its cg $x/3$ from right end as follows.

$$\Sigma M = 0, \qquad M + \frac{5x \times x}{2} \times \frac{x}{3} = 0$$

or, $\quad M = -5x^3/6 = -6.67$ kN m at $x = 2$ m

–ve sign means M is actually acting opposite to the direction assumed earlier in the free body diagram. Figure 5.8(iv) shows third degree curve for bending moment.

Note: Slope of S.F. diagram dV/dx is –ve as $w(x)$ is +ve downward. As $w(x)$ increases in x-direction then $d^2V/dx^2 = -dw/dx < 0$ and S.F. diagram will concaves downward.

The shear force is –ve, the bending moment diagram will slope downwards. As V decreases in x-direction, i.e. $d^2M/dx^2 = -dV/dx < 0$ and the bending moment diagram will concave downwards.

Example 5.4 Draw shearing force and bending moment diagrams for the cantilever beam shown in Fig. 5.9.

Solution Consider a section of the beam at a distance x from the free end as shown in Fig. 5.9(ii). Shear force V is given by the equilibrium equation as follows,

$$\Sigma F = 0, \quad V = 20 <x-1>^0 + 5 <x-2>^1 \text{ kN}$$

where the pointed brackets are called *singularity functions* which are zero if the quantity within the pointed brackets is –ve.

For $0 < x < 1$ m, $\quad V = 0$
For $1 < x < 2$ m, $\quad V = 20$ kN
For $2 < x < 4$ m, $\quad V = 20 + 5(x - 2)$
$\quad = 30$ kN at $x = 4$ m

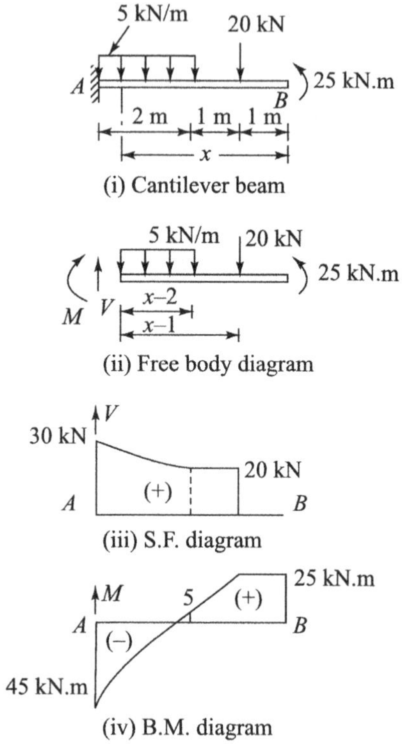

Fig. 5.9

The shearing force diagram shown in Fig. 5.9(iii) is constant 20 kN for $1 < x < 2$ m and varies linearly from 20 to 30 kN for $2 < x < 4$ m.

Similarly, the bending moment function is given by taking moment of the force in the free body diagram about left end. The distributed load 5 kN/m over the length $< x - 2 >$ acts at mid-point $< x - 2 >/2$ and its moment about left end will be $5 < x - 2 >^2/2$ as follows.

$$\Sigma M = 0, \qquad M = 25 <x>^0 - 20 <x-1>^1 - 5 <x-2>^2/2$$

For $0 < x < 1$ m, $\quad M = 25$ kN m
For $1 < x < 2$ m, $\quad M = 25 - 20(x - 1)$ kNm
For $2 < x < 4$ m, $\quad M = 25 - 20(x - 1) - 5 <x-2>^2/2$
$\quad = -45$ kN m at $x = 4$ m

The bending moment diagram shown in Fig. 5.9 (iv) is constant 25 kN m for $0 < x < 1$ m, varies linearly from 25 to 5 kN m for $1 < x < 2$ m and varies as second degree curve from 5 kN m to -45 kN m for $2 < x < 4$ m.

Example 5.5 Draw shear force and bending moment diagram for a simply supported beam carrying a concentrated load W at mid span as shown in Fig. 5.10

Solution Support reactions R_1 and R_2 are given by equilibrium equations are follows.

$\Sigma F = 0, \qquad R_1 + R_2 = W \qquad$ (i)

Taking moment about end B,

$\Sigma M = 0, \qquad R_1 \times L = W \times L/2$

or, $\qquad R_1 = W/2 \qquad$ (ii)

From, Eqs (i) and (ii), $R_2 = W/2$

Considering equilibrium of a section of the beam at any distance x as shown in Fig. 5.10(ii),

$\Sigma F = 0,$ for $0 < x < L/2$, $\quad V = R_1 = W/2$

for $L/2 < x < L$, $\quad V = R_1 - W = -W/2$

$\Sigma M = 0,$ for $0 < x < L/2$ $\quad M = R_1 \times x = Wx/2$

for $L/2 < x < L$, $\quad M = R_1 \times x - W \times (x - L/2)$

$= W(L - x)/2$

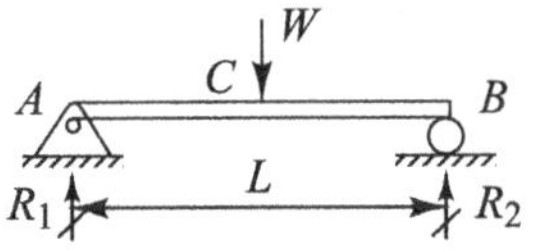

(i) Simply supported beam

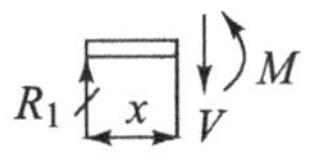

(ii) Free body diagram

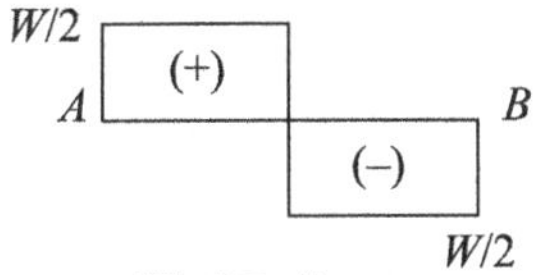

(iii) S.F. diagram

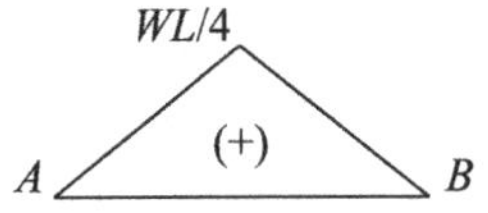

(iv) B.M. diagram

Fig. 5.10

V vs. x and M vs. x is shown in Fig. 5.10(iii) and (iv), respectively. Maximum bending moment $WL/4$ will occur at mid span where shearing force is zero.

Example 5.6 Draw shear force and bending moment diagrams for a simply supported beam of span L carrying a uniformly distributed load w as shown in Fig. 5.11.

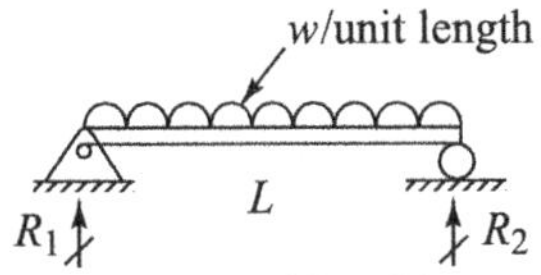

(i) Beam with u.d.l

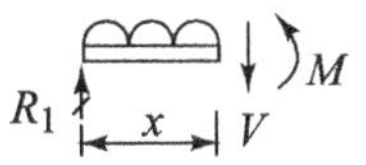

(ii) Free body diagram

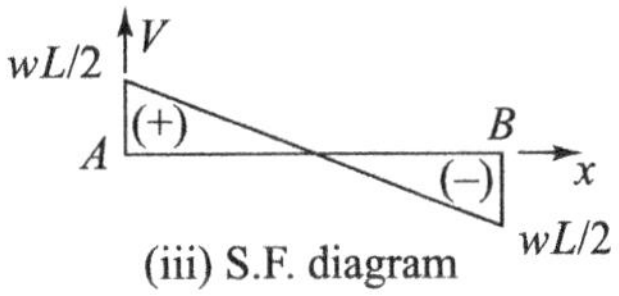

(iii) S.F. diagram

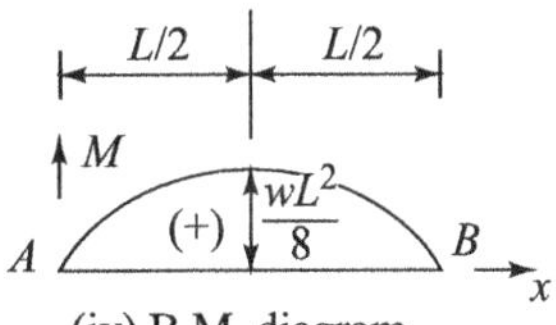

(iv) B.M. diagram

Fig. 5.11

Solution Support reactions R_1 and R_2 are given by equilibrium equation as follows.

$\Sigma F = 0, \qquad R_1 + R_2 = wL \qquad$ (i)

Taking moment about end B (assume *udl* $w \times L$ at its cg $L/2$),

$\Sigma M = 0, \qquad R_1 \times L = (w \times L) \times L/2$

or, $\qquad R_1 = wL/2 \qquad$ (ii)

From Eqs (i) and (ii), $R_2 = wL/2$

Considering equilibrium of a section of the beam at any distance x as shown in Fig. 5.11(ii),

$\Sigma F = 0, \qquad V = R_1 - w \times x = w(L/2 - x)$

at $\quad x = 0, \qquad V = wL/2$

at $\quad x = L/2, \qquad V = 0$

and at $x = L$, $V = -wL/2$

$\Sigma M = 0$, $M = R_1 \times x - wx \times x/2$

$= wx(L - x)/2$

$= 0$ at $x = 0$ and L

$= wL^2/8$ at $x = L/2$

V vs. x and M vs. x is shown in Fig. 5.11(iii) and (iv), respectively.

Note: $w(x)$ is +ve (downward), $dV/dx = -w$, i.e. slope of S.F. diagram is –ve (downward). As $w(x)$ is constant in x-direction then $d^2V/dx^2 = -dw/dx = 0$, and S.F. diagram will be linear.

The shear force is +ve from $0 < x < L/2$, then B.M. diagram will slope upward. The shear force is –ve from $L/2 < x < L$, then B.M. diagram will slope downwards. As V decreases in x-direction, then $d^2M/dx^2 = dV/dx < 0$ and the bending moment diagram will concave downward.

Example 5.7 Draw shear force and bending moment diagrams for a simply supported beam of span 6 m carrying a linearly varying uniformly distributed load $w(x) = 5x$ kN/m as shown in Fig. 5.12.

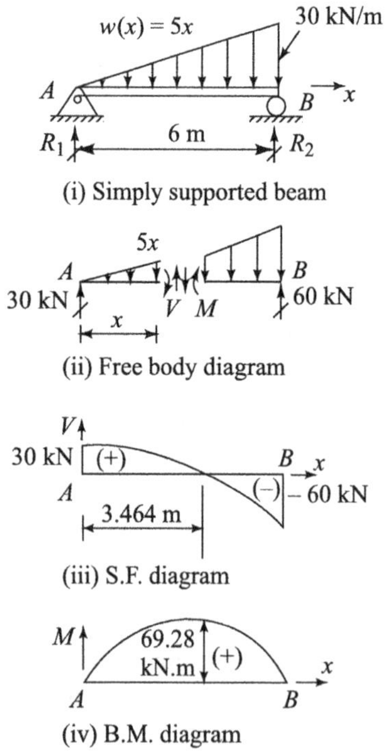

Fig. 5.12

Solution Using equilibrium equations,

$$\Sigma F = 0, \qquad R_1 + R_2 = \frac{30 \times 6}{2} \qquad \text{(i)}$$

Take moment about B assuming triangular load to act at its cg,

$$\Sigma M = 0, \qquad R_1 \times 6 = \frac{30 \times 6}{2} \times \frac{6}{3}$$

$$\text{or,} \qquad R_1 = 30 \text{ kN m} \qquad \text{(ii)}$$

From Eqs (i) and (ii), $R_2 = 60$ kN m

Considering equilibrium of a section of the beam at any distance x as shown in Fig. 5.12 (ii),

$$\Sigma F = 0, \quad V = 30 - \frac{5x \times x}{2}$$

$V = 30$ kN at $x = 0$

$= 0$ at $x = \sqrt{30 \times 2/5}$ $= 3.464$ m from A

$= -60$ kN at $x = 6$ m

$$\Sigma M = 0, \quad M = 30 \times x - \frac{5x \times x}{2} \times \frac{x}{3}$$

$= 30x - 5x^3/6$

$= 0$ at $x = 0$ and L

M will be maximum where shear force is zero (Eq. 5.3).

$$M_{max} = 30 \times \sqrt{12} - 5(\sqrt{12})^3/6 = 69.28 \text{ kN m}$$

V vs. x and M vs. x is shown in Fig. 5.12(iii) and (iv). For rate of change of slope of S.F. and B.M. diagrams, see the note in Example 5.3 and 5.6.

Example 5.8 Draw shear force and bending moment diagrams for a simply supported beam of span 5 m carrying a concentrated anticlockwise moment 100 kN m at mid span as shown in Fig. 5.13(i). Also draw the deflected shape of the beam.

Solution Using equilibrium equations,

$\Sigma F = 0$, $R_1 + R_2 = 0$ (i)

Taking moment about B, $R_1 \times 5 = 100$

or, $R_1 = 20$ kN m (ii)

Considering equilibrium of a section of the beam at any distance x from end A,

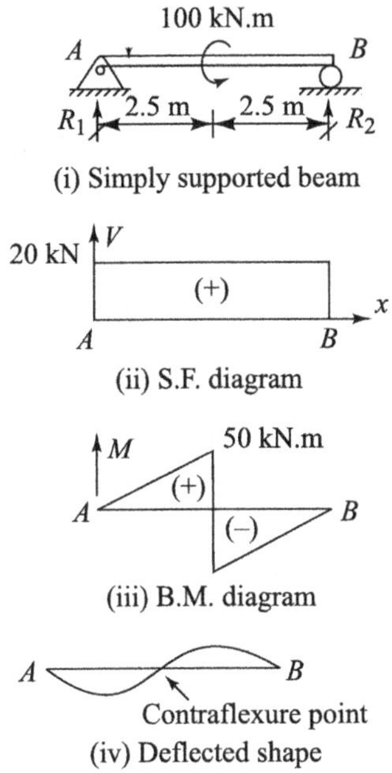

(i) Simply supported beam
(ii) S.F. diagram
(iii) B.M. diagram
(iv) Deflected shape

Fig. 5.13

$$\Sigma F = 0, \quad V = R_1 = 20 \text{ kN} \quad \text{for } 0 < x < 5 \text{ m}$$
$$\Sigma M = 0, \quad M = R_1 \times x = 20x \quad \text{for } 0 < x < 2.5 \text{ m}$$
$$= 100 - 20x \quad \text{for } 2.5 < x < 5 \text{ m}$$

V vs. x and M vs. x is shown in Fig. 5.13(ii) and (iii), respectively. Figure 5.13(iv) shows the deflected shape of the beam. The slope the deflected shape changes at the point of contraflexure where the B.M. diagram changes from +ve to –ve, i.e. from sagging to hogging.

Example 5.9 Draw shear force and bending moment diagram for a simply supported beam of span 8 m carrying loads as shown in Fig. 5.2(i).

Solution Using equilibrium equations,

$$\Sigma F = 0, \quad R_1 + R_2 = 6 \times 4 + 8 = 32 \text{ kN} \qquad \text{(i)}$$

Taking moment about R_2,

$$R_1 \times 8 = 6 \times 4 \times 6 - 20 + 8 \times 2 = 140 \text{ kN m}$$

or, $$R_1 = 17.5 \text{ kN m} \qquad \text{(ii)}$$

Considering equilibrium of a section of the beam at any distance x from end R_1,

$$\Sigma F = 0, \quad V = 17.5 - 6x \text{ kN} \quad \text{for } 0 < x < 4 \text{ m}$$
$$= 17.5 \text{ kN} \quad \text{at } x = 0$$
$$= 0 \quad \text{at } x = 2.917 \text{ m}$$
$$= -6.5 \text{ kN} \quad \text{at } x = 4 \text{ m}$$
$$V = 17.5 - 6 \times 4 \quad \text{for } 4 \text{ m} < x < 6 \text{ m}$$
$$= -6.5 \text{ kN}$$
$$V = 17.5 - 24 - 8$$
$$= -14.5 \text{ kN} \quad \text{for } 6 \text{ m} < x < 8 \text{ m}$$
$$\Sigma M = 0, \quad M = 17.5x - 6x^2/2 \quad \text{for } 0 < x < 4 \text{ m}$$
$$= 0 \quad \text{at } x = 0$$
$$= 25.52 \text{ kN.m} \quad \text{at } x = 2.917 \text{ m}$$
$$= 22 \text{ kN.m} \quad \text{at } x = 4 \text{ m}$$
$$M = 17.5x - 6 \times 4\ (x - 2) + 20 \quad \text{for } 4 < x < 6 \text{ m}$$
$$M = 17.5x - 6 \times 4\ (x - 2) + 20 - 8\ (x - 6) \quad \text{for } 6 < x < 8 \text{ m}$$

V vs. x and M vs. x is shown in Fig. 5.14(i) and (ii), respectively.

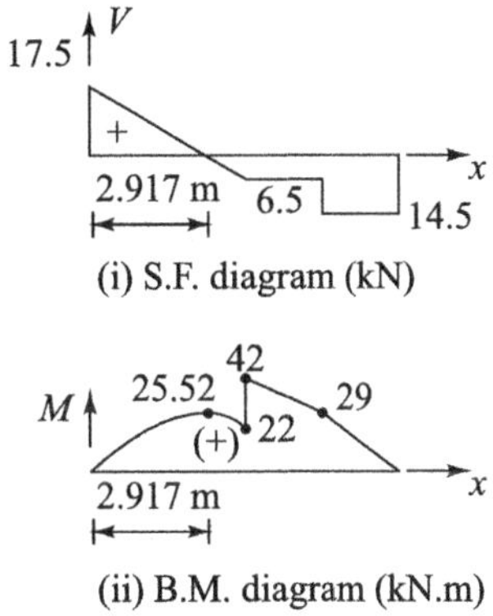

(i) S.F. diagram (kN)
(ii) B.M. diagram (kN.m)

Fig. 5.14

Example 5.10 Draw bending moment diagram for the beam shown in Fig. 5.15(i) using principle of superposition.

Solution (i) When the beam is under 40 and 80 kN m. as in Fig. 15.15(ii).

Taking moment about end B,

$$R_A \times 4 = 80 - 40$$

or, $$R_A = 10 \text{ kN}$$

Moment at any point in the beam is then given by,

$$M = -40 - 10x$$

Figure 5.15(ii) shown the B.M. diagram for the beam under 40 and 80 kN m.

(ii) When the beam is under udl 40 kN/m: Figure 5.15(iii) shown the B.M. diagram according to the results of Example 5.6.

(iii) The combined B.M. diagram is shown in Fig. 5.15(iv) by superposition of Fig. 5.15(ii) and (iii).

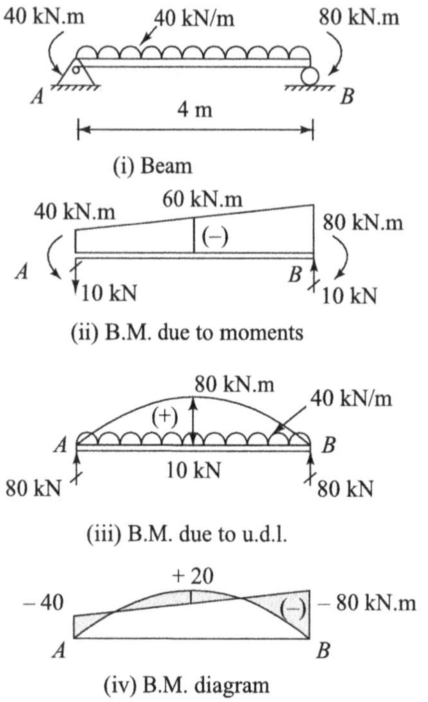

Fig. 5.15

Example 5.11 Write a computer program to calculate the shearing force and bending moment for a simply supported beam. It should also calculate the maximum bending moment in between two nodes.

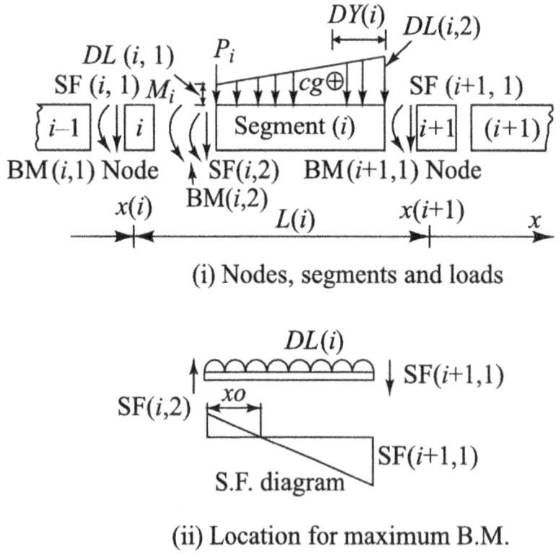

Fig. 5.16

Solution A simply supported beam is divided into different segments between nodes where external loads M_i, P_i, DL $(i, 1)$ and DL $(i, 2)$ are applied as shown in Fig. 5.16(i). SF $(i, 1)$, BM $(i, 1)$ are the shear forces and bending moments on left side of a node i and SF $(i, 2)$, BM $(i, 2)$ are the shear forces and bending moments on left side of a segment i. Counter i is used to count either nodes or segment.

Sign Convention The forces are in positive sense as shown in Fig. 5.16. The results are printed as per the previous sign convention shown in Fig. 5.3.

Procedure The program first calculates the support reactions R_1 and R_2 by taking moment of the forces about R_2. Distributed load on a segment i is assumed to act at its cg at a distance DY (i) from $i + 1$ node as follows.

$$DY\,(i) = \frac{2\,DL\,(i,1) + DL\,(i,2)}{DL\,(i,1) + DL\,(i,2)} \times \frac{L\,(i)}{3}$$

Moment of distributed load on a segment i with respect to node $i + 1$

$$\text{MR} = \frac{DL\,(i,1) + DL\,(i,2)}{2} \times DY\,(i)$$

This trapezoidal distributed load takes care of all types of variable/constant distributed load except while finding maximum bending moment, the udl case had to be checked separately.

The program next calculates the shear force and bending moment at each node. Finally the program checks the location of maximum bending moment within two nodes by locating points of zero shear force, e.g. in case of udl DL (i) on a segment i as shown in Fig. 5.16(ii), the location of zero shear force is,

$$X_0 = \text{SF}\,(i, 2)/DL\,(i)$$

Reader should verify the expression for X_0 in case of trapezoidal distributed load in the program (QBASIC) given below.

```
INPUT "NUMBER OF NODES = ?", N
PRINT " INPUT X-COORDINATES OF EACH NODE
_WITH ZERO AT LEFT END"
FOR I = 1 TO N
INPUT X(I)
```

```
NEXT I
FOR I = 1 TO N - 1
L (I) = X (I + 1) - X (I)
NEXT I
INPUT “NUMBER OF POINT LOAD = ?”, NP
FOR I = 1 TO NP
INPUT “NODE NUMBER, POINT LOAD (DOWN +ve)
_= ?”, NODE, P (NODE)
NEXT I
INPUT “NUMBER OF EXTERNAL MOMENTS = ?”, NM
FOR I = 1 TO NM
INPUT “NODE NUMBER, EXTERNAL MOMENT
_(ANTICLOCKWISE +ve) = ?”, NODE, M(NODE)
NEXT I
INPUT “NUMBER OF DISTRIBUTED LOAD (DOWN
_+ve) = ?”, NDS
FOR I = 1 TO NDS
INPUT “SEGMENT NUMBER, LOADLEFT, LOADRIGHT
_= ?”, NS, DL (NS, 1), DL (NS, 2)
DF(NS) = (DL(NS, 1) + DL (NS, 2))
DY(NS)=(DL(NS,1)+DF(NS))*L(NS)/(3*DF(NS))
NEXT I
INPUT “NODE NUMBER OF SIMPLE SUPPORTS, N1,
_N2 = ”, N1, N2
“**** CALCULATE SIMPLE SUPPORT REACTIONS
_R1 AND R2 ****”
SUM = P (1)
MR2= M (1) + P (1) * (X (N2) - X (1))
FOR I = 1 TO N - 1
IF I > N2 THEN
SUM = SUM + P (I + 1) + DF (I) * L (I)/2
MD=(DF(I)/2)*L(I)*(X(I+1)-X(N2)-DY(I))
MR2=MR2+M(I+1)-P(I+1)*(X(I+1)-X(N2))-MD
GOTO 100
END IF
SUM = SUM + P (I + 1) + DF (I) * L (I)/2
MD = (DF(I)/2*L(I)*(DY(I)+X(N2)-X(I+1))
MR2=MR2+M(I+1)-P(I+1)*(X(I+1)-X(N2))+MD
100 NEXT I
R1 = MR2/(X (N2) - X (N1))
R2 = SUM - R1
PRINT “SUPPORT REACTION R1 = ”, R1
PRINT “SUPPORT REACTION R2 = ”, R2
P (N1) = P (N1) - R1
P (N2) = P (N2) - R2
“**** CALCULATE SHEAR FORCE AND BENDING
_MOMENTS ****”
SF (1, 2) = P (1)
BM (1, 2) = M (1)
FOR I = 1 TO N - 1
SF (I + 1, 1) = SF (I, 2) + DF (I) * L (I)/2
SF (I + 1, 2) = SF (I + 1, 1) + P (I + 1)
MR = (DL (I, 1) + DF (I)) * L (I) * L (I)/6
BM (I+1,1)=BM(I,2)+SF(I, 2)*L(I)+MR
BM (I + 1, 2) = BM (I + 1, 1) + M (I + 1)
NEXT I
PRINT “SEGMENT”, “SHEAR LEFT”, “SHEAR
_RIGHT”, “MOMENT LEFT”, “MOMENT RIGHT”
FOR I = 1 TO N -1
PRINT I,-SF(I,2),-SF(I+1,1),-BM(I,2),-BM
_(I+1,1)
NEXT I
“** SEARCH FOR MAX BM. WITHIN A SEGMENT
_WHERE SF. IS ZERO**”
FOR I = 1 TO N - 1
IF SF (I, 2) * SF (I + 1, 1) > = 0 THEN 200
IF DL (I, 1) = DL (I, 2) THEN
X0 = - SF (I, 2)/DL(I, 1)
GOTO 300
END IF
A = (DL (I, 2) - DL (I, 1))/L(I)
D=4*DL(I,1)*DL(I,1)-8*SF(I,2)*A
X1 = (-2 * DL (I, 1) + SQR (D))/(2 * A)
X2 = (-2 * DL (I, 1) - SQR (D))/(2 * A)
IF X1 > 0 AND X1 < L (I) THEN X0 = X1 ELSE
IF X2 > 0 AND X2 < L (I) THEN X0 = X2
300 DD = DL (I, 2) - DL (I, 1)
BMAX (I) = BM (I, 2) + SF (I, 2) * X0 + (3
_*DL (I, 1) + DD*X0/L (I)) *X0*X0/6
PRINT “SEGMENT”, “MAX. BM.”, “DIST. FROM
_LEFT”
PRINT I, -BMAX (I), X0
200 NEXT I
END
```

Underscore indicates that the line should be typed in continuation with the previous line. Qbasic does not support underscore.

Example 5.12 Calculate shear force and bending moment for the beam shown in Fig. 5.17 using the computer program of Ex. 5.11.

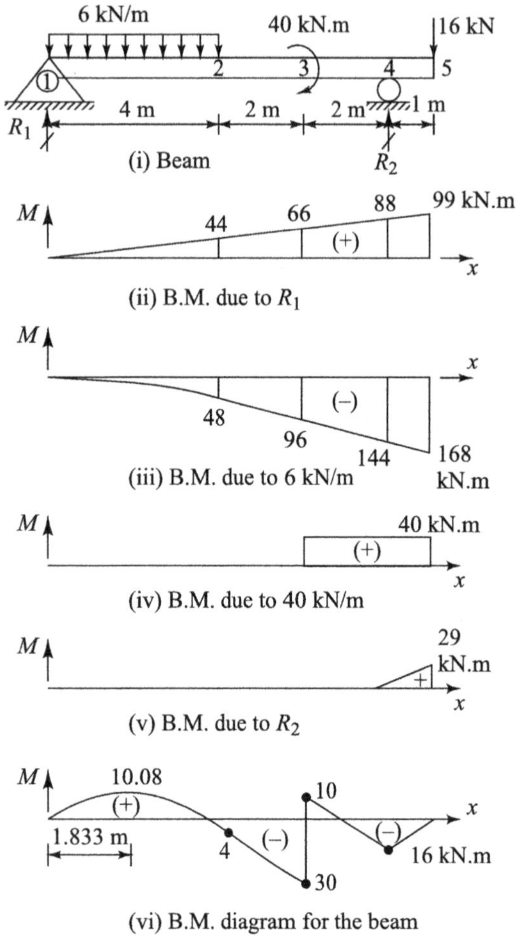

Fig. 5.17

Solution Input/output of the computer program from screen for the beam of Example 5.11 is as follows.

```
NUMBER OF NODES = ? 5
INPUT X-COORDINATES OF EACH NODE WITH ZERO
AT LEFT END
? 0
? 4
? 6
? 8
? 9
NUMBER OF POINT LOAD = ? 1
NODE NUMBER, POINT LOAD (DOWN +ve) = ? 5, 16
NUMBER OF EXTERNAL MOMENT = ? 1
NODE NUMBER, EXTERNAL MOMENT (ANTICLOCKWISE
+ve) = ? 3, -40
NUMBER OF DISTRIBUTED LOAD (DOWN +ve) = ? 1
SEGMENT NUMBER, LOADLEFT, LOADRIGHT = ? 1,
6, 6
NODE NUMBER OF SIMPLE SUPPORTS, N1, N2 =
1, 4
SUPPORT REACTION R1 = 11
SUPPORT REACTION R2 = 29
```

SEGMENT	SHEAR LEFT	SHEAR RIGHT	MOMENT LEFT	MOMENT RIGHT
1	11	-13	0	-4
2	-13	-13	-4	-30
3	-13	-13	10	-16
4	16	16	-16	0

SEGMENT	MAX. BM.	DIST. FROM LEFT
1	10.08333	1.833333

Example 5.13 Draw bending moment for the beam shown in Fig. 5.17(i).

Solution Support reactions are determined using static equilibrium equations,

$$\Sigma F = 0, \quad R_1 + R_2 = 6 \times 4 + 16 = 40 \text{ kN} \qquad \text{(i)}$$

Taking moment about R_2,

$$\Sigma M = 0, \quad R_1 \times 8 + 40 + 16 \times 1 = (6 \times 4) \times 6 \text{ kN m}$$

or, $$R_1 = 11 \text{ kN} \qquad \text{(ii)}$$

From Eqs (i) and (ii), $R_2 = 29$ kN

Bending moment due to individual loads in x-direction is as follows.

$M(R_1) = 11 \times x = 99$ kNm at $x = 9$ m

$M(6) = -(6 \times x) \times \dfrac{x}{2}$ for $0 < x < 4$ m

$= -(6 \times 4) \times (x - 2)$ for $4 < x < 9$ m

$M(40) = 40$ kN m for $6 < x < 9$ m

$M(R_2) = 29 \times (x - 8)$ for $8 < x < 9$ m

Figure 5.17(ii) through (v) show bending moment diagram for the individual loads and their sum is shown in Fig. 5.17(iv).

Example 5.14 Calculate shear force and bending moment for the beam shown in Fig. 5.18 using the computer program of Example 5.11.

Solution The vertical component of the inclined load 80 kN will cause bending moment in the beam and its horizontal component will cause axial compression in the beam as shown in the axial force diagram in Fig. 5.18(ii).

Input/output of the computer program from screen for the beam of Ex. 5.14 is as follows.

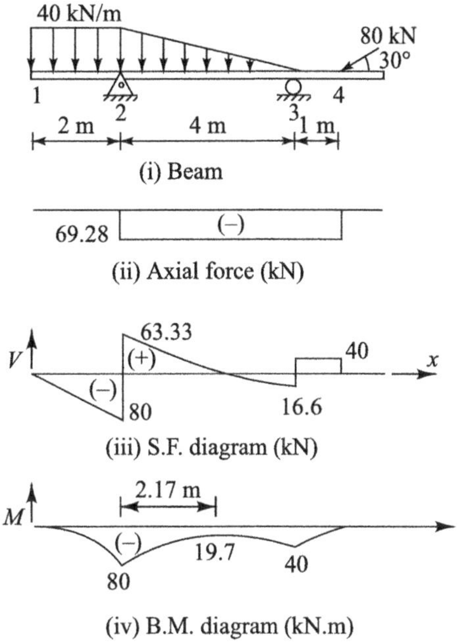

Fig. 5.18

```
NUMBER OF NODES = ? 4
INPUT X-COORDINATES OF EACH NODE WITH ZERO
AT LEFT END
0
2
6
7
NUMBER OF POINT LOAD = ? 1
NODE NUMBER, POINT LAOD (DOWN +ve) = ? 4, 40
NUMBER OF EXTERNAL MOMENTS = ? 0
NUMBER OF DISTRIBUTED LOAD (DOWN +ve) = ? 2
SEGMENT NUMBER, LOADLEFT, LOADRIGHT=? 1,40,40
SEGMENT NUMBER, LOADLEFT, LOADRIGHT=? 2,40,0
NODE NUMBER OF NUMBER SUPPORTS, N1, N2 = 2, 3
SUPPORT REACTION R1 = 143.3333
SUPPORT REACTION R2 = 56.66666

SEGMENT     SHEAR LEFT    SHEAR RIGHT    MOMENT LEFT    MOMENT RIGHT
   1        0             -80            0              -80
   2        63.33334      -16.66666      -80            -39.99995
   3        40            40             -39.99995      4.577637E-05
SEGMENT     MAX. BM.      DIST. FROM LEFT
   2        -19.71396     2.174259
```

Figure 5.18 shows these values in the shear force and bending moment diagram for the beam.

Example 5.15 Draw the bending moment diagram for the beam shown in Fig. 5.19 by the method of superposition.

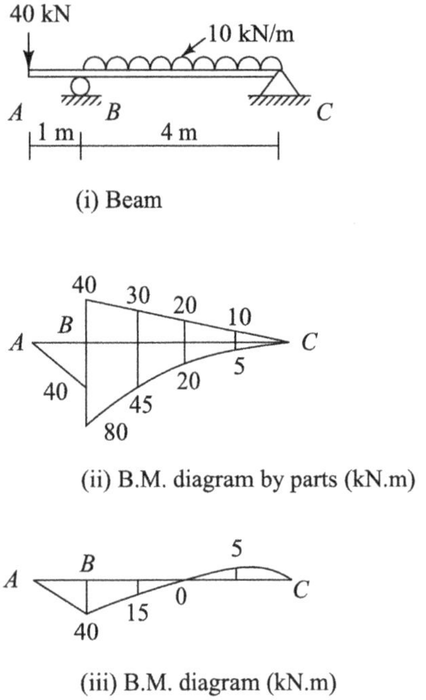

Fig. 5.19

Solution Support reaction R_c is given by taking moment about B as follows.

$\Sigma M = 0, \quad R_c \times 4 + 40 \times 1 = 10 \times 4 \times 4/2$

or, $\quad R_c = 10$ kN

Figure 5.19(ii) shows the bending moment diagram by parts for 40 kN concentrated load, udl 10 kN/m and R_c about B. The final bending moment diagram is obtained by adding the three bending moment diagrams as shown in Fig. 5.19(iii).

Example 5.16 Draw the shear force for bending moment diagram for the beam shown in Fig. 5.20.

Solution Figure 5.20 shows the free body diagram by cutting hinge at C. Support reaction R_a is given by taking moment about C as follows.

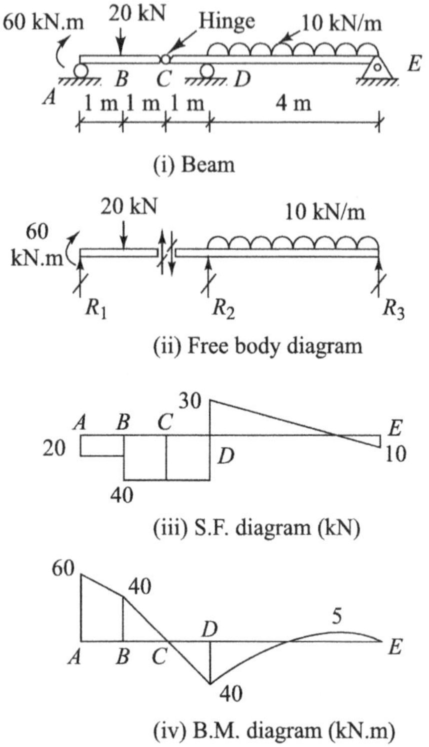

Fig. 5.20

$\Sigma M = 0, \quad R_a \times 2 + 60 = 20 \times 1$

or, $\quad R_a = -20$ kN

–ve sign means that R_a is acting in reverse direction.

Shear force V at hinge C is given by,

$\Sigma F = 0, \quad V = 20 + R_a = 20 + 20 = 40$ kN

Figure 5.20(iii) and (iv) shows the force and bending moment diagrams for the beam. The right portion of the free body diagram is same as the beam of Example 5.15. The hinge at C will transfer the shear force from BC to CD but the bending moment will be zero at the hinge.

Example 5.17 Figure 5.21(i) show a portal frame of an industrial shed under wind load and a suspended crane load. Analyse the frame if its legs are fixed to concrete base through anchor bolts which allow some lateral movement.

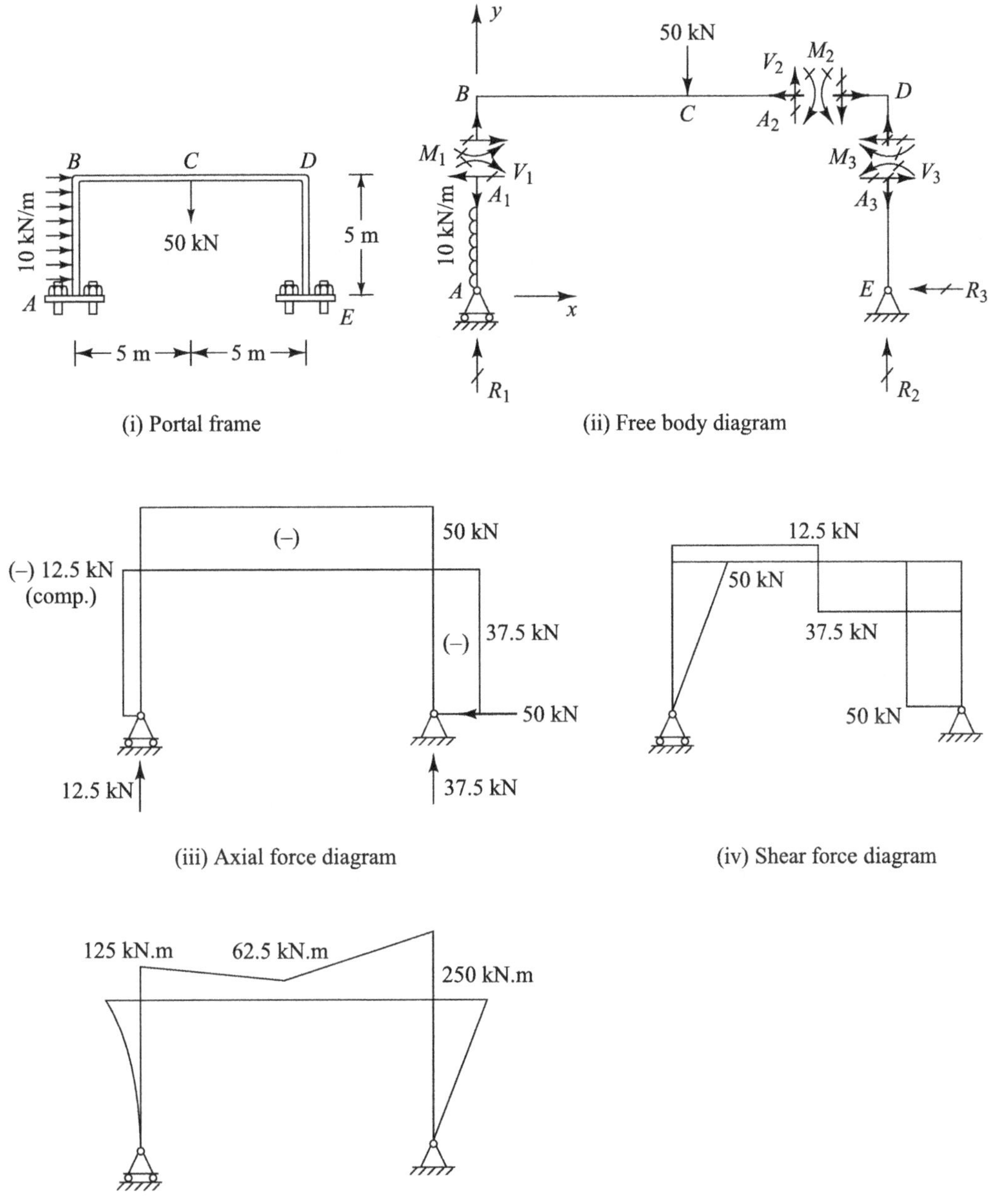

(i) Portal frame

(ii) Free body diagram

(iii) Axial force diagram

(iv) Shear force diagram

(v) Bending moment diagram

Fig. 5.21

Solution Assuming the base reactions as shown in Fig. 5.21(ii) for which the portal frame is determinate. Taking moment of the forces about joint *E*,

$\Sigma M_E = 0, \quad R_1 \times 10 + 10 \times 5 \times 5/2 - 50 \times 5 = 0,$

or, $\quad R_1 = 12.5$ kN

$\Sigma F_x = 0, \quad R_3 = 10 \times 5 = 50$ kN

$\Sigma F_y = 0, \quad R_2 = 50 - 12.5 = 37.5$ kN

Taking sections at 3 locations very near the joints and applying equilibrium equations on left sides.

Section at 1,

$\Sigma F_x = 0, \quad V_1 = 10 \times 5 = 50$ kN

$\Sigma F_y = 0, \quad A_1 = R_1 = 12.5$ kN

$\Sigma M_1 = 0, \quad M_1 - 10 \times 5 \times 5/2 = 0,$

or, $\quad M_1 = 125$ kNm

Section at 2,

$\Sigma F_x = 0, \quad A_2 = 10 \times 5 = 50$ kN

$\Sigma F_y = 0, \quad V_2 = 50 - R_1 = 37.5$ kN

$\Sigma M_2 = 0, \quad M_2 - 10 \times 5 \times 5/2 - 50 \times 5 + 12.5 \times 10 = 0,$

or, $\quad M_2 = 250$ kNm

Section at 3,

$\Sigma F_x = 0, \quad V_3 = 10 \times 5 = 50$ kN

$\Sigma F_y = 0, \quad A_3 = 50 - R_1 = 37.5$ kN

$\Sigma M_3 = 0, \quad M_3 - 10 \times 5 \times 5/2 - 50 \times 5 + 12.5 \times 10 = 0,$

or, $\quad M_3 = 250$ kNm

Figure 5.21 (iii), (iv) and (v) shows the axial force, shear force and bending moment diagrams. B.M. diagram is drown on the tension face.

EXERCISES

PROBLEMS

5.1 through 5.13. Analyse the beams shown in Fig. 5.22 through 5.34 and draw their shear force, bending moment diagram.

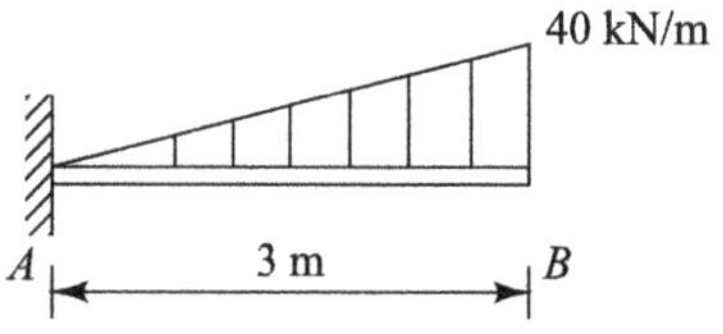

Fig. 5.22 *Prob. 5.1*

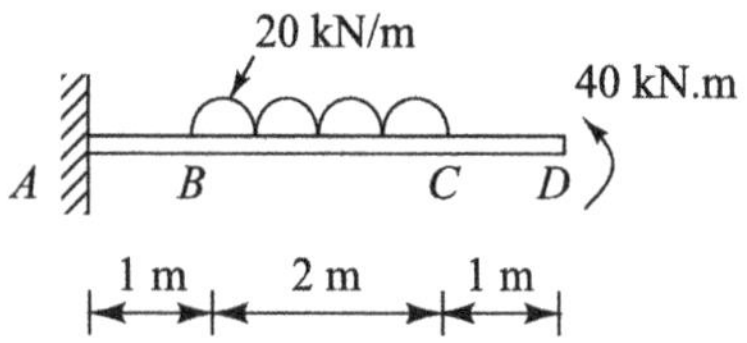

Fig. 5.23 *Prob. 5.2*

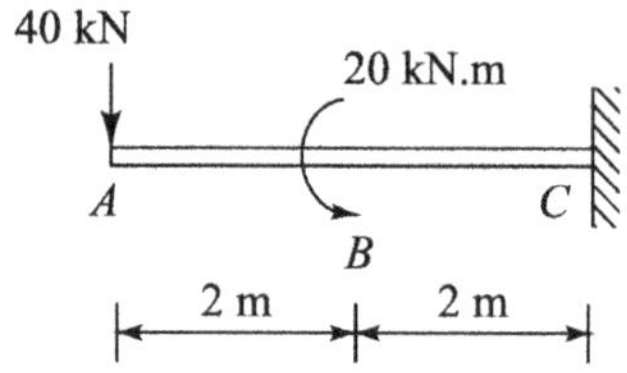

Fig. 5.24 *Prob. 5.3*

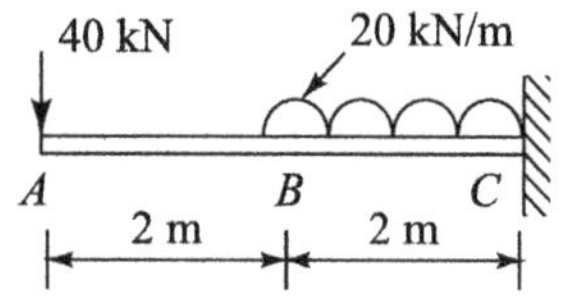

Fig. 5.25 *Prob. 5.4*

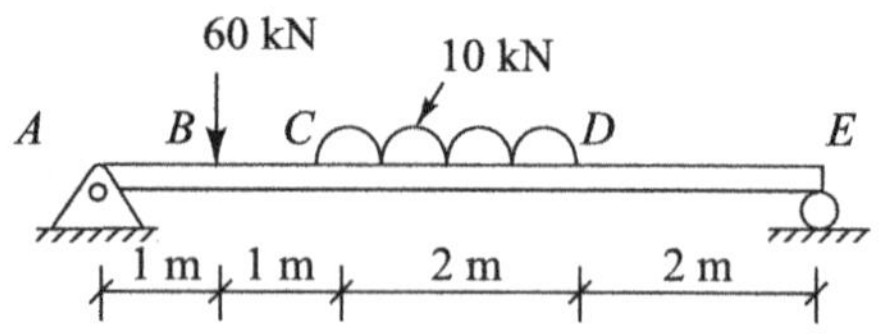

Fig. 5.26 *Prob. 5.5*

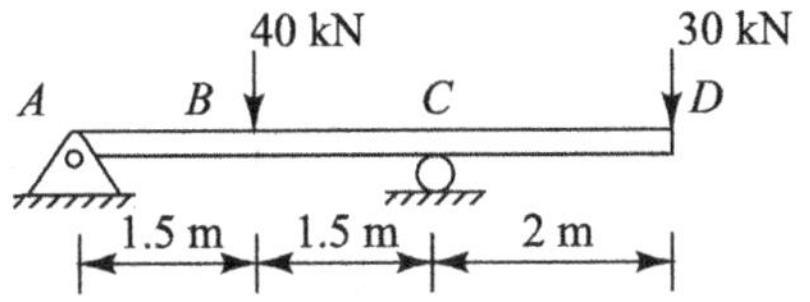

Fig. 5.27 *Prob. 5.6*

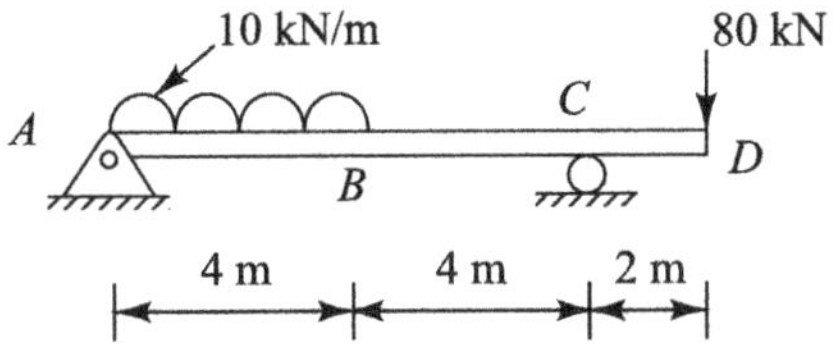

Fig. 5.28 *Prob. 5.7*

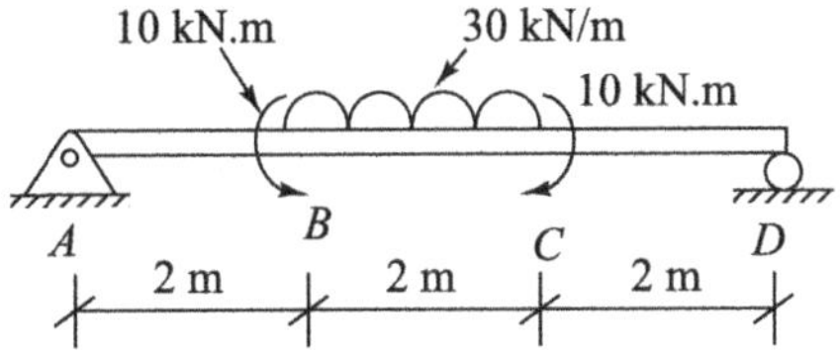

Fig. 5.29 *Prob. 5.8*

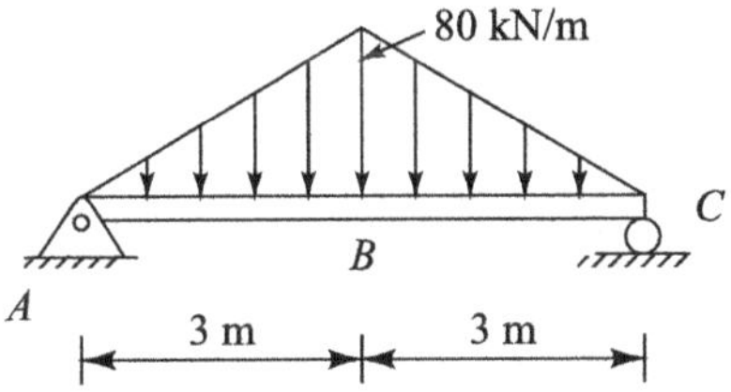

Fig. 5.30 *Prob. 5.9*

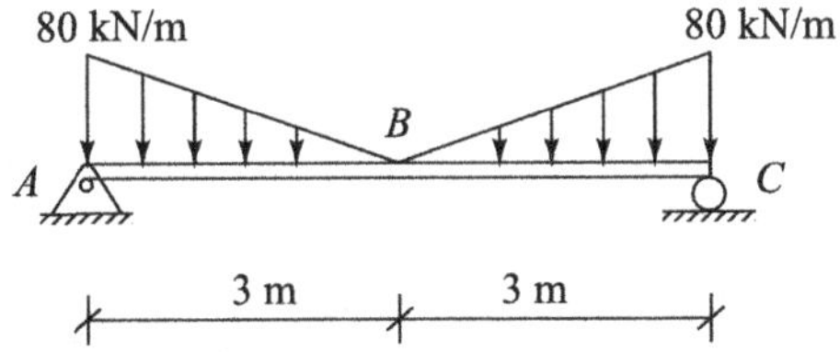

Fig. 5.31 *Prob. 5.10*

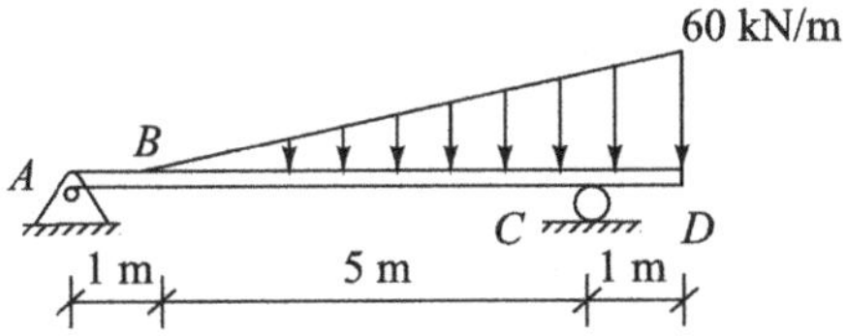

Fig. 5.32 *Prob. 5.11*

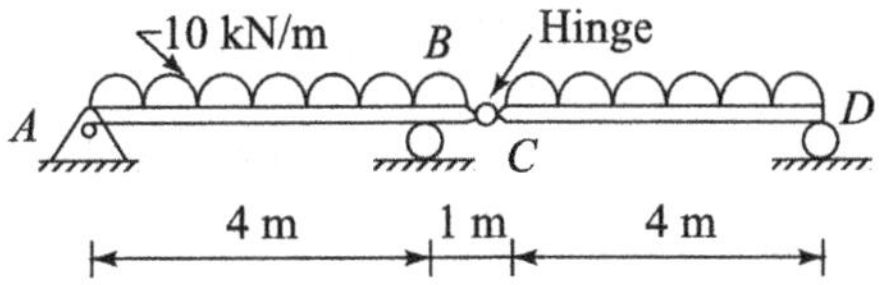

Fig. 5.33 *Prob. 5.12*

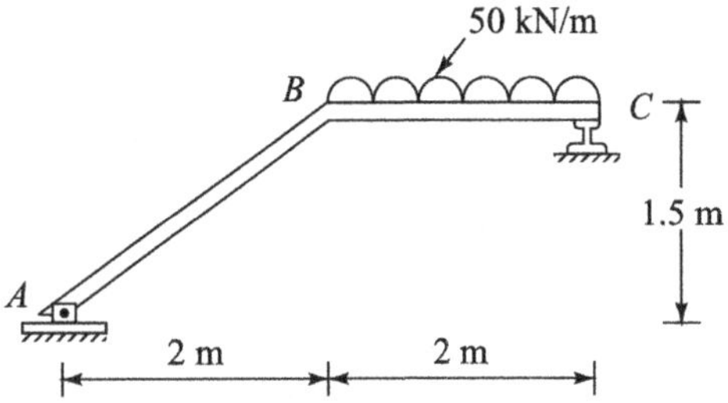

Fig. 5.34 *Prob. 5.13*

5.14 Check the computer program of Example 5.11 to analyse the beam shown in Fig. 5.35 and mark the correction if any in the program.

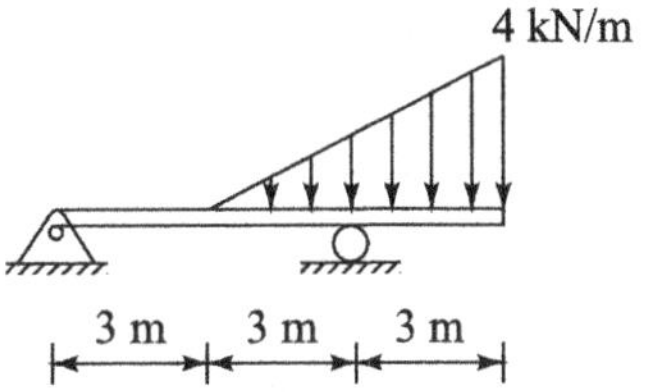

Fig. 5.35 *Prob. 5.14*

5.15 Analyse the beam shown in Fig. 5.36 with the computer program.

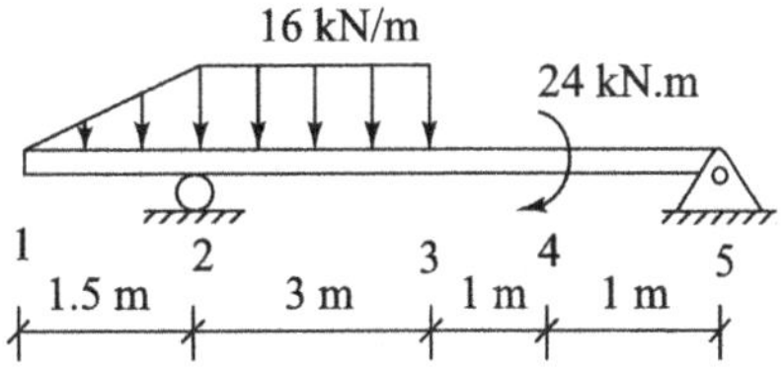

Fig. 5.36 *Prob. 5.15*

OBJECTIVE QUESTIONS

5.16 Consider the following statements:
If a simply supported beam of uniform cross-section is subjected to a clockwise moment at the left support and an equal anticlockwise moment at the right support, then the

1. B.M.D. will be in the shape of a rectangle.
2. S.F.D. will be a straight line coinciding with the base.
3. Deflection curve will be in the shape of a circular arc.

Of these statements
(a) 1, 2 and 3 are correct
(b) 1 and 2 are correct
(c) 1 and 3 are correct
(d) 2 and 3 are correct

5.17 In a cantilever beam, the bending moment at any section of the beam is equal to the
(a) area of the shear force diagram between the free end of the beam and that section.
(b) area of the bending moment diagram between the free end of the beam and that section.
(c) area of shear force diagram between the fixed end of the beam and that section.
(d) vertical ordinate at that section in the shear force diagram.

5.18 Match List I with List II and select the correct answer.

List I ***(Type of beam with loading)*** — ***List II*** ***(Maximum bending moment value)***

A. 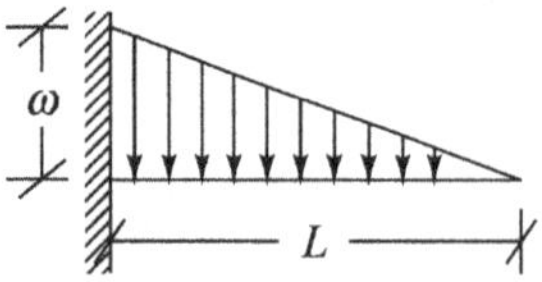

1. $\frac{\omega L^2}{12}$

B. 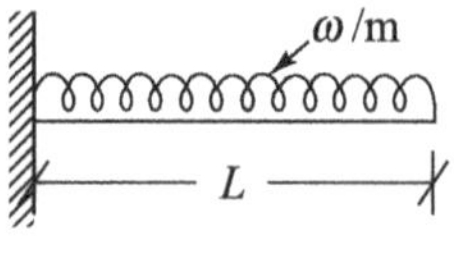

2. $\frac{\omega L^2}{6}$

C. 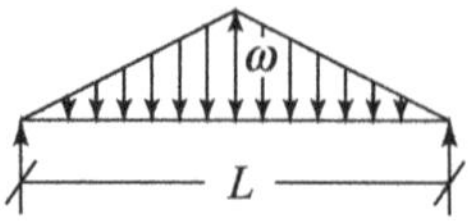

3. $\frac{\omega L^2}{2}$

D. 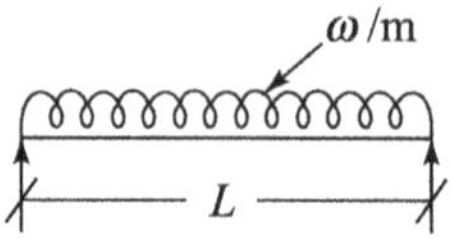

4. $\frac{\omega L^2}{8}$

	A	B	C	D		A	B	C	D
(a)	4	3	2	1	(b)	1	3	2	4
(c)	2	3	1	4	(d)	2	4	1	3

5.19 Match List I with List II and select the correct answer using the codes given below the lists:

List I ***(Type of beam with type of loading)*** — ***List II*** ***(S.F. diagram)***

A.  1. 

B. 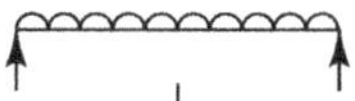 2. 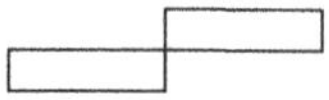

C. 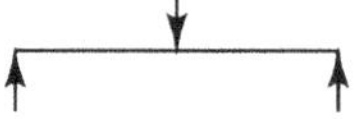 3. 

D. 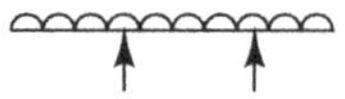 4. 

Codes :	A	B	C	D
(a)	4	1	3	2
(b)	4	3	2	1
(c)	3	4	1	2
(d)	3	4	2	1

5.20 Match List I with List II and select the correct answer

List I (Type and position of force on cantilever)	***List II (Shape of bending moment diagram for a cantilever)***
A. Carrying linearly varying load from zero at its free end and maximum at the fixed end	1. Parabola

B. Subject to uniformly distributed load	2. Rectangle
C. Carrying concentrated load at its free end	3. Cubic parabola
D. Whose free end is subjected to a couple	4. Triangle

	A	B	C	D		A	B	C	D
(a)	1	2	3	4	(b)	4	3	2	1
(c)	3	1	4	2	(d)	1	3	4	2

5.21 The given sketch show a beam of supported length '*l*' and overhang 0.1*l* carrying a concentrated load *W* at the end of the overhang. Which one of the following figures would represent the correct shear force diagram for the beam?

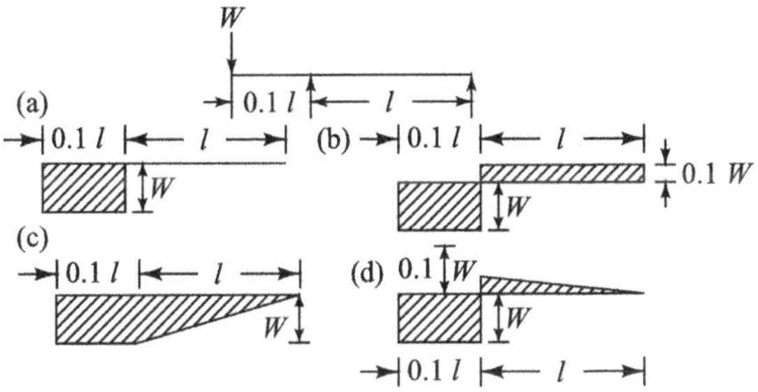

5.22 For the beam shown in the given figure, the maximum positive bending moment is equal to negative bending moment. The value of L_1 is

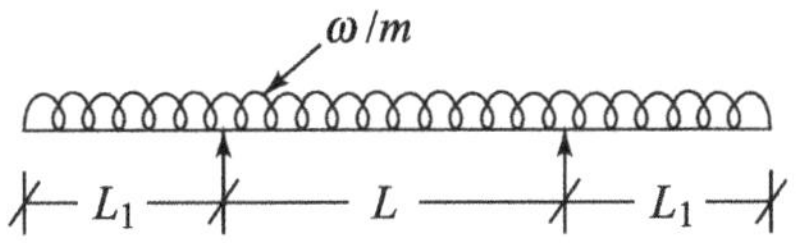

(a) $\frac{L}{\sqrt{2}}$ (b) $\frac{L}{\sqrt{3}}$

(c) $\frac{L}{2}$ (d) $\frac{L}{2\sqrt{2}}$

5.23 A long construction member of uniform section is to be lifted using ropes at *C* and *D* (see figure). The causes bending moments due to self-weight as shown. To minimise the peak value of bending moment, the overhang '*b*' shall be such that

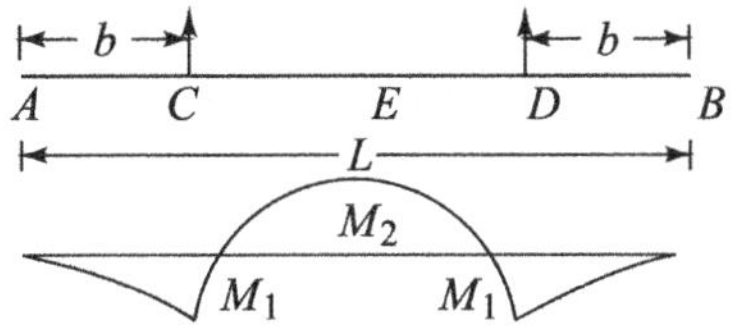

(a) $M_2 = 0$ (b) $M_1 = M_2$
(c) $M_2 = 2\,M_1$ (d) $b = L/4$

5.24 For the shear force diagram shown in the given figure the loaded beam will be

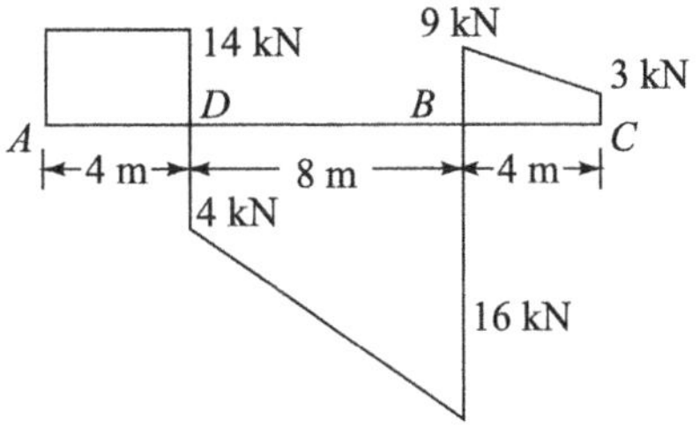

(a)

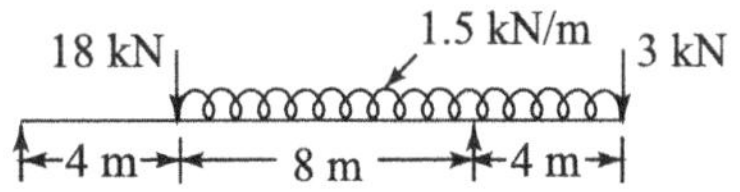

(b)

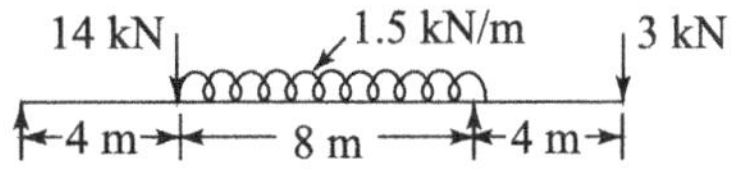

(c)

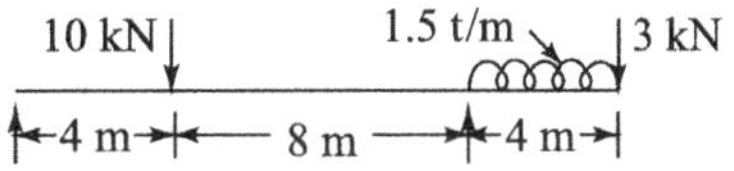

(d)

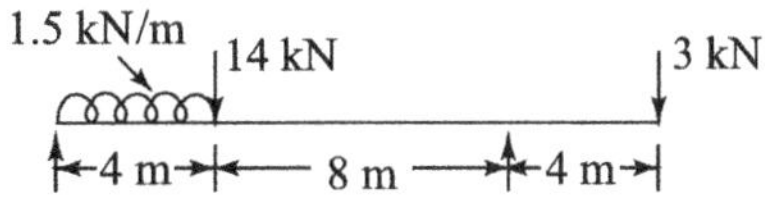

5.25 If the second derivative of bending moment, *M* with respect to length, *x*, i.e. d^2M/dx^2 is positive, then,

(1) the bending moment diagram will concave downward,
(2) the bending moment diagram will concave upwards
(3) the shear force diagram will slope downwards
(4) the shear force diagram will slope upwards

Of these statements

(a) only 1 is correct
(b) only 2 is correct
(c) 2 and 3 are correct
(d) 2 and 4 are correct

5.26 The given figure shows a cantilever beam loaded with a varying uniformly distributed load, 12 kN/m and a concentrated load of 10 kN. The shear force diagram will be,

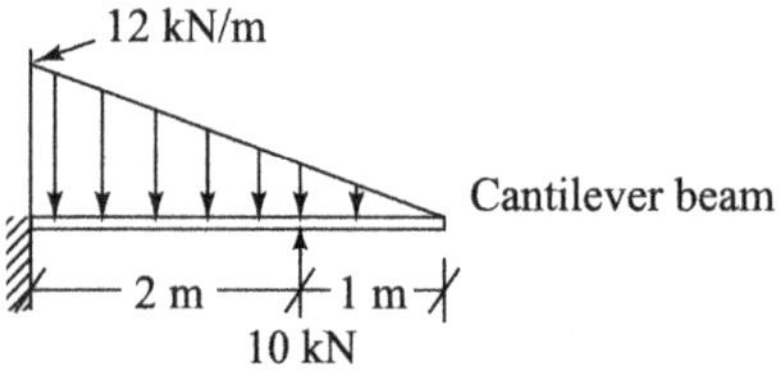

(a)
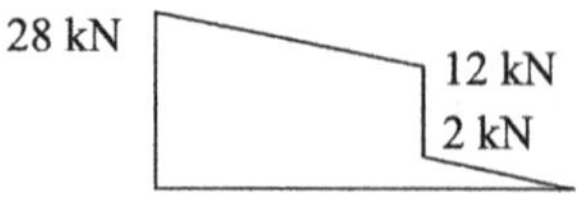

(b)
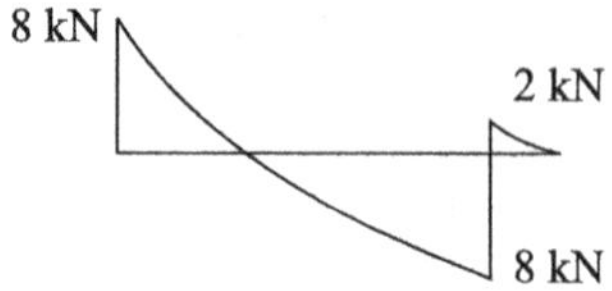

(c)
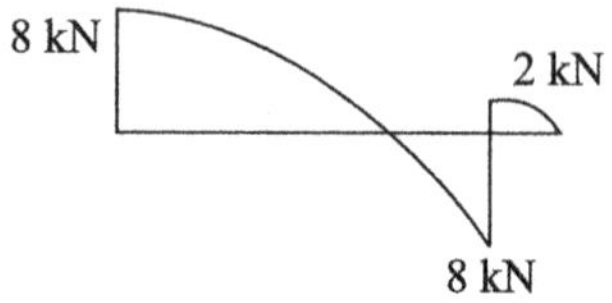

(d)
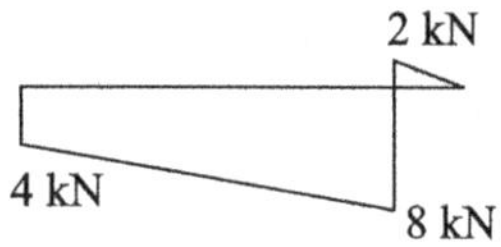

6

Centroid, Second Moment of Area, and Product Moment of Area

CONCEPT REVIEW

6.1 INTRODUCTION

In many problems of engineering mechanics, the shape and disposition of a plane area relative to some reference convey a certain meaning. Some standard results are derived here which shall also be used later in other chapters.

6.2 CENTROID AND FIRST MOMENT OF AREA

The coordinates of the centroid of a plane area shown in Fig. 6.1 are given by integrating the moments of elementary areas dA about x or y axis and equating it to the moment of the entire area about the respective axis as follows,

$$\bar{x} = \frac{\int x\,dA}{A}, \quad \bar{y} = \frac{\int y\,dA}{A} \qquad (6.1\,\text{a,b})$$

where, $\bar{x}$ and $\bar{y}$ = coordinates of centroid.

The numerators of Eq. (6.1) are called the first moments of area.

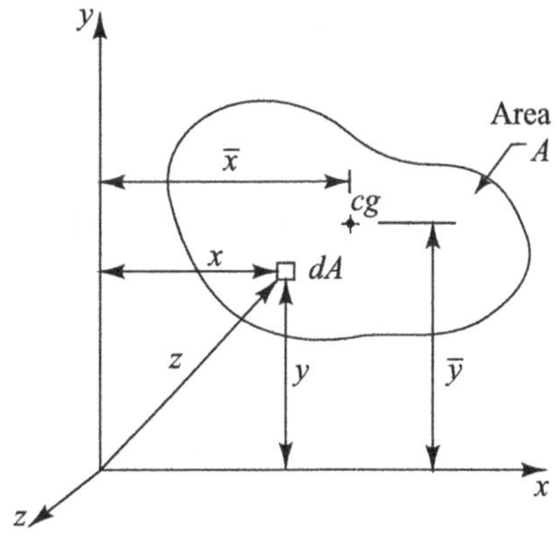

Fig. 6.1

6.3 SECOND MOMENT AND THE PRODUCT MOMENT OF AN AREA

Moments of inertia or the second moments of a plane area A shown in Fig. 6.1 about x and y axes are given by,

$$I_x = \int y^2\, dA\,,\ I_y = \int x^2\, dA \qquad (6.2\text{a,b})$$

The product moment of an area is defined as,

$$I_{xy} = \int xy\, dA \qquad (6.2\text{c})$$

I_x and I_y are always +ve while I_{xy} may vary from +ve to −ve values.

Radius of gyration about x and y axes are defined as,

$$r_x = \sqrt{I_x / A}\,,\ r_y = \sqrt{I_y / A} \qquad (6.3\text{a,b})$$

6.4 PARALLEL AXIS THEOREM

Consider the plane area A shown in Fig. 6.2. The axes x_G and y_G pass through the centroid of the plane surface. The x and y axes are located at distances x_1 and y_1 respectively from the axes through the centroid. The moment of inertia about x-axis for the element of area dA is given by,

$$dI_x = (y_1 + y')^2\, dA$$

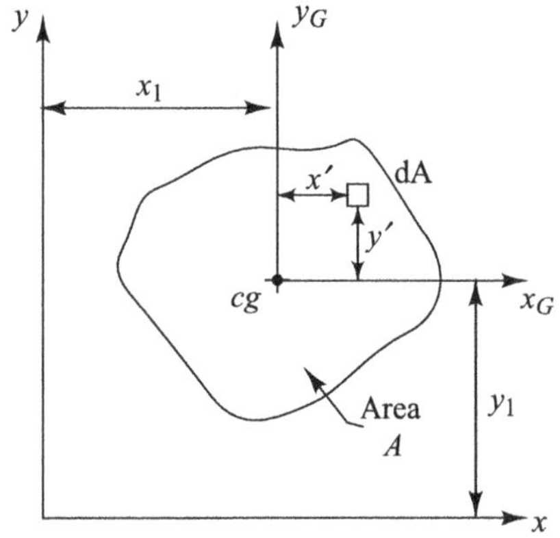

Fig. 6.2

For entire area A, the second moment of area about x-axis is,

$$I_x = \int dI_x = \int (y' + y_1)^2 dA$$

$$= \int y'^2\, dA + \int y_1^2\, dA + \int 2y' y_1\, dA$$

From Eq 6.1.b, $\int y'\, dA = 0$, since the x_G–y_G axes pass through the centroid, i.e. $\bar{y} = 0$ w.r.t. x_G axis.

therefore, $$I_x = I_{xG} + Ay_1^2 \qquad (6.4\text{a})$$

Similarly, $$I_y = I_{yG} + Ax_1^2 \qquad (6.4\text{b})$$

The parallel axis theorem for product moment of area is derived similarly as follows,

$$I_{xy} = \int dI_{xy} = \int (x' + x_1)(y' + y_1) dA$$

$$= \int x' y' dA + \int x_1 y_1 dA + \int x_1 y' dA + \int x' y_1 dA$$

or, $I_{xy} = I_{xGyG} + Ax_1 y_1 + 0 + 0$

or, $$I_{xy} = I_{xGyG} + Ax_1 y_1 \qquad (6.4\text{c})$$

6.5 PRINCIPAL AXES AND PRINCIPAL SECOND MOMENT OF AREA

At any point in the plane of an area there exist two perpendicular axes about which the second moments of area are maximum and minimum for that point. These maximum and minimum values of second moments of area are termed as Principal second moments of the area. Consider two sets of reference axes x–y and x_1–y_1 in a plane area A as shown in Fig. 6.3. The second moment of area with respect to x_1 axis is,

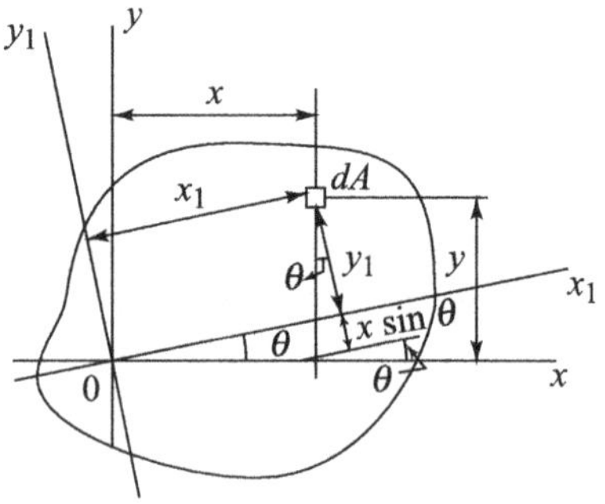

Fig. 6.3

$$I_{x1} = \int y_1^2 dA = \int (y\cos\theta - x\sin\theta)^2 dA$$

$$= \cos^2\theta \int y^2 dA + \sin^2\theta \int x^2 dA$$

$$-2\sin\theta\cos\theta \int xy dA$$

$$= I_x \cos^2\theta + I_y \sin^2\theta - 2I_{xy}\sin\theta\cos\theta$$

$$\therefore \quad I_{x1} = \left(\frac{I_x + I_y}{2}\right) + \left(\frac{I_x - I_y}{2}\right)\cos 2\theta - I_{xy}\sin 2\theta \quad (6.5a)$$

Similarly the second moment of area about y_1 axis can be written by replacing θ by $\theta + \pi/2$ as follows.

$$I_{y1} = \left(\frac{I_x + I_y}{2}\right) - \left(\frac{I_x - I_y}{2}\right)\cos 2\theta + I_{xy}\sin 2\theta \quad (6.5b)$$

Differentiating Eq. (6.5a) wrt θ, for $I_{max/min}$,

$$\frac{dI_{x1}}{d\theta} = (I_x - I_y)\sin 2\theta - 2I_{xy}\cos 2\theta = 0$$

$$\text{or,} \quad \tan 2\theta_p = \frac{-I_{xy}}{\left(\frac{I_x - I_y}{2}\right)} \quad (6.6)$$

where, θ_p = orientation of principal axes. Eq. (6.6) gives two values of θ_p which are 90° apart.

The maximum or minimum values of I_{x1} is given by substituting the value of θ_p from Eq.(6.6) in Eq.(6.5a) as follows.

$$(I_{x1})_{max/min} = \frac{I_x + I_y}{2} \pm \sqrt{\left(\frac{I_x - I_y}{2}\right)^2 + I_{xy}^2} \quad (6.7)$$

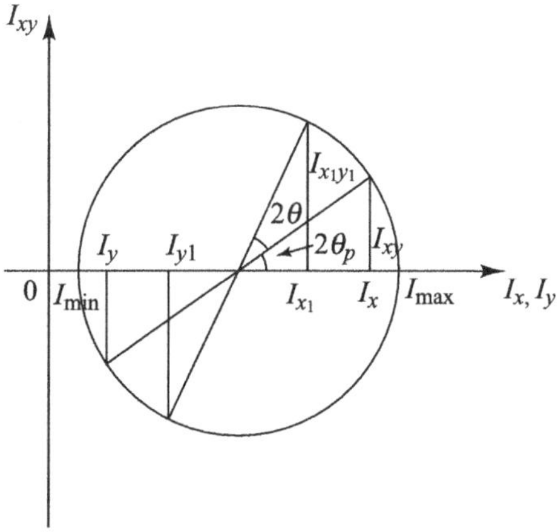

Fig. 6.4 *Mohr's circle*

6.6 PRODUCT MOMENT OF AREA FOR PRINCIPAL AXES

The product moment of area about $x_1 - y_1$ axes is given by (Fig. 6.3),

$$I_{x_1y_1} = \int x_1 y_1 dA$$

$$= \int (x\cos\theta + y\sin\theta)\,(y\cos\theta - x\sin\theta)dA$$

$$= (\cos^2\theta - \sin^2\theta)\int xy dA + \sin\theta\cos\theta \int (y^2 - x^2)dA$$

$$= I_{xy}(\cos^2\theta - \sin^2\theta) + (I_x - I_y)\sin\theta\cos\theta$$

$$\text{or,} \quad I_{x_1y_1} = \left(\frac{I_x - I_y}{2}\right)\sin 2\theta + I_{xy}\cos 2\theta \quad (6.8)$$

$I_{x_1y_1} = 0$ corresponding to the value of θ_p from Eq. (6.6) for principal axes.

Equation (6.5) through (6.8) be represented graphically by Mohr's circle shown in Fig. 6.4. It shows that the maximum value of $I_{x_1y_1}$

$$(I_{x_1y_1})_{max/min} = \frac{I_{max} - I_{min}}{2} = \sqrt{\left(\frac{I_x - I_y}{2}\right)^2 + I_{xy}^2} \quad (6.9)$$

6.7 PERPENDICULAR AXIS THEOREM: POLAR MOMENT OF INERTIA

Consider the elementary area dA in Fig. 6.1. Its second moment of area about the perpendicular axis z, also known as Polar moment of inertia I_z is given by,

$$I_z = \int z^2 dA = \int (x^2 + y^2)\,dA$$

$$\text{or,} \quad I_z = I_x + I_y \quad (6.10)$$

The polar moment of inertia is an important parameter in torsion problems.

EXAMPLES

Example 6.1 Locate the centroid of a triangle *abc* with reference to its base *ab* as shown in Fig. 6.5.

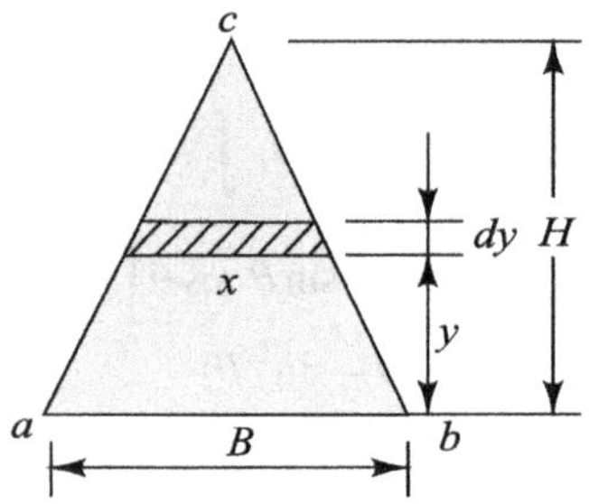

Fig. 6.5

Solution Centroid distance is given by Eq. (6.1) as

$$\bar{y} = \frac{\int y\,dA}{A}$$

where, $A = \dfrac{BH}{2}$

and $dA = x \times dy$, area of elementary strip

or
$$\bar{y} = \frac{\int_0^H xy\,dy}{BH/2} \qquad \text{(i)}$$

From similar triangles, $\dfrac{x}{B} = \dfrac{H-y}{H}$

or
$$\bar{y} = \frac{2}{BH}\int_0^H \left(By - \frac{By^2}{H}\right)dy$$

or
$$\bar{y} = \frac{2}{H}\left|\frac{y^2}{2}\right|_0^H - \frac{2}{H^2}\left|\frac{y^3}{3}\right|_0^H = \frac{H}{3}$$

Example 6.2 Locate the centroid of a trapezium *abcd* with reference to its base *ab* as shown in Fig. 6.6.

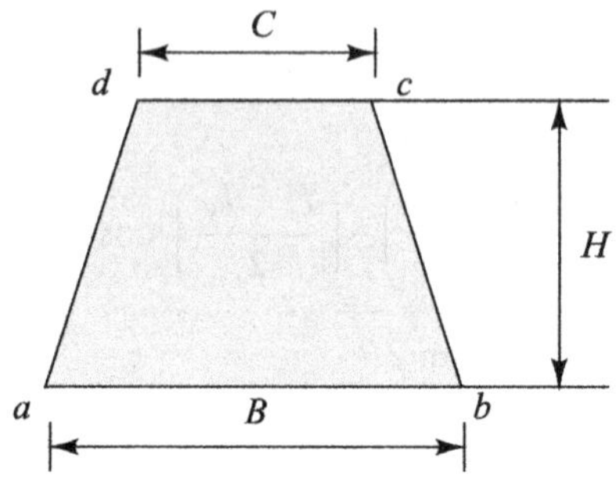

Fig. 6.6

Solution The area of trapezoid consists of two parts

(i) a rectangle of area $C \times H$,

(ii) a triangle of area $\frac{1}{2} H \times (B - C)$,

Taking moment area of both parts about base *ab*,

$$\bar{y} = \frac{\text{Moment of area}}{\text{Total area}}$$

$$= \frac{(C \times H) \times H/2 + [(B-C) \times H/2] \times H/3}{C \times H + (B-C) \times H/2}$$

$$= \left(\frac{B+2C}{B+C}\right) \times \frac{H}{3}$$

Example 6.3 Locate centroid of a semicircle of radius *R*, with respect to its diameter as shown in Fig. 6.7.

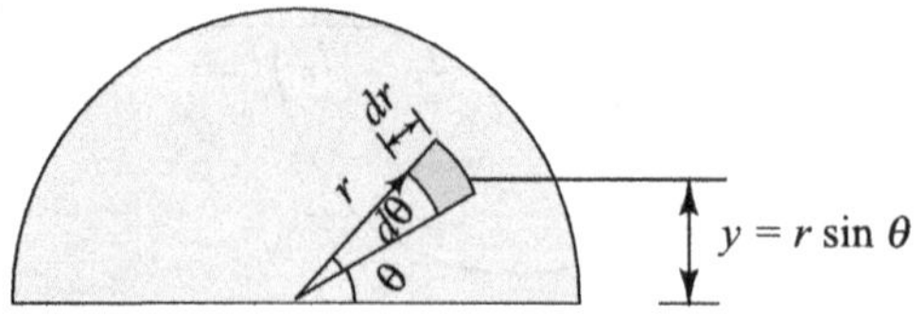

Fig. 6.7

Solution Consider an elementary area,

$$dA = r\,d\theta \times dr$$

then centroid distance,

$$\bar{y} = \frac{\int y\,dA}{A}$$

or, $$\bar{y} = \frac{\int_0^{\pi}\int_0^{R}(r\sin\theta)(r\,d\theta\,dr)}{\pi R^2/2}$$

$$= \frac{\frac{R^3}{3}\int_0^{\pi}\sin\theta\,d\theta}{\pi R^2/2} = \frac{2R}{3\pi}\left|-\cos\theta\right|_0^{\pi}$$

or, $$\bar{y} = \frac{4R}{3\pi}$$

Example 6.4 Locate the centroid of quarter-parabolic area shown in Fig. 6.8.

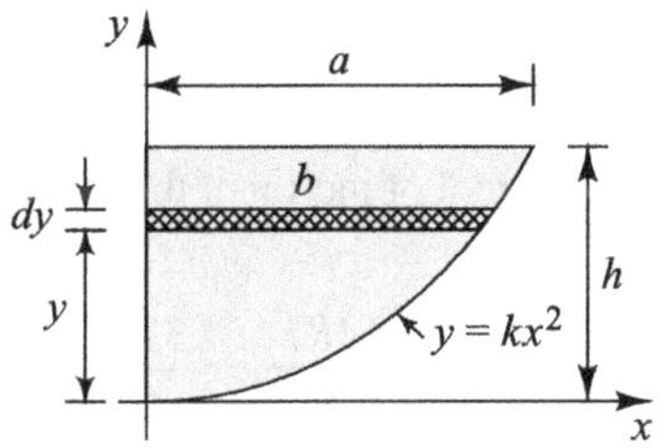

Fig. 6.8

Solution Substituting boundary condition $y = h$ at $x = a$ in $y = kx^2$ gives, $k = h/a^2$.

Area of an element,

$$dA = b \times dy = \frac{y^{1/2}}{k^{1/2}}dy = \frac{a}{h^{1/2}}y^{1/2}dy$$

Area of quarter parabola,

$$A = \frac{a}{h^{1/2}}\int_0^{h} y^{1/2}dy = \frac{a}{h^{1/2}}\left|\frac{y^{3/2}}{3/2}\right|_0^{h}$$

or, $$A = \frac{2}{3}ha$$

Taking moment of area about x-axis,

$$\bar{y} = \frac{\int y\,dA}{A} = \frac{\frac{a}{h^{1/2}}\int_0^{h} y^{3/2}dy}{\frac{2}{3}ha}$$

or, $$\bar{y} = \frac{3}{5}h$$

Similarly $\bar{x} = \frac{3}{8}a$ can be derived by taking moment of area about y-axis.

Example 6.5 Determine the second moments of area of the triangle shown in Fig. 6.5 about an axis coinciding with the base, and about an axis parallel to the base and passing through centroid.

Solution The second moment of area of the triangle shown in Fig. 6.5, about base *ab* is given by,

$$I_x = \int y^2 dA = \int_0^{H} y^2 x\,dy$$

$$= \int_0^{H}\left(By^2 - \frac{By^3}{H}\right)dy$$

or, $$I_x = B\left|\frac{y^3}{3}\right|_0^{H} - \frac{B}{H}\left|\frac{y^4}{4}\right|_0^{H} = \frac{BH^3}{12}$$

Using parallel axis theorem for moment of inertia about centroid,

$$I_x = I_{cg} + A\,\bar{y}^2$$

where $\bar{y} = \frac{H}{3}$ as derived in Example 6.1

$$\therefore \quad I_{cg} = \frac{BH^3}{12} - \frac{BH}{2} \times \frac{H^2}{9}$$

or, $$I_{cg} = \frac{BH^3}{36}$$

Example 6.6 Determine the second moment of inertia of the rectangle shown in Fig. 6.9 about

(i) x-axis coinciding with base,

(ii) an axis parallel to x-axis and passing through centroid.

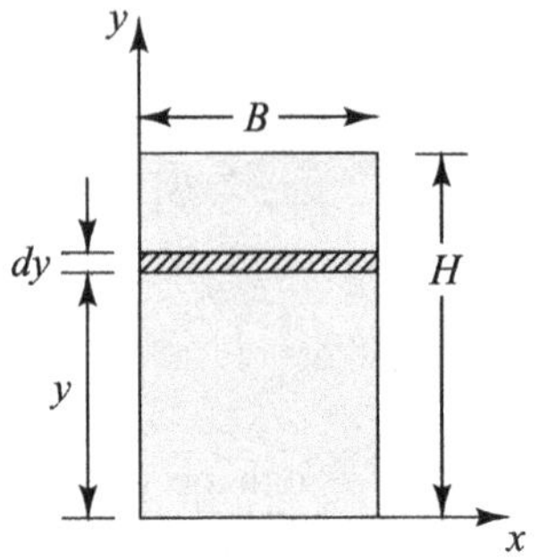

Fig. 6.9

Solution The second moment of the rectangle about x-axis is given by,

$$I_x = \int y^2 dA = \int_0^H y^2 \times B\,dy = \frac{BH^3}{3}$$

The second moment of area about centroid is given by the parallel axis theorem of follows.

$$I_x = I_{cg} + Ay^2$$

$$\text{or} \quad I_{cg} = \frac{BH^3}{3} - BH \times \left(\frac{H}{2}\right)^2 = \frac{BH^3}{3} - \frac{BH^3}{4}$$

$$\text{or} \quad I_{cg} = \frac{BH^3}{12}$$

Example 6.7 Determine the second moment of area of a circle of radius R about its diameter.

Solution Consider an elementary area $dA = dr \times rd\theta$ as shown in Fig. 6.10. Its second moment of area about x-y axes is given by,

$$dI_x = y^2\,dA = (r \sin\theta)^2 \times dr \times rd\theta$$

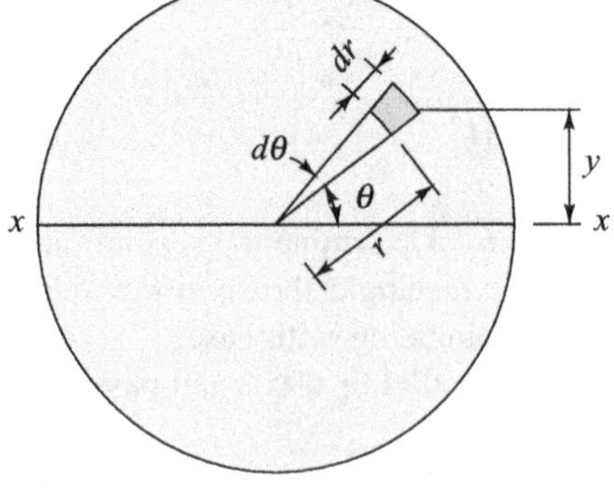

Fig. 6.10

Integrating for second moment of the entire area,

$$I_x = \int_0^R \int_0^{2\pi} r^3 \sin^2\theta\, d\theta\, dr$$

$$\text{or,} \quad I_x = \left|\frac{r^4}{4}\right|_0^R \int_0^{2\pi} \sin^2\theta\, d\theta$$

$$\text{or,} \quad I_x = \frac{\pi R^4}{4}$$

Example 6.8 Determine the moment of inertia of the area shown in Fig. 6.11 about x-axis.

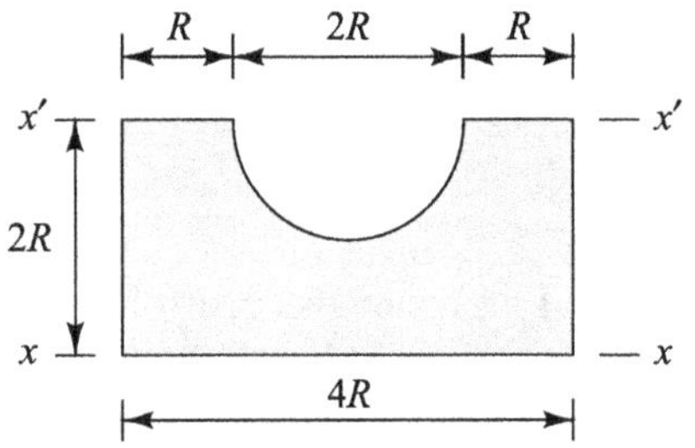

Fig. 6.11

Solution Moment of inertia of the rectangle about x-axis,

$$I_x = \frac{4R \times (2R)^3}{3} = \frac{32R^4}{3}$$

Moment of inertia of semicircle about its diameter,

$$I_{x'} = \frac{\pi R^4}{8}$$

(It can be derived by integrating dI_x of Example 6.7 for circle within the limits 0 to π). Moment of inertia of semicircle about its centroid,

$$I_{cg} = I_{x'} - A\,\bar{y}^2$$

$$= \frac{\pi R^4}{8} - \frac{\pi R^2}{2}\left(\frac{4R}{3\pi}\right)^2$$

$$= \frac{\pi R^4}{8} - \frac{8R^4}{9\pi}$$

Moment of inertia of semicircle about x-axis,

$$I_x = I_{cg} + A\left(2R - \frac{4R}{3\pi}\right)^2$$

or, $$I_x = \frac{\pi R^4}{8} - \frac{8R^4}{9\pi} + \frac{\pi R^2}{2} \times 4R^2 + \frac{\pi R^2}{2} \times \frac{16R^2}{9\pi^2} - \frac{\pi R^2}{2} \times \frac{16R^2}{3\pi}$$

or, $$I_x = R^4\left(\frac{\pi}{8} + 2\pi - \frac{8}{3}\right)$$

The moment of inertia of the area shown in Fig. 6.11 is given by subtracting moment of inertia of semicircle about x-axis from the moment of inertia of the rectangle about x-axis as follows,

$$I_x = \frac{32R^4}{3} - R^4\left(\frac{\pi}{8} + 2\pi - \frac{8}{3}\right)$$

Example 6.9 Determine the second moment of area of a T-section shown in Fig. 6.12 about horizontal axis passing through centroid.

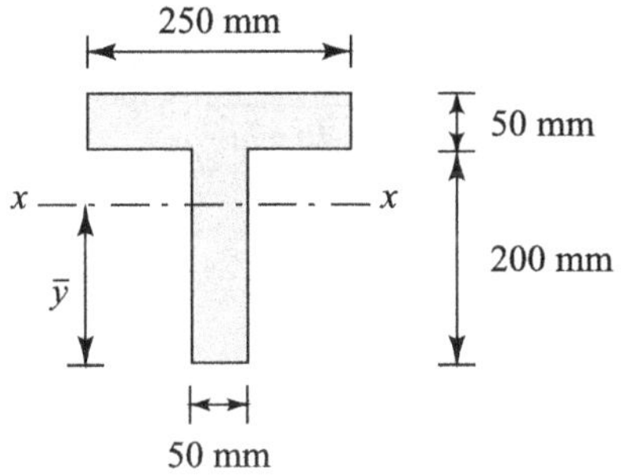

Fig. 6.12

Solution Taking moment of area about bottom fibre, the location of centroid is given by,

$$\bar{y} = \frac{(250 \times 50) \times 225 + (50 \times 200) \times 100}{250 \times 50 + 50 \times 200}$$
$$= 169.44 \text{ mm}$$

The second moment of area of flange and web is first taken about horizontal axis through their own centroid and then transferred to centroid of the T-section as follows.

$$I_x = \frac{250 \times 50^3}{12} + 250 \times 50 \times (225 - 169.44)^2 + \frac{50 \times 200^3}{12} + 50 \times 200 \times (169.44 - 100)^2$$
$$= 122.74 \times 10^6 \text{ mm}^4$$

Example 6.10 Determine moment of inertia and radius of gyration of the unequal I-section shown in Fig. 6.13 about horizontal and vertical axis passing through centroid.

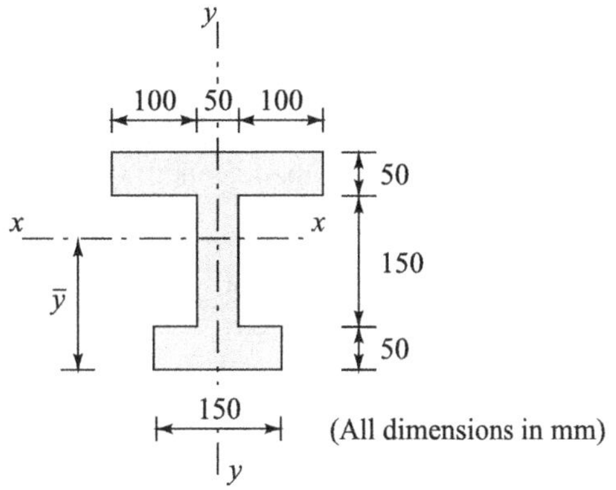

Fig. 6.13

Solution Area of the section,

$$A = 250 \times 50 + 150 \times 50 + 150 \times 50$$
$$= 27500 \text{ mm}^2$$

Taking moment of the area about bottom fibre, the location of centroid is given by,

$$\bar{y} = \frac{(250 \times 50) \times 225 + (50 \times 150) \times 125 + (150 \times 50) \times 25}{27500}$$

or, $\bar{y} = 143.18$ mm

The moment of inertia of flanges and web is first taken about their own centroid and then transferred to horizontal axis through centroid by parallel axis theorem as follows.

$$I_x = \frac{250 \times 50^3}{12} + 250 \times 50 \times (225 - 143.18)^2 + \frac{50 \times 150^3}{12} + 50 \times 150 \times (125 - 143.18)^2 + \frac{150 \times 50^3}{12} + 150 \times 50 \times (143.18 - 25)^2$$
$$= 209.14 \times 10^6 \text{ mm}^4$$

Radius of gyration

$$r_x = \sqrt{\frac{I_x}{A}} = \sqrt{\frac{209.14 \times 10^6}{27500}}$$

$$= 87.2 \text{ mm}$$

Both flanges and web are symmetric about y-axis which passes through centroid of the section. Sum of moment inertia of both flanges and web about vertical axis gives the moment of inertia of I-section as follows.

$$I_y = \frac{50 \times 250^3}{12} + \frac{150 \times 50^3}{12} + \frac{50 \times 150^3}{12}$$

$$= 80.73 \times 10^6 \text{ mm}^4$$

$$r_y = \sqrt{\frac{I_y}{A}} = \sqrt{\frac{80.73 \times 10^6}{27500}} = 54.18 \text{ mm}$$

Example 6.11 Determine the product moment of inertia the rectangle of size $B \times H$ about

(i) x-y axes shown in Fig. 6.14, and

(ii) their parallel axes through the centroid.

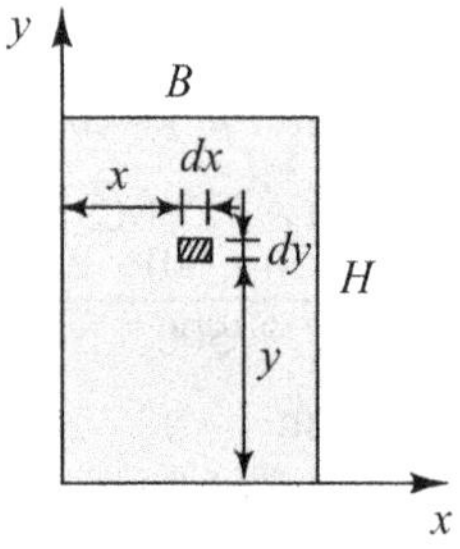

Fig. 6.14

Solution (i) Integrating the product moment of inertia of a small elementary area,

$$I_{xy} = \int_{y=0}^{y=H} \int_{x=0}^{x=B} xy\, dx\, dy$$

$$= \left[\frac{x^2}{2}\right]_0^B \int_{y=0}^{y=H} y\, dy = \frac{B^2}{2}\left[\frac{y^2}{2}\right]_0^H$$

$$\text{or,} \quad I_{xy} = \frac{B^2 H^2}{4}$$

(ii) The product moment of inertia about centroid is given by parallel axis theorem,

$$I_{xy} = I_{xgyg} + A\,\bar{x}\,\bar{y}$$

$$\text{or } I_{xgyg} = \frac{B^2H^2}{4} - BH \times \frac{B}{2} \times \frac{H}{2} = 0$$

The product moment of inertia about centroid is zero as $x'-y'$ axes through centroid are the principal axes of the rectangle.

Example 6.12 Determine the maximum and minimum radii of gyration of the angle section shown in Fig. 6.15 about the centroid.

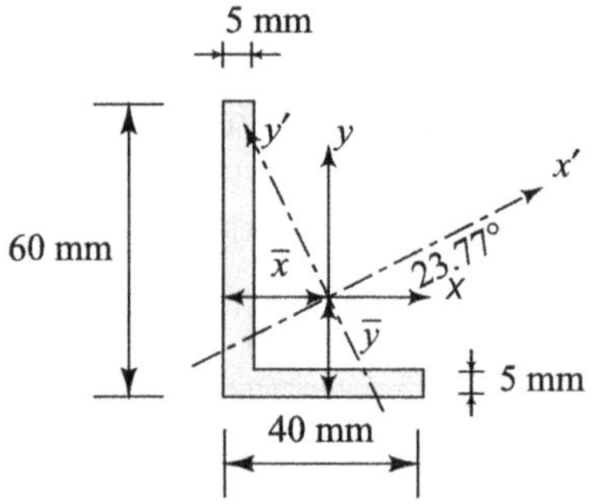

Fig. 6.15

Solution Taking moment of area about bottom and left and fibre, the location of centroid is given by,

$$\bar{y} = \frac{60 \times 5 \times 30 + 35 \times 5 \times 2.5}{60 \times 5 + 35 \times 5} = 19.868 \text{ mm}$$

$$\bar{x} = \frac{55 \times 5 \times 2.5 + 40 \times 5 \times 20}{60 \times 5 + 35 \times 5} = 9.868 \text{ mm}$$

Moments of inertia of both legs are calculated about their own centroid and then transferred to the centroid of the angle section by parallel axis theorem as follows.

$$I_x = \frac{5 \times 60^3}{12} + 5 \times 60 \times (30 - 19.868)^2$$

$$+ \frac{35 \times 5^3}{12} + 35 \times 5 \times (19.868 - 2.5)^2$$

$$= 173950 \text{ mm}^4$$

$$I_y = \frac{5 \times 40^3}{12} + 5 \times 40 \times (20 - 9.868)^2$$

$$+ \frac{55 \times 5^3}{12} + 55 \times 5 \times (9.868 - 2.5)^2$$

$$= 62700 \ \text{mm}^4$$

The product moment of inertia of both legs is zero about axes parallel to *x*-*y* axes and passing through their centroid since they are the principal axes. The product moment of inertia of the angle about a centroid is then calculated using parallel axis theorem as follows.

$$I_{xy} = 0 + 60 \times 5 \times (30 - 19.868) \times (2.5 - 9.868)$$

$$+ 0 + 35 \times 5 \times (2.5 - 19.868) \times \left(\frac{35}{2} + 5 - 9.868\right)$$

$$= -60789.5 \ \text{mm}^4$$

Principal values of moment of inertia are given by Eq. (6.7) as follows.

$$I_{\text{max/min}} = \frac{I_x + I_y}{2} \pm \sqrt{\left(\frac{I_x - I_y}{2}\right)^2 + I_{xy}^2}$$

$$= \frac{173950 + 62700}{2} \pm \sqrt{\left(\frac{173950 - 62700}{2}\right)^2 + 60789.5^2}$$

$$= 200723.5,\ 35926.7 \ \text{mm}^4$$

$$r_{\max} = \sqrt{\frac{I_{\max}}{A}} = \sqrt{\frac{200723.5}{475}} = 20.6 \ \text{mm}$$

$$r_{\min} = \sqrt{\frac{I_{\min}}{A}} = \sqrt{\frac{35926.7}{475}} = 8.7 \ \text{mm}$$

Inclination of principal axes by Eq. (6.6),

$$\theta_p = \frac{1}{2}\tan^{-1}\left(\frac{-I_{xy}}{\frac{I_x - I_y}{2}}\right)$$

$$= \frac{1}{2}\tan^{-1}\left(\frac{60789.5}{\frac{173950 - 62700}{2}}\right)$$

or $\theta_p = 23.77°$

The calculations can be done for any angle using a C++ computer program as follows,

```
#include<iostream.h>
#include<math.h>
#include<conio.h> // to clear output screen
void main ()
{
  clrscr ();
cout<< " Calculate centroid x, y and Ix,"
<< "Iy, Ixy for angle" << endl;
float x, y, a, b, t, area, Ix, Iy, Ixy,
Imax, Imin, rmax, rmin, theeta;
float pie = 355./113.;
  cout << " input leg sizes: a, b and"
  <<"thickness: t";
  cin >> a >> b >> t;
  area = a*t+(b-t)*t;
  y = (a*t*a/2+ (b-t)*t* t /2)/area;
  x = ((a-t)*t*t/2 + b*t*b/2)/area;
Ix = t*pow(a,3)/12+t*a*pow(a/2-y,2)+
  (b-t)*pow(t,3)/12+(b-t)*t*pow(y-t/2,2);
Iy = t*pow(b,3)/12+t*b*pow(b/2-x,2)+
  (a-t)*pow(t,3)/12+(a-t)*t*pow(x-t/2,2);
Ixy = a*t*(a/2-y)*(t/2-x)+(b-t)*t*
  (t/2-y)*(t+(b-t)/2-x);
Imax = (Ix+Iy)/2+sqrt(pow((Ix-Iy)/
  2,2)+Ixy*Ixy);
Imin = (Ix+Iy)/2-sqrt(pow((Ix-Iy)/
  2,2)+Ixy*Ixy);
rmax = sqrt (Imax/area);
rmin = sqrt (Imin/area);
theeta = (180/pie)*0.5*atan(-Ixy/((Ix-Iy)/
2));
cout  << "    area = " <<area<<endl;
cout  << "       y = " <<y<<endl;
cout  << "       x = " <<x<< endl;
cout  << "      Ix = " << Ix<< endl;
cout  << "      Iy = " << Iy<< endl;
```

```
cout   << "       Ixy  = " << Ixy<< endl;
cout   << "      Imax  = " << Imax<< endl;
cout   << "      Imin  = " << Imin<< endl;
cout   << "      rmax  = " << rmax>> endl;
cout   << "      rmin  = " << rmin<< endl;
cout   << "    theeta  = " << theeta<<
                          " ( in degrees)";
getch ();
}}
```

The output of this program for 60 × 40 × 5 mm angle section is as follows.

```
      area = 475
         y = 19.868422
         x = 9.868421
        Ix = 173950.109375
        Iy = 62700.109375
       Ixy = -60789.472656
      Imax = 200723.53125
      Imin = 35926.683594
      rmax = 20.55665
      rmin = 8.696846
    theeta = 23.770065 (in degrees)
```

The principal $x' - y'$ axes at 23.77° is shown in Fig. 6.15.

EXERCISES

PROBLEMS

6.1 through 6.4 Locate the centroid of the plane areas shown in Fig. 6.16 through 6.19.

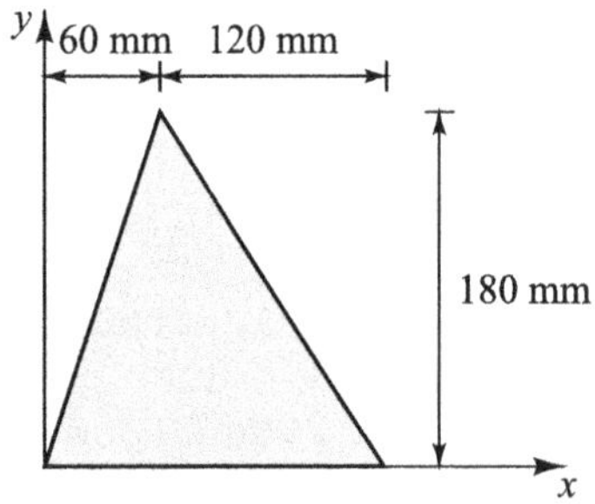

Fig. 6.16 *Prob 6.1*

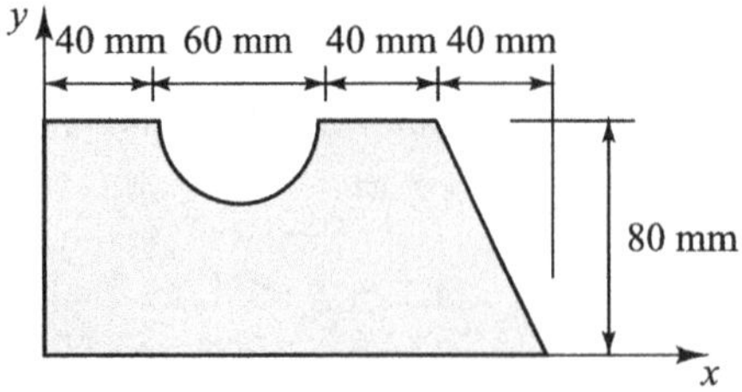

Fig. 6.17 *Prob 6.2*

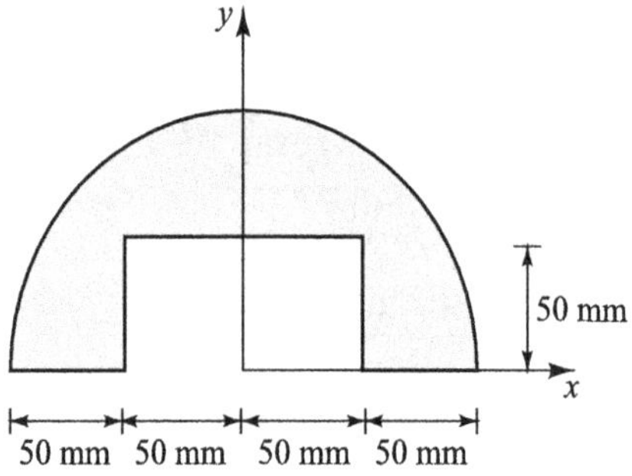

Fig. 6.18 *Prob 6.3*

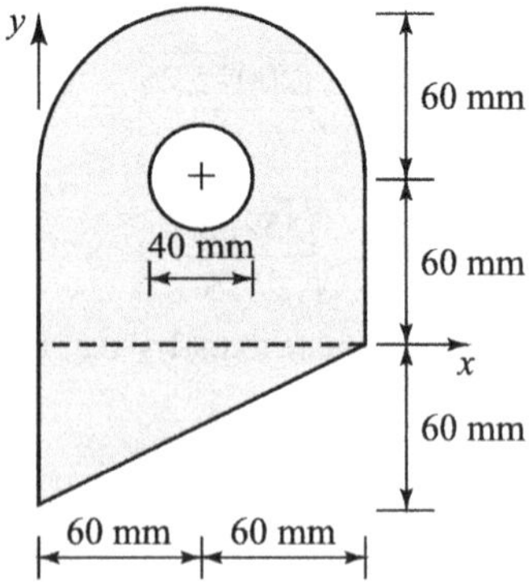

Fig. 6.19 *Prob 6.4*

6.5 through 6.7 Calculate the second moments of area for the plane areas shown in Fig. 6.20 through 6.22 about x and y axes.

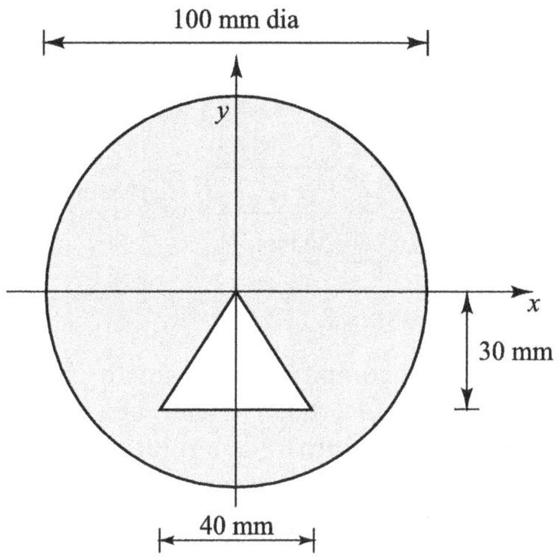

Fig. 6.20 *Prob 6.5*

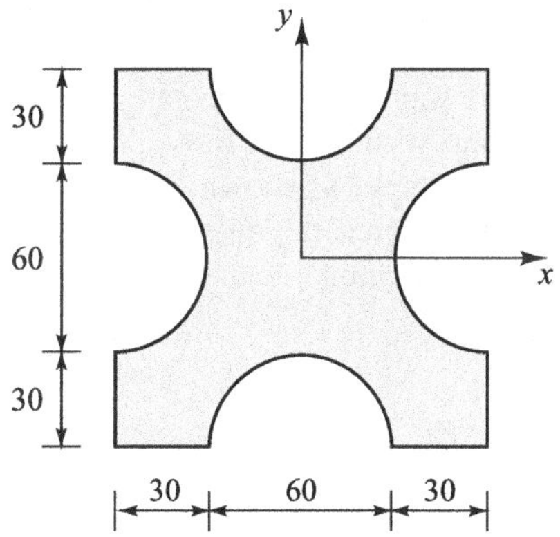

Fig. 6.21 *Prob 6.6*

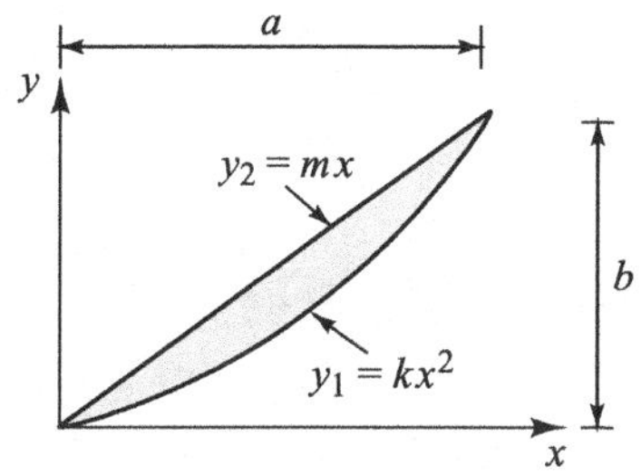

Fig. 6.22 *Prob 6.7*

6.8 through 6.10 Calculate the second moments of area for the plane areas shown in Fig. 6.23 through 6.25 about horizontal and vertical axes passing through centroid.

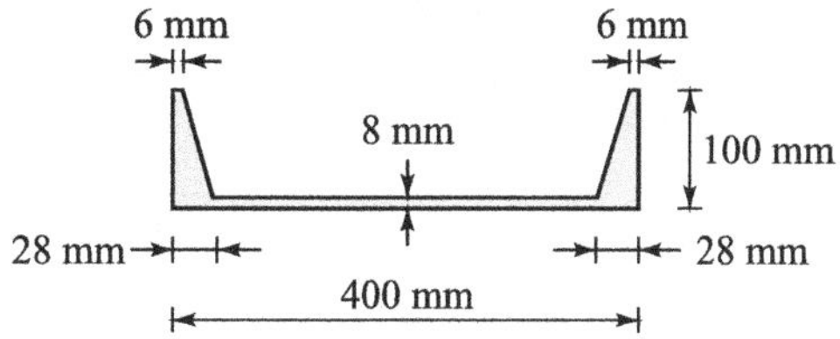

Fig. 6.23 *Prob 6.8*

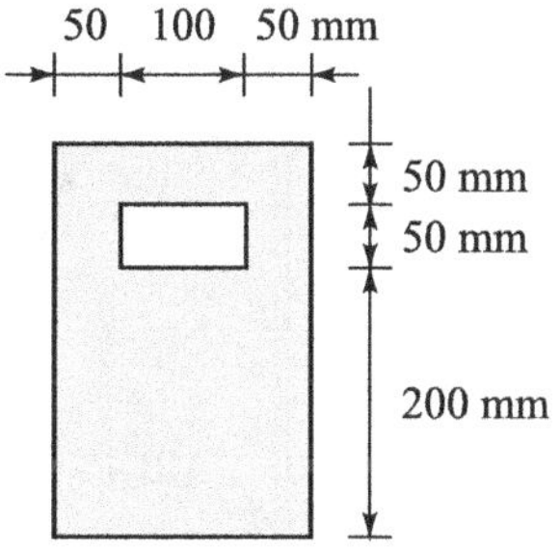

Fig. 6.24 *Prob 6.9*

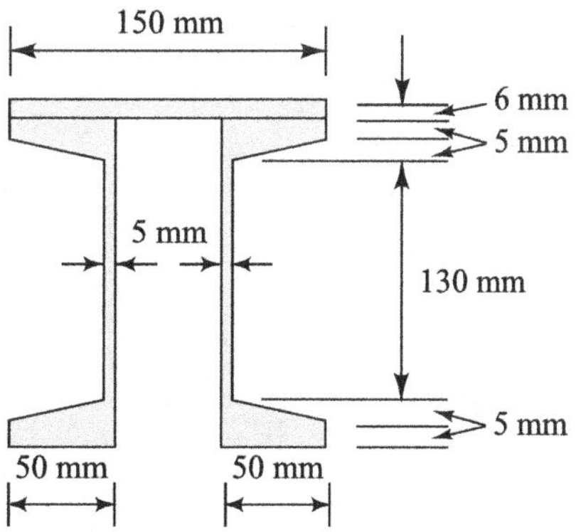

Fig. 6.25 *Prob 6.10*

6.11 A compound beam is built up with a channel section ISLC 200 attached to the top flange of I-section MB450 as shown in Fig. 6.26. Calculate the second moments of area for the compound beam about horizontal and vertical axes passing through centroid if the properties of ISLC 200 and MB 450 are as follows.

ISLC 200: Area = 2626 mm^2 Centroid distance,

c_y = 23.6 mm

$I_x = 146.8 \times 10^4$ mm^4,

$I_y = 17.29 \times 10^6$ mm^4

Web thickness = 5.5 mm.

MB450: Area = 9230 mm^2

Depth = 450 mm

$I_x = 303.908 \times 10^6$ mm^4,

$I_y = 8.34 \times 10^6$ mm^4

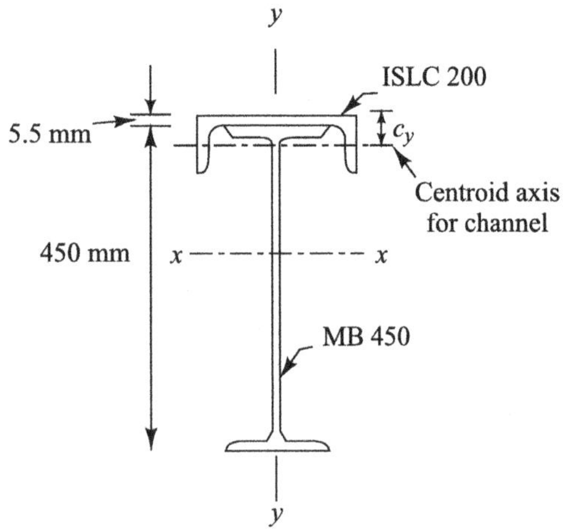

Fig. 6.26 *Prob 6.11*

6.12 and 6.13 Determine the maximum and minimum radii of gyration for the plane areas in Fig. 6.27 and 6.28 about centroid.

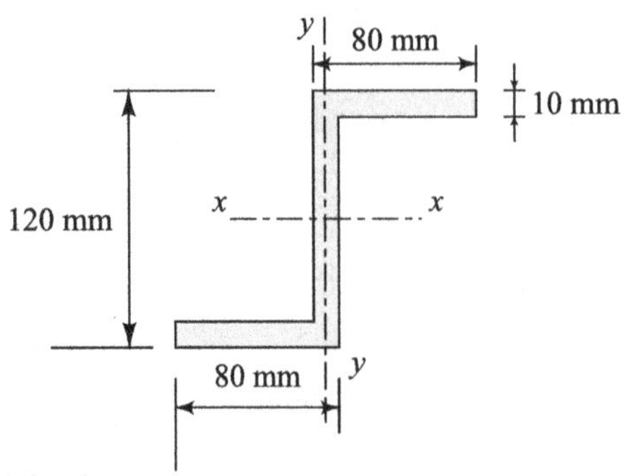

Fig. 6.27 *Prob 6.12*

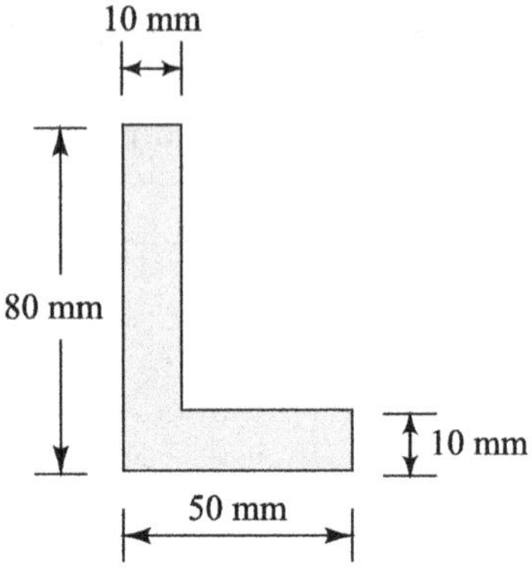

Fig. 6.28 *Prob 6.13*

6.14 Write a computer program to calculate moment of inertia of a plane area which can be divided into rectangular areas about horizontal and vertical axis through centroid. It should be extended to calculate maximum and minimum radius of gyration and inclination of principal axes through centroid about horizontal axis.

(**Hint:** Divide a plane area into horizontal rectangles. The moments of inertia is the sum of moment of inertia of each rectangle about horizontal and vertical axes through bottom and left hand corner as shown in Fig. 6.29. It is then transferred to centroidal axis by parallel axis theorem.)

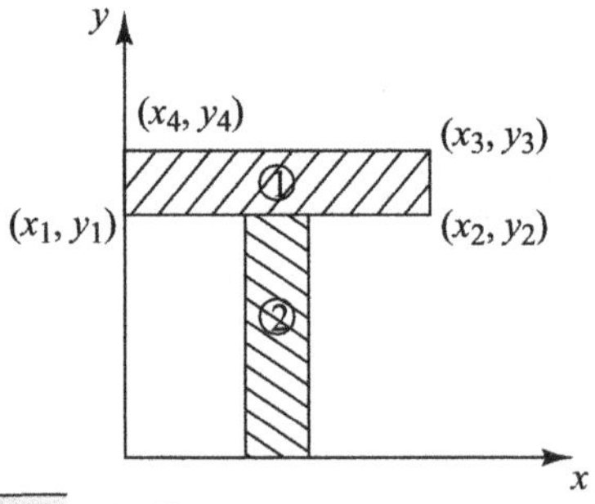

Fig. 6.29 *Prob 6.14*

OBJECTIVE QUESTIONS

6.15 Match the shapes in List I with their respective centroid distance in List II and answer using code.

List I		*List II*
(1) Quarter circle	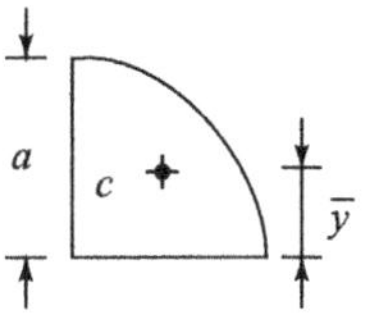	(A) 0.3 a
(2) Semi Ellipse	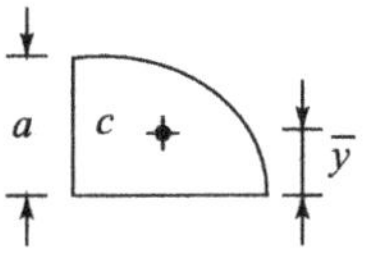	(B) 0.424 a
(3) Parabolic Spandrel	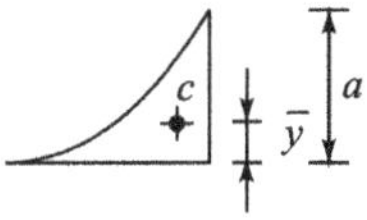	(C) 0.4 a
(4) Semi Parabolic area	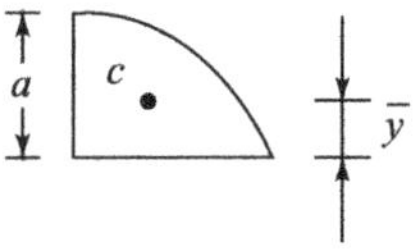	(D) 0.424 a

Answer Code	1	2	3	4
(a)	A	B	C	D
(b)	B	C	A	D
(c)	B	D	A	C
(d)	D	C	B	A

6.16 Consider the moments of inertia about an axis perpendicular to web through centroid I_x and about the axis parallel to web through centroid, I_y for a channel section, and an *I*-section having same depth of web, thickness and cross-sectional area, then

(1) I_x of channel = I_x of I-section
(2) I_x of channel > I_x of I-section
(3) I_y of channel = I_y of I-section
(4) I_y of channel > I_y of I-section
(5) I_y of channel < I_y of I-section

Of these statements,

(a) 1 and 5 are correct
(b) 2 and 3 are correct
(c) 1 and 4 are correct
(d) 2 and 4 are correct

6.17 Which of the axis shown in Fig. 6.30 has minimum radius of gyration for an angle section.

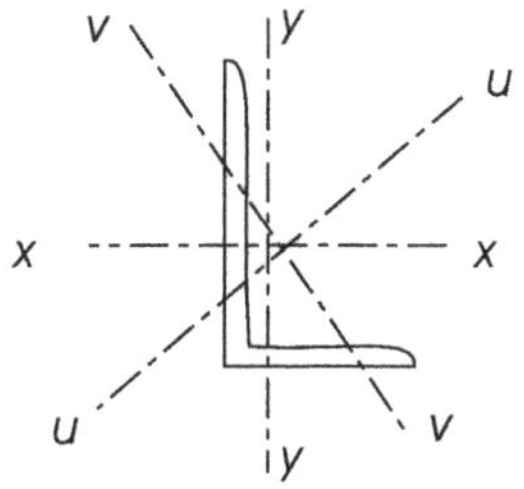

Fig. 6.30

(a) *x*-*x* axis (b) *y*-*y* axis
(c) *u*-*u* axis (d) *v*-*v* axis

6.18 Which of the compound sections found by two equal angles has highest value of minimum radius of gyration.

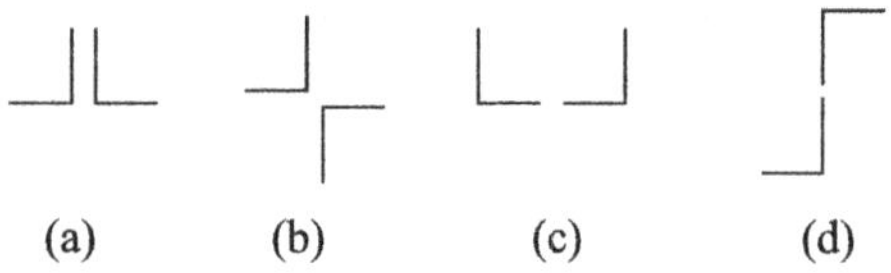

Fig. 6.31

7

Stresses in Beams

CONCEPT REVIEW

7.1 BENDING STRESSES

Consider a beam section under constant bending moment M as shown in Fig. 7.1(a). It is assumed that,

(1) a plane section in $y - z$ plane will remain plane and normal to the axis after bending,

(2) the material follows Hooke's law and the moduli of elasticity in tension and compression are equal.

As per these assumptions a vertical plane is displaced to an inclined plane such that the strain varies linearly from zero at neutral axis to a maximum tension at bottom and a maximum compression at top. Variation of stress in similar to strain as stress is proportional to strain.

Strain at a distance y from neutral axis EF is given by,

$$\varepsilon = \frac{GH - EF}{EF} = \frac{(R + y)\theta - R\theta}{R\theta} = \frac{y}{R}$$

Stress at a distance y from neutral axis,

$$\sigma = E\varepsilon = \frac{Ey}{R}$$

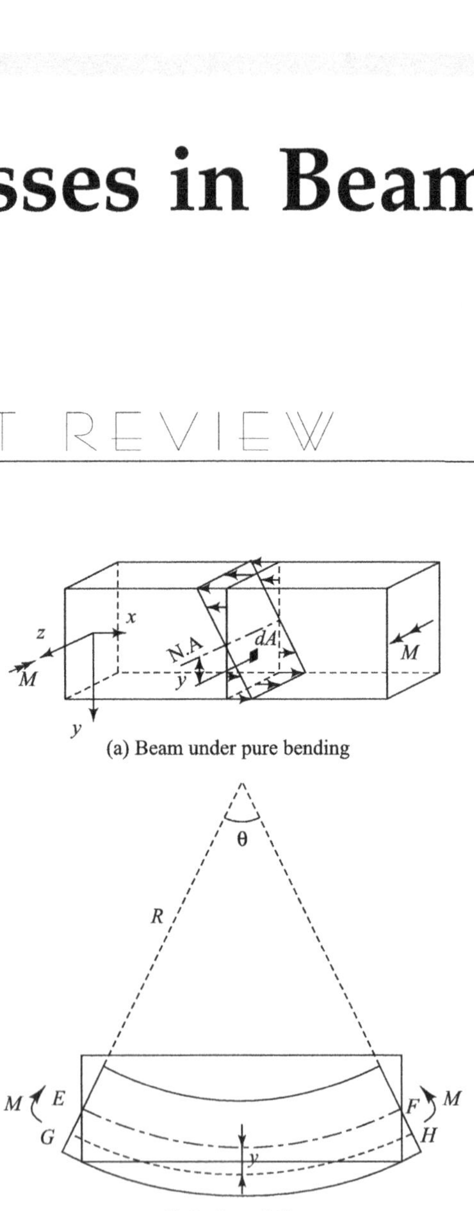

(a) Beam under pure bending

(b) Deflected Shape

Fig. 7.1

or, $\dfrac{\sigma}{y} = \dfrac{E}{R}$ (i)

where, $1/R$ = curvature of neutral axis. The applied moment M produces interval resisting moment given by moment of internal force about neutral axis as follows.

$$M = \int (\sigma \, dA) \times y = \frac{E}{R} \int y^2 \, dA$$

or, $M = \dfrac{EI}{R}$ (ii)

where, dA = elementary area at a distance y from neutral axis

and, I = second moment of area about neutral axis

Equations (i) and (ii) are generally known as moment equation as follows.

$$\frac{M}{I} = \frac{\sigma}{y} = \frac{E}{R} \quad (7.1)$$

Location of neutral axis

The summation of the force on all the elementary areas is zero as there is no axial force on the beam.

Thus, $\int \sigma \, dA = 0$

or, $\dfrac{E}{R} \int y dA = \dfrac{E}{R} \; \bar{y} \; A = 0$

$\dfrac{EA}{R} \neq 0$ and therefore $\bar{y} = 0$,

As $\bar{y}$ is the distance of centroid from neutral axis, $\bar{y} = 0$ means neutral coincides with centroid of the plane section.

7.2 SHEARING STRESS IN BEAMS

The first assumption in the theory of bending that plane section remains plane after bending is strictly true in case of pure bending. It will not hold good in case the shear force is also acting on the beam due to shear deformations. However the magnitude of shear stress is generally low in beams so that the assumption is used with reasonable accuracy.

Consider a beam element of length dx under bending moment M and $M + dM$ and there is shear force V due to varying moment. The shear stress at a distance y_o from neutral axis is evaluated by considering equilibrium of an element *abcd* shaded in Fig. 7.2(a).

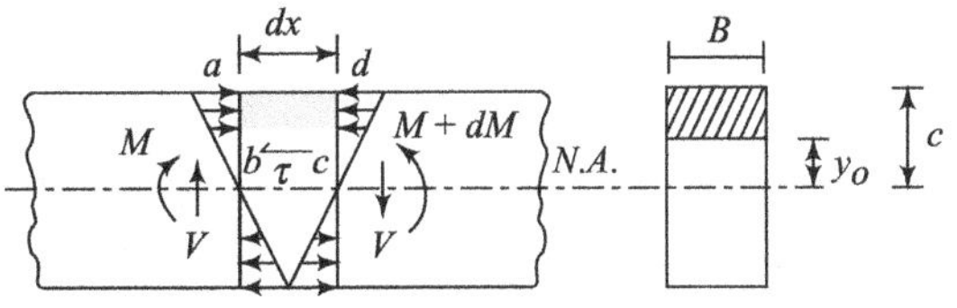

(a) Beam (b) Cross-section

Fig. 7.2

The difference of force due to bending on face *ab* and *cd* is resisted by shear force on face *bc* as follows.

$$\tau B \, dx = \int_{y_0}^{c} \frac{(M + dM)}{I} \, ydA - \int_{y_0}^{c} \frac{My \, dA}{I}$$

or $\tau = \dfrac{1}{IB} \dfrac{dM}{dx} \displaystyle\int_{y_0}^{c} y \, dA$

or $\tau = \dfrac{VQ}{IB}$ (7.2)

where,

$V = \dfrac{dM}{dx}$, shear force due to varying moment

$Q = A\bar{y}$ = first moment of area beyond distance y_o from neutral axis

It should be noted that vertical shear stress is complementary to horizontal shear as derived in equilibrium Eq. (1.18).

EXAMPLES

Example 7.1 A circular steel cantilever beam of 80 mm diameter and 5 m length, carries a concentrated load of 1 kN at the free end. Plot the variation of bending stress over the cross-section of the beam at the critical section.

Solution Maximum bending moment

$= 1 \times 5 = 5$ kN.m

Maximum bending stress at $y = R$ as per Eq. (7.1)

$$\sigma = \frac{My}{I} = \frac{5 \times 10^6 \times 40}{\pi \times 40^4/4} = 99.47 \text{ MPa}$$

The bending stress variation is linear as per Eq. (7.1) as shown in Fig. 7.3.

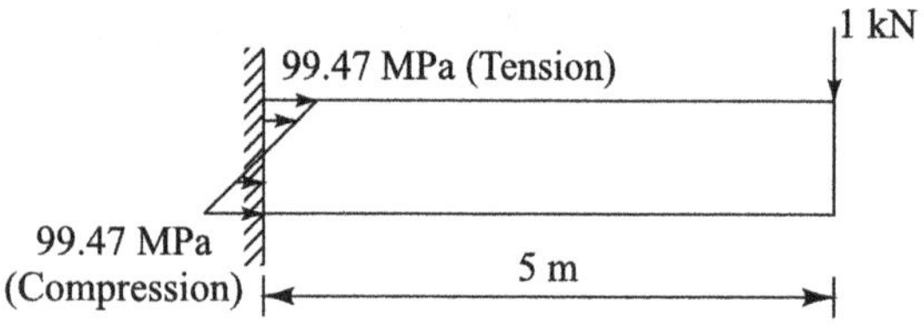

Fig. 7.3

Example 7.2 Compare moment carrying capacity of the beam sections shown in Fig. 7.4.

Solution If the material of the beams in same then σ = permissible stress in bending = constant, then the moment carrying capacity, $M = \sigma \times I/y$ will depend on $I/y = Z$ which is called sectional modulus.

(i) Sectional modulus of rectangular section,

$$Z = \frac{I}{y} = \frac{bd^3}{12 \times d/2} = \frac{bd^2}{6}$$

or, $$Z = \frac{300 \times 400^2}{6} = 8 \times 10^6 \text{ mm}^3$$

(ii) Sectional modulus of circular section

$$Z = \frac{\pi R^4}{4 \times R} = \frac{\pi \times 200^3}{4} = 6.28 \times 10^6 \text{ mm}^3$$

(iii) Sectional modulus of I-section

$$Z = \frac{BD^3/12 - bd^3/12}{D/2}$$

$$= \frac{400 \times 400^3/12 - 300 \times 200^3/12}{400/2}$$

or $Z = 9.67 \times 10^6$ mm^3

The sectional area of I-section is lowest but its moment resistance capacity is highest.

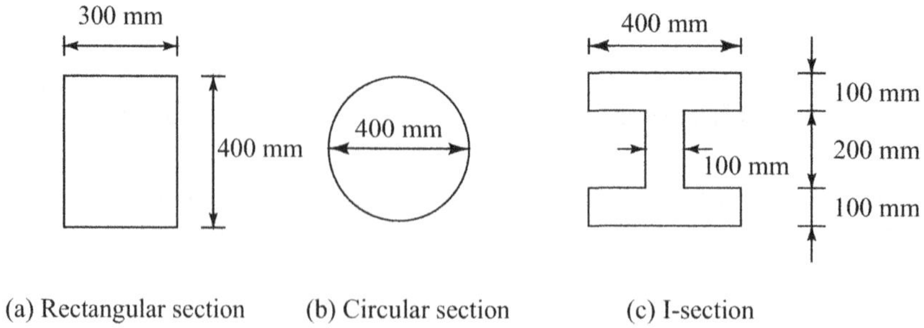

Fig. 7.4

Example 7.3 A rectangular beam 200 mm wide and 300 mm deep carries an uniformly distributed load of 10 kN per m over a simply supported span of 6 m. Determine,

(a) the maximum stress in the beam due to bending

(b) the radius of curvature for the section where bending is maximum if the modulus of elasticity is 200 GPa.

Solution Maximum bending moment,

$$M = \frac{wl^2}{8} = \frac{10 \times 6^2}{8} = 45 \text{ kN m}$$

Maximum bending stress,

$$\sigma = \frac{M}{Z} = \frac{45 \times 6^2}{200 \times 300^2/6} = 15 \text{ MPa}$$

The radius of curvature corresponding to these values in given by,

$$R = \frac{Ey}{\sigma} = \frac{200 \times 10^3 \times (300/2)}{15} = 2 \times 10^6 \text{ mm}$$

Example 7.4 Determine the longest span over which the *I*-beam cross-section shown in Fig. 6.13 can carry 15 kN/m if the permissible stress in bending for the beam is 165 MPa. Also determine the bending stress at the junction of the top flange and the web.

Solution The sectional properties are calculated in Example 6.10.

The *I*-section has two values of sectional modulus and the smallar value is critical

$$Z_1 = I/y_1 = \frac{209.14 \times 10^6}{143.18} = 1.46 \times 10^6 \text{ mm}^3$$

$$Z_2 = I/y_2 = \frac{209.14 \times 10^6}{250 - 143.18} = 1.96 \times 10^6 \text{ mm}^3$$

$$M = \sigma Z = 165 \times 1.46 \times 10^6 = 241 \times 10^6 \text{ N.mm}$$

Equating the applied moment to moment of resistance,

$$\frac{wl^2}{8} = M$$

or, span, $l = \sqrt{8M/w} = \sqrt{\dfrac{8 \times 241 \times 10^6}{15}}$

$= 11337.5$ mm

For stress at a depth of 50 mm, i.e.

$y = 200 - 143.18 = 56.82$ mm from centroid,

$$\sigma = \frac{My}{I} = \frac{241 \times 10^6 \times 56.82}{209.14 \times 10^6} = 65.47 \text{ MPa}$$

Example 7.5 Determine the maximum increase possible in moment of resistance of a square beam when bent about its diagonal by cutting its top and bottom edges as shown in Fig. 7.5.

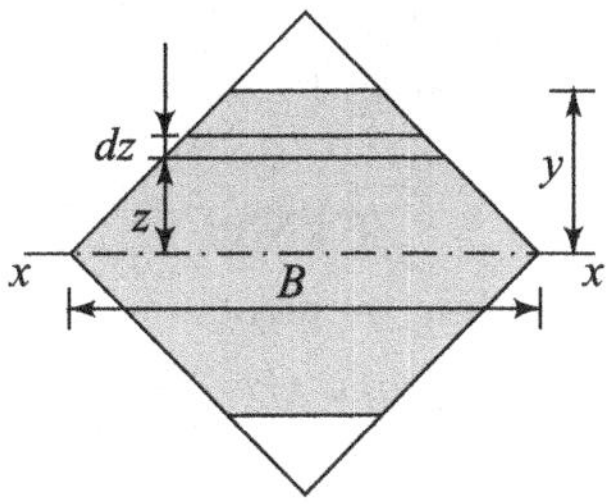

Fig. 7.5

Solution For square section,

$$I_x = \frac{2bh^3}{12} = \frac{B^4}{48}$$

$$Z_1 = \frac{I_x}{B/2} = \frac{B^3}{24}$$

For truncated section,

$$I_x = 2\int_0^y z^2 (B - 2z)\, dz$$

$$= 2\left[\frac{By^3}{3} - \frac{2y^4}{4}\right]$$

$$Z_2 = 2\left[\frac{By^2}{3} - \frac{y^3}{2}\right]$$

For maximum Z_2

$$\frac{dZ_2}{dy} = 0 = 2\left[\frac{2By}{3} - \frac{3y^2}{2}\right]$$

or, $y = \dfrac{4B}{9}$

$\therefore$ Maximum $Z_2 = \dfrac{32B^3}{729}$

Increase in moment of resistance

$$= \frac{Z_2 - Z_1}{Z_1} \times 100 = 5.35\%$$

Example 7.6 Design a timber joist for a simply supported span of 4 m at a spacing of 1.5 m centre to centre. The dead load of the roof covering is 2 kN/m^2 and live load in 3 kN/m^2. Take the permissible stress in bending as 16.8 N/mm^2 and specific weight of timber to be 8 kN/m^3.

Solution Load on beam per metre span,

$$\text{Dead laod} = 2 \times 1.5 \times 1 = 3.0 \text{ kN/m}$$

$$\text{Live load} = 3 \times 1.5 \times 1 = 4.5 \text{ kN/m}$$

Self weight, assuming 15 × 30 cm beam

$$= 8 \times 0.15 \times 0.3 \times 1 = 0.36 \text{ kN/m}$$

Total udl, $w = 7.86$ kN/m

Maximum bending moment,

$$M = \frac{wl^2}{8} = \frac{7.86 \times 4^2}{8} = 15.72 \text{ kN m}$$

Sectional modulus required,

$$Z = \frac{M}{\sigma} = \frac{15.72 \times 10^6}{16.8} = 935714.3 \text{ mm}^3$$

Taking depth by width ratio, $d/b = 2$

$$Z = \frac{bd^2}{6} = \frac{4b^3}{6} = 935714.3 \text{ mm}^3$$

Thus, width of beam, $b = 112$ mm

and, depth of beam, $d = 224$ mm

Example 7.7 Determine the dimensions of a rectangular timber beam which can be cut from a circular log of diameter 800 mm so that it has maximum strength for pure bending.

Solution For maximum strength in bending, the sectional modulus, $Z = bh^2/6$ should be maximum, where b and h are the width and height of the rectangular beam as shown in Fig. 7.6.

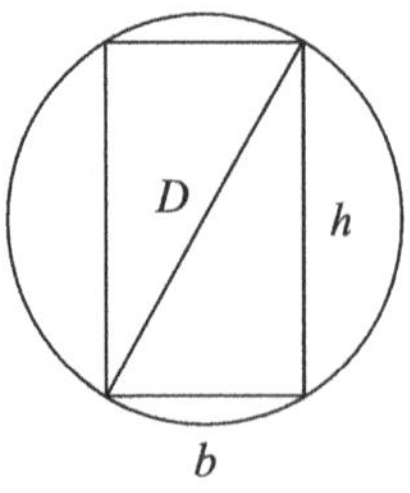

Fig. 7.6

For maximum Z,

$$\frac{dZ}{db} = 0 = \frac{d}{db}(bh^2) = \frac{d}{db}(b \times (D^2 - b^2))$$

which gives, $3b^2 = D^2$

or, $b = 800/\sqrt{3} = 462$ mm

and, $d = \sqrt{D^2 - b^2} = 653$ mm

Example 7.8 Determine the maximum stress in the unsymmetric *I* beam shown in Fig. 7.7.

Solution Taking moment of area about bottom fibre, the neutrual axis is,

$$\bar{y} = \frac{300 \times 50 \times 275 + 200 \times 50 \times 150 + 50 \times 200 \times 25}{300 \times 50 + 200 \times 50 + 50 \times 200}$$

$$= 167.86 \text{ mm}$$

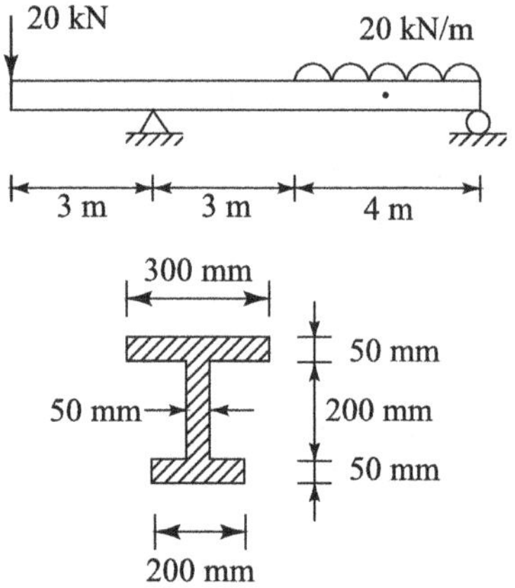

Fig. 7.7

Using parallel axis theorem,

$$I_{x,cg} = \frac{300 \times 50^3}{12} + 300 \times 50\,(275 - 167.86)^2 + \frac{50 \times 200^3}{12} + 50 \times 200\,(150 - 167.86)^2 + \frac{200 \times 50^3}{12} + 200 \times 50\,(167.86 - 25)^2$$

$$= 4.18 \times 10^8 \text{ mm}^4$$

The maximum negative bending moment is – 60 kN.m at left support and the corresponding stress are,

$$\text{at bottom fibre, } \sigma = \frac{My}{I} = \frac{60 \times 10^6 \times 167.86}{4.18 \times 10^8}$$

$$= 24.09 \text{ MPa (Compression)}$$

$$\text{at top fibre, } \sigma = \frac{60 \times 10^6 \times (300 - 167.86)}{4.18 \times 10^8}$$

$$= 18.97 \text{ MPa (Tension)}$$

The maximum positive bending moment is 58.98 kN m at 2.4285 m from right support and the corresponding stress are,

$$\text{at bottom fibre, } \sigma = \frac{58.98 \times 10^6 \times 167.86}{4.18 \times 10^8}$$

$$= 23.68 \text{ MPa (Tension)}$$

$$\text{at top fibre, } \sigma = \frac{58.98 \times 10^6 \times (300 - 167.86)}{4.18 \times 10^8}$$

$$= 18.64 \text{ MPa (Compression)}$$

Thus maximum tension is 23.68 MPa at 2.4285 m from right support and the maximum compression is 24.09 MPa at left support at bottom fibre in both cases.

Example 7.9 A simply supported steel beam of channel type cross-section is to carry loads as shown in Fig.7.8. Determine the maximum bending stresses in the beam.

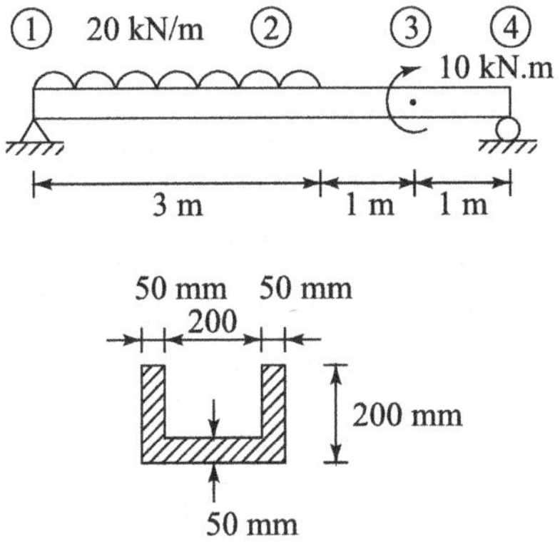

Fig. 7.8

Solution The beam is analysed using the computer program of Example 5.11. The input data is as follows.

```
NUMBER OF NODES = ? 4
INPUT X-COORDINATES OF EACH NODE-ZERO AT
LEFT END
  0.0
  3.0
  4.0
  5.0
NUMBER OF POINT LOAD = ?
  0
NUMBER OF EXTERNAL, MOMENTS = ?
  1
MODE NUMBER, EXTERNAL, MOMENT (ANTICLOCKWISE
+ve) = ?
  3    - 10.0
NUMBER OF DISTRIBUTED LOAD (DOWN +ve = ?
  1
SEGEMENT NUMBER, LOADLEFT, LOADRIGHT = ?
  1    20.0    20.0
NODE NUMBER OF SIMPLE SUPPORTS, N1, N2 =
  1    4
```

The output data is as follows:

```
 SUPPORT REACTION R1 = 40.0000000
 SUPPORT REACTION R2 = 20.0000000
SEGEMENT NO. = 1
 SHEAR LEFT = 40.000, SHEAR RIGHT = -20.00
 MOMENT LEFT = .000, MOMENT RIGHT = 30.000
SEGEMENT NO. = 2
 SHEAR LEFT = -20.000, SHEAR RIGHT -20.000
 MOMENT LEFT = 30.000, MOMENT RIGHT = 10.000
SEGEMENT NO. = 3
 SHEAR LEFT = -20.000, SHEAR RIGHT = -20.000
 MOMENT LEFT = 20.000, MOMENT RIGHT = .000
 SEGEMENT NO. MAX.BM. DIST.FROM LEFT
 1             40.000        2.000
```

The channel section is to carry maximum bending moment of 40 kN m at 2 m from left. The sectional properties of the channel section are calculated using C++ program of Problem 6.12. The input/output data is as follows

```
input Number of rectangles = 3
input for 1 rectangle
```

```
input x and y coordinates of right top
  corner 300 200
input x and y coordinates of left top corner
  250 0
input for 2 rectangle
input x and y coordinates of right top
  corner 50 200
input x and y coordinates of left bottom
  corner 0 0
input for 3 rectangle
input x and y coordinates of right top
  corner 250 50
input x and y coordinates of left bottom
  corner 50 0
area = 3000
  y = 75
  x = 150
  Qx = 2250000
  Qy = 4500000
Ix = 2.75e+08
Iy = 1.024999e+09
Ixcg = 1.0625e+08
Iycg = 3.499999e+08
Ixy = 0
Imax = 3.499999e+08
Imin = 1.0625e+08
rmax = 108.012337
rmin = 59.511902
theeta (in degrees) = 0
```

Substituting the values in Eq. (7.1), stress at the bottom fibre,

$$\sigma = \frac{My}{I} = \frac{40 \times 10^6 \times 75}{1.0625 \times 10^8}$$

$$= 28.23 \text{ MPa}$$

Similarly the bending stress at top fibre,

$$\sigma = \frac{40 \times 10^6 \times (200 - 75)}{1.0625 \times 10^8}$$

$$= 47.06 \text{ MPa}$$

Thus maximum bending stress = 47.06 MPa

The readers could join the two programmes and incorporate the bending stress formula to calculate the bending stresses at various points in a beam.

Example 7.10 A flitched beam shown in Fig. 7.9(a) consists of 150 × 300 mm timber section bonded with two 10 × 300 mm steel plates. Determine the stresses in timber and steel when the beam is subjected to a bending moment of 60 kNm if the ratio of modulus of elasticity of steel and timber is 20.

Solution Steel and timber will stretch together
strain in steel = strain in timber

or, $$\left(\frac{\sigma_b}{E}\right)_{\text{steel}} = \left(\frac{\sigma_b}{E}\right)_{\text{timber}}$$

or, $$\frac{\sigma_{\text{steel}}}{\sigma_{\text{timber}}} = \frac{E_{\text{steel}}}{E_{\text{timber}}} = 20$$

Steel is replaced by an equivalent timber section as shown in Fig. 7.9(b). The width of equivalent timber section is 20 times the width of steel section in a direction parallel to neutral axis since steel is 20 times more effective in resisting the applied load. The moment of inertia of the transformed section is,

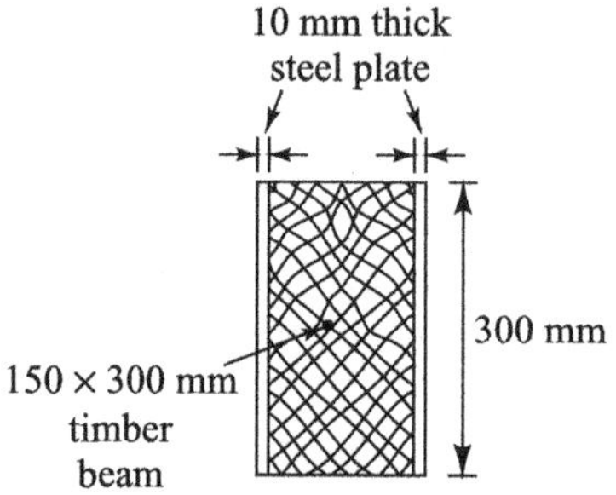

(a) Composite beam

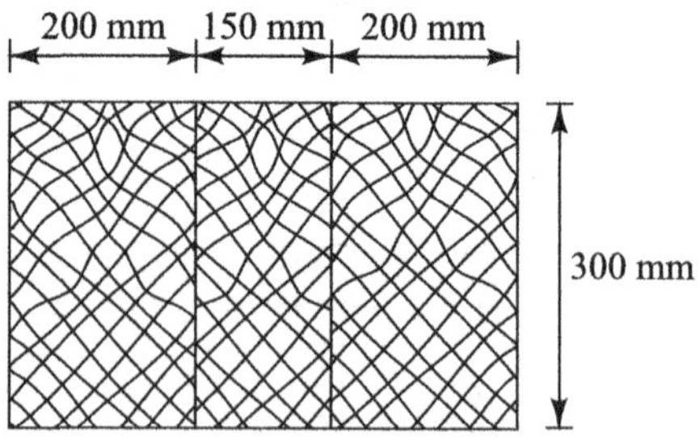

(b) Equivalent beam

Fig. 7.9

$$I = \frac{bd^2}{12} = \frac{550 \times 300^3}{12}$$
$$= 1.2375 \times 10^9 \text{ mm}^4$$

The maximum bending stress in the transformed section,

$$\sigma = \frac{My}{I} = \frac{60 \times 10^6 \times 150}{1.2375 \times 10^9}$$
$$= 7.273 \text{ MPa}$$

Thus stress in timber = 7.273 MPa

and stress in steel = 20 × 7.273 = 145.45 MPa

Example 7.11 A flitched beam is made of a timber joist 200 mm wide 250 mm deep with a 15 mm × 150 mm steel plate fixed in a groove as shown in Fig. 7.10 (a). Calculate the moment of resistance of the beam if the permissible bending stress in timber is 7.5 MN/m^2. Also calculate the maximum stress in steel if the ratio of modulii of steel and timber is 20.

Solution As in the previous example, the steel portion is replaced by equivalent timber portion in a direction parallel to neutral axis.

Width of beam at transformed section

$$= (200 - 15) + 20 \times 15 = 485 \text{ mm}$$

Second moment of area of the transformed section,

$$= 200 \times \frac{250^3}{12} + \frac{(485 - 200) \times 150^3}{12}$$
$$= 3.4 \times 10^8 \text{ mm}^4$$

Moment of resistance of the transformed section,

$$M = \sigma \times \frac{I}{y} = \frac{7.5 \times 3.4 \times 10^8}{(250/2)}$$
$$= 20.4 \text{ kN.m}$$

Maximum stress at 150/2 = 75 mm from neutral axis,

$$\sigma = \frac{My}{I} = \frac{20.4 \times 10^6 \times 75}{3.4 \times 10^8}$$
$$= 4.5 \text{ MPa}$$

Since steel has 20 times more stress, the maximum stress in steel = 4.5 × 20 = 90 MPa.

Alternate solution Second moment of area of steel,

$$I_s = \frac{15 \times 150^3}{12} = 4.22 \times 10^6 \text{ mm}^4$$

Second moment of area of timber,

$$I_t = \frac{200 \times 250^3}{12} - \frac{15 \times 150^3}{12}$$
$$= 256.2 \times 10^6 \text{ mm}^4$$

From strain diagram (Fig. 7.10(c)), maximum streain in steel and timber are related as follows.

$$\frac{\varepsilon_t}{\varepsilon_s} = \frac{125}{75} = \frac{5}{3}$$

or, $$\varepsilon_t = \frac{5}{3}\varepsilon_s$$

Stress in timber, $$\sigma_t = \varepsilon_t E_t = \frac{5}{3}\varepsilon_s E_t$$

$$= \frac{5}{3}\frac{E_t}{E_s}\sigma_s = \frac{\sigma_s}{12}$$

or, $$\sigma_s = 12\,\sigma_t$$

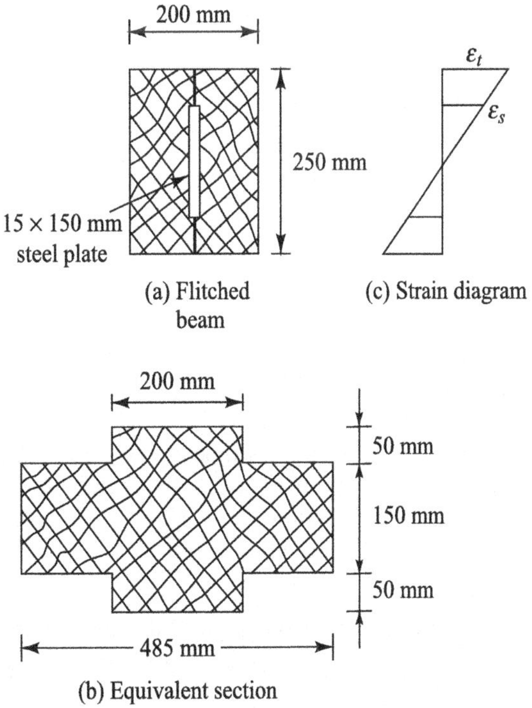

Fig. 7.10

Moment of resistance of the composite beam is the sum of moment shared by timber and steel as follows,

$$M = M_t + M_s = \frac{\sigma_t I_t}{125} + \frac{\sigma_s I_s}{75}$$

$$= \frac{7.5 \times 256.2 \times 10^6}{125} + \frac{12 \times 7.5 \times 4.22 \times 10^6}{75}$$

$$= 20.44 \times 10^6 \text{ N.mm} = 20.44 \text{ kN.m}$$

Example 7.12 Determine the load carrying capacity including self weight of the compound beam shown in Fig.7.11(a) over an effective span of 6 m. The permissible stress in steel and timber are 165 MPa and 17.2 MPa respectively and the modulii of elasticity of steel and timber are 200 GPa and 11.2 GPa. Also calculate the maximum stress in timber.

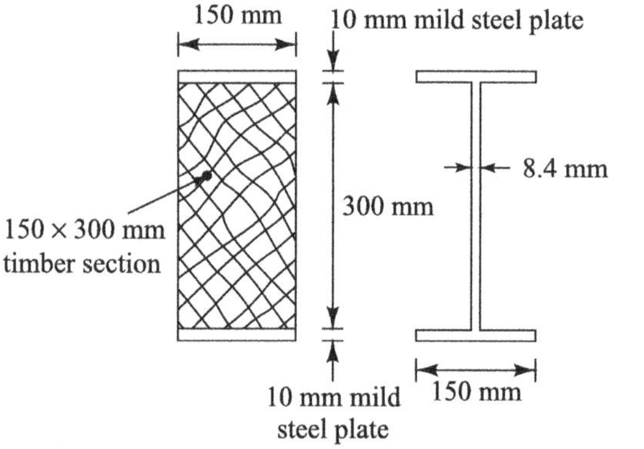

Fig. 7.11

Solution Transforming timber section into steel section as shown in Fig. 7.11(b), thickness of equivalent steel section

$$= \frac{150 \times 11.2}{200} = 8.4 \text{ mm}$$

Second moment of area of the transformed section,

$$I = \frac{150 \times 320^3}{12} - \frac{(150 - 8.4) \times 300^3}{12}$$

$$= 9.1 \times 10^7 \text{ mm}^4$$

Moment of resistance,

$$M = \frac{\sigma I}{y} = \frac{165 \times 9.1 \times 10^7}{160}$$

$$= 93.84 \text{ kN.m}$$

Equating moment of resistance and bending moment

$$\frac{wl^2}{8} = 93.84 \text{ kN.m}$$

udl the beam can carry,

$$w = \frac{93.84 \times 8}{6 \times 6} = 20.85 \text{ kN/m}$$

Maximum stress at 150 mm from neutral axis in the transformed steel section,

$$\sigma = \frac{My}{I} = \frac{93.8 \times 10^6 \times 150}{9.1 \times 10^7} = 154.68 \text{ MPa}$$

Maximum stress in timber

$$= 154.68 \times \frac{11.2}{200} = 8.66 \text{ MPa} \ngtr 17.2 \text{ MPa}$$

which is all right

(**Note:** Moment of resistance should be such that the stress in the timber section does not exceed the permissible value.)

Example 7.13 Plot shear stress distribution on a rectangular beam section of width B and depth D due to shear force V.

Solution The shear stress variation across the section is given by Eq. (7.2),

$$\tau = \frac{VQ}{IB}$$

where, Q = first moment of shaded area beyond distance y_0 of from neutral axis as shown in Fig. 7.12(a)

$$= B \times (D/2 - y_0) \times \left(y_0 + \frac{D/2 - y_0}{2} \right)$$

$$= \frac{B\,(D^2/4 - y_0^2)}{2}$$

and $I = BD^3/12$

$$\therefore \quad \tau = \frac{6V}{BD^3}(D^2/4 - y_0^2)$$

The plot of τ wrt y_0 is parabolic as shown in Fig. 7.12(b).

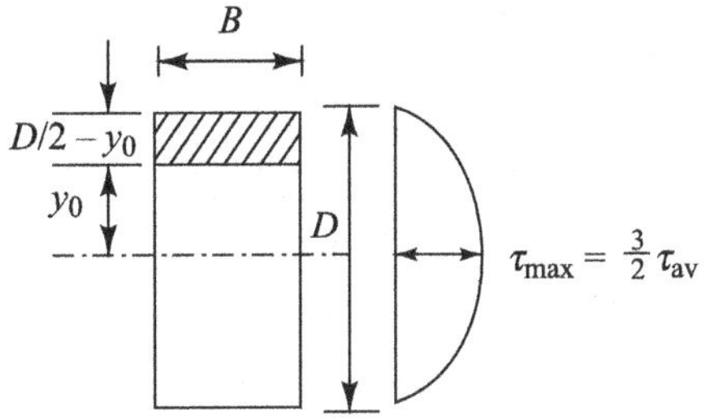

(a) Rectangle (b) Shear stress distribution section

Fig. 7.12

at $\quad y_0 = D/2, \quad \tau = 0$

at $\quad y_0 = 0, \quad \tau_{max} = \dfrac{1.5V}{BD} = 1.5\ \tau_{average}$

Example 7.14 Plot shear stress distribution on a circular beam section of radius R due to shear force V.

Solution The shear stress variation is given by Eq. (7.2).

$$\tau = \frac{V}{IB}A\bar{y}$$

where,
$$I = \frac{\pi R^4}{4}$$

and
$$A\,\bar{y} = \int_{y_0}^{R} y\,z\,dy$$

substitute,
$$z = 2\sqrt{R^2 - y^2}$$

or,
$$z^2 = 4(R^2 - y^2)$$

or
$$2z\,dz = -4 \times 2y\,dy$$

or
$$y\,dy = -\frac{1}{4}z\,dz$$

$\therefore$
$$\tau = \frac{V}{IB}\int_{y_0}^{R} yz\,dy$$
$$= \frac{V}{IB}\int_{B}^{0}\frac{-z^2}{4}dz$$

or
$$\tau = \frac{V}{4IB}\int_{0}^{B} z^2 dz = \frac{VB^2}{12I}$$

Substituting $(B/2)^2 = R^2 - y_0^2$

$\therefore$
$$\tau = \frac{V(R^2 - y_0^2)}{3I}$$
$$= \frac{4}{3}\frac{V(R^2 - y_0^2)}{\pi R^4}$$

The variation of τ wrt various values of y_0 is parabolic as shown in Fig.7.13(b) and the maximum value of shear stress,

at $\quad y_0 = 0, \quad \tau_{max} = \dfrac{4}{3}\dfrac{V}{\pi R^2} = \dfrac{4}{3}\tau_{average}$

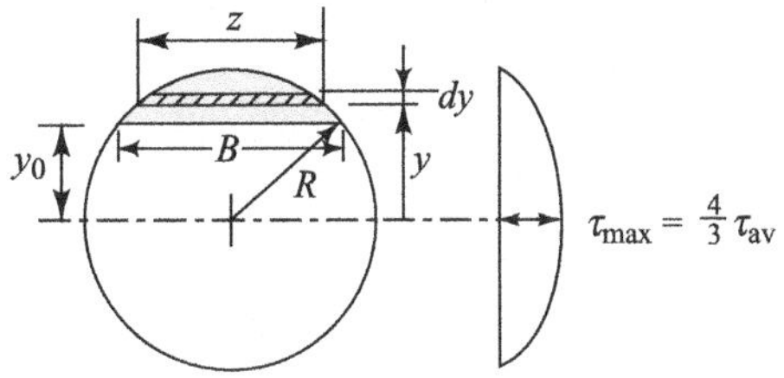

(a) Circular section (b) Shear stress distribution

Fig. 7.13

Example 7.15 A hollow circular cantiliver beam 4 m long has external diameter 200 mm and internal diameter 100 mm is to carry 20 kN at its free end. Plot bending and shearing stress near fixed end and also determine the ratio of maximum and average shear stress.

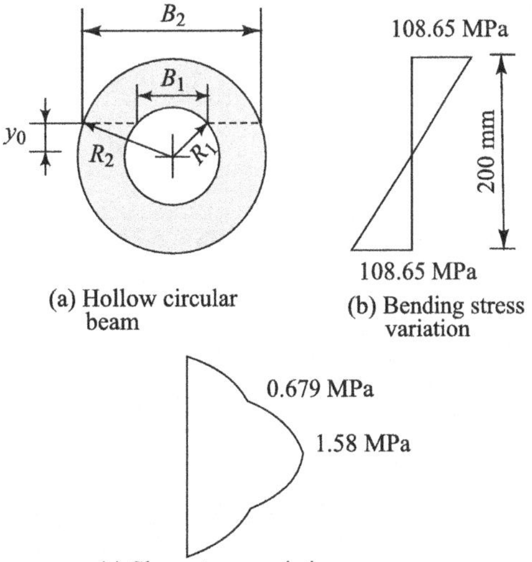

(a) Hollow circular beam (b) Bending stress variation

(c) Shear stress variation

Fig. 7.14

Solution

$$I = \frac{\pi}{4}(R_2^4 - R_1^4) = \frac{\pi(100^4 - 50^4)}{4}$$
$$= 73.63 \times 10^6 \text{ mm}^4$$

Bending stress variation is linear as per Eq. (7.1) and it is maximum at $y = 100$ mm as follows.

$$\sigma = \frac{My}{I} = \frac{(20 \times 4 \times 10^6) \times 100}{73.63 \times 10^6}$$
$$= 108.65 \text{ MPa}$$

Shear stress at $y_0 = 50$ mm is given by the expression derived in the previous example as follows,

$$\tau = \frac{V(R_2^2 - y_0^2)}{3I} = \frac{4}{3}\frac{V(R_2^2 - y_0^2)}{\pi(R_2^4 - R_1^4)}$$

$$= \frac{4}{3} \times \frac{20000(100^2 - 50^2)}{\pi(100^4 - 50^4)}$$

or, $\tau = 0.679$ MPa

Shear stress at $y_0 = 25$ mm,

$$\tau = \frac{VA\bar{y}}{IB}$$

or, $$\tau = \frac{V\left[\int_0^{B_2} z^2 dz - \int_0^{B_1} z^2 dz\right]}{4I(B_2 - B_1)}$$

$$= \frac{V(B_2^3 - B_1^3)}{12I(B_2 - B_1)} = \frac{V(B_2^2 + B_1^2 + B_1 B_2)}{12I}$$

where,

$$B_2 = 2\sqrt{R_2^2 - y_0^2} = 2\sqrt{100^2 - 25^2}$$
$$= 193.65 \text{ mm}$$

$$B_1 = 2\sqrt{R_1^2 - y_0^2} = 2\sqrt{50^2 - 25^2}$$
$$= 86.6 \text{ mm}$$

or, $$\tau = \frac{20000(193.65^2 + 86.6^2 + 193.65 \times 86.6)}{12 \times 73.63 \times 10^6}$$
$$= 14 \text{ MPa}$$

Shear stress at $y_0 = 0$,

$$\tau_{max} = \frac{20000(200^2 + 100^2 + 200 \times 100)}{12 \times 73.63 \times 10^6}$$
$$= 1.58 \text{ MPa}$$

Average shear stress,

$$\tau_{av} = \frac{V}{\pi(R_2^2 - R_1^2)} = \frac{20000}{\pi(100^2 - 50^2)}$$
$$= 0.85 \text{ MPa}$$

$$\frac{\tau_{max}}{\tau_{av}} = \frac{1.58}{0.85} = 1.86.$$

Example 7.16 An equilateral triangular beam of sides 300 mm and span 4 m is to carry a uniformly distributed load of 20 kN/m. Plot the variation of bending and shearing stress at a section where they are maximum.

Solution Maximum bending moment at mid span,

$$M = \frac{wl^2}{8} = \frac{20 \times 4^2}{8} = 40 \text{ kN.m}$$

Maximum shear force at supports,

$$V = \frac{wl}{2} = \frac{20 \times 4}{2} = 40 \text{ kN}$$

Moment of inertia

$$= \frac{BH^3}{36} = \frac{300 \times (300 \sin 60°)^3}{36}$$
$$= 146 \times 10^6 \text{ mm}^4$$

Location of neutral axis

$$= \frac{H}{3} = \frac{300 \sin 60}{3} = 86.6 \text{ mm}$$

Bending compressive stress,

$$\sigma_{bc} = \frac{My}{I} = \frac{40 \times 10^6 \times 86.6}{146 \times 10^6}$$
$$= 23.72 \text{ MPa}$$

Bending tensile stress,

$$\sigma_{bt} = \frac{40 \times 10^6 \times 173.2}{146 \times 10^6} = 47.45 \text{ MPa}$$

For variation of shear stress below neutral axis, consider the shaded area of width b, below neutral axis as shown in Fig.7.15(a). The shear stress is given by,

$$\tau = \frac{VA\bar{y}}{Ib}$$

where, $$A\,\bar{y} = \frac{bh}{2}\left(\frac{2H}{3} - \frac{2h}{3}\right)$$

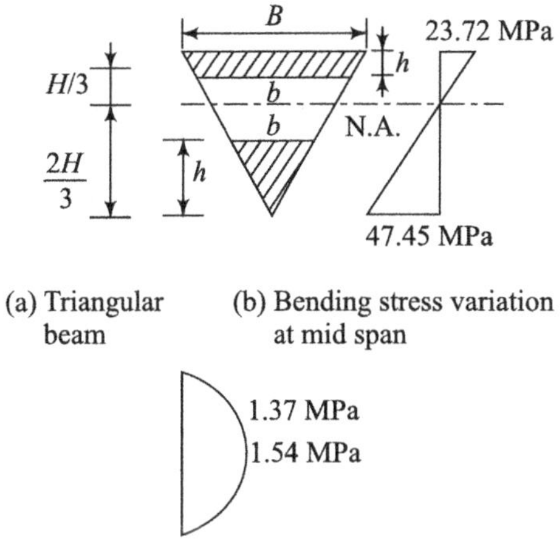

(a) Triangular beam (b) Bending stress variation at mid span

(c) Shear stress variation at end supports

Fig. 7.15

or, $$\tau = \frac{V(hH - h^2)}{3I}$$

The variation is parabolic and for maximum τ,

$$\frac{\partial \tau}{\partial h} = 0 = H - 2h$$

or $$h = H/2$$

At $h = 0$, $$\tau = 0$$

At $h = H/2$, $$\tau_{max} = \frac{V(H^2/2 - H^2/4)}{3BH^3/36}$$

$$= \frac{V}{BH/3} = \frac{3}{2}\tau_{average}\,.$$

$$= \frac{40000}{300 \times 259.8/3} = 1.54\text{ MPa}$$

At neutral axis,

$$\tau = \frac{V(2H^2/3 - 4H^2/9)}{3\,BH^3/36} = \frac{8V}{3BH}$$

$$= \frac{8 \times 40000}{3 \times 300 \times 259.8} = 1.37\ \text{MPa}$$

For variation of shearing stress above neutral axis consider the shaded area of width b as shown in Fig.7.15(a), for which b and $A\bar{y}$ are as follow.

$$b = \frac{B}{H}(H - h)$$

$$A\bar{y} = \frac{(B+b)}{2} \times h\left[\frac{H}{3} - \left(\frac{B+2b}{B+b}\right)\frac{h}{3}\right]$$

$$= \frac{h}{6}\left[B(H-h) + b(H-2h)\right]$$

$$= \frac{hB(H-h)^2}{3H}$$

or, $$\tau = \frac{VA\bar{y}}{Ib} = \frac{V\,hB(H-h)^2}{3H \times \dfrac{BH^3}{36} \times \dfrac{B}{H}(H-h)}$$

$$= \frac{12Vh(H-h)}{BH^3}$$

At top fiber, $h = 0$, $\tau = 0$

At neutral axis $h = H/3$, $\tau = \dfrac{8V}{3BH} = 1.37$ MPa

The linear bending stress variation at mid span and parabolic shear stress variation near end supports is shown in Fig. 7.15.

Example 7.17 Plot the shear stress distribution on an I-beam section of flange width B, overall depth D and thickness t as shown in Fig. 7.16, due to a shear force V.

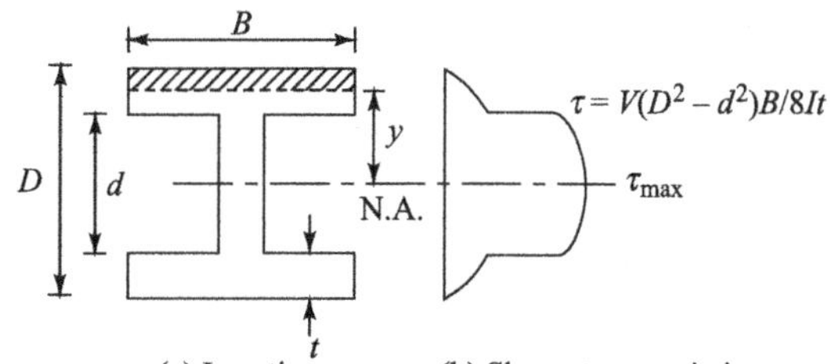

(a) I-section (b) Shear stress variation

Fig.7.16

Solution ***Shear stress in the flange*:** At a distance y from neutral axis the shear stress is given by,

$$\tau = \frac{VA\bar{y}}{IB} = \frac{V}{IB}B\left(\frac{D}{2}-y\right)\left(y+\frac{1}{2}\left(\frac{D}{2}-y\right)\right)$$

$$= \frac{V}{2I}\left(\frac{D^2}{4}-y^2\right)$$

This variation is parabolic and $\tau = 0$ at top/bottom fibre.

At the junction with web, $y = d/2$, and

$$\tau = \frac{V}{8I}(D^2 - d^2)$$

Shear stress in the web: At any distance y from neutral axis,

$$A\bar{y} = B\left(\frac{D}{2}-\frac{d}{2}\right)\left(\frac{d}{2}+\frac{1}{2}\left(\frac{D}{2}-\frac{d}{2}\right)\right)$$

$$+ t\left(\frac{d}{2}-y\right)\left(y+\frac{1}{2}\left(\frac{d}{2}-y\right)\right)$$

$$= \frac{B}{8}\left(D^2-d^2\right)+\frac{t}{8}\left(d^2-4y^2\right)$$

$$\tau = \frac{V}{It}A\bar{y} = \frac{V}{8It}\left(B(D^2-d^2)+t(d^2-4y^2)\right)$$

At the junction with the flange, $y = d/2$,

$$\tau = \frac{V}{8I}(D^2-d^2)\frac{B}{t}$$

Thus, at the junction, the shear stress suddenly jumps to B/t times the shear stress in the flange.

At neutral axis, $y = 0$ and,

$$\tau = \tau_{max} = \frac{V}{8I}\left[\frac{B}{t}(D^2-d^2)+d^2\right]$$

Example 7.18 Plot the shear stress intensity on a T-beam section shown in Fig. 7.17 if the shear force at the section is 80 kN.

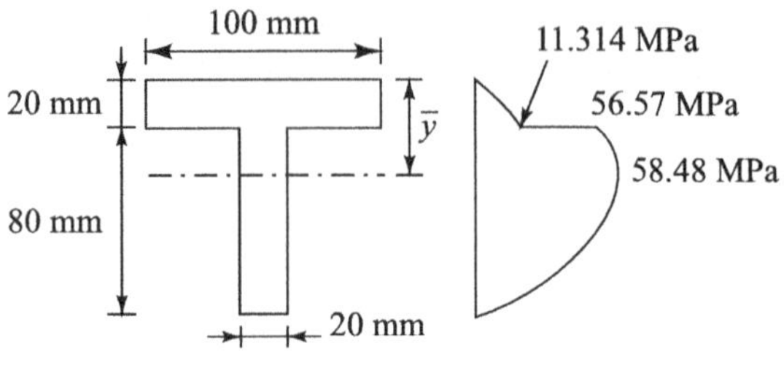

(a) T-section (b) Shear stress distribution

Fig. 7.17

Solution Location of neutral axis by taking moment of area about top.

$$\bar{y} = \frac{100\times20\times10+80\times20\times(40+20)}{100\times20+80\times20}$$

$$= 32.22 \text{ mm}$$

$$I_{cg} = \frac{100\times20^3}{12}+100\times20\times(32.22-10)^2$$

$$+\frac{20\times80^3}{12}+20\times80\times(60-32.22)^2$$

$$= 3142222 \text{ mm}^4$$

Shear stress in the flange at the junction,

$$\tau = \frac{V\,A\bar{y}}{IB} = \frac{80000\times100\times20\times22.22}{3142222\times100}$$

$$= 11.314 \text{ MPa}$$

Shear stress in the web at the junction,

$$\tau = \frac{11.314\times100}{20} = 56.57 \text{ MPa}$$

Maximum shear stress at neutral axis,

$$\tau_{max} = \frac{80000\times(100-32.22)^2\times20}{3142222\times20\times2}$$

$$= 58.48 \text{ MPa}$$

Example 7.19 Plot the shear stress distribution over a beam cross-section shown in Fig.7.18(a) due to a shear stress of 80 kN.

Solution

$$I_{cg} = \frac{20 \times 100^3}{12} + \frac{80 \times 20^3}{12}$$
$$= 1.72 \times 10^6 \text{ mm}^4$$

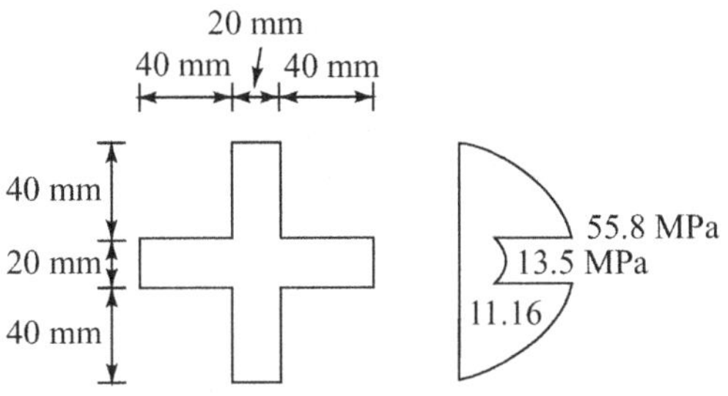

(a) Beam section (b) Shear stress variation

Fig. 7.18

For sections at a distance y from neutral axis such that 10 mm < y < 50 mm, the shear stress is given by,

$$\tau = \frac{VA\bar{y}}{IB}$$

$$= \frac{80000 \times 20 \times (50 - y)\left[y + \frac{1}{2}(50 - y)\right]}{1.72 \times 10^6 \times 20}$$

$$= 0.02325 \times (50^2 - y^2)$$

$$= 55.8 \text{ MPa at } y = 10 \text{ mm}$$

Similarly the variation of shear stress for y < 10 mm is given by,

$$\tau = \frac{80000 \times \left(24000 + 50(100 - y^2)\right)}{1.72 \times 10^6 \times 100}$$

$$= 11.16 \text{ MPa at } y = 10 \text{ mm}$$
$$= 13.5 \text{ MPa at } y = 0 \text{ mm}$$

Example 7.20 Plot the shear stress variation for a square beam cross-section of side B if one of its diagonal is horizontal as shown in Fig. 7.19.

Solution Depth of the beam, $D = \sqrt{2}\,B$

$$I_{cg} = 2 \times \frac{D}{12} \times \left(\frac{D}{2}\right)^3 = \frac{D^4}{48}$$

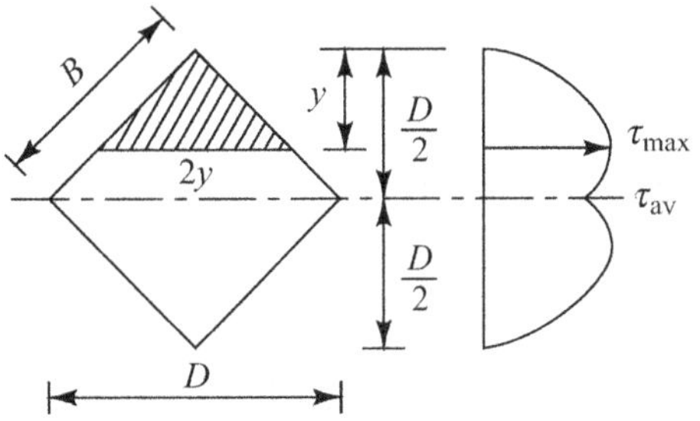

(a) Square bear section (b) Shear stress variation

Fig. 7.19

Shear stress at a depth y from the top

$$\tau = \frac{V\,A\bar{y}}{I \times 2y} = \frac{V\,y^2}{2y\,I}\left(\frac{D}{2} - \frac{2}{3}y\right)$$

$$= \frac{4Vy}{D^4}(3D - 4y)$$

This variation is parabolic as shown in Fig. 7.19(b). and $\tau = 0$, at $y = 0$.

At $y = \frac{D}{2}$, $\tau = \frac{2V}{D^2} = \tau_{average}$

For maximum shear stress,

$$\frac{d\tau}{dy} = 0$$

or $3D - 8y = 0$

which gives, $y = \frac{3}{8}D$

and,

$$\tau_{max} = \frac{4V}{D^4}\left(\frac{3}{8}D\right)\left(3D - \frac{12}{8}D\right)$$

$$= \frac{9}{8}\frac{V}{B^2}$$

Example 7.21 A symmetrical I-section 30 mm thick, 600 mm deep, has 200 mm wide flanges. Its top flange is reinforced with 400 mm × 40 mm cover plate as shown in Fig. 7.20. The cover plate and the flange of the I-section are joined by two rows of rivets which can resist a shear force of

35 kN per rivet. Determine the pitch of the rivets if the beam is to carry a maximum vertical shear force of 250 kN.

Solution Area of I-section = 2 × 200 × 30 + 540 × 30 = 28200 mm^2

For I-section,

$$I_{cg} = \frac{200 \times 600^3}{12} - \frac{(200-30)(600-60)^3}{12}$$

$$= 1.369 \times 10^9 \text{ mm}^4$$

Location of neutral axis of the built-up beam by taking moment of area about top,

$$\bar{y} = \frac{400 \times 40 \times 20 + 28200 \times (300 + 40)}{400 \times 40 + 28200}$$

$$= 224.16 \text{ mm}$$

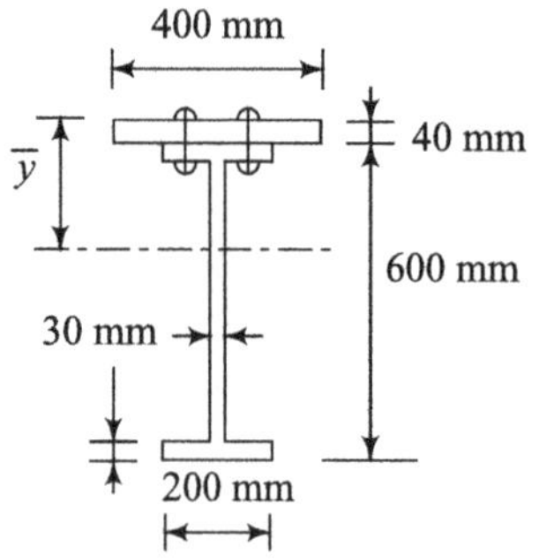

Fig. 7.20

For built-up beam,

$$I_{cg} = 1.369 \times 10^9 + 28200\,(340 - 224.16)^2$$

$$+ \frac{400 \times 40^3}{12} + 400 \times 40\,(224.16 - 20)^2$$

$$= 2.4 \times 10^9 \text{ mm}^4$$

The horizontal shear force per unit length of the beam is

$$\tau \times b \times 1 = \frac{V}{I} \times A\bar{y}$$

If, p = pitch of the rivet, horizontal shear force at the junction of cover plate and flange per pitch,

$$= \frac{V \times A\bar{y} \times p}{I}$$

$$= \frac{250000 \times 400 \times 40 \times (224.16 - 20)\, p}{2.4 \times 10^9}$$

$$= 340.27\, p$$

Equating horizontal shear force per pitch with the strength of 2 rivets per pitch,

$$2 \times 35000 = 340.24\, p$$

$$p = 205.7 \text{ mm (say 205 mm).}$$

Example 7.22 A beam 150 mm wide, 240 mm deep is to carry an uniformly distributed load of 20 kN per metre over an effective span of 8 m. Determine the principal stresses and their direction 20 mm above the neutral axis at mid span and 0.25 m from end support.

Solution $I_{cg} = \frac{150 \times 240^3}{12} = 172.8 \times 10^6 \text{ mm}^4$

At mid span:

Bending moment $= \frac{wl^2}{8} = \frac{20 \times 8^2}{8} = 160$ kN. m

Shear stress = 0

Bending stress at mid span 20 mm above neutral axis,

$$\sigma = \frac{My}{I} = \frac{160 \times 10^6 \times 20}{172.8 \times 10^6}$$

$$= 18.52 \text{ MPa (Compressive)}$$

It is the principal stress since no other stress is acting at this point.

At 0.25 m from end support.

Bending moment $= 80 \times 0.25 - \frac{20 \times 0.25^2}{2}$

$= 19.375$ kN. m

Shear force = 80 – 20 × 0.25 = 75 kN

Bending stress 20 mm above neutral axis,

$$\sigma = \frac{M \times y}{I} = \frac{19.375 \times 10^6 \times 20}{172.8 \times 10^6}$$

$$= 13.78 \text{ MPa (Compressive)}$$

Shear stress 20 mm above neutral axis,

$$\tau = \frac{V A\bar{y}}{I} = \frac{75\times10^3\times150\times100\times(20+50)}{172.8\times10^6}$$

$= 18.52$ MPa

Principal stresses,

$$\sigma_{\text{max/min}} = \frac{-13.78}{2} \pm \sqrt{\left(\frac{13.78}{2}\right)^2 + 18.52^2}$$

$= -26.65$ MPa, 12.86 MPa

Inclination of principal plane,

$$\theta = \frac{1}{2}\tan^{-1}\left(\frac{-18.52}{13.78/2}\right) = -34.8° \text{ w.r.t } \sigma$$

The principal stress in a simply supported beam carrying uniformly distributed load lies along the principal stress lines as shown in Fig. 7.21.

Concrete and masonry beams are weak in tension and therefore they sometimes develop diagonal tension cracks in the absence of proper tension reinforcement as the orientation of principal tension stress is diagonally upward near the end supports. In steel I-beam ripples develop perpendicular to the direction of principal compression stress near supports due to buckling.

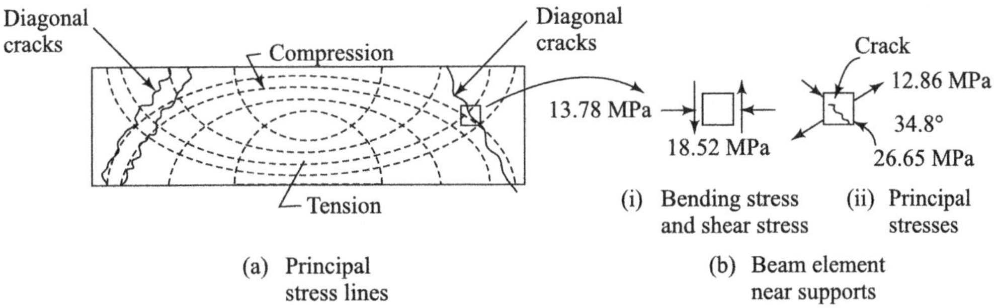

Fig. 7.21

EXERCISES

PROBLEMS

7.1 An I-beam 50 mm × 100 mm has thickness 5 mm as shown in Fig. 7.16. Determine the maximum bending moment which can be applied about the principal axes of the beam if the bending stress is not to exceed 150 MPa.

7.2 A 100 mm × 100 mm T beam has thickness 10 mm and length 5 m. It is simply supported over a span of 3 m and overhangs 1 m over both sides as shown in Fig. 7.22. If the maximum permissible bending stress is 165 MPa, determine the maximum uniformly distributed load that can be applied over the beam.

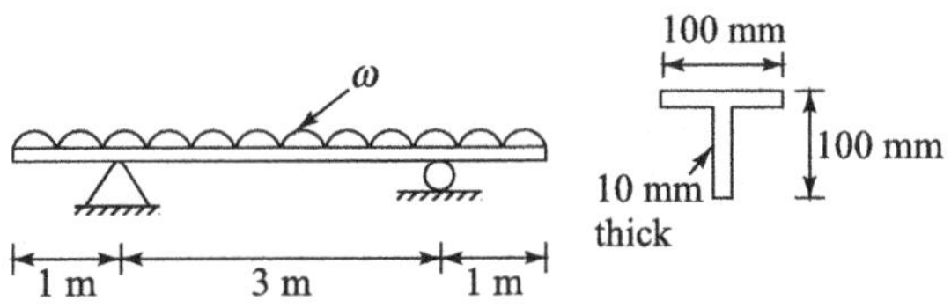

Fig. 7.22

7.3 Determine the maximum bending and shearing stress in the simply supported rectangular beam 100 mm wide, 200 mm deep carrying loads as in Fig. 5.29.

7.4 Figure 7.23 shows the cross-section of a cast iron beam. Find the greatest sagging moment that can be applied safely and the corresponding radius of curvature of the beam if the permissible bending stresses are 100 MPa in compression and 30 MPa in tension. Take $E = 200$ GPa.

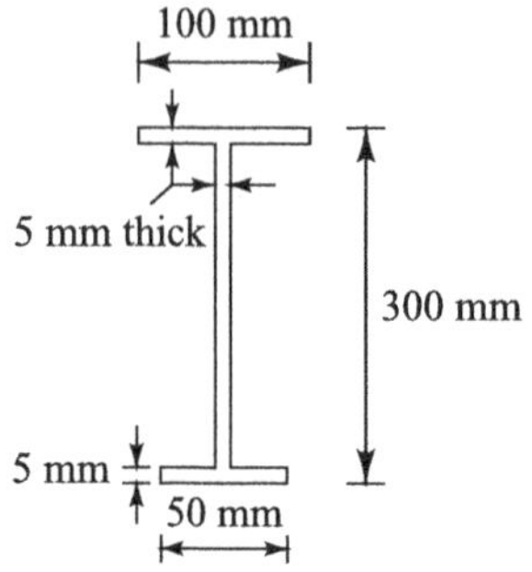

Fig. 7.23

7.5 A 300 mm square concrete beam is 6 m long. It is to be lifted horizontally by two slings. Determine the maximum bending and shearing stress due to self weight for that location of the slings which will cause minimum bending of the beam. The unit weight of concrete is 25 kN/m^3.

7.6 A simply supported I-beam has depth 500 mm and second moment of area 40000 cm^4 with flanges reinforced by 8 mm thick plates. The beam is to carry a uniformly distributed load 120 kN/m over a span of 5 m. If the allowable bending stress is 160 MPa, determine the width of reinforcing plates.

7.7 A cantilever timber beam 50 mm wide 100 mm deep is reinforced with 5 mm thick steel plates on top and bottom faces. A concentrated load of 3 kN is applied at the free end. Find maximum bending stress in steel and timber at 0.5 m from the free end if $E_s/E_t = 17$. Also find shear stress at the interface of steel and timber. If the steel plates are attached to timber by double row of 3 mm diameter nails for which maximum allowable shear stress is 100 MPa, find the maximum distance the nails should be fixed along the beam.

7.8 A timber beam 150 mm by 75 mm is strengthened by two 100 mm wide, 12 mm thick steel plates as shown in Fig. 7.24. If the ratio of Young's moduli of steel and timber $E_s/E_t = 17$ and the allowable bending stress for timber is 7 MPa, determine the maximum simply supported span for the beam to support a load of 4 kN/m.

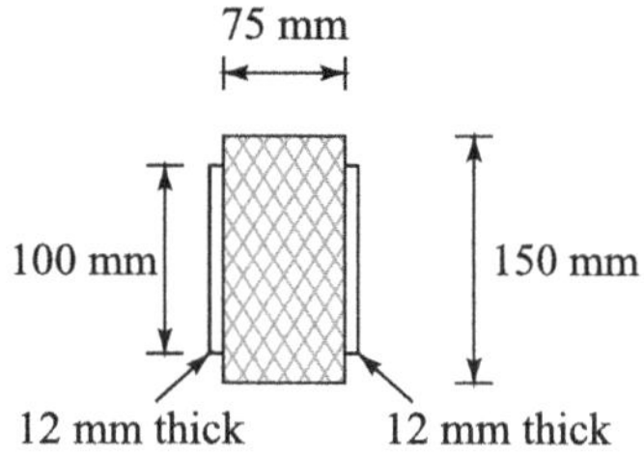

Fig. 7.24

7.9 A steel bar of diameter 100 mm is encased in a copper tube of outer diameter 150 mm. It is subjected to a bending moment of 15 kN.m. Determine the maximum stress in each material due to bending if the ratio of Young's moduli of steel and copper is 2.

7.10 A 200 × 200 mm T-section 10 mm thick bracket is to carry an inclined load of 5 kN as shown in Fig. 7.25. Determine the principal stresses at a distance 1 m from free end and 20 mm from bottom edge.

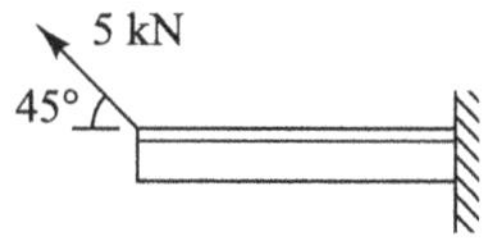

Fig. 7.25

7.11 A timber beam has a box section made by nailing together 50 mm thick planks as shown in Fig. 7.26. Determine the spacing of the nails if the beam is to withstand a shear force of 7 kN and the nails are capable of resisting 0.4 kN shear force.

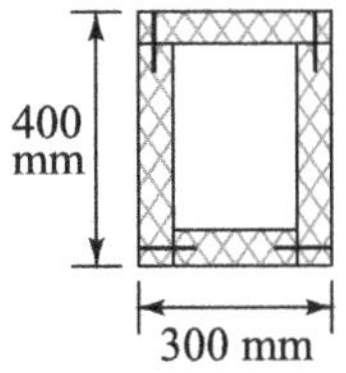

Fig. 7.26

OBJECTIVE QUESTIONS

7.12 The ratio of flexural strength of two beams of square cross-section, the first beam being placed with its top and bottom sides horizontally and the second beam being placed with one diagonal horizontally is,

(a) $\sqrt{3}$ (b) $1/\sqrt{3}$

(c) $1/\sqrt{2}$ (d) $\sqrt{2}$

7.13 For a square-section beam bent as shown in the Fig. 7.27, the exaggerated view of the deformed cross-section is,

Fig. 7.27

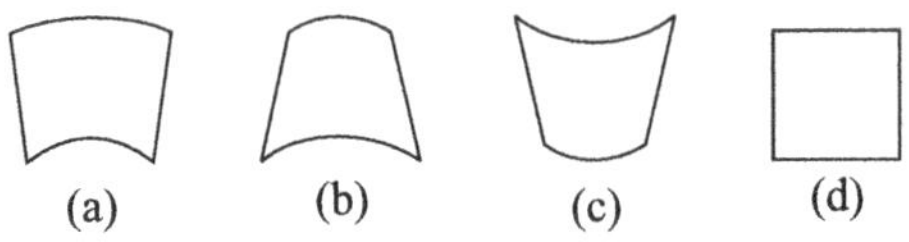

7.14 In a simply supported rectangular beam under uniformly distributed load, a hole has to be made in the breadthwise direction at mid span. From structural strength point of view, it would be advisable to have the hole made at,

(a) the bottom
(b) the top
(c) mid depth
(d) 1/4 depth either from the top or the bottom.

7.15 A flitched beam shown in Fig. 7.28, is subjected to a bending moment. The bending strain variation across the section will be as in,

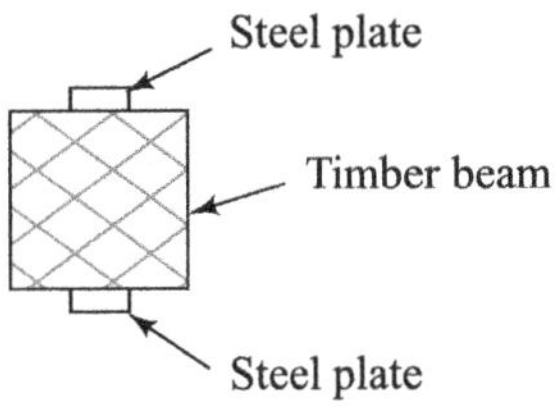

Fig. 7.28

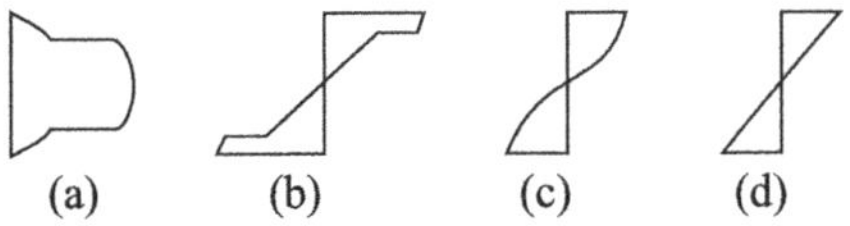

7.16 A simply supported beam of uniform strength along the span of length L, carries a point laod W at mid span. If the breadth b of the beam is constant and the permissible stress in bending is σ, then the depth of the beam at a distance x from the support will be,

(a) $\dfrac{6\,Wx}{\sigma b}$ (b) $\sqrt{\dfrac{6\,Wx}{\sigma b}}$

(c) $\dfrac{3\,Wx}{\sigma b}$ (d) $\sqrt{\dfrac{3\,Wx}{\sigma b}}$

7.17 Three smooth wooden battens are placed one above the other and are subjected to bending moment M as shown in Fig. 7.29. The bending stress distribution is given by,

Fig. 7.29

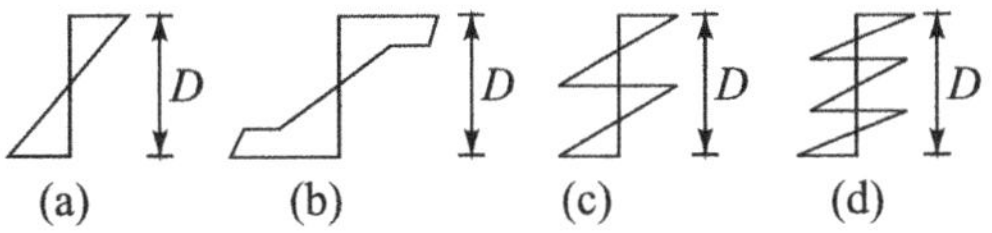

7.18 A T-section is used as a simply supported beam with uniformly distributed load. The maximum bending stress will occur at the,

(a) top of the flange
(b) centroid of the section

(c) mid point of the web
(d) extreme point in the web

7.19 A steel I-beam has over all depth 30 cm. If the bending stresses developed at the top and bottom of the beam are 120 MPa and 30 MPa, then the depth of the neutral axis from the top of the beam will be,

(a) 18 cm (b) 20 cm
(c) 24 cm (d) 25 cm

7.20 A wooden beam of width B and depth D is strengthened by two steel plates of the thickness t and depth D on both sides of the beam. The allowable bending stress is wood is σ and modular ratio of steel to wood is m. The allowable bending moment is,

(a) $\frac{\sigma}{6} D^2 (B + mt)$ (b) $\frac{\sigma}{6} D^2 (B + 2\,mt)$

(c) $\frac{\sigma}{6} D^2 (mB + 2\,t)$ (d) $\frac{\sigma}{6} D^2 (mD + t)$

7.21 A simply supported beam of span L carries a concentrated load W at its mid-span. If the width b of the beam is constant throughout the span, then, when the permissible bending stress is σ, the beam's mid-span depth will be,

(a) $\frac{3\,WL}{2\,b\sigma}$ (b) $\sqrt{\frac{3\,WL}{2\,b\sigma}}$

(c) $\frac{6\,WL}{b\sigma}$ (d) $\sqrt{\frac{6\,WL}{b\sigma}}$

7.22 A cantilever of constant depth carries a uniformly distributed load over the whole span. To make the maximum stress at all the sections same, the breadth of the section at a distance x from the free end should be proportional to,

(a) x (b) $\sqrt{x}$
(c) x^2 (d) x^3

7.23 The ratio of moment carrying capacity of a circular beam of diameter D and a square beam of side D is

(a) $3\pi/4$ (b) $2\pi/3$
(c) $3\pi/8$ (d) $3\pi/16$

7.24 Given that m = modular ratio E_s/E_t in a flitched beam, I_t = moment of inertia of the timber portion and I_s = moment of intertia of the steel portion, the equivalent moment of inertia is given by,

(a) $m\,I_t + I_s$ (b) $I_t/m + I_s$
(c) $I_s/m + I_t$ (d) $m\,I_s + I_t$

7.25 A beam of square cross-section is used to carry a shear force under a certain system of loading with one side of the square kept vertical. The beam is now rotated to make one of its diagonal vertical under the same system of loading. The ratio of maximum shear stress intensity per unit width at nautral axis over the cross-secion between former and the later beam is,

(a) 0.5 (b) 1.0
(c) 1.5 (d) 2.0

7.26 A beam has a triangular section, having base b and altitute h. If a section of the beam is subjected to a shear force F, the shear stress at the level of neutral axis is the cross-section is given by,

(a) $\frac{4}{3}\frac{F}{bh}$ (b) $\frac{3}{4}\frac{F}{bh}$

(c) $\frac{3}{8}\frac{F}{bh}$ (d) $\frac{8}{3}\frac{F}{bh}$

7.27 The ratio of maximum shear stress and average shear stress in a circular beam is equal to,

(a) 3/2 (b) 4/3
(c) 2 (d) $\pi/2$

8

Trusses

CONCEPT REVIEW

8.1 INTRODUCTION

Trusses are generally used for large spans. They are made by joining slender members by rivets or bolts or weld. The joints are considered to be pin joints which are capable of transmitting axial forces only. This assumption is not true and the truss members will have some distortion/bending due to the unaccounted bending/torsion which may be neglected. The truss member may not transmit these forces as they are slender. The pin joints in trusses were found to be unsatisfactory due to their cost and maintenance. An exact analysis is required when the forces are large and deflections cause serious problems. Trusses are broadly classified based on methods of their analysis as follows:

1. Determinate truss.
2. Indeterminate truss.
3. Unstable truss or mechanism.

Figure 8.1(a) shows an unstable truss *ABCD*. An additional member *BC* makes the structure stable as shown in Fig. 8.1(b). It is also called "Determinate truss" as it can be analysed by the equations of static equilibrium, i.e., $\Sigma M = 0$ and $\Sigma F = 0$. A determinate truss will have minimum constrains to render the structure stable. Any additional constrain like an additional member *AC* as in Fig. 8.1(c) will make the truss "Indeterminate". Indeterminate trusses are analysed using displacement equations which is kept beyond the scope of this book. A plane truss is determinate if Eq. (8.1) is statisfied.

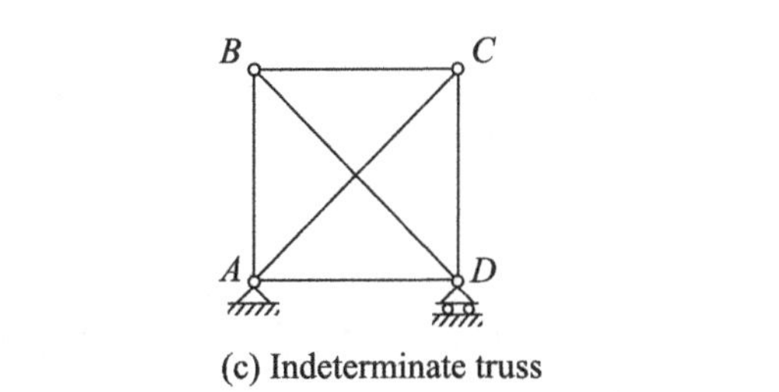

(a) Unstable frame

(b) Determinate truss

(c) Indeterminate truss

Fig. 8.1

$$m + r = 2j \qquad (8.1)$$

where, m = number of members
r = number of support reactions
j = number of joints

Each joint of a plane truss gives two equilibrium equation $\Sigma F_x = 0$ and $\Sigma F_y = 0$. The total number of unknowns are the forces in m number of members and r number of support reactions and it should not exceed $2j$ number of equations. A truss will be indeterminate if $m + r > 2j$ and a truss will be unstable if $m + r < 2j$.

8.2 ASSUMPTIONS

Following assumptions are made in analysing trusses,

1. All members are joined at ends by frictionless pin joints.
2. All loads are applied in the plane of the truss at the joints only.
3. Weight of the member can be neglected.
4. The centroid of various members coincide at the joint.
5. Bending and torsion are neglected.

8.3 METHODS OF ANALYSIS

These methods of analysis are used as follows:

(1) *Method of joints:* Two equations of equilibrium $\Sigma F_x = 0$ and $\Sigma F_y = 0$ are applied at each joint to determine unknown forces in two members at a joint. If more than two members are meeting at a joint, then equations from other joints will give the solution as explained in Example 8.4.

(2) *Method of sections:* This method is more convenient if it is required to determine forces in a few members only. A section is cut along the members where forces are required to be determined. The equations of equilibrium $\Sigma F_x = 0$, $\Sigma F_y = 0$, $\Sigma M = 0$ for the free body diagram will give the solutions as explained in Example. 8.2.

(3) *Graphical method:* Consider the joint 6 in Fig. 8.2(a). According to the principle of polygon of forces, if a number of forces acting at a joint are in equilibrium, then the vector diagram of the forces will be a closed polygon as shown in Fig. 8.3(b). The closing side of the force polygon is the force which the joint exerts on the members. This property is utilized in funicular diagram with a modification to determine the support reactions of a structure, since if the external forces are parallel, the closed polygon will not form. The procedure to form funicular diagram is explained in Example 8.7. Member forces can also be determined graphically by making force polygons. If force polygon is made for the entire structure at a time in one diagram, then it is known as Maxwell's diagram as explained in Ex. 8.8 and 8.9.

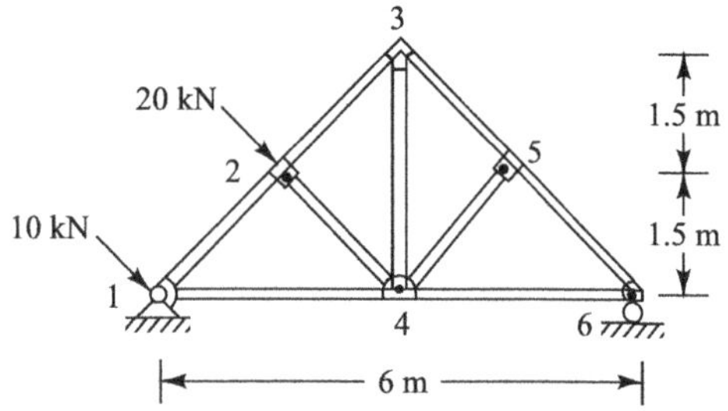

Fig. 8.2 *(a) Truss*

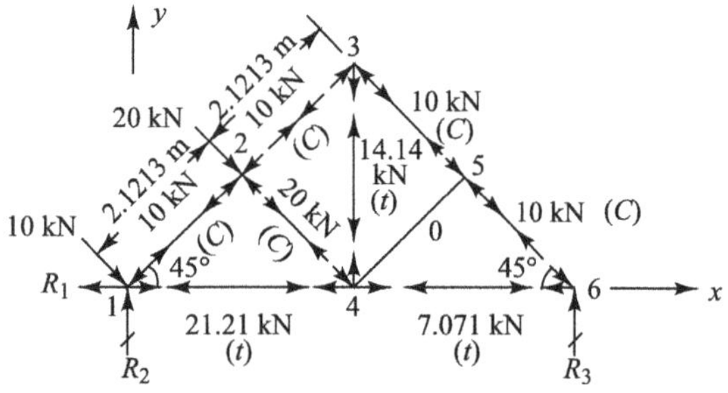

Fig. 8.2 *(b) Free Body Diagram*

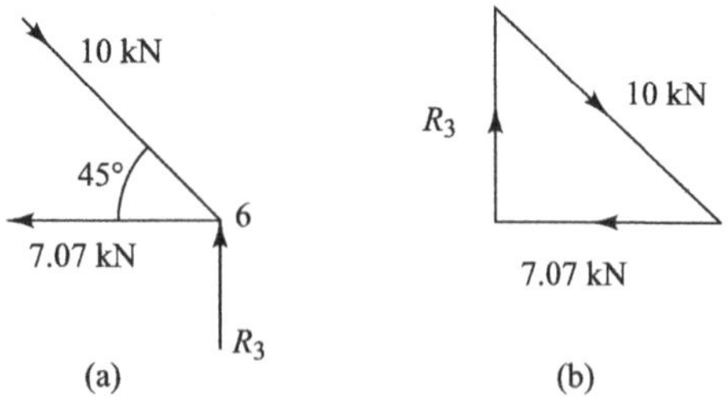

Fig. 8.3 *(a) Space Diagram (b) Force Diagram*

EXAMPLES

Example 8.1 Analyze the truss shown in Fig 8.2(a) by the method of joints, assuming all the joints are hinged.

Solution $m + r = 2j$

$9 + 3 = 2 \times 6$, the truss is determinate.

Support reactions: Taking moment of all forces about joint 1 and joint 6,

$\Sigma M_1 = 0, \quad R_3 \times 6 = 20 \times 2.1213$

or, $\quad R_3 = 7.071$ kN

$\Sigma M_6 = 0, \quad R_2 \times 6 = 20 \times 2.1213 + 10 \times 4.2426$

or, $\quad R_2 = 14.142$ kN

Equating horizontal component of all forces with R_1,

$\Sigma F_x = 0, \quad R_1 = 30 \times \sin 45°$

$= 21.2132$ kN

Member forces: Apply equilibrium equations at all joints by considering the free body diagram of Fig. 8.2(b). It should be seen that the members in tension (bottom chord) tend to pull their end nodes axially inward and compression members push their end nodes outward. Since no external axial force is acting on any members, the two end forces will be equal.

Joint 6

$\Sigma F_y = 0, \quad F_{65} \sin 45° = 7.071$ kN

or, $\quad F_{65} = 10$ kN

$\Sigma F_x = 0, \quad F_{65} \cos 45° = F_{64}$

or, $\quad F_{64} = 7.071$ kN

Joint 5: Member 45 is zero force member since members 35 and 56 cannot have any force component in its perpendicular direction.

$\therefore \quad F_{35} = F_{56} \quad$ or, $F_{35} = 10$ kN

Joint 3

$\Sigma F_x = 0, \quad F_{32} \cos 45° = F_{35} \cos 45°$

or, $\quad F_{32} = 10$ kN

$\Sigma F_y = 0, \quad F_{34} = F_{35} \sin 45° + F_{32} \sin 45°$

or, $\quad F_{34} = 14.1421$ kN

Joint 2: Again by the same logic as for joint 5.

$F_{24} = 20$ kN

and, $\quad F_{21} = F_{23} = 10$ kN

Joint 1

$\Sigma F_x = 0, \quad F_{14} = F_{12} \cos 45° + R_1 - 10 \cos 45°$

$= 21.2132$ kN

Example 8.2 Determine the forces in the members $p-q$, $r-t$ and $s-t$ of the truss shown in Fig. 8.4(a).

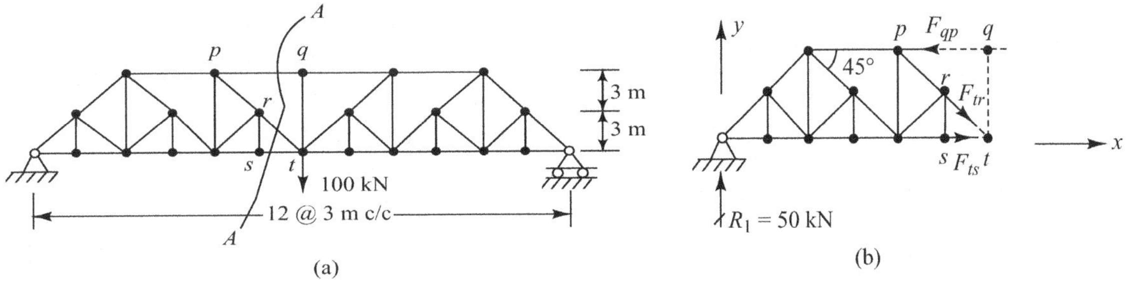

Fig. 8.4

Solution Method of section is advantageous in such cases. By symmetry, $R_1 = R_2 = 100/2 = 50$ kN. Consider the equilibrium of the free body diagram by taking the section along $A - A$ as shown in Fig. 8.4(b).

$\Sigma F_x = 0, \qquad F_{ts} + F_{tr} \cos 45° = F_{qp}$

$\Sigma F_y = 0, \qquad F_{tr} \sin 45° = 50$ kN

or, $\qquad F_{tr} = 70.7$ kN

Taking moment about joint P,

$\Sigma M_p = 0, \qquad F_{ts} \times 6 = 50 \times 12,$

or, $\qquad F_{ts} = 100$ kN

$\therefore \qquad F_{qp} = 100 + 70.7 \cos 45°,$

or, $\qquad F_{qp} = 150$ kN

Example 8.3 Analyse the truss shown in Fig. 8.5(a) by the method of joints. Also find force in member 35 using method of section.

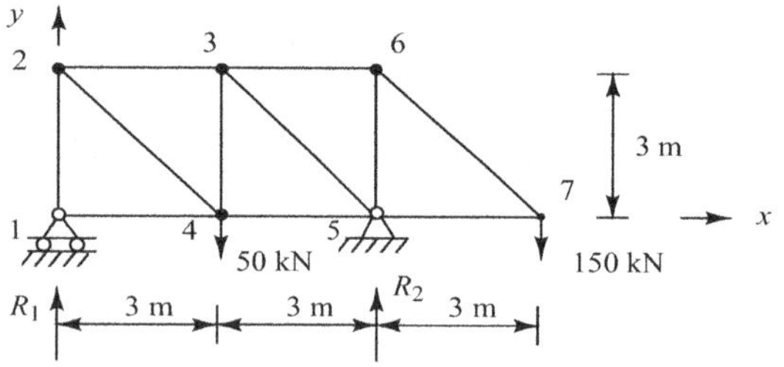

(a)

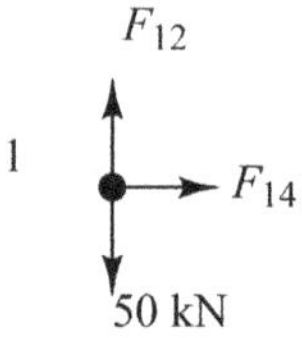

(b) Joint 1

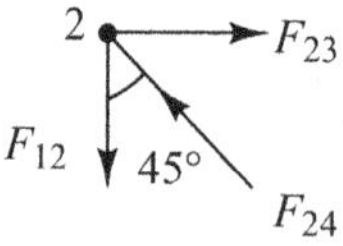

(c) Joint 2

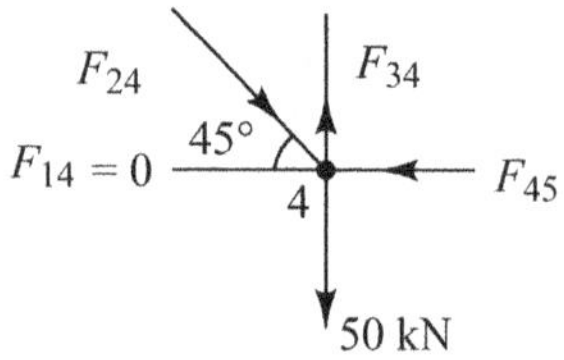

(d) Joint 4

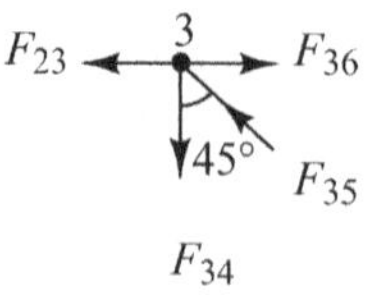

(e) Joint 4

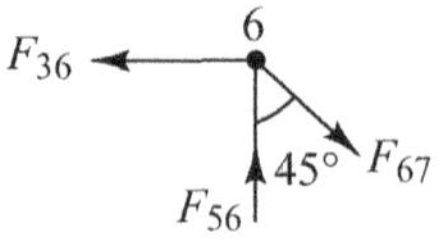

(f) Joint 6

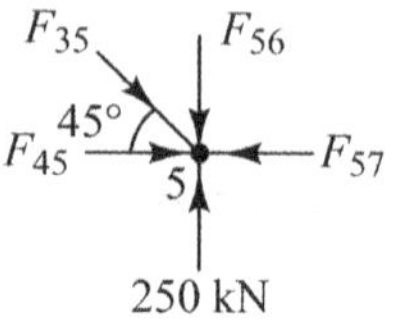

(g) Joint 5

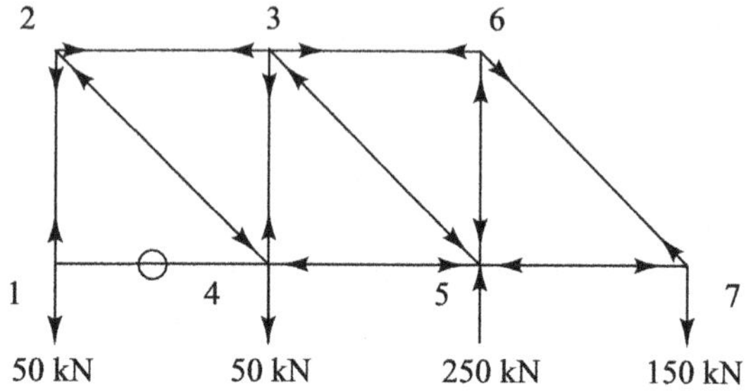

(h)

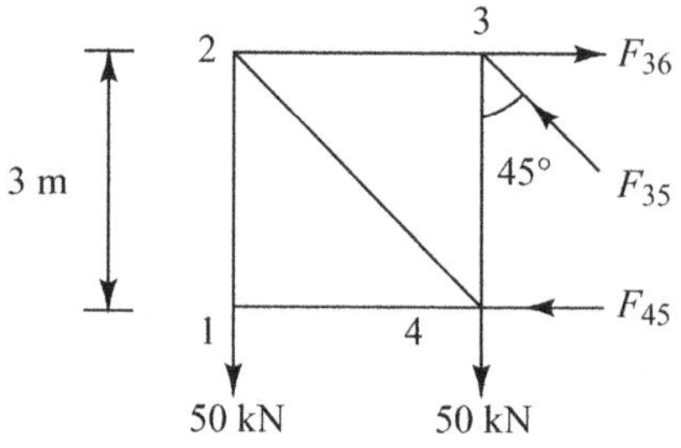

(i)

Fig. 8.5 (a)–(i)

Solution $m + r = 11 + 3 = 14 = 2j$, the truss is determinate.

For support reactions, R_1 and R_2 consider equilibrium equations as follows.

$\Sigma F_y = 0, \qquad R_1 + R_2 = 50 + 150 = 200$ kN

Taking moment about reaction R_1,

$R_2 \times 6 = 150 \times 9 + 50 \times 3$

or, $\qquad R_2 = 250$ kN

and $\qquad R_1 = -50$ kN $\qquad$ (–ve sign indicate R_1 is downward)

Joint 1: Consider free body diagram of joint 1 in Fig. 8.5(b). Direction of F_{12} is assumed upward since the downward 50 kN should be resisted by upward force.

$\Sigma F_x = 0, \quad F_{14} = 0$

$\Sigma F_y = 0, \quad F_{12} = 50$ kN

Mark the direction of F_{12} at joint 1 in Fig. 8.5(h). Also mark F_{12} of opposite direction at joint 2.

Joint 2: Direction of F_{12} was obtained from joint 1. Direction of F_{24} is assumed upward to counter downward force F_{12} at joint 2. If the assumed direction is wrong then it will come out –ve mathematically. Similarly assume directions of all the forces in Fig. 8.5(h). Apply equilibrium equation at joint 2 as follows (Fig. 8.5(c)).

$\Sigma F_y = 0, \quad F_{12} = F_{24} \cos 45°$

$F_{24} = 50\sqrt{2}$ kN

$\Sigma F_x = 0, \quad F_{24} \sin 45° = F_{23}$

or $\quad F_{23} = 50$ kN

Joint 4: Consider equilibrium of joint 4 in Fig. 8.5(d),

$\Sigma F_x = 0, \quad F_{45} = F_{24} \cos 45°$

or $\quad F_{45} = \dfrac{50\sqrt{2}}{\sqrt{2}} = 50$ kN

$\Sigma F_y = 0, \quad F_{34} = F_{24} \sin 45° + 50 = 100$ kN

Joint 3: Consider equilibrium of joint 3 in Fig. 8.5(e),

$\Sigma F_y = 0, \quad F_{34} = F_{35} \cos 45°$

$F_{35} = 100\sqrt{2}$ kN

$\Sigma F_x = 0, \quad F_{36} = F_{35} \sin 45° + F_{23}$

$= \dfrac{100\sqrt{2}}{\sqrt{2}} + 50$

or $\quad F_{36} = 150$ kN

Joint 6: Consider equilibrium of joint 6 in Fig. 8.5(f),

$\Sigma F_x = 0, \quad F_{36} = F_{67} \sin 45°$

or, $\quad F_{67} = \dfrac{F_{36}}{\sin 45°} = 150\sqrt{2}$

$\Sigma F_y = 0, \quad F_{56} = F_{67} \cos 45° = \dfrac{150\sqrt{2}}{\sqrt{2}}$

$= 150$ kN

Joint 5: Consider equilibrium of joint 5 in Fig. 8.5(g),

$\Sigma F_x = 0, \quad F_{57} = F_{45} + F35 \cos 45°$

$= 50 + \dfrac{100\sqrt{2}}{\sqrt{2}}$

or, $\quad F_{57} = 150$ kN

The forces in all members are as follows. The arrow in Fig. 8.5(h) show the force a member exerts on a joint. Thus a force indicates tension in the member.

Member	*Force*	
14	0	
12	50 kN	Tension
23	50 kN	Tension
24	$50\sqrt{2}$ kN	Compression
34	100 kN	Tension
45	50 kN	Compression
35	$100\sqrt{2}$ kN	Compression
36	150 kN	Tension
56	150 kN	Compression
57	150 kN	Compression
67	$150\sqrt{2}$ kN	Tension

Method of section: Consider the free body diagram in Fig. 8.5(i) and take the moment of forces about joint 3 as follows.

$\Sigma M = 0, \quad F_{45} \times 3 = 50 \times 3$

or, $\quad F_{45} = 50$ kN

Similarly

$\Sigma F_y = 0, \quad F_{35} \cos 45° = 50 + 50 = 100$ kN

or $\quad F_{35} = 100\sqrt{2}$ kN

Also apply

$\Sigma F_x = 0, \quad F_{36} = F_{45} + F_{35} \sin 45° = 150$ kN

Example 8.4 Analyse the truss shown in Fig. 8.6.

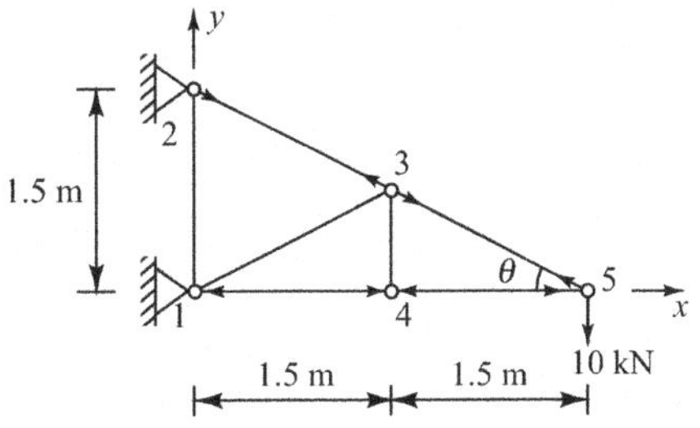

Fig. 8.6

Solution $m + r = 7 + 4 = 11 \neq 2j$

Apparently the truss is indeterminate, but member 1 – 2 is zero force member as it is between fixed supports and hence it is a determinate truss.

Member 34 is also a zero force member as it has no other force in its direction at joint 4.

Similarly member 13 is also a zero force member as member 23 and 35 are in one line and there is no other force across them except 13.

$$\theta = \tan^{-1}\frac{1.5}{3} = 26.565°$$

Joint 5: Consider equilibrium at joint 5 as follows.

$$\Sigma F_y = 0, \quad F_{35}\sin\theta = 10$$

$$F_{35} = \frac{10}{\sin 26.565°} = 22.36 \text{ kN}$$

$$\Sigma F_x = 0, \quad F_{45} = F_{35}\cos\theta$$

$$= 22.36 \times \cos 26.565°$$

or, $$F_{45} = 20 \text{ kN}$$

Equilibrium of joints 3 and 4 gives,

$$F_{23} = F_{35} = 22.36 \text{ kN}$$

$$F_{14} = F_{45} = 20 \text{ kN}$$

These forces are in a line and there is no other force in their direction.

Example 8.5 Analyse the truss shown in Fig. 8.7.

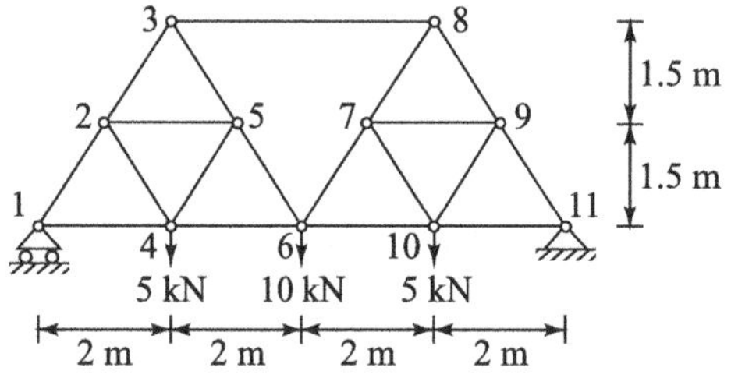

Fig. 8.7(a)

Solution $m + r = 19 + 3 = 22 = 2j$, determinate truss

For support reaction, consider equilibrium equations as follows.

$$\Sigma F_y = 0, \quad R_1 + R_2 = 5 + 10 + 5 = 20 \text{ kN}$$

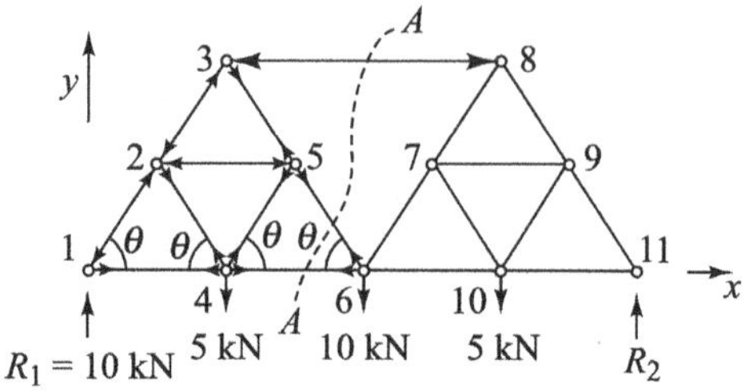

Fig. 8.7(b)

Taking moment about R_2,

$$\Sigma M = 0, \quad R_1 \times 8 = 5 \times 6 + 10 \times 4 + 5 \times 2$$

or, $$R_1 = 10 \text{ kN} = R_2$$

$$\theta = \tan^{-1} 1.5/1 = 56.31°$$

Joint 1: Consider equilibrium at joint 1 as follows.

$$\Sigma F_y = 0, \quad F_{12}\sin\theta = 10$$

or, $$F_{12} = \frac{10}{\sin 56.31°} = 12 \text{ kN}$$

$$\Sigma F_x = 0, \quad F_{14} = F_{12}\cos\theta$$

or, $$F_{14} = 12 \times \cos 56.31°$$

$$= 6.67 \text{ kN}$$

Both joints 2 and 4 has more than 3 unknown forces. Method of sections can be used to find forces in 38, 56 and 46 by taking section along $A - A$ as shown in Fig. 8.7(b). Consider the equilibrium of left side free body of truss in Fig. 8.7(c) as follows,

$$\Sigma F_x = 0, \quad F_{38} = F_{56}\cos\theta + F_{46}$$

or, $$F_{38} = 0.5547\, F_{56} + F_{46} \quad \text{(i)}$$

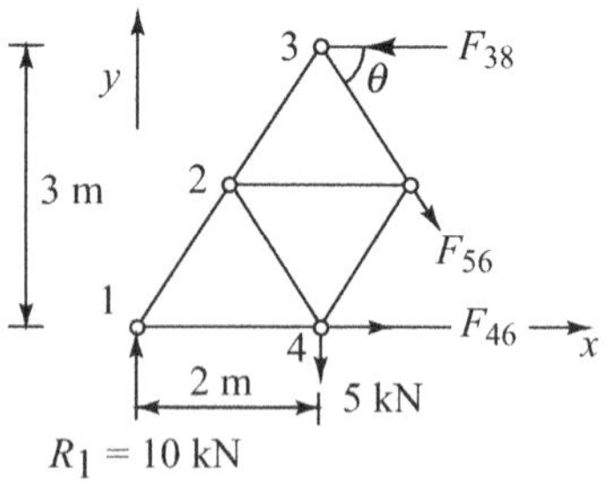

Fig. 8.7(c)

$\Sigma F_y = 0, \quad F_{56} \sin \theta = R_1 - 5$

or, $$F_{56} = \frac{5}{\sin 56.31°} = 6 \text{ kN}$$

Taking moment of forces about joint 3,

$\Sigma M = 0, \quad F_{46} \times 3 = 10 \times 2$

or, $F_{46} = 6.67$ kN

From Eq. (i), $F_{38} = 10$ kN

Joint 3: Fig. 8.7(d)

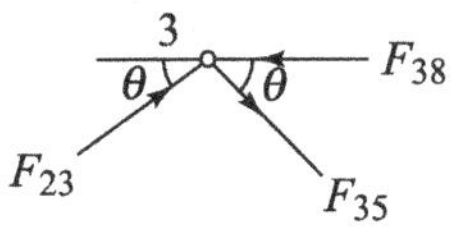

Fig. 8.7(d)

$\Sigma F_y = 0, \quad F_{23} \sin \theta = F_{35} \sin \theta$

or, $F_{23} = F_{35}$ (ii)

$\Sigma F_x = 0, \quad F_{23} \cos \theta + F_{35} \cos \theta = F_{38}$

or, $$F_{23} + F_{35} = \frac{10}{\cos 56.31°} = 18 \quad \text{(iii)}$$

From Eqs (ii) and (iii),

$F_{23} = F_{35} = 9$ kN

Joint 2: Fig. 8.7(e),

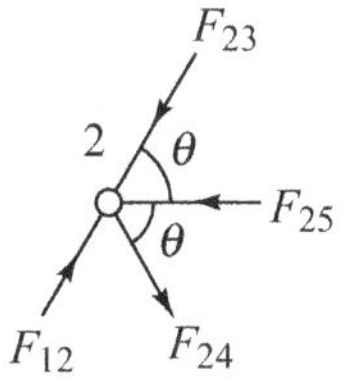

Fig. 8.7(e)

$\Sigma F_x = 0, F_{12} \cos \theta + F_{24} \cos \theta = F_{23} \cos \theta + F_{25}$

or, $$\frac{F_{25}}{\cos \theta} - F_{24} = F_{12} - F_{23} = 12 - 9 = 3$$

or, $1.8\, F_{25} - F_{24} = 3$ (iv)

$\Sigma F_y = 0, \quad F_{12} \sin \theta = F_{23} \sin \theta + F_{24} \sin \theta$

or, $F_{24} = F_{12} - F_{23}$

$= 12 - 9 = 3$ kN

Substitute in Eq. (iv)

$$F_{25} = \frac{3-3}{1.8} = 0$$

Joint 4: Fig. 8.7(f),

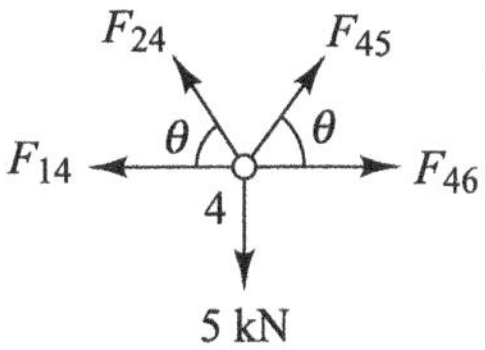

Fig. 8.7(f)

$\Sigma F_x = 0, \quad F_{14} + F_{24} \cos \theta = F_{46} + F_{45} \cos \theta$

or, $F_{24} \cos \theta - F_{45} \cos \theta = F_{46} - F_{14}$

$= 6.67 - 6.67 = 0$

or, $F_{24} = F_{45} = 3$ kN

$\Sigma F_y = 0, \quad F_{24} \sin \theta + F_{45} \sin \theta = 5$

or, $2 \times 3 \times \sin 56.13 = 5$ kN

Hence the calculations are verified. The forces in left half of the truss are tabulated as follows. The forces in other half will be symmetrical.

Member	*Force*	
12	12 kN	Compression
14	6.67 kN	Tension
38	10 kN	Compression
23	9 kN	Compression
35	9 kN	Tension
25	0	
24	3 kN	Tension
45	3 kN	Tension
56	6 kN	Tension
46	6.67 kN	Tension

Example 8.6 Solve the truss shown in Fig. 8.8(a).

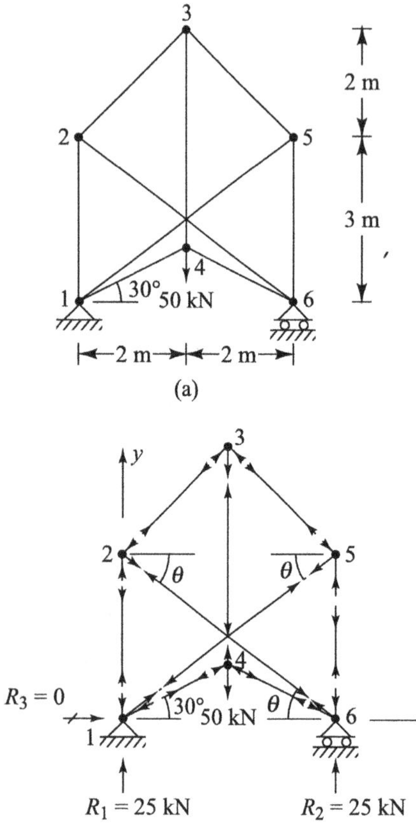

Fig. 8.8

Solution $m + r = 2j$

or, $9 + 3 = 2 \times 3$ determinate

Support reactions: By symmetry, $R_1 = R_2 = 25$ kN and since there is no external horizontal force, $R_3 = 0$.

Member forces: Applying equilibrium equations at the joints using the free body diagram in Fig. 8.8(b).

Joint 6: Since there are three unknown member forces and two equilibrium equations, assume member force $F_{26} = T$ kN (tension).

$\Sigma F_x = 0,\quad F_{64} \cos 30° = T \cos \theta$

or, $F_{64} = 0.924T$

$\Sigma F_y = 0,\quad F_{56} + F_{64} \sin 30° = T \sin \theta + 25$

or, $F_{56} = 0.138T + 25$

Joint 5

$\Sigma F_x = 0,\quad F_{51} \cos \theta = F_{53} \cos 45°$

or, $F_{51} = 0.884\, F_{53}$

$\Sigma F_y = 0,\quad F_{53} \sin 45° + F_{51} \sin \theta = F_{56}$

or, $F_{53} = 0.111T + 20.2$

Joint 3

$\Sigma F_x = 0,\quad F_{35} \cos 45° = F_{32} \cos 45°$

or, $F_{32} = 0.111T + 20.2$

$\Sigma F_y = 0,\quad F_{34} = F_{35} \sin 45° + F_{32} \sin 45°$

$= 0.157T + 28.6$

Joint 4

$\Sigma F_x = 0,\quad F_{46} \cos 30° = F_{41} \cos 30°$

or, $F_{41} = 0.924T$

$\Sigma F_y = 0,\quad F_{34} + F_{46} \sin 30° + F_{41} \sin 30° = 50$

or, $0.157T + 28.6 + 2 \times 0.924T \sin 30°$

$= 50$

or, $T = 19.8$ kN

Substituting back the value of T, we get

$F_{26} = 19.8$ kN (tension)

$F_{64} = 18.3$ kN (compression)

$F_{56} = 27.73$ kN (compression)

$F_{53} = 22.4$ kN (compression)

$F_{51} = 19.8$ kN (tension)

$F_{34} = 31.7$ kN (tension)

Since the truss is symmetric, the other half of the truss will have symmetric member forces.

Example 8.7 Determine graphically the end reactions for the simply supported beam carrying 40 kN concentrated load inclined at 30° at the middle of a 5 m span.

Solution *Bow's notation*: Spaces between the forces are named like *A*, *B* and *C* in the space diagram (Fig. 8.9 (a))

Force diagram: Force *AB* is drawn in the force diagram (Fig. 8.9(b)) to some scale = 40 kN. Any point *O* is chosen as pole and joined with *A* and *B*. Forces between spaces *A*, *B* and *C* in space diagram are represented by the line *AB*, *BC* and *CA*, respectively in the force diagram.

Funicular polygon: It is possible to draw Funicular polygon if the location of one of the reactions and the line of action of the other is known. Start from the reaction of known location, i.e. R_A in Fig. 8.9(a). Draw a line *oa* parallel to *OA*, to intersect line of action of 40 kN force at *p*. Draw *ob* parallel to *OB*, to intersect the line of action of reaction R_B at *q*. The closing side *oc* gives the direction of *OC* in the force diagram. Draw *BC* parallel to the known direction of R_B to intersect *OC* at *C*.

Sides *BC* and *AC* in the force diagram give the magnitude and direction of the reactions in cyclic order.

R_B = 10 kN (vertical, upward)

and R_A = 37 kN (30° upward)

Instrumental error is inherent in graphical method. R_A can also be resolved into horizontal and vertical components R_{Ax} and R_{Ay} respectively.

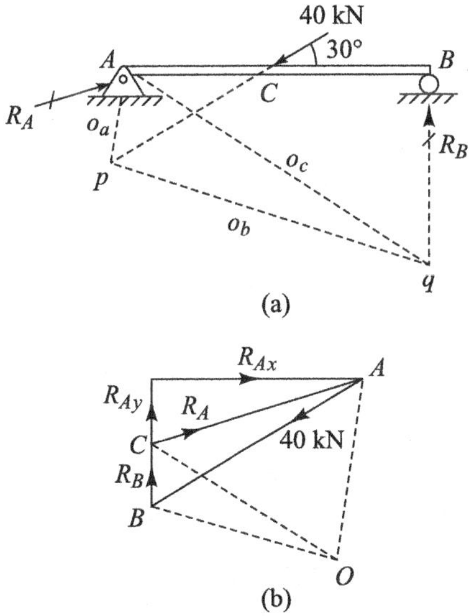

Fig. 8.9 *(a) Space Diagram and Funicular Polygon (b) Force Polygon*

Example 8.8 Solve Example 8.1 by graphical method.

Solution *Bow's notation:* Spaces between member forces and external forces are marked from *A* to *H* as shown in Fig. 8.10(a).

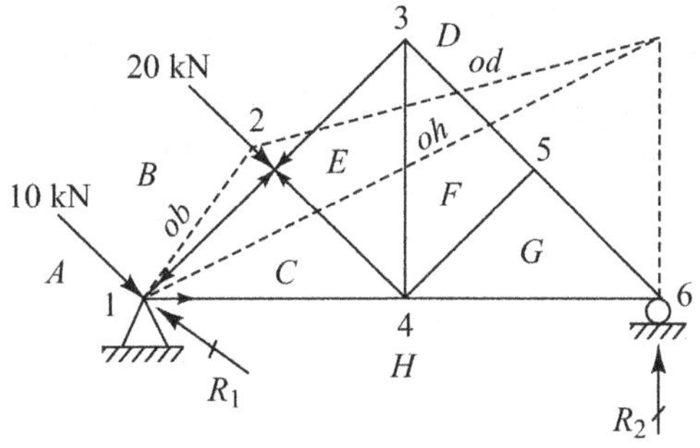

Fig. 8.10 *(a) Truss and Funicular Diagram*

Force diagram: Draw forces *AB* and *BD* in the force diagram (Fig. 8.10(b)) to some scale = 10 and 20 kN, respectively. Any point *O* is chosen as pole and joined with *B* and *D*.

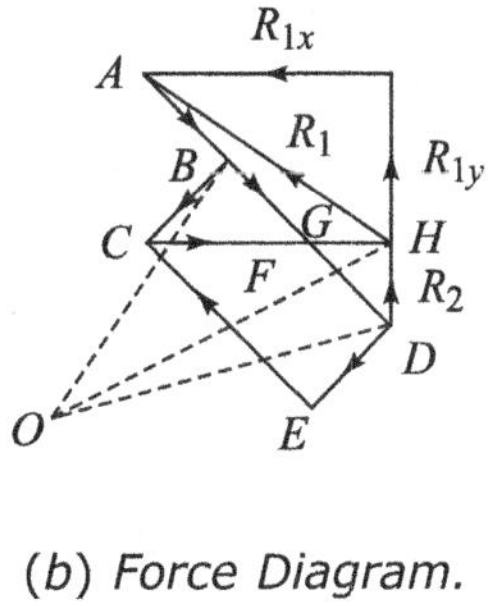

Fig. 8.10 *(b) Force Diagram.*

Funicular polygon: Start from the reaction of a known location, i.e. R_A. Draw lines *ob* and *od* parallel to *OB* and *OD*. *oh* is drawn from the point of intersection of *od* and the line of action of R_2 to the other end of space *H*.

Force diagram: Draw *OH* parallel to *oh*. From *D* draw *DH* parallel to R_2 to intersect *OH* at *H*. Join *AH*.

$$DH = R_2 = 7 \text{ kN}$$

and, $$HA = R_1 = 25.5 \text{ kN},$$

$$R_{1y} = 14 \text{ kN}, R_{1x} = 21 \text{ kN}.$$

From *B*, draw *BC* and from *H*, draw *HC* parallel their respective members to intersect at the point *C* in the force diagram. From *C*, draw *CE* and from *D*, draw *DE*. To locate *G* in the force diagram, draw *HG* from *H* and *DG* from *D*. Similarly *F* is located by drawing *DF* from *D* and *GF* from *G*. *G* and *F* coincide at one point indicating a zero force in member *FG*.

To find the direction of force, move in cyclic order in the closed polygon for a join say for joint 2, start *BD* downward, *DE* downward, *EC* and upward. *CB* will be downward in the force polygon for joint 1.

Example 8.9 Determine member forces in the cambered Fink truss carrying wind load as shown in Fig. 8.11(a).

Solution

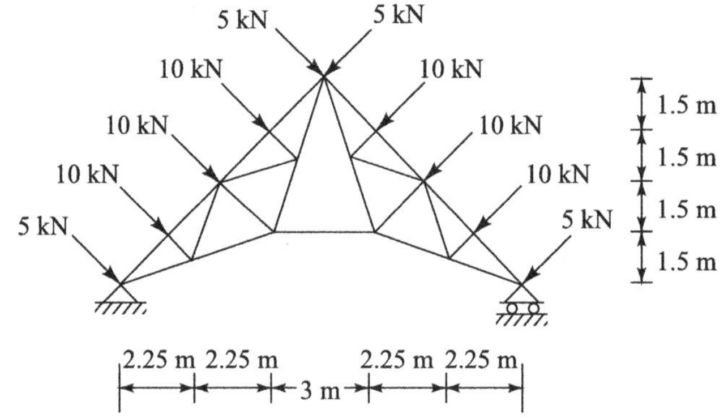

Fig. 8.11 (a) *Fink Truss*

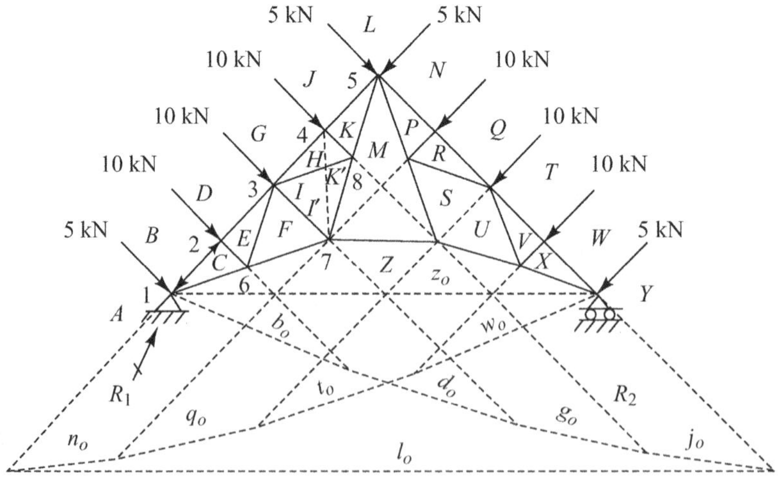

Fig. 8.11 (b) *Space and Funicular Diagram*

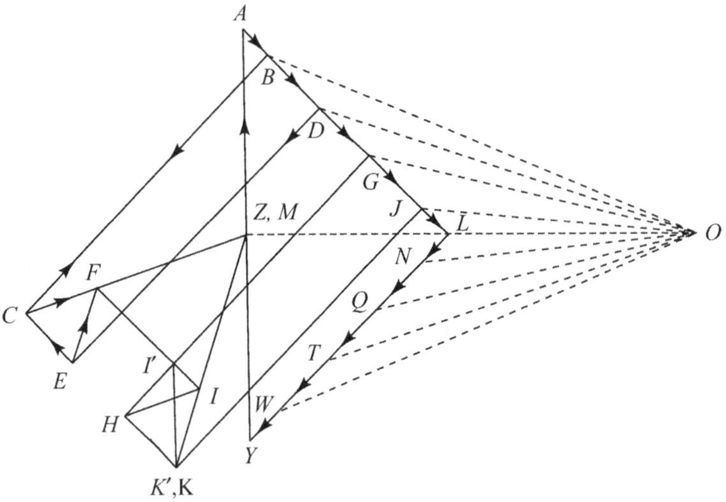

Fig. 8.11 *(c) Maxwell's Diagram*

Force diagram: Plot external forces, *AB*, *BD*, *DG* ... etc. as in Fig. 8.11(c) to some scale. Take a pole *O* and join *BO*, *DO*, etc.

Funicular diagram: Starting from support reaction of a known location R_1, draw *bo*, *do*, *go*, ..., *wo* between the line of action of their respective spaces *B*, *D*, *G*, ... etc. in the space diagram (Fig. 8.11(b)). The last line *wo* intersect line of action of R_2 at the support itself. Draw the closing line *zo*.

Force diagram: Draw *ZO* parallel to *zo*, *YZ* parallel to R_2 intersecting at Z. $YZ = R_2 = ZA = R_1 = 28.5$ kN.

Strating from joint 1, draw force diagram for each joint. From point *B* in the force diagram, draw *BC* parallel to member 12 and from *Z* draw *ZC* parallel to member 16 to intersect *BC* at *C*. *ABCZA* is the closed force polygon for joint 1. Direction of forces is given by moving from *A*, *B*... to *Z* in the cyclic order. *BC* will have opposite sign in force polygons of joints 1 and 2. Force polygons for joints 2 and 6 are made by this procedure.

Joint 3 has more than 2 unknown forces and this procedure will not work. To overcome this difficulty subsitute member 47 shown by dotted line is provided in place of member 38 and 48. This substitution does not effect the stability of the truss nor does it alter the force in the members outside the panel of 3478 since there is no change if the method of section is used to determine the forces of members in the adjacent panels.

Spaces formed by substitute member are marked *I′* and *K′*. Now force polygons for joints 3 and 4 are made by the usual procedure to locate point *K′* in the force diagram. The location of *K′* coincides with *K* in the original truss, as in either case the forces in members 45 and 58 remain unchanged. Now return to the original truss and starting with *K* (or *K′*) in the force polygon, locate *H* and *I* by forming force polygons for joints 4 and 3.

Combined force diagram also known as Maxwell's diagram is drawn for only one half of the truss since the member forces are symmetrical. From the force diagram,

$BC = DE = GH = JK = 50$ kN, $EF = IH = 11$ kN,

$CZ = 33$ kN, $MZ = 0$ kN, $CE = HK = 10$ kN,

$KM = 33$ kN, $FI = 20$ kN, $FZ = IM = 23$ kN

EXERCISES

PROBLEMS

8.1 through 8.20 Analyse the trusses shown in Fig. 8.12 through 8.31 by all three methods (i) method of joints, (ii) method of sections and (iii) grahpical method.

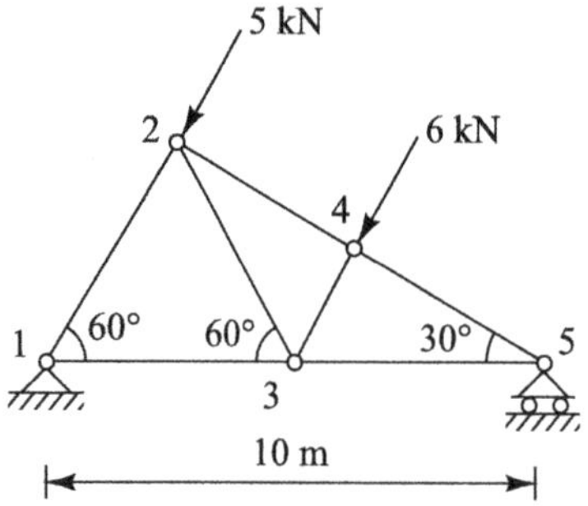

Fig. 8.12 *Prob 8.1*

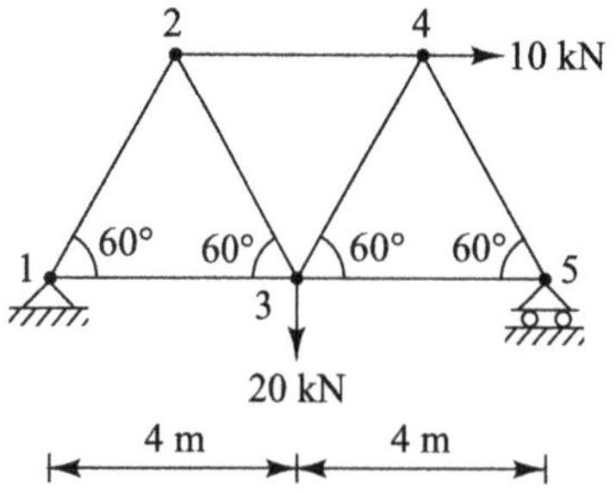

Fig. 8.13 *Prob 8.2*

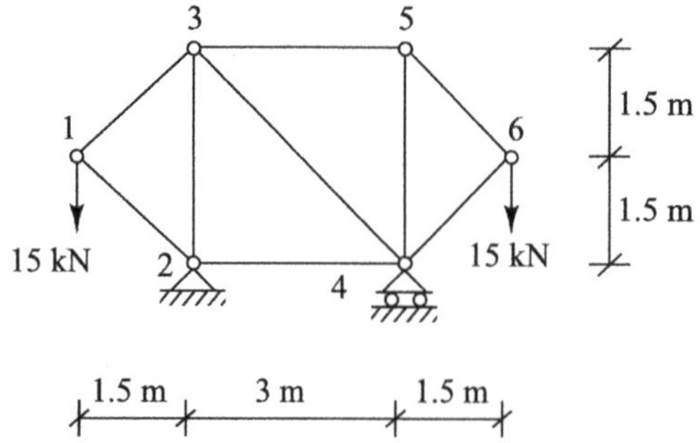

Fig. 8.14 *Prob 8.3*

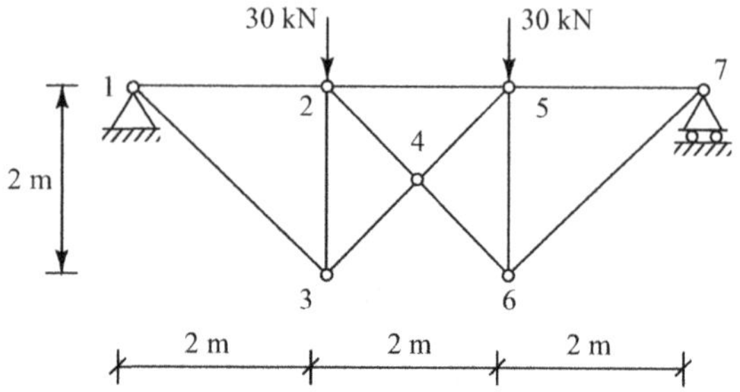

Fig. 8.15 *Prob 8.4*

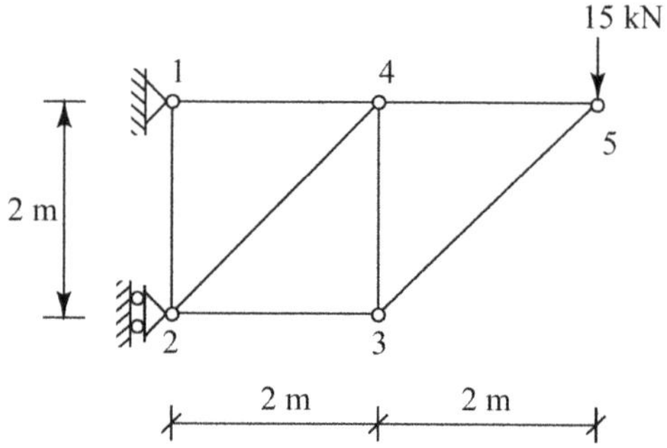

Fig. 8.16 *Prob 8.5*

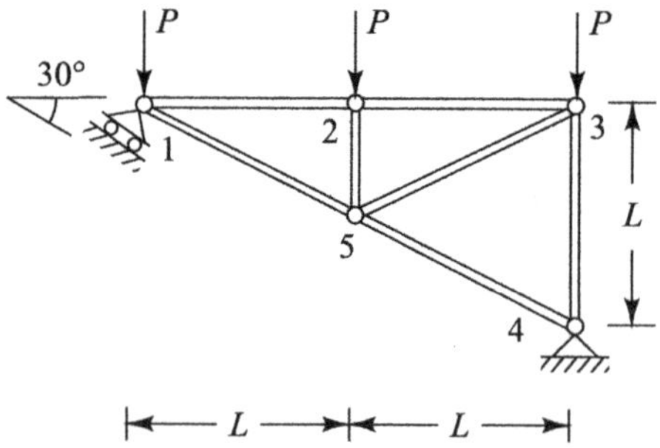

Fig. 8.17 *Prob 8.6*

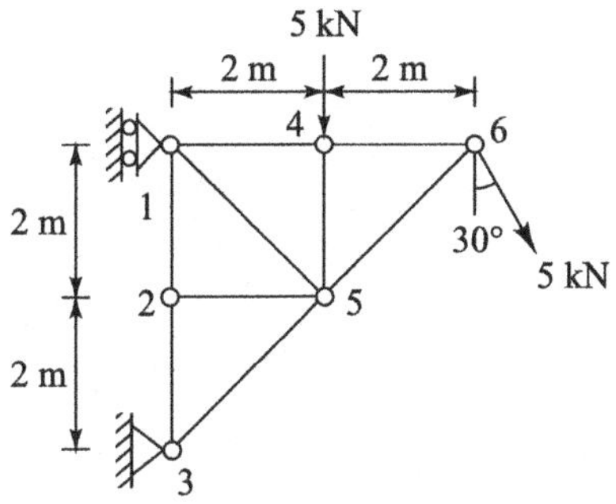

Fig. 8.18 *Prob 8.7*

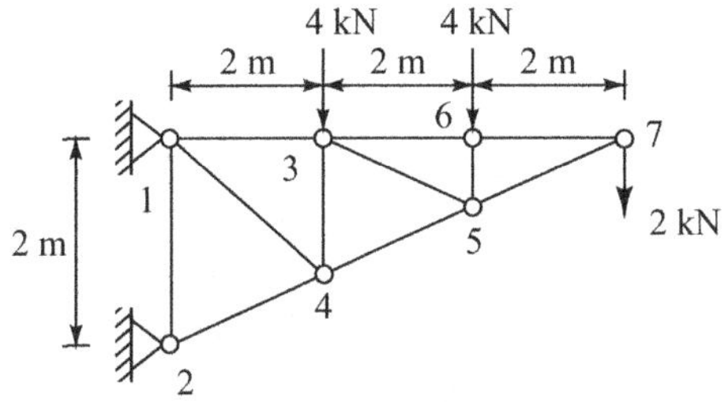

Fig. 8.19 *Prob 8.8*

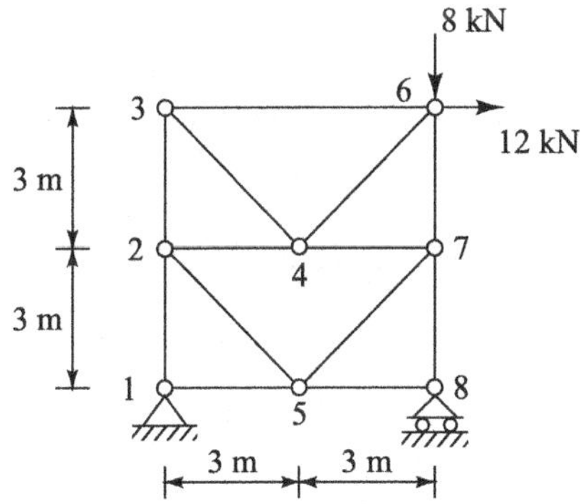

Fig. 8.20 *Prob 8.9*

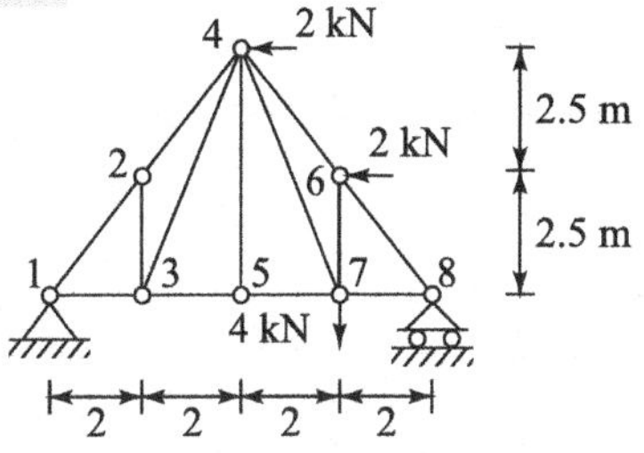

Fig. 8.21 *Prob 8.10*

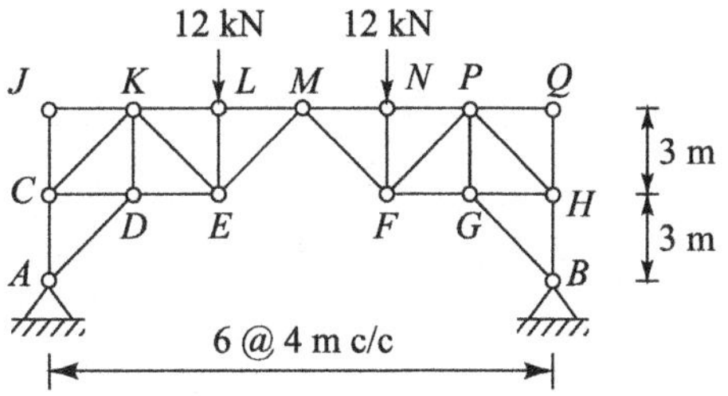

Fig. 8.22 *Prob 8.11*

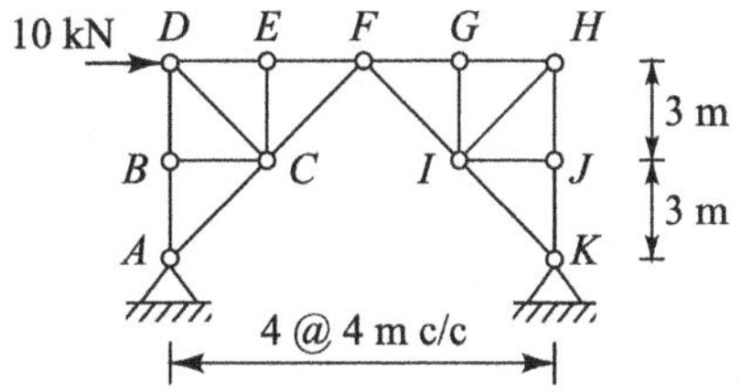

Fig. 8.23 *Prob 8.12*

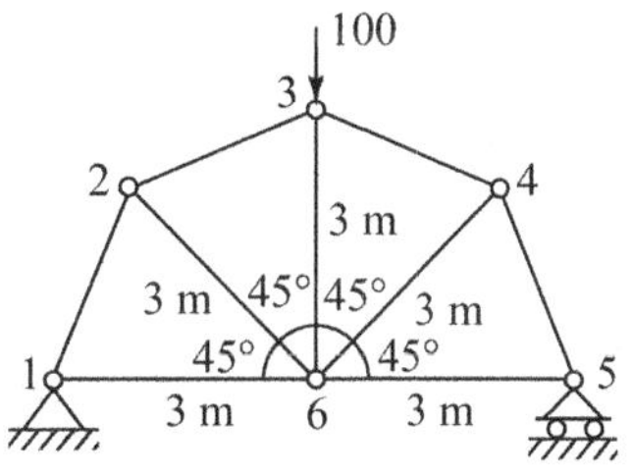

Fig. 8.24 *Prob 8.13*

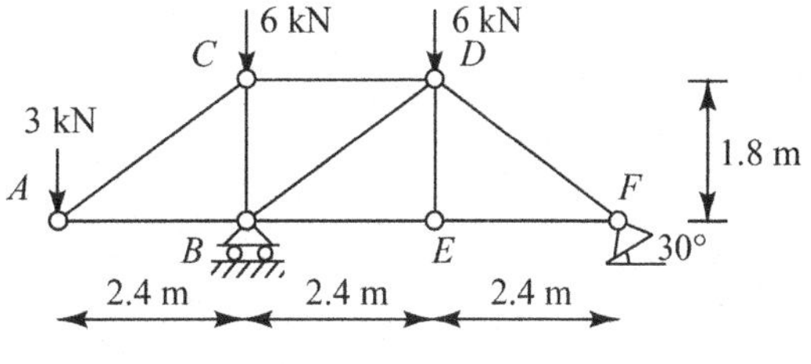

Fig. 8.25 *Prob 8.14*

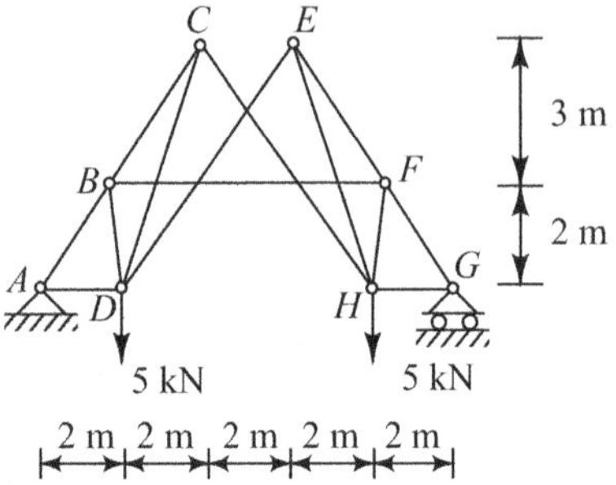

Fig. 8.26 *Prob 8.15*

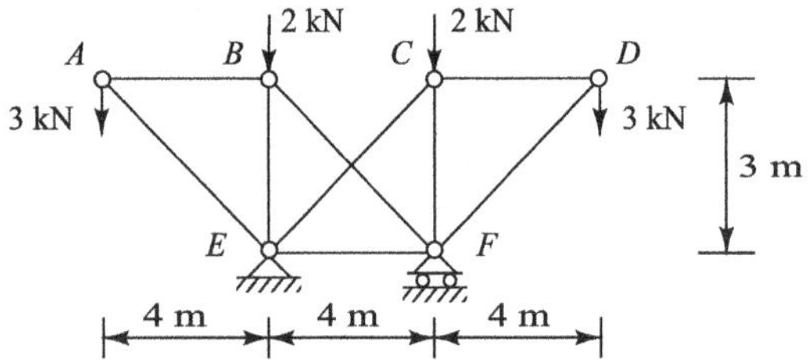

Fig. 8.27 *Prob 8.16*

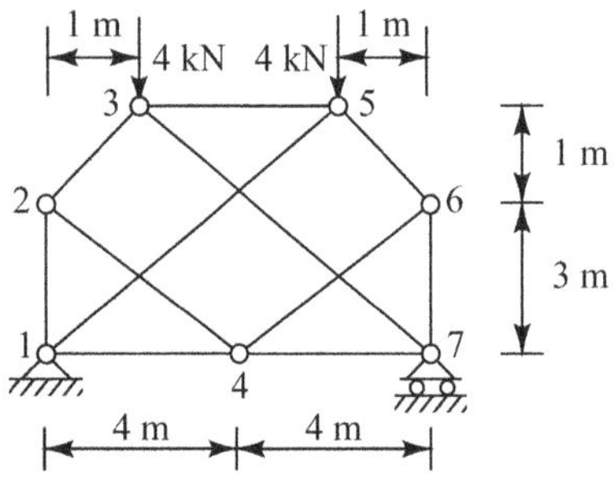

Fig. 8.28 *Prob 8.17*

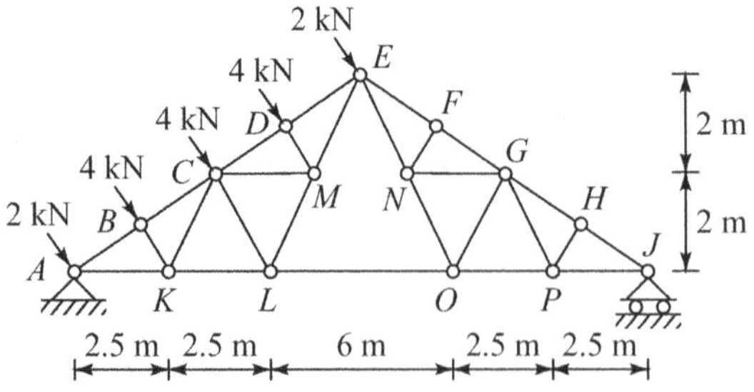

Fig. 8.29 *Prob 8.18*

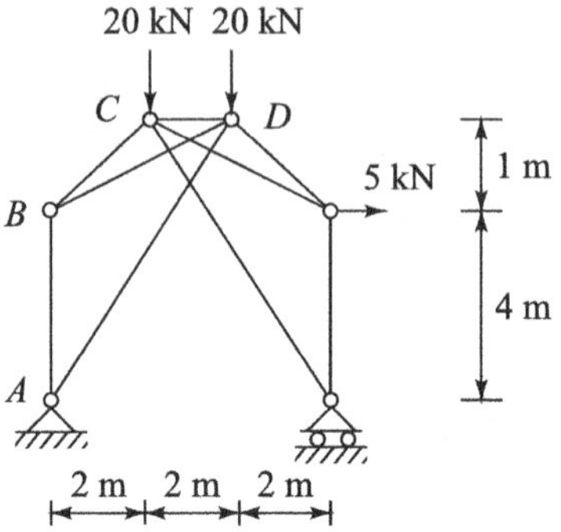

Fig. 8.30 *Prob 8.19*

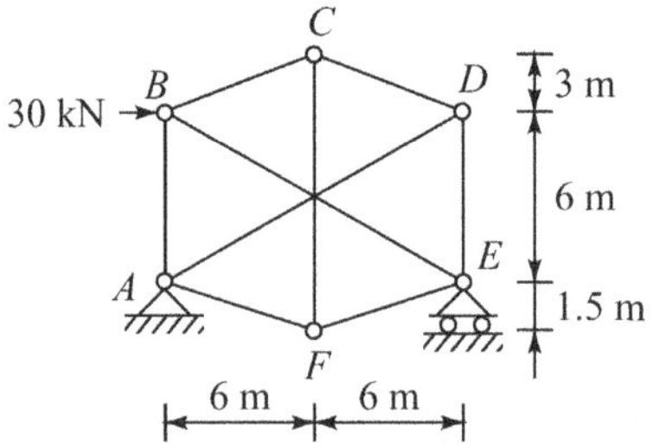

Fig. 8.31 *Prob 8.20*

8.21 Write computer programs to analyse a truss by forming equilibrium equation for each member by the method of joints.

OBJECTIVE QUESTIONS

8.22 The force in the member *AB* of the truss shown in Fig. 8.32 is,

(a) 25 kN (compressive)
(b) $25\sqrt{2}$ kN (tensile)
(c) $25\sqrt{2}$ kN (compressive)
(d) 25 kN (tensile)

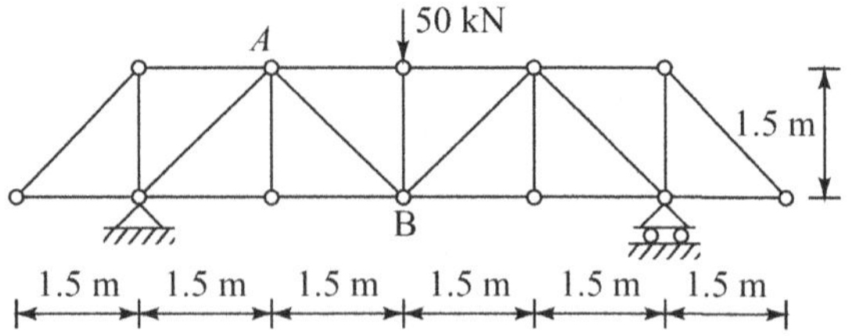

Fig. 8.32

8.23 The force in member *AB* of the truss shown in Fig. 8.33 is

(a) 4 kN (compression)
(b) 36 kN (compression)
(c) 16 kN (tensile)
(d) 20 kN (tensile)

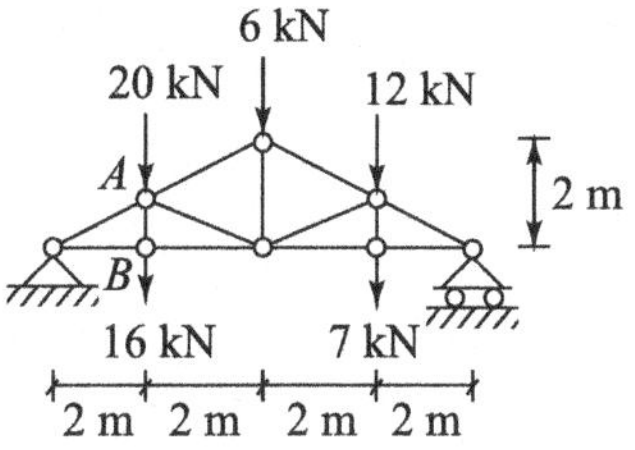

Fig. 8.33

8.24 The force in member *BD* of the truss shown in Fig. 8.34 is,

(a) 5 kN (tensile)
(b) 5 kN (compression)
(c) $5\sqrt{2}$ kN (compression)
(d) zero

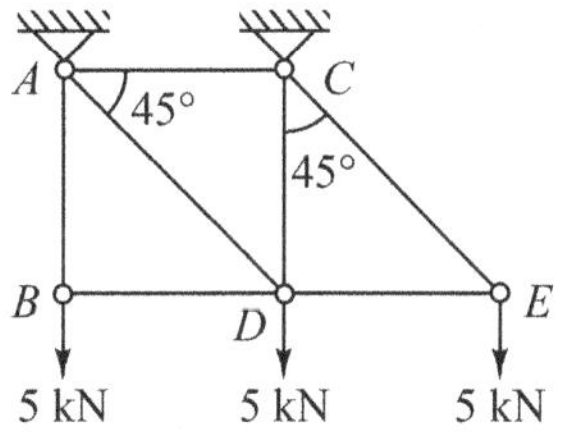

Fig. 8.34

8.25 Force in the member *AB* of the frame shown in Fig. 8.35 will be

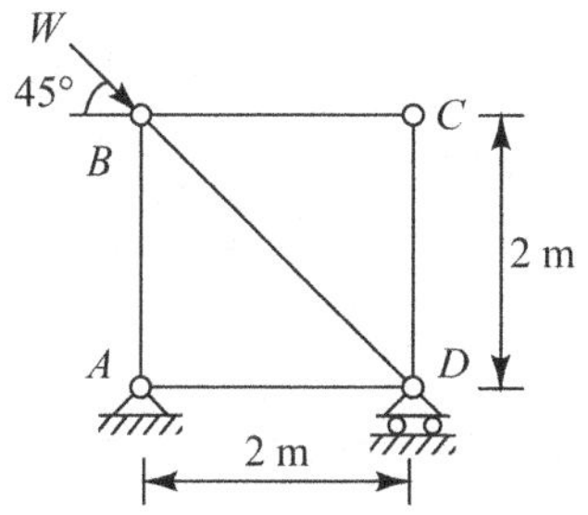

Fig. 8.35

(a) zero (b) W

(c) $W\sqrt{2}$ (d) $\dfrac{W}{\sqrt{2}}$

8.26 Force in the member *AD* of the frame shown in Fig. 8.36 will be

(a) zero
(b) $5\sqrt{2}$ kN (compression)
(c) $5\sqrt{2}$ kN (tension)
(d) $10\sqrt{2}$ kN (tension)

(**Hint:** Consider the equilibrium of the free body above a horizontal section through *AC*)

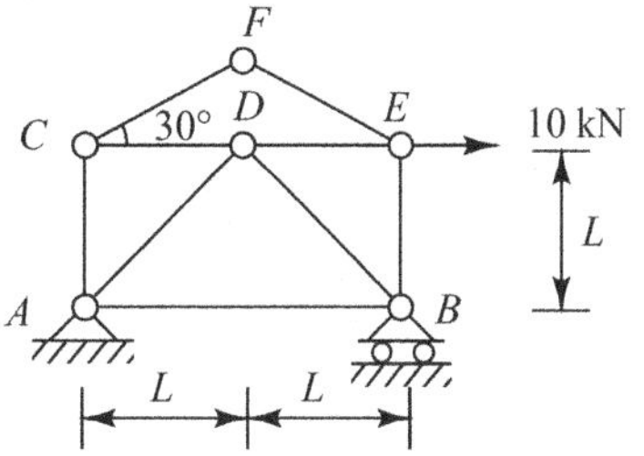

Fig. 8.36

8.27 Force in the member *AC* of the truss shown in Fig. 8.37 is,

(a) 50 kN (compression)
(b) 50 kN (tension)
(c) $\dfrac{50\sqrt{3}}{2}$ kN (tension)
(d) zero

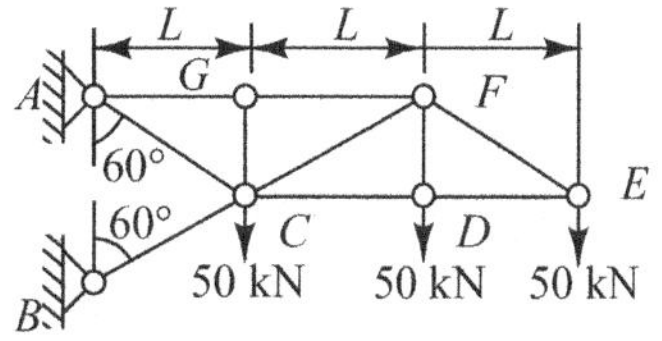

Fig. 8.37

8.28 Force in the member *ED* of the truss shown in Fig. 8.38 is,

(a) 40 kN (compression)
(b) 80 kN (tension)
(c) 80 kN (compression)
(d) 120 kN (compression)

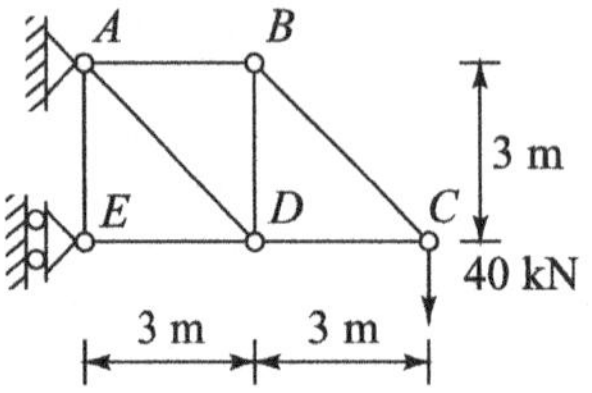

Fig. 8.38

8.29 Consider the truss shown in Fig. 8.39 and match list I of members and list II of forces and select the correct answer.

List I	*List II*
A. PR	1. 40 kN (tension)
B. RS	2. 40 kN (compression)
C. SU	3. 60 kN (tension)
D. RT	4. $50\sqrt{2}$ kN (compression)

Answer code	*A*	*B*	*C*	*D*
(a)	3	2	1	4
(b)	4	1	2	3
(c)	3	1	2	4
(d)	4	2	1	3

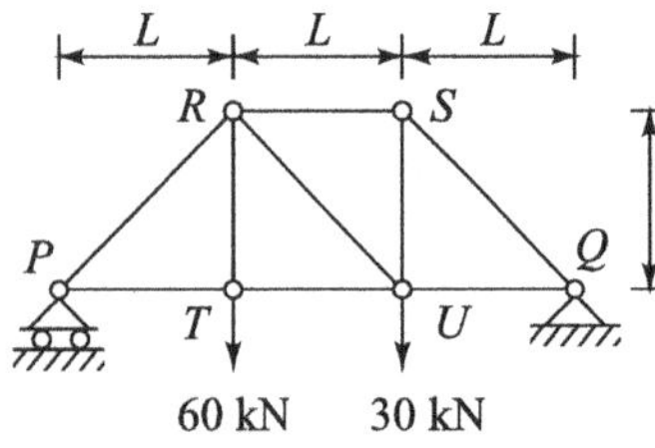

Fig. 8.39

8.30 Axial force in the member *BC* of the truss shown in Fig. 8.40 is,

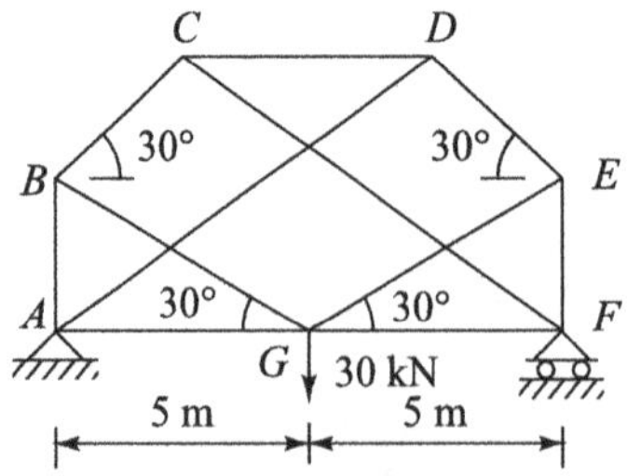

Fig. 8.40

(a) 15 kN (b) $10\sqrt{3}$ kN

(c) $15\sqrt{3}$ kN (d) 30 kN

(**Hint:** Force in *BG* = *EG* = 30/2 sin 30° due to symmetry)

8.31 The force in member *AC* of the pin jointed truss shown in Fig. 8.41 is,

(a) $10\sqrt{2}$ kN (tension)

(b) $10\sqrt{2}$ kN (compression)

(c) zero

(d) 10 kN (tension)

(**Hint:** Force in diagonal members will be equal. GH and HE are zero force members)

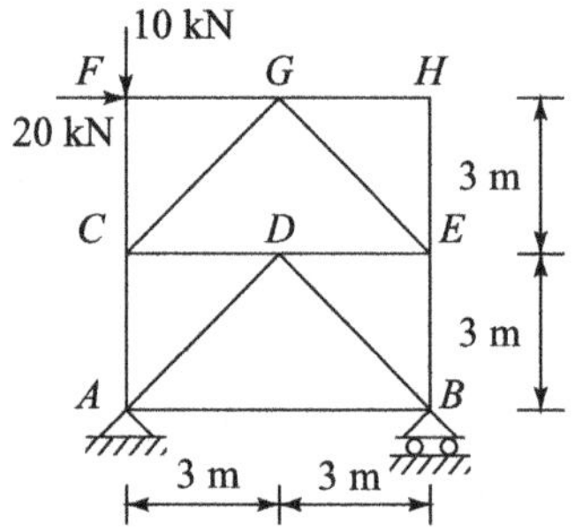

Fig. 8.41

8.32 A pin jointed truss is loaded as shown in Fig. 8.42. Match List I with List II and select the correct answer.

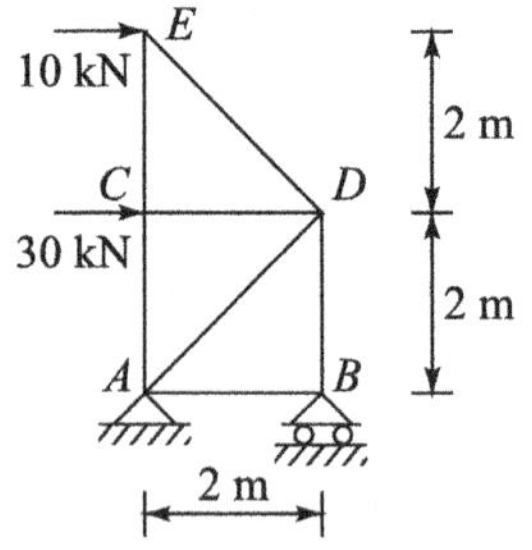

Fig. 8.42

List I *(member)*	*List II* *(force)*
A. *AB*	1. 30 kN
B. *AC*	2. 50 kN
C. *BD*	3. 0
D. *CD*	4. 10 kN

Answer code	A	B	C	D
(a)	3	2	1	4
(b)	4	3	2	1
(c)	3	4	2	1
(d)	4	3	1	2

8.33 The member(s) of the truss shown in Fig. 8.43 which carry zero force is/are,
(a) *EC* only
(b) *EC* and *AB*
(c) *EC* and *AC*
(d) *EC*, *AC* and *AB*

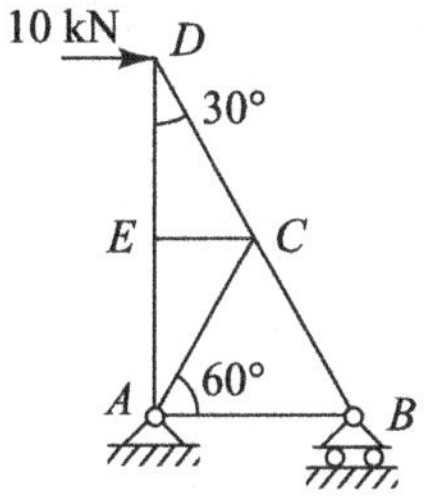

Fig. 8.43

8.34 Which is the ratio of the forces in the members *AB*, *BE* and *AE* of the pin jointed truss shown in Fig. 8.44?
(a) 5 : 4 : 3 (b) 4 : 3 : 5
(c) $\frac{1}{4}:\frac{1}{3}:\frac{1}{5}$ (d) None of the above

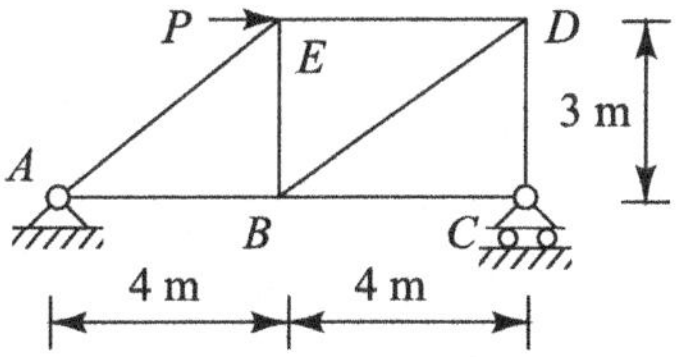

Fig. 8.44

8.35 The force in members *DF* of the truss shown in Fig. 8.45 is,
(a) 1.5 kN (tension)
(b) 4.5 kN (tension)
(c) 1.5 kN (compression)
(d) 4.5 kN (compression)

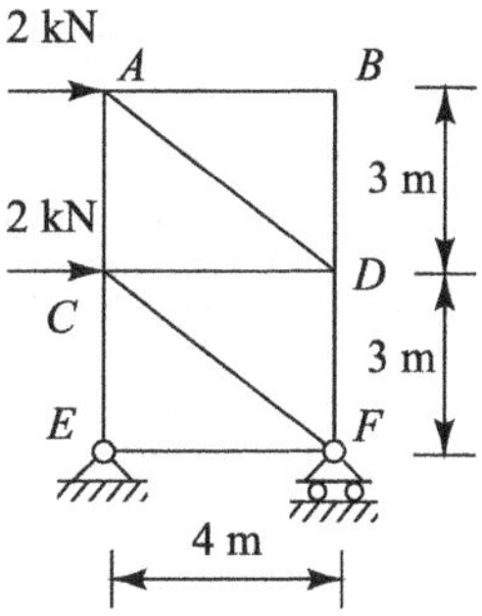

Fig. 8.45

8.36 The force in member *EF* of the truss shown in Fig. 8.27 is,
(a) zero (b) 2.5 kN
(c) 5 kN (d) 8 kN

8.37 In the system of coplaner concurrent forces shown in Fig. 8.46, if *R* is the known force, then to find the unknown force F_1, all forces are to be resolved in the direction of,
(a) force F_1
(b) force F_2
(c) force *R*
(d) normal F_1

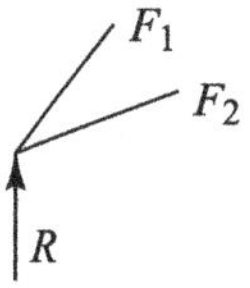

Fig. 8.46

8.38 The force in member *DF* of the pin jointed truss shown in Fig. 8.47 is,
(a) 5 kN (b) $5\sqrt{2}$ kN
(c) 10 kN (d) $10\sqrt{2}$ kN

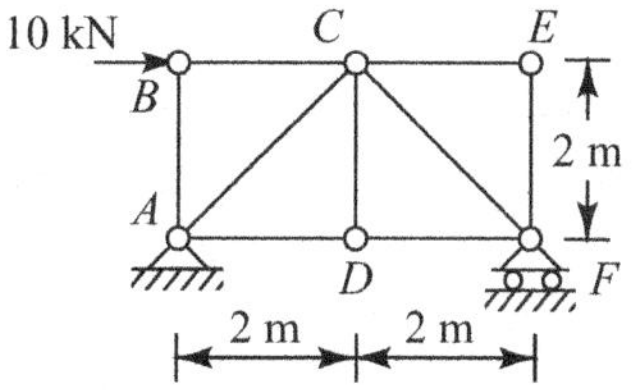

Fig. 8.47

8.39 In a beam *AB* shown in Fig. 8.48, the direction of support reaction at *A* and *B* is given by,

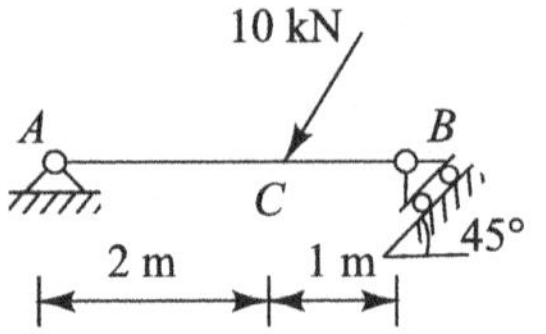

Fig. 8.48

(a)

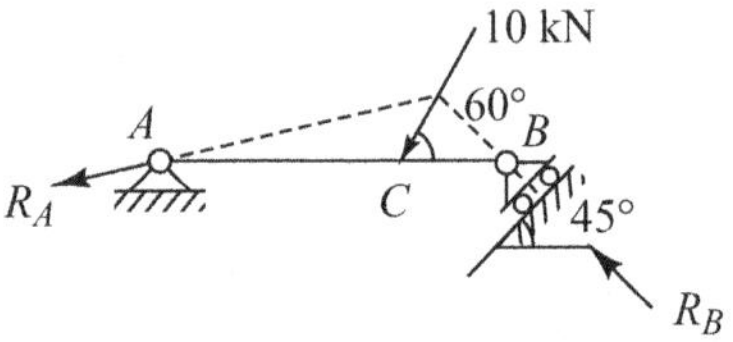

(b)

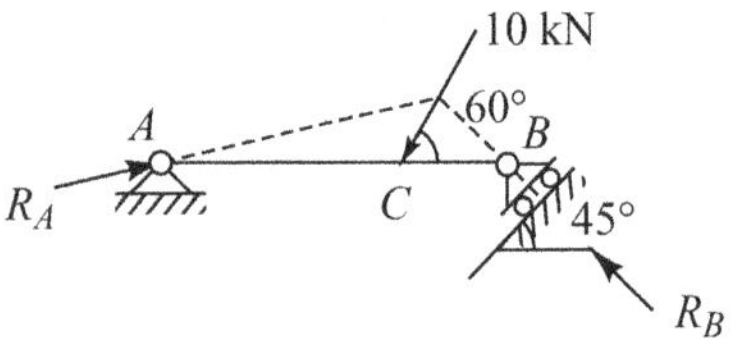

(c)

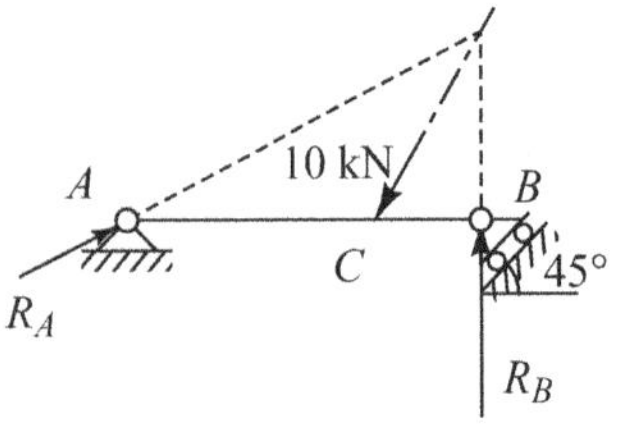

(d)

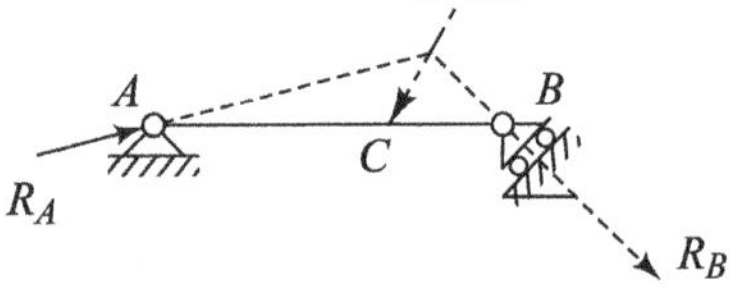

8.40 The force diagram for joint 2 of the pin jointed cantilever truss shown in Fig. 8.49 will be,

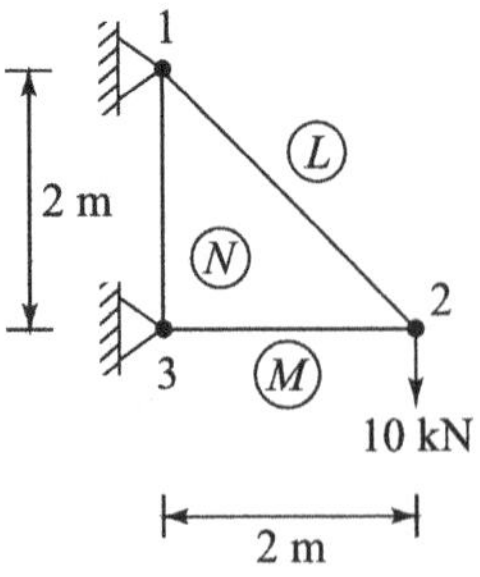

Fig. 8.49

(a)

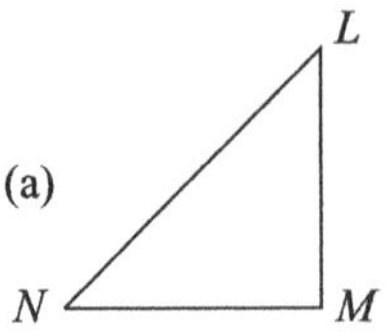

(b)

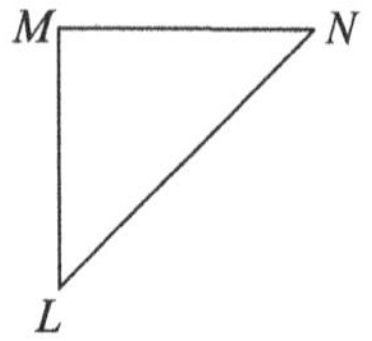

(c)

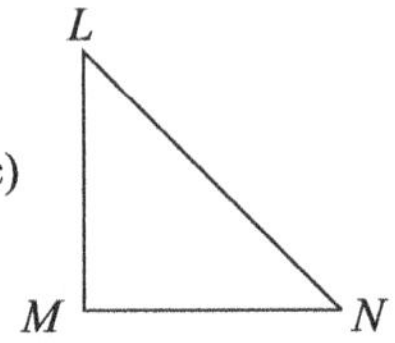

(d)

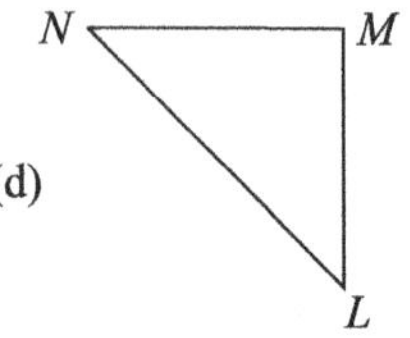

8.41 The force diagram for joint 1 of the truss shown in Fig. 8.50 will be,

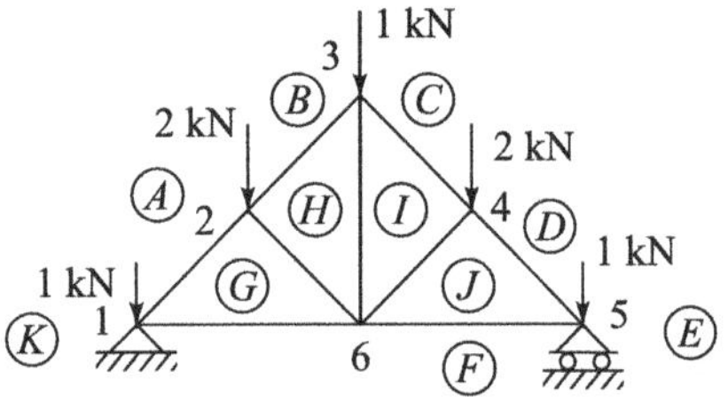

Fig. 8.50

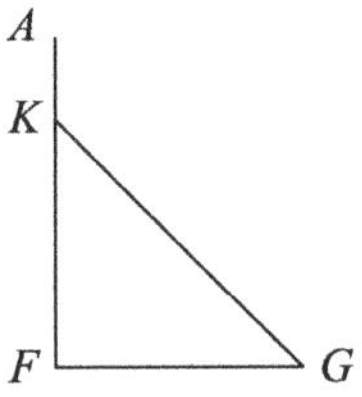

(a)

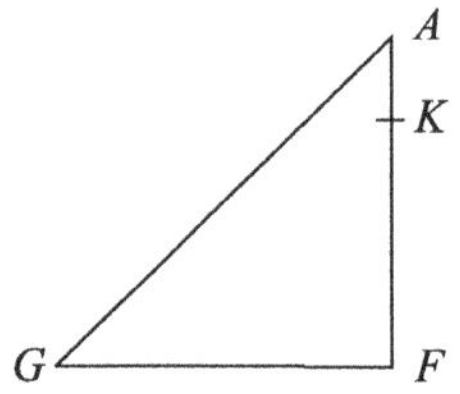

(b)

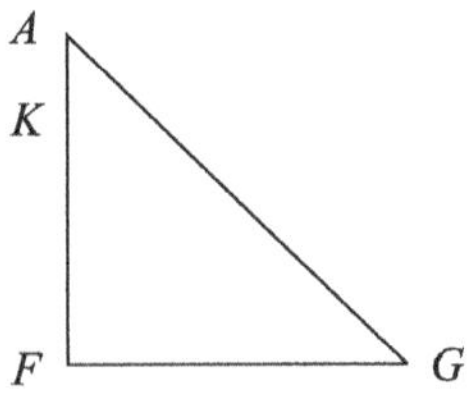

(c)

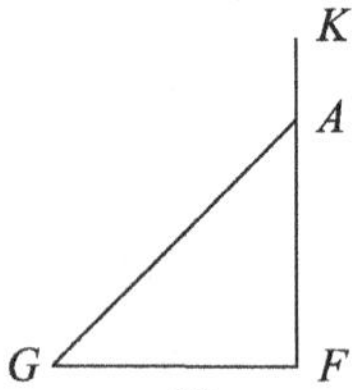

(d)

9

Compression Members

CONCEPT REVIEW

9.1 INTRODUCTION

A short column may fail by crushing if the applied load exceeds its yield stress, but a long column fails by buckling at a stress much smaller than its yield stress. The permissible compressive load for slender members is reduced due to instability. Such an instability can cause failure at all sections under compressive load, e.g. I-beam web crippling at supports and under concentrated loads, thin shafts under torsion, torsional buckling of deep and narrow beams besides long columns.

9.2 STABILITY OF LONG COLUMNS—EULER'S FORMULA

The failure of an axially loaded compression member takes place by bending resulting from a number of effects as follows:

(i) no member is perfectly straight,
(ii) material is not perfectly homogeneous,
(iii) load is not at centre perfectly,
(iv) some occasional transverse load may act.

If a column attains neutral equilibrium in a slightly deformed configuration under an axial load P, it is called the *critical load.*

9.2.1 Both Column Ends are Hinged (Fig. 9.1)

As earlier in Art. 7.1, the deformed shape under bending is given by,

$$\frac{M}{EI} = \frac{1}{R} \approx -\frac{d^2y}{dx^2}$$

Bending moment induced due to the deflected shape,

$$M = P \times y$$

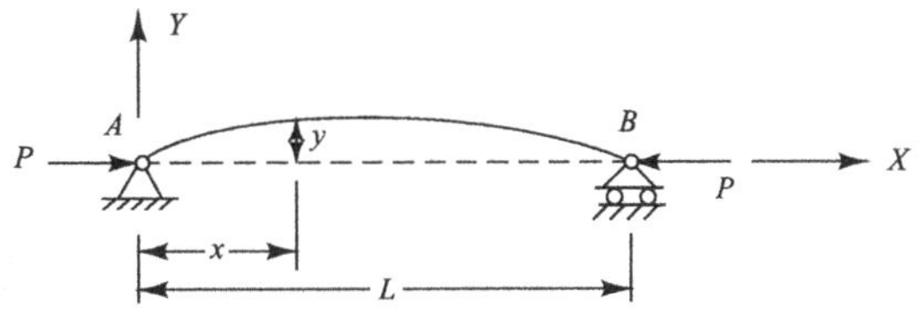

Fig. 9.1

or, $$\frac{d^2y}{dx^2} + \frac{P \times y}{EI} = 0$$

The solution of this differential equation is of the form,

$$y = C \sin\sqrt{\frac{P}{EI}}\, x + D \cos\sqrt{\frac{P}{EI}}\, x$$

Boundary conditions: At $x = 0$, $y = 0 \Rightarrow D = 0$

$$\text{At } x = L,\ y = 0 \Rightarrow C \sin\sqrt{\frac{P}{EI}}\, L = 0,$$

either, $C = 0$, then $y = 0$ everywhere (straight bar). It is a trivial solution of no use.

$\therefore$ $$\sin\sqrt{\frac{P}{EI}} \times L = 0,$$

or $$\sqrt{\frac{P}{EI}} \times L = n\pi \text{ rad } (n = 1, 2, 3, ...)$$

or, $$P = \frac{n^2\pi^2 EI}{L^2}$$

The first mode of buckling occurs at $n = 1$, when P has the smallest value called critical load.

$$P_{cr} = \frac{\pi^2 EI}{L^2} \tag{9.1}$$

This is called the Euler's buckling load for hinged end column. The corresponding deflected shape is given by,

$$y = C \sin\sqrt{\frac{P}{EI}}\, x = C \sin\frac{\pi x}{L}$$

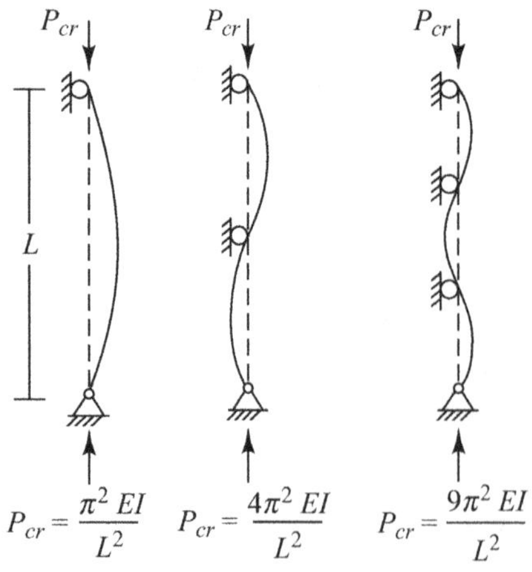

Fig. 9.2

It is not possible to determine C, i.e. amplitude of the deflected shape because of the approximation in the expression for $1/R$.

Theoretically, higher modes of buckling are, also possible at higher P_{cr} by applying external restraint at the point of inflection as shown in Fig. 9.2.

9.2.2 One End Fixed, Other End Free

Usual bending equation gives the deformed shape shown in Fig. 9.3.

$$-\frac{d^2y}{dx^2} = \frac{M}{EI} = -\frac{P \times (a - y)}{EI}$$

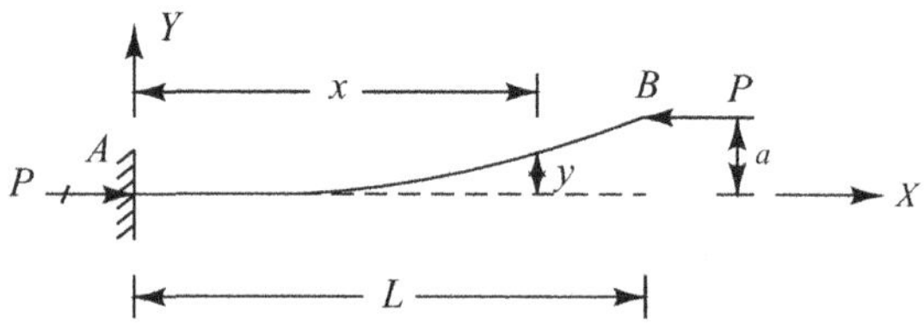

Fig. 9.3

The solution of this differential equation is of the form,

$$y = C \sin\sqrt{\frac{P}{EI}}\, x + D \cos\sqrt{\frac{P}{EI}}\, x + a$$

Boundary conditions:

At $x = 0$, $y = 0 \Rightarrow D = -a$

At $x = 0$, $\frac{dy}{dx} = 0 \Rightarrow C = 0$

or, $$y = a\left(1 - \cos\sqrt{\frac{P}{EI}}\, x\right)$$

At $x = L$, $$y = a = a\left(1 - \cos\sqrt{\frac{P}{EI}}\, L\right)$$

or, $$\sqrt{\frac{P}{EI}}\, L = \frac{\pi}{2}, \frac{3\pi}{2}, \frac{5\pi}{2} ...$$

For least significant value,

$$P_{cr} = \frac{\pi^2 EI}{4L^2} \tag{9.2}$$

9.2.3 Both Ends Fixed

The end reaction M_O is induced due to buckling of the column under an axial load P. Bending equation gives the deformed shape shown in Fig. 9.4 for moment at a distance x and y from the origin A.

$$-EI\frac{d^2y}{dx^2} = -M_o + P \times y$$

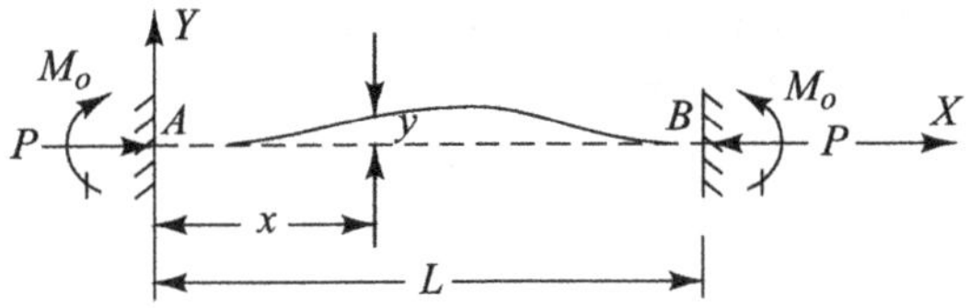

Fig. 9.4

The solution of this differential equation is of the form,

$$y = C\sin\sqrt{\frac{P}{EI}}x + D\cos\sqrt{\frac{P}{EI}}x + \frac{M_o}{P}$$

Boundary conditions:

At $x = 0, y = 0 \Rightarrow D = -\frac{M_o}{P}$

At $x = 0, \frac{dy}{dx} = 0 \Rightarrow C = 0,$

$$y = \frac{M_o}{P}\left(1 - \cos\sqrt{\frac{P}{EI}}x\right)$$

At $x = L$, $$y = 0 = \frac{M_o}{P}\left(1 - \cos\sqrt{\frac{P}{EI}}L\right)$$

i.e. either $\frac{M_o}{P} = 0$ (trivial solution)

or, $1 - \cos\sqrt{\frac{P}{EI}}L = 0$

$$\sqrt{\frac{P}{EI}}L = 0, 2\pi, 4\pi, ...$$

Also at $x = L$,

$$\frac{dy}{dx} = 0 = \frac{M_o}{P}\sqrt{\frac{P}{EI}}\sin\sqrt{\frac{P}{EI}}L$$

or, $$\sqrt{\frac{P}{EI}}L = 0, \pi, 2\pi, 3\pi, ...$$

To satisfy both,

$$\sqrt{\frac{P}{EI}}L = 0, 2\pi, 4\pi, ...$$

For least significant value,

$$P_{cr} = \frac{4\pi^2 EI}{L^2} \qquad (9.3)$$

9.2.4 One End Fixed, Other End Hinged

The end reaction R is induced due to buckling of the column under an axial load P. Bending equation gives the deformed shape shown in Fig. 9.5 for moment at any section at a distance x and y from the origin A.

$$EI\frac{d^2y}{dx^2} = R(L - x) - P \times y$$

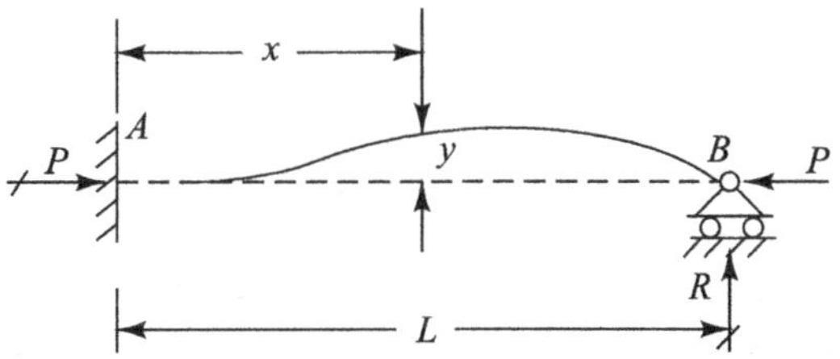

Fig. 9.5

The solution of this differential equation is of the form,

$$y = C\sin\sqrt{\frac{P}{EI}}x + D\cos\sqrt{\frac{P}{EI}}x + \frac{R(L - x)}{P}$$

Boundary conditions:

At $x = 0, \quad y = 0 \Rightarrow D = -\frac{R \times L}{P}$

and, at $x = 0, \frac{dy}{dx} = 0 \Rightarrow C = \sqrt{\frac{EI}{P}} \times \frac{R}{P}$

which gives,

$$y = \frac{R}{P}\sqrt{\frac{EI}{P}} \times \sin\sqrt{\frac{P}{EI}}\, x - \frac{RL}{P} \times$$

$$\cos\sqrt{\frac{P}{EI}}\, x + \frac{R(L-x)}{P}$$

Also at $x = L$, $y = 0$, which gives

$$\tan\sqrt{\frac{P}{EI}}L = \sqrt{\frac{P}{EI}}L$$

The smallest root of this equation gives,

$$\sqrt{\frac{P_{cr}}{EI}}L = 4.4934 \text{ rad}$$

or,

$$P_{cr} \simeq \frac{2\pi^2 EI}{L^2} \tag{9.4}$$

9.2.5 Effective Length of a Column

The basic formula in column buckling is the Euler's formula Eq. (9.1) which assumes that,

(1) The column is initially straight,
(2) The load is concentric,
(3) Both ends of the column are hinged,
(4) The material is not stressed beyond the proportional limit,

and such a column is called an **Ideal** or **Euler Column**. The load required to buckle such a column is known as Euler load P_e. Therefore rewriting Eq. (9.1) as,

$$P_e = \frac{\pi^2 EI}{L^2} \tag{9.5a}$$

Critical load for different end conditions is expressed in terms of Euler loads as follows:

$$P_{cr} = \frac{P_e}{K^2} = \frac{\pi^2 EI}{(KL)^2} = \frac{\pi^2 EI}{L_e^2} \tag{9.5b}$$

where L_e = effective length or equivalent to a column hinged at both ends

K = effective length factor.

Table 9.1 *Effective Length, L_e*

End condition	*Effective length L_e*
Both ends hinged	L
One end fixed, other end free	$2L$
Both ends fixed	$L/2$
One end fixed, other end hinged	$L/\sqrt{2}$

9.2.6 Limitations of Euler's Formula

Critical stress corresponding to critical load is given by,

$$\sigma_{cr} = \frac{P_{cr}}{A} = \frac{\pi^2 EI}{L^2 A} = \frac{\pi^2 E}{(L/r)^2} \tag{9.6}$$

where, $r^2 = I/A$ radius of gyration

It should be noted that Critical Euler's stress s_{cr} depends on E and L/r only and takes no account of yield stress σ_y of the material. For steels taking $E = 200$ GPa, σ_{cr} exceeds σ_y for slenderness ratio $L/r < 90$, as shown in Fig. 9.7. Therefore Euler's buckling load can represent column behaviour at higher values of L/r.

Euler's formula takes no account of eccentricity of load and factor of safety. In any case, perfect centric loading is not possible and the end conditions assumed in above four cases are also not perfectly possible.

9.3 SECANT FORMULA—ECCENTRIC LOADING

Consider a column AB hinged at both ends with an eccentric load P as shown in Fig. 9.6. At section C, bending moment,

$$M = P \times y + M_A = P(y + e)$$

Substituting in bending equation,

$$-EI\frac{d^2y}{dx^2} = M = P(y+e)$$

or,

$$\frac{d^2y}{dx^2} = \frac{-P}{EI}(y+e) = -\mu^2(y+e)$$

where, $\mu^2 = P/EI$

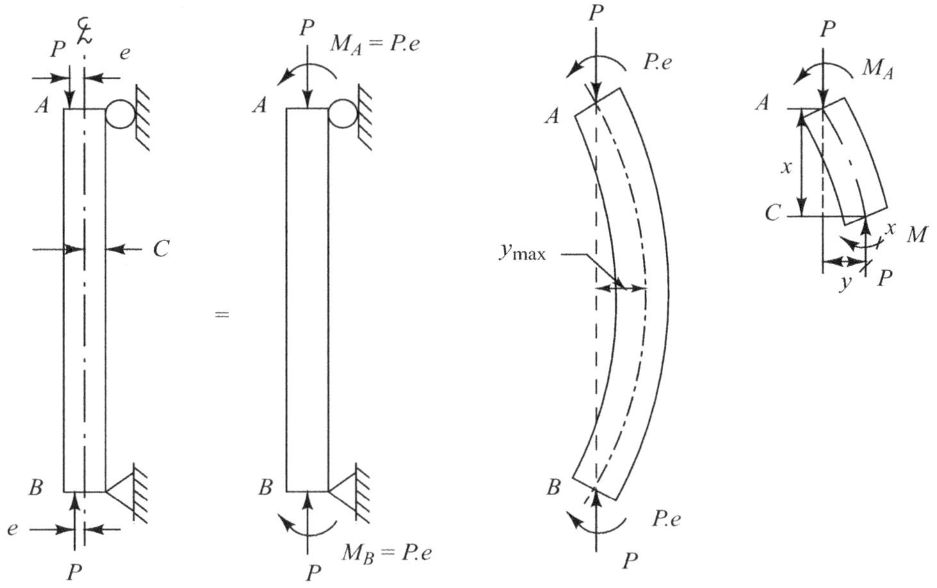

Fig. 9.6

The solution of this differential equation is of the form,

$$y = C \sin \mu x + D \cos \mu x - e$$

Boundary conditions:

At $\quad x = 0, y = 0 \Rightarrow D = e$

At $\quad x = L, y = 0$, which gives,

$$C \sin \mu L = e(1 - \cos \mu L)$$

or, $\quad C(2 \times \sin \mu L/2 \times \cos \mu L/2) = e(2 \sin^2 \mu L/2)$

or, $\quad C = e \tan \mu L/2$

or, $\quad y = e(\tan(\mu L/2) \times \sin \mu x + \cos \mu x - 1)$

For maximum deflection:

At $\quad x = L/2, y = y_{max}$

$$y_{max} = e(\tan(\mu L/2) \times \sin(\mu L/2) + \cos \mu L/2 - 1)$$

or, $\quad y_{max} = e(\sec \mu L/2 - 1) \qquad (9.7)$

Load P becomes critical at $\sqrt{\dfrac{P}{EI}} \times \dfrac{L}{2} = \dfrac{\pi}{2}$, for which y_{max} tends to infinity.

$$P_{cr} = \frac{\pi^2 EI}{L^2}$$

(same as Euler's load for hinged column)

Maximum stress under direct compression and bending at $y = L/2$ is given by,

$$\sigma_{max} = \frac{P}{A} + \frac{M_{max} \times c}{I}$$

where $I = A \times r^2$ and $M_{max} = P(y_{max} + e)$

$$\sigma_{max} = \frac{P}{A}\left[1 + \frac{e \times c}{r^2} \sec \mu L/2\right] \qquad (9.8)$$

These derivations for P_{cr} and σ_{max} can be used for any end conditions if appropriate values of L_e are used in place of L. The value σ_{max} does not vary linearly with P and therefore principle of superposition will not be valid for more than one load acting on the column. Figure 9.7 shows variation of P/A vs. L_e/r for Euler's and Secant formula, taking $\sigma_{max} = \sigma_y = 250$ MPa and $E = 200$ GPa for mild steel. Eccentricity ratio $e \times c/r^2$ is taken equal to 0.25 for the design of steel structures in the absence of any accurate value of e.

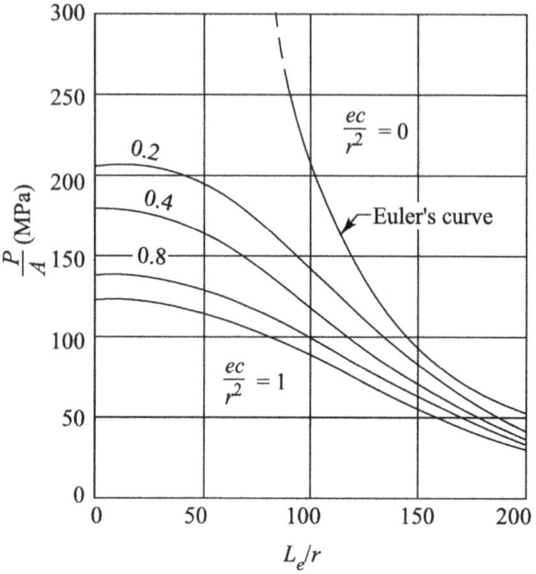

Fig. 9.7 *Yield Stress vs. Slenderness Ratio for Mild Steel*

9.4 OTHER FORMULAE

Real column falls short of idealizations assumed in theoretical formulae, which do not have proper correlation with the experimental data. Since no single formula is adequate for all materials with different end conditions and L_e/r ratios, various empirical formulae are used in different codes for column strength.

9.4.1 Rankine-Gordan Formula

This empirical relation gives critical axial load P_{cr} as follows:

$$\frac{1}{P_{cr}} = \frac{1}{P_c} + \frac{1}{P_e} \tag{9.9}$$

where,

P_c = crushing compressive load = $\sigma_y \times A$

P_e = Euler's critical load = $\pi^2 EI/L_e^2$

$$\text{or, } P_{cr} = \frac{P_c P_e}{P_c + P_e} = \frac{\sigma_y A}{1 + C(L_e/r)^2} \tag{9.10}$$

where, C = constant

Indian Standard IS-800-1984 recommends Merchant–Rankine formula for permissible compressive stress σ_{ac} using 0.6 for safety factor as follows:

$$\sigma_{ac} = 0.6 \times \sigma_{cr}$$

$$= 0.6 \times \frac{\sigma_e \sigma_y}{\left[\sigma_e^{1.4} + \sigma_y^{1.4}\right]^{1/1.4}} \tag{9.11}$$

where, σ_e = Euler's stress and σ_y = yield stress

9.4.2 AISC Formula

American Institute of Steel Construction specifies a parabolic variation from σ_y to $\sigma_y/2$ along with Euler's formula as shown in Fig. 9.8 and the allowable stress σ_{all} is given as follows:

For $L_e/r < C_c$

$$\sigma_{all} = \frac{\sigma_y}{\text{F.S.}}\left(1 - \frac{(L_e/r)^2}{2C_c^2}\right) \quad \text{(Parabolic) (9.12a)}$$

and for $C_c \le L_e/r \le 200$

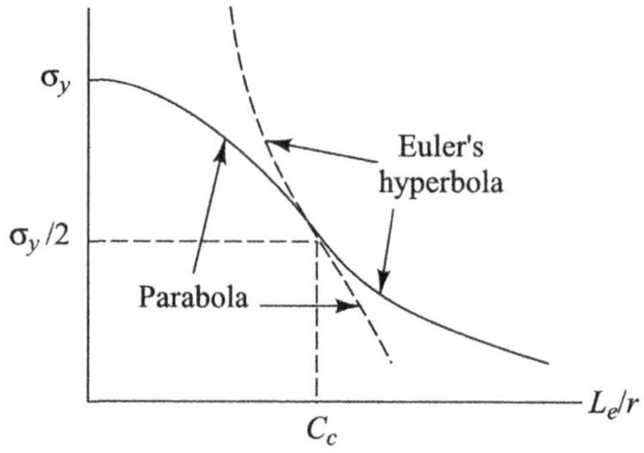

Fig. 9.8

$$\sigma_{all} = \frac{\pi^2 E}{\text{F.S.}(L_e/r)^2} \quad \text{(Euler's Formula) (9.12b)}$$

where,

F.S = Factor of safety

$$C_c = \sqrt{\frac{2\pi^2 E}{\sigma_y}}$$

9.4.3 Engesser Inelastic Column Theory

This is also known as modified Euler's formula since the derivation is similar. Euler assumed E as constant and therefore not applicable beyond limit of proportionality. Engesser proposed tangent modulus E_T to be used instead of E and therefore,

$$P_{cr} = \frac{\pi^2 E_T I}{L_e^2} \tag{9.13}$$

9.4.4 Straight Line Formula

Chicago Building code gives linear expression for safe working stress σ_w, for $30 < L_e/r < 120$,

$$\sigma_w = 112 - 0.49(L_e/r) \text{ MPa} \tag{9.14}$$

9.5 PARTIAL END CONSTRAINTS

Consider a column shown in Fig. 9.9(a), having one end constrained partially by a spring of stiffness k (N.m/rad) and the other end simply supported on a roller. End reaction R is induced due to buckling of the column under critical axial load P_{cr}. Bending equation gives the deformed shape shown in Fig. 9.9(b), for moment at any section at a distance x and y from the origin A.

$$-EI\frac{d^2y}{dx^2} = P_{cr} \times y - R \times x = P_{cr} \times y - \frac{k\theta}{L} \times x$$

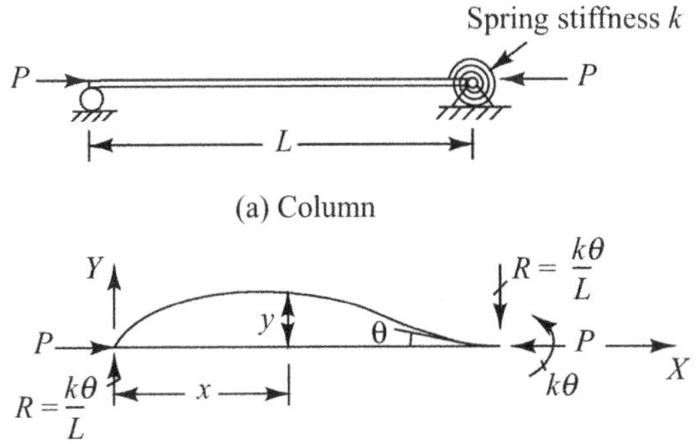

(b) Deflected shape

Fig. 9.9

The solution of this differential equation is of the form,

$$y = C\sin\sqrt{\frac{P_{cr}}{EI}}x + D\cos\sqrt{\frac{P_{cr}}{EI}}x + \frac{k\theta}{P_{cr}L} \times x \tag{9.15}$$

Critical load P_{cr} for different end conditions is expressed in term of Euler load in the form of Eq. (9.5b) as follows:

$$P_{cr} = \frac{P_e}{K^2} = \frac{\pi^2 EI}{(KL)^2}$$

$$\text{and let,} \quad k' = \frac{k}{(EI/L)} \tag{9.16}$$

Substituting Eq. (9.5) and (9.16) in Eq.(9.15),

$$y = C\sin\frac{\pi x}{KL} + D\cos\frac{\pi x}{KL} + \frac{k'\theta x}{(\pi/K)^2} \tag{9.17}$$

Boundary conditions:

At $x = 0$, $y = 0 \Rightarrow D = 0$

At $x = L$, $y = 0$

$$\Rightarrow \quad C = \frac{-k'\theta L}{(\pi/K)^2 \sin(\pi/K)}$$

and

At $x = L$, $\dfrac{dy}{dx} = -\theta$

which gives,

$$k' = \frac{(\pi/K)}{\cot(\pi/K) - 1/(\pi/K)} \tag{9.18}$$

Equation (9.18) gives effective length factor K in terms of relative rotational stiffness factor k'.

If both the ends of the column have some constrain against rotation, then there are two rotational stiffness factors k_1' and k_2' and the equation work out to be,

$$\left[\left(\frac{\pi}{K}\cos\frac{\pi}{K} - \sin\frac{\pi}{K}\right)k_1' - \left(\frac{\pi}{K}\right)^2 \sin\frac{\pi}{K}\right] \left[\left(\frac{\pi}{K}\cos\frac{\pi}{K} - \sin\frac{\pi}{K}\right)k_2' - \left(\frac{\pi}{K}\right)^2 \sin\frac{\pi}{K}\right] - \left(\frac{\pi}{K} - \sin\frac{\pi}{K}\right)^2 k_1'k_2' = 0 \tag{9.19}$$

The solution for Eq. (9.19) is given in Fig. 9.10 for effective length factor K verses k_1' and k_2'.

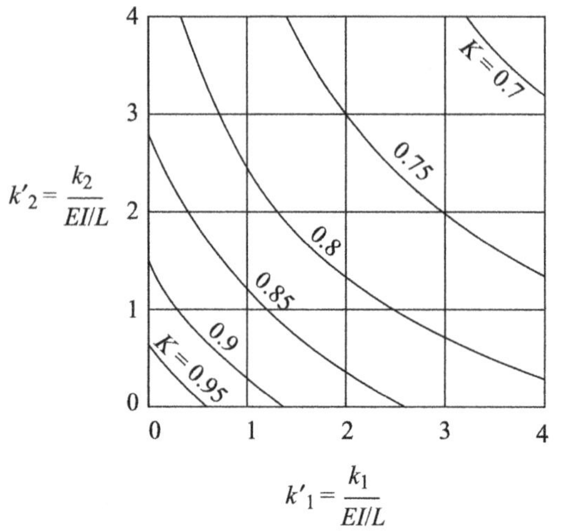

Fig. 9.10

9.6 VIRTUAL WORK METHOD

Virtual work method is widely used due to its simplicity. It gives an approximate value of the critical load and is particularly useful in cases where exact differential equation becomes too complicated.

Consider a compression member shown in Fig. 9.11 to carry an axial load P and a lateral load Q. The beam is given a virtual axial displacement δ, and the consequent strain energy U, stored due to bending is given by,

$$U = \int_O^L \frac{M^2 dx}{2EI} = \int_O^L \frac{(-EI \times d^2y/dx^2)^2 dx}{2EI}$$

$$= \frac{EI}{2}\int_O^L \left(\frac{d^2y}{dx^2}\right)^2 dx \qquad (9.20)$$

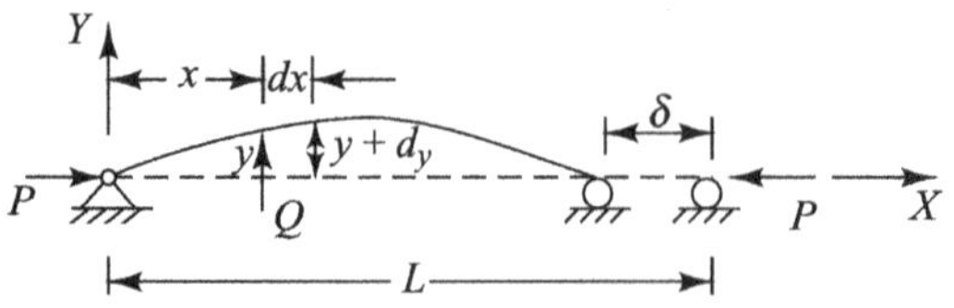

Fig. 9.11

Work done by external forces,

$$V = P \times \delta + Q \times y \qquad (9.21)$$

Equating, Work done = Energy stored,

$$P \times \delta + Q \times y = \frac{EI}{2}\int_O^L \left(\frac{d^2y}{dx^2}\right)^2 dx \qquad (9.22)$$

The strain energy due to bending is constant for a particular deflected shape for different combinations of P and Q. The value of Q will go on decreasing with increase in the value of P. The critical value of P is such the bending energy stored for a particular deflected shape is due to axial load P_{cr} alone and any lateral load Q, how-so-ever small will cause instability.

$$P_{cr} \times \delta = \frac{EI}{2}\int_O^L \left(\frac{d^2y}{dx^2}\right)^2 dx \qquad (9.23)$$

where $\delta = \int_O^L (ds - dx)$

and, ds = length of the column along elastic curve corresponding to length dx

$$ds = \sqrt{(dx^2 + dy^2)} = \left[1 + \left(\frac{dy}{dx}\right)^2\right]^{1/2} dx$$

$$\approx dx + \frac{1}{2}\left(\frac{dy}{dx}\right)^2 dx$$

or, $$\delta = \int_O^L \frac{1}{2}(dy/dx)^2 dx \qquad (9.24)$$

Substituting δ in Eq. (9.23)

$$P_{cr} = \frac{EI\int_O^L (d^2y/dx^2)^2 dx}{\int_O^L (dy/dx)^2 dx} \qquad (9.25)$$

To begin with, the deflected shape $y(x)$ is not known. Rayleigh's method for determination of critical frequencies of vibration of rods is used to solve this problem. An approximate deflected shape is assumed and approximate P_{cr} is obtained which will always be greater than the true value.

Rayleigh–Ritz method has slightly different solution procedure. The deflected shape is assumed as follows:

$$y = a_1 f_1(x) + a_2 f_2(x) + ... + a_n f_n(x)$$

where $f_1(x), f_2(x), ..., f_n(x)$ are the functions of x, such that each term satisfies the boundary conditions. The total potential energy $(U + V)$ is calculated by this assumed deflected shape function using Eqs. (9.20) and (9.21). The value of the coefficients $a_1, a_2, ..., a_n$ should be such that the total energy does not change, i.e.,

$$\frac{\partial(U+V)}{\partial a_1} = 0, \quad \frac{\partial(U+V)}{\partial a_2} = 0 .... \frac{\partial(U+V)}{\partial a_n} = 0$$

9.7 BEAM COLUMNS

Sometimes the compression members are subjected to bending moments due to lateral loads. Such members are called beam-columns. The two actions are not treated separately and the results are not superimposed because the deflection due to bending moment has an influence on the column buckling under axial load that would be ignored in the solution by superimposition.

9.7.1 Beam Column Subjected to Concentrated Lateral Load

Consider a column with a concentrated load W at its centre. Bending equation gives the deformed shape shown in Fig. 9.12 for moment at any section at a distance x and y from the origin C.

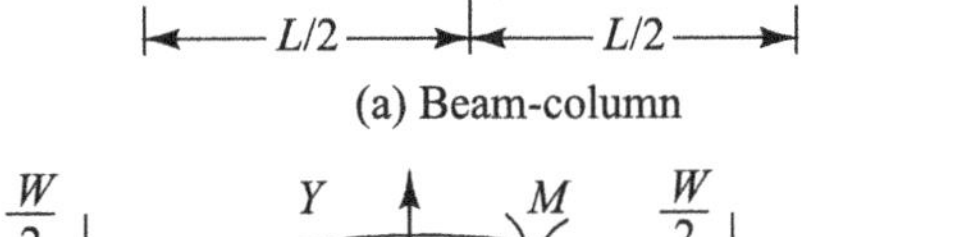

(a) Beam-column

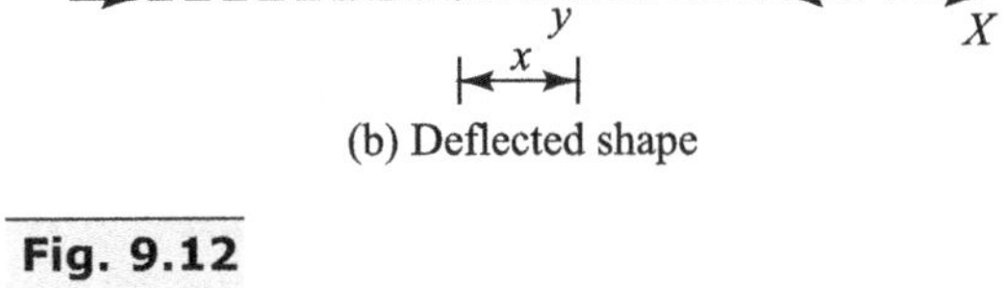

(b) Deflected shape

Fig. 9.12

For $0 \le x \le \frac{L}{2}$

$$-EI\frac{d^2y}{dx^2} = M = P \times y + \frac{W}{2}\left(\frac{L}{2} - x\right)$$

or

$$\frac{d^2y}{dx^2} + \mu^2 y = \frac{-W}{2EI}\left(\frac{L}{2} - x\right)$$

where $\mu^2 = P/EI$

The solution of this differential equation is of the form,

$$y = C\sin\mu x + D\cos\mu x - \frac{W}{2P}\left(\frac{L}{2} - x\right)$$

Boundary conditions:

At $x = 0, \frac{dy}{dx} = 0 \quad \Rightarrow C = \frac{-W}{2\mu P}$

At $x = \frac{L}{2}, y = 0 \quad \Rightarrow D = \frac{W}{2\mu P}\tan\mu\frac{L}{2}$

Then,

$$y = \frac{W}{2P}\left[\frac{1}{\mu}\left(\tan\frac{\mu L}{2}\cos\mu x - \sin\mu x\right) - \left(\frac{L}{2} - x\right)\right] \tag{9.26}$$

Maximum deflection y at $x = 0$,

$$y_{max} = \frac{W}{2P}\left[\frac{1}{\mu}\tan\mu L/2 - L/2\right] \tag{9.27}$$

Maximum bending moment at $x = 0$,

$$M_{max} = P \times y_{max} + WL/4$$

$$= \frac{W}{2\mu}\tan\mu L/2$$

If Z is the relevant section modulus, then maximum compressive stress is given by,

$$\sigma_{max} = \frac{P}{A} + \frac{M}{Z} = \frac{P}{A} + \frac{W}{2\mu Z}\tan\mu L/2 \tag{9.28}$$

9.7.2 Beam Column Subjected to Distributed Lateral Load

Consider a column subjected to a distributed lateral load w as shown in Fig. 9.13. Bending equation gives the deformed shape for moment at any section at a distance x and y from the origin at C.

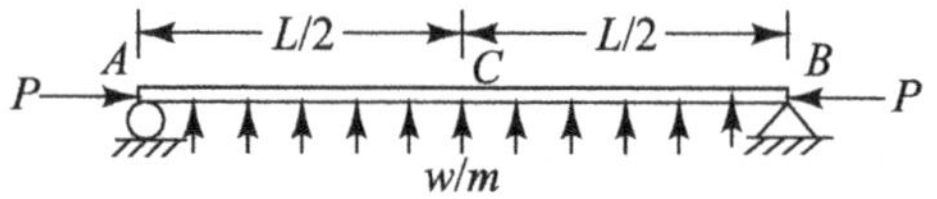

(a) Beam-column

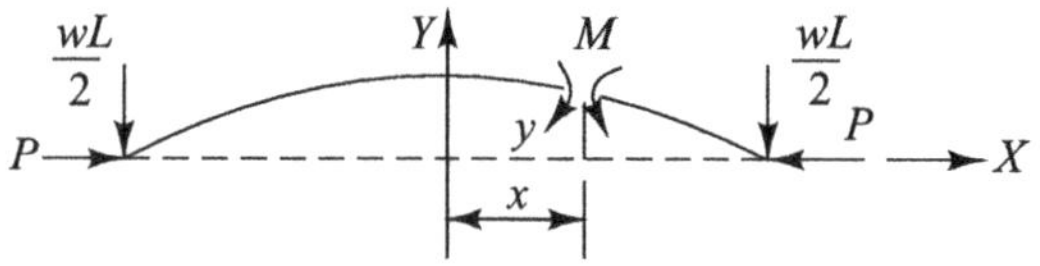

(b) Deflected shape

Fig. 9.13

$$-EI\frac{d^2y}{dx^2} = M = P\times y + w\left(\frac{L^2}{8} - \frac{x^2}{2}\right)$$

or,
$$\frac{d^2y}{dx^2} + \mu^2 y = \frac{-w}{EI}\left(\frac{L^2}{8} - \frac{x^2}{2}\right)$$

where $\mu^2 = P/EI$

The solution of this differential equation is of the form,

$$y = C\sin\mu x + D\cos\mu x + \frac{w}{2P}x^2 - \frac{wEI}{P}\left(\frac{L^2}{8EI} + \frac{1}{P}\right)$$

Boundary conditions:

$$\text{At } x = 0,\ \frac{dy}{dx} = 0 \Rightarrow C = 0$$

$$\text{At } x = \frac{L}{2},\ y = 0 \Rightarrow D = \frac{wEI}{P^2\cos\mu L/2}$$

Then,
$$y = \frac{w}{\mu^2 P}\times\frac{\cos\mu x}{\cos\mu L/2} + \frac{wx^2}{2P} - \frac{w}{\mu^2}\left(\frac{L^2}{8EI} + \frac{1}{P}\right) \quad (9.29)$$

Maximum deflection y at $x = 0$,

$$y_{max} = \frac{w}{\mu^2 P\cos\mu L/2} - \frac{w}{\mu^2}\left(\frac{L^2}{8EI} + \frac{1}{P}\right) \quad (9.30)$$

Maximum bending moment at $x = 0$,

$$M_{max} = P\times y_{max} + wL^2/8$$

$$= \frac{w}{\mu^2}(\sec\mu L/2 - 1) \quad (9.31)$$

If Z is the relevant section modulus, then maximum compressive stress is given by,

$$\sigma_{max} = \frac{P}{A} + \frac{M}{Z} = \frac{P}{A} + \frac{w}{\mu^2 Z}(\sec\mu L/2 - 1) \quad (9.32)$$

9.8 COLUMNS WITH INITIAL CURVATURE

In practice, a column cannot be made perfectly straight. Consider a pin ended column having its centroidal axis initially curved as shown in Fig. 9.14. For a simpler analysis, assume initial curvature given by a sinusoidal form as follows:

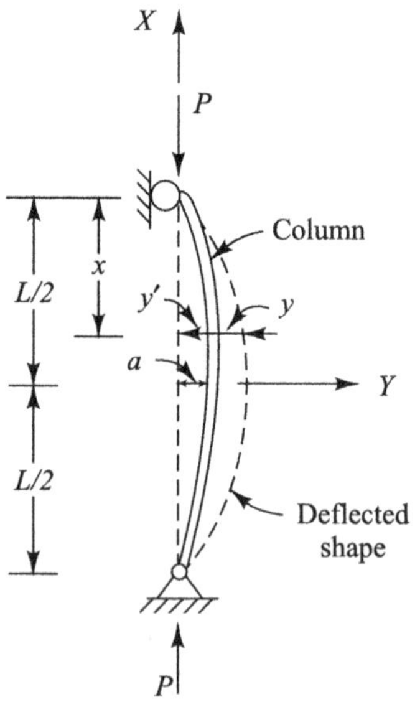

Fig. 9.14

$$y' = a \sin \frac{\pi x}{L}$$

where a = maximum initial curvature at the centre of the column.

Let y be the additional lateral displacement when some axial load P is applied on the column. At any section, bending moment,

$$M = P \times (y + y')$$

Substituting in bending equation,

$$-EI \frac{d^2 y}{dx^2} = M = P \times (y + y')$$

or, $$\frac{d^2 y}{dx^2} + \mu^2 y = -\mu^2 a \sin \frac{\pi x}{L}$$

where $\mu^2 = P/EI$

The solution of this differential equation is of the form,

$$y = C \sin \mu x + D \cos \mu x + \frac{\mu^2 a}{(\pi / L)^2 - \mu^2} \sin \frac{\pi x}{L}$$

Boundary conditions:

At $x = 0$, $y = 0$ $\Rightarrow D = 0$

At $x = L$, $y = 0$ $\Rightarrow C = 0$

$$y = \frac{\mu^2 a}{(\pi / L)^2 - \mu^2} \sin \frac{\pi x}{L} = \frac{a \sin(\pi x / L)}{(P_e / P) - 1} \tag{9.33}$$

where $P_e = \pi^2 EI/L^2$ is the Euler's load given by Eq. (9.5a)

Maximum deflection y at $x = L/2$,

$$y_{max} = \frac{a}{(P_e / P) - 1} \tag{9.34}$$

Maximum bending moment at $x = L/2$,

$$M_{max} = P \times (y_{max} + a) = Pa\left(\frac{P_e}{P_e - P}\right) \tag{9.35}$$

If Z is the relevant section modulus, then maximum compressive stress is given by,

$$\sigma_{max} = \frac{P}{A} + \frac{M}{Z} = \frac{P}{A} + \frac{aPP_e}{Z(P_e - P)} \tag{9.36}$$

EXAMPLES

Example 9.1 Determine the critical load P_{cr} for a column AB, free at end A and restrained against rotation by a spring of stiffness k at other end B as shown in Fig. 9.15.

Solution Equating moment of the system about end B,

$$P_{cr} L \sin \theta = k\theta$$

For small θ, $\sin \theta \approx \theta$ or, $P_{cr} = k/L$

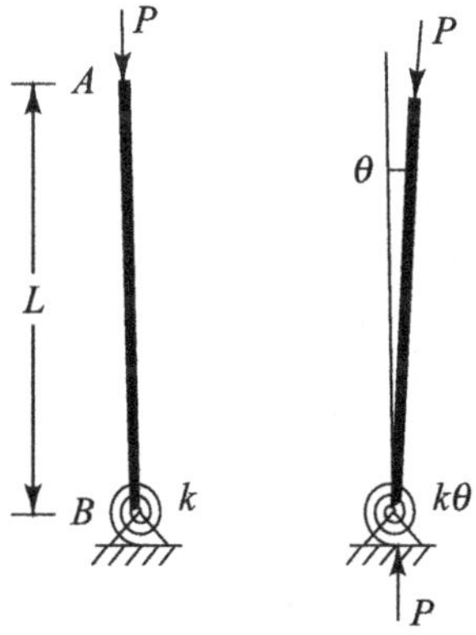

Fig. 9.15

Example 9.2 A rectangular timber column AB is restrained at end B in the ground. End A is restricted to move in x–z plane by two ropes as shown in Fig. 9.16. Neglecting tension in the cables, determine the most efficient cross-section of the column under an axial load of 20 kN. Take E = 12.5 GPa and factor of safety = 2.5.

Solution Column in x–z plane:

Effective length, $L_e = L/\sqrt{2}$ (Table 9.1)

Rad. of gyration, $r^2 = \frac{I}{A} = \frac{db^3}{12bd} = \frac{b^2}{12}$

Slenderness ratio, $\frac{L_e}{r} = \frac{L}{\sqrt{2}} \times \frac{\sqrt{12}}{b} = \frac{L\sqrt{6}}{b}$

Column in y–z plane

Effective length, $L_e = 2L$ (Table 9.1)

Rad. of gyration, $r^2 = \frac{I}{A} = \frac{bd^3}{12bd} = \frac{d^2}{12}$

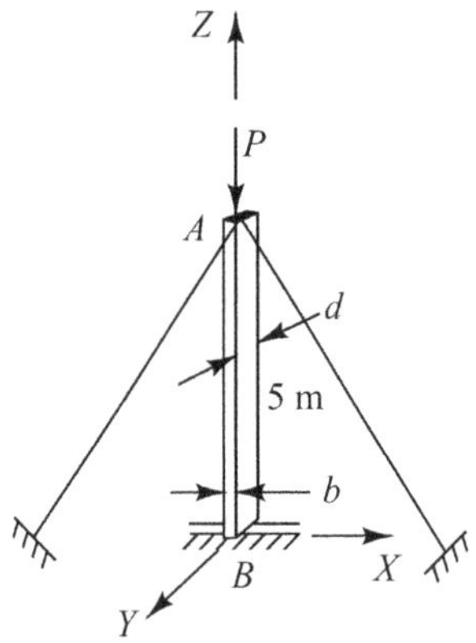

Fig. 9.16

Slenderness ratio, $\frac{L_e}{r} = 2L \times \frac{\sqrt{12}}{d} = \frac{L\sqrt{48}}{d}$

For most efficient section, critical loads in two modes of buckling should be equal for which slenderness ratio in two directions should be equal.

or, $\frac{L\sqrt{6}}{b} = \frac{L\sqrt{48}}{d}$

or, $d = 2.83b$ (i)

Design for given data F.S. = 2.5

$$P_{cr} = \text{F.S.} \times P = 2.5 \times 20 = 50 \text{ kN}$$

where, $P_{cr} = \frac{\pi^2 EI}{L_e^2}$

For bending in y–z plane,

$$50 = \frac{\pi^2 \times 12.5 \times 10^3 \times bd^3/12}{(2 \times 5000)^2} \quad \text{(ii)}$$

From Eqs (i) and (ii), $b = 68$ mm and $d = 192.6$ mm

Example 9.3 A steel column section SC 250 has sectional area = 109 cm^2 and second moment of area, $I_x = 12500$ cm^4 and $I_y = 3260$ cm^4. If yield stress of steel is 250 MPa and $E = 200$ GPa, determine the minimum length for which Euler's equation may be used for buckling load if,

(a) both ends of the column are hinged,
(b) both ends of the column are fixed rigidly.

Solution Minimum $I = 3260$ cm^4

$$\text{Minimum } r = \sqrt{\frac{I_{min}}{A}} = \sqrt{\frac{3260}{109}} = 5.469 \text{ cm}$$

Critical stress, $\sigma_{cr} = \frac{\pi^2 E}{(L_e/r)^2}$

or, $250 = \frac{\pi^2 \times 200 \times 10^3}{(L_e/5.469)^2}$

or, $L_e = 485.96$ cm

Therefore minimum length for which Euler's equation is applicable,

(a) when both ends are hinged,
$L = L_e = 4859.6$ mm (Table 9.1)
(b) when both ends are fixed,
$L = 2 \times L_e = 9719.2$ mm.

Example 9.4 Determine the critical stress for the column in Example 9.3 if the effective length of the column is 3 m by Euler, and Merchant–Rankine's formula.

Solution *Critical stress by Euler's formula,*

(a) when both ends are hinged, effective length $L = 3000$ mm
Critical stress,

$$\sigma_{cr} = \frac{\pi^2 E}{(L_e/r)^2} = \frac{\pi^2 \times 200 \times 10^3}{(3000/54.69)^2} = 656 \text{ MPa}$$

(b) when both ends are fixed, $L_e = 0.5 \times 3000 = 1500$ mm

Critical stress,

$$\sigma_{cr} = \frac{\pi^2 E}{(L_e / r)^2} = 4 \times 656 = 2624 \text{ MPa}$$

Euler's formula gives very high theoretical value of critical stress when the length of this column is less than 4859.6 mm. Euler's formula is therefore not applicable to short columns which do not buckle.

Critical stress by Merchant–Rankine formula,

(a) when both ends are hinged, Euler's stress $\sigma_e = 656$ MPa

Critical stress,

$$\sigma_{cr} = \frac{\sigma_e \times \sigma_y}{\left(\sigma_e^{1.4} + \sigma_y^{1.4}\right)^{1/1.4}}$$

$$\text{or, } \sigma_{cr} = \frac{656 \times 250}{\left(250^{1.4} + 250^{1.4}\right)^{1/1.4}} = 212 \text{ MPa}$$

(b) when both ends are fixed, Euler's stress, $\sigma_e = 2624$ MPa

$$\text{or, } \sigma_{cr} = \frac{2624 \times 250}{(2624^{1.4} + 250^{1.4})^{1/1.4}} = 243.56 \text{ MPa}$$

Note: Allowable stress in axial compression, $\sigma_{ac} = 127.2$ MPa in case (a) and 146.136 MPa in case (b).

Example 9.5 A steel bar 5 mm × 10 mm × 1000 mm is hinged at both ends. Determine its Euler's critical buckling load and estimate the maximum critical deflection before the material attains its yield points at 250 MPa. Take $E = 200$ GPa.

Solution Minimum moment of inertia,

$$I_{\min} = \frac{bd^3}{12} = \frac{10 \times 5^3}{12} = 104.17 \text{ mm}^4$$

Euler's critical load,

$$P_e = \frac{\pi^2 EI}{L^2} = \frac{\pi^2 \times 200 \times 10^3 \times 104.17}{1000^2}$$

$$= 205.6 \text{ N}$$

Direct axial compressive stress,

$$\sigma_c = \frac{205.6}{5 \times 10} = 4.112 \text{ MPa}$$

If δ is the maximum central deflection at mid length, the corresponding maximum moment $M = P \times \delta = \sigma_b \times I/y$

Maximum bending stress,

$$\sigma_b = \frac{P \times \delta}{I / y} = \frac{205.6 \times \delta}{104.17 / 2.5}$$

$$= 4.93 \times \delta \text{ MPa}$$

Total stress = $\sigma_c + \sigma_b \le$ yield stress σ_y

or, $4.112 + 4.93 \times \delta = 250$ or, $\delta = 49.87$ mm

Example 9.6 An I-section steel column it to carry load, applied at the face of the flange. It is rigidly fixed at the base and hinged at the top. Determine the allowable load if the length of the column, $L = 5.374$ m, factor of safety = 2.5, $E = 200$ GPa, and $\sigma_y = 250$ MPa. Properties of the I-section (SC 200) are as follows:

Depth of the section, $d = 200$ mm,

Area, $A = 7680$ mm^2, $I_x = 5530$ cm^4, $r_x = 84.8$ mm.

Solution Effective length,

$$L_e = \frac{L}{\sqrt{2}} = \frac{5.375}{\sqrt{2}} = 3.8 \text{ m}$$

Using secant formula,

$$\sigma_{\max} = \frac{P}{A}\left[1 + \frac{ec}{r^2} \sec \sqrt{\frac{P}{EI}} \times \frac{L_e}{2}\right]$$

where $\sigma_{\max} = 250$ MPa, $e = c = 200/2 = 100$ mm

$$250 = \frac{P}{7680}\left[1 + \frac{100 \times 100}{84.8^2} \sec \sqrt{\frac{P}{200000 \times 5530 \times 10^4}} \times \frac{3800}{2}\right]$$

Solving by hit and trial, $P \approx 7.45 \times 10^5$ N

$$\text{Allowable } P = \frac{7.45 \times 10^5}{2.5} = 298000 \text{ N}$$

Example 9.7 A 40 mm square strut of aluminium is 0.8 m long. It is rigidly fixed at the base and free to rotate or sway at the top. If the eccentricity of the load is 5 mm along an axis perpendicular to a face, determine the load for which horizontal deflection of free end is 10 mm. Also determine the maximum stress in the bar. Take $E = 70$ GPa.

Solution Effective length,

$L_e = 2L = 2 \times 800 = 1600$ mm (Table 9.1)

$$I = \frac{bd^3}{12} = \frac{40 \times 40^3}{12} = 2.13 \times 10^5 \text{ mm}^4$$

$$y_{\max} = e \times \left(\sec \sqrt{\frac{P}{EI}} \times \frac{L}{2} - 1 \right)$$

$$\text{or, } 10 = 5 \times \left(\sec \sqrt{\frac{P}{70000 \times 2.13 \times 10^5}} \times \frac{1600}{2} - 1 \right)$$

Solving by hit and trial, $P = 35400$ N

$$\sigma_{\max} = \frac{P}{A}\left[1 + \frac{ec}{r^2} \sec \sqrt{\frac{P}{EI}} \times \frac{L_e}{2}\right]$$

where $e = 5$ mm, $c = 40/2 = 20$ mm, $r^2 = I/A$ $= 1600/12$

$$\sigma_{\max} = \frac{35400}{1600}\left[1 + \frac{5 \times 20}{1600/12} \sec \sqrt{\frac{35400}{70000 \times 2.13 \times 10^5}} \times \frac{1600}{2}\right] = 72 \text{ MPa}$$

Example 9.8 Determine the critical load P for the portal frame shown in Fig. 9.17(a), if the second moment of area of the columns is 255×10^4 mm^4 and $E = 200$ GPa. The columns and the beam are welded rigidly at the joint.

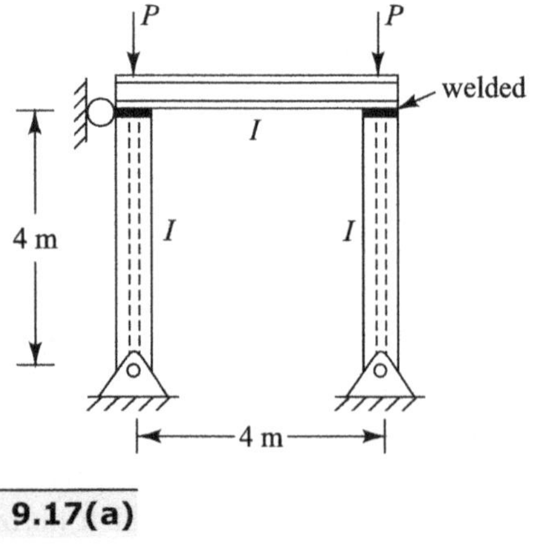

Fig. 9.17(a)

Solution Figure 9.17(b) shows two possible alternates for the deflected shape. The relation between rotation θ and moment M at the beam end is obtained by appendix Table F as follows:

Case 1, $$\theta = \frac{1}{3}\frac{ML}{EI} + \frac{1}{6}\frac{ML}{EI} = \frac{1}{2}\frac{ML}{EI}$$

Case 2, $$\theta = \frac{1}{3}\frac{ML}{EI} - \frac{1}{6}\frac{ML}{EI} = \frac{1}{6}\frac{ML}{EI}$$

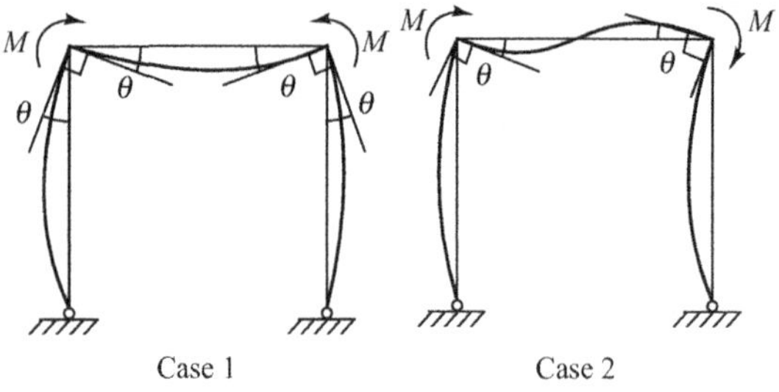

Fig. 9.17(b) *Alternate Modes of Buckling*

The restrain at column top against rotation

$$= k = M/\theta = \frac{2EI}{L} \quad \text{Case 1}$$

$$= \frac{6EI}{L} \quad \text{Case 2}$$

If end restrain k is less, the effective length factor K is more as per Fig. 9.10. If K is more, the effective length L_e (= KL) is more and it will be the critical case for buckling. Therefore Case 1 is more critical.

From Fig. 9.10, $K = 0.87$, corresponding to

$$k_1' = \frac{2EI/L}{EI/L} = 2 \text{ and, } k_2' = 0 \quad \text{(hinged)}$$

$$L_e = KL = 0.87 \times 4000 \text{ mm}$$

$$P_{cr} = \frac{\pi^2 EI}{(KL)^2} = \frac{\pi^2 \times 200000 \times 255 \times 10^4}{(0.87 \times 4000)^2}$$

$$= 4.156 \times 10^5 \text{ N}$$

Example 9.9 Derive an expression for the buckling load P on the portal frame shown in Fig. 9.18(a).

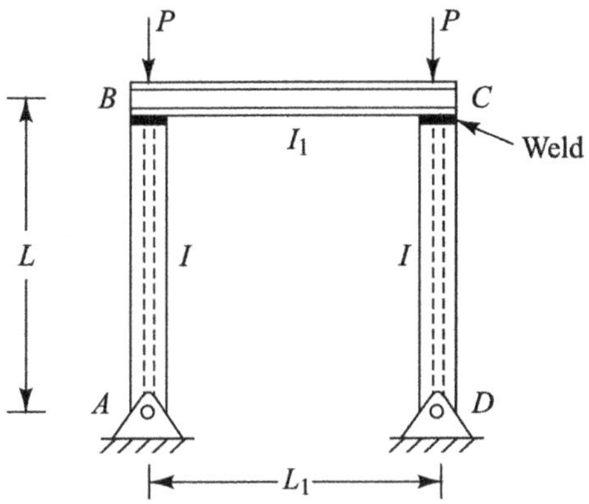

Fig. 9.18(a)

Solution Figure 9.18(b) shows the deflected shape of the portal frame. Taking hinged end as the origin, the bending equation is given by,

$$-EI\frac{d^2y}{dx^2} = P \times y$$

As earlier in Art. 9.2.1, the solution of this equation is of the form,

$$y = C \sin \mu x + D \cos \mu x \quad \text{where } \mu^2 = P/EI$$

Boundary conditions:

At $x = 0$, $y = 0 \Rightarrow D = 0$

At $x = L$, $\quad y = \delta \Rightarrow C = \dfrac{\delta}{\sin \mu L}$

At $x = L$, $\quad \dfrac{dy}{dx} = \theta = \delta\mu \cot \mu L \qquad$ (i)

Since joint B is rigid, the column end slope will be equal to the beam end slope. The beam end slope is equal to the sum of rotations due to the moments at both ends.

or, $\quad \theta = \dfrac{1}{3}\dfrac{ML_1}{EI_1} - \dfrac{1}{6}\dfrac{ML_1}{EI_1} = \dfrac{1}{6}\dfrac{ML_1}{EI_1} \qquad$ (ii)

Equating (i) and (ii), $\dfrac{1}{6}\dfrac{ML_1}{EI_1} = \delta\mu \cot \mu L \qquad$ (iii)

Taking moment of the reaction force P at joint A, about joint B,

$$P \times \delta = M$$

Substituting in Eq. (iii),

$$\frac{P\delta L_1}{6EI_1} = \delta\mu \cot \mu L$$

or, $\quad \mu L \tan \mu L = \dfrac{6I_1L}{IL_1}$

Solving by hit and trial method for smallest root if

$I_1 = I$ and $L_1 = L$, then $\mu L = 1.35$

or $\quad P_{cr} = 1.8225\ EI/L^2$

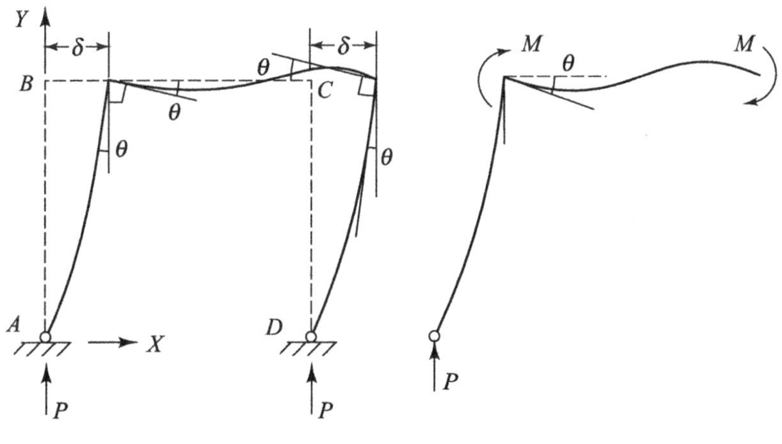

Fig. 9.18(b) *Deflected Shape*

Example 9.10 A column of length L is restrained at both ends by springs of stiffness k (N m/rad). Show that the critical load P is given by the expression, $\mu + (k/EI) \tan (\mu L/2) = 0$ where, $\mu^2 = P/EI$ and EI = flexural rigidity of the column.

Solution End reaction moments M develop due to buckling of the column as shown in Fig. 9.19. Bending equation defines the deflected shape as follows:

$$\frac{d^2y}{dx^2} = \frac{M - P \times y}{EI}$$

Its solution is given by,

$$y = C \sin \mu x + D \cos \mu x + M/P$$

Boundary conditions:

At $x = 0, y = 0 \Rightarrow D = -M/P$

At $x = \frac{L}{2}, \frac{dy}{dx} = 0 \Rightarrow C = \frac{-M}{P} \tan \mu L/2$

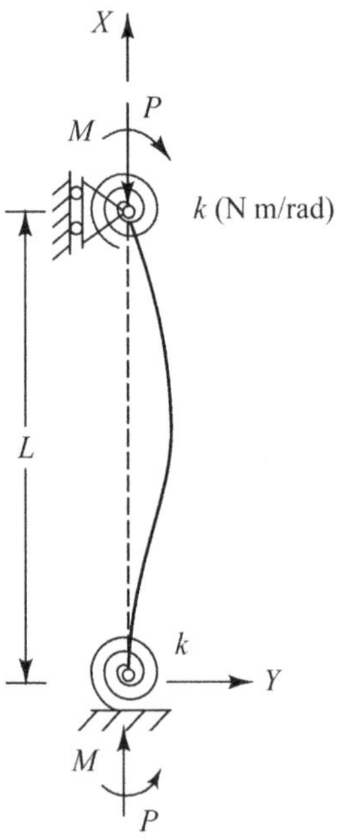

Fig. 9.19

At $x = 0$, $\frac{dy}{dx} = \theta = \frac{-M}{P} \mu \tan \mu L/2$

Substituting $M = k\theta$,

$$\theta = \frac{-k\theta}{P} \mu \tan \mu L/2$$

or, $P = -\mu k \tan \mu L/2$

or, $\mu^2 = -\frac{\mu k}{EI} \tan \mu L/2$

or, $$\mu + \frac{k}{EI} \tan \mu L/2 = 0$$

Example 9.11 A horizontal compression member consists of a 100 mm diameter (outside) tube of wall thickness 6 mm. It is 7 m long and carries a load of 50 kN. Calculate the maximum stress in the pipe if the self-weight of the pipe is 139 N/m. Take E = 200 GPa.

Solution Area

$$A = \pi(100^2 - 88^2)/4 = 1772 \text{ mm}^2$$

Moment of inertia,

$$I = \pi(100^4 - 88^4)/64 = 1.965 \times 10^6 \text{ mm}^4$$

Sectional modulus,

$$Z = \frac{I}{(d/2)} = \frac{1.965 \times 10^6}{50}$$
$$= 3.93 \times 10^4 \text{ mm}^3$$

$$\mu^2 = \frac{P}{EI} = \frac{50000}{200000 \times 1.965 \times 10^6}$$
$$= 1.27 \times 10^{-7}$$

$$\mu = 3.567 \times 10^4$$

UDL, $w = 0.139$ N/mm

From Eq. (9.32),

$$\sigma_{max} = \frac{P}{A} \pm \frac{w}{\mu^2 Z}\left(\sec \frac{\mu L}{2} - 1\right)$$

$$= \frac{50000}{1772} \pm \frac{0.139}{1.27 \times 10^{-7} \times 3.93 \times 10^4}$$

$$\left(\sec \frac{3.567 \times 10^{-4} \times 7000}{2} - 1\right)$$

$$= 28.22 \pm 60.05 = 88.26 \text{ N/mm}^2, \quad -31.83 \text{ N/mm}^2$$

Example 9.12 A 400 mm × 400 mm concrete column is 4 m long. It is to carry an axial load of 200 kN in addition to a lateral load of 12.5 kN. Determine the maximum stress in the column if the ends of the column are rigidly fixed against rotation but not against sway. Take $E = 2.55 \times 10^4$ N/mm^2.

Solution This problem can be solved by expressions derived for hinged end condition in Art. 9.7.1, using equivalent hinged end column with lateral load. $W = 2 \times 12.5$ kN and length, $L = 4$ m, as explained in Fig. 9.20.

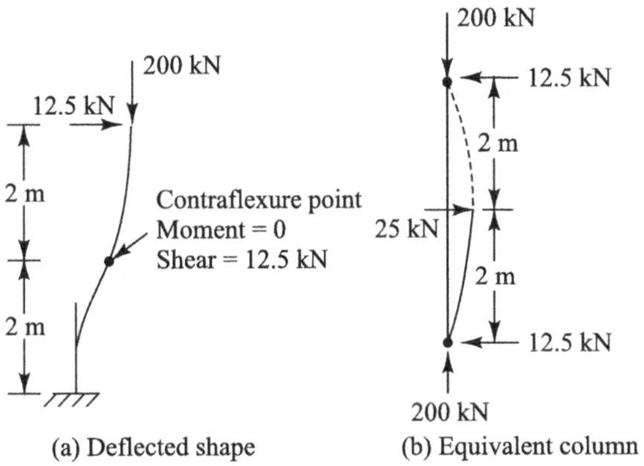

Fig. 9.20

Area, $A = 400 \times 400 = 160000$ mm^2

Moment of inertia,

$$I = \frac{bd^3}{12} = \frac{400 \times 400^3}{12}$$
$$= 2.133 \times 10^9 \text{ mm}^4$$

Sectional modulus,

$$Z = \frac{I}{(d/2)} = \frac{2.133 \times 10^9}{200}$$
$$= 1.067 \times 10^7 \text{ mm}^3$$

$$\mu^2 = \frac{P}{EI} = \frac{200000}{25500 \times 2.133 \times 10^9}$$
$$= 3.677 \times 10^{-9}$$
$$\mu = 6.06 \times 10^{-5}$$

From Eq. (9.28),

$$\sigma_{\max} = \frac{P}{A} \pm \frac{W}{2\mu Z} \tan(\mu L/2)$$
$$= \frac{200000}{160000} \pm \frac{25000}{2 \times 6.06 \times 10^{-5} \times 1.067 \times 10^7}$$
$$\tan \frac{6.06 \times 10^{-5} \times 4000}{2}$$
$$= 1.25 \pm 2.356 = 3.6 \text{ N/mm}^2, -1.1 \text{ N/mm}^2$$

Example 9.13 Determine the critical bulking load for a column with hinged ends using strain energy method.

Solution Assume half sine wave curve for the deflected shape of the column as shown in Fig. 9.21. It satisfies the boundary conditions as follows:

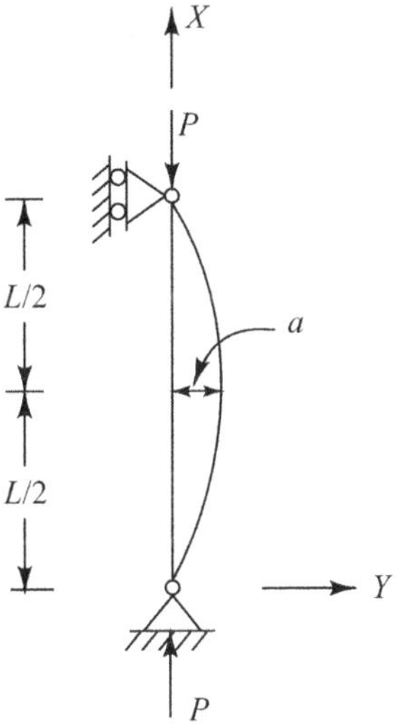

Fig. 9.21

$$y = a \sin \pi x/L$$

whereas, $y = 0$ at $x = 0$ and $x = L$ and $dy/dx = 0$ at $x = L/2$

Substituting in Eq. (9.25),

$$P_{cr} = \frac{EI\int_0^L (d^2y/dx^2)^2 dx}{\int_0^L (dy/dx)^2 dx}$$

$$= \frac{\dfrac{a^2\pi^4 EI}{L^4}\displaystyle\int_O^L \sin^2(\pi x/L)dx}{a^2\dfrac{\pi}{L^2}\displaystyle\int_O^L \cos^2(\pi x/L)dx} = \frac{\pi^2 EI}{L^2}$$

Example 9.14 A long slender cantilever column is rigidly fixed at the base and free at the other. Determine its self weight under which it will buckle using strain energy method.

Solution Let self weight/unit length = w.

Figure 9.22 shows the deflected shape of the column. Let the deflected shape be given by,

$$y = a(1 - \cos \pi x/2L)$$

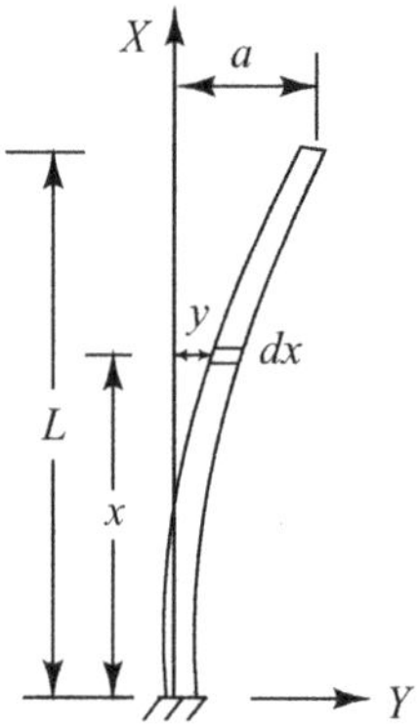

Fig. 9.22

Strain energy due to bending (Eq. 9.20)

$$U_b = \frac{EI}{2}\int_O^L \left[\left(\frac{d^2y}{dx^2}\right)^2\right] dx$$

$$= \frac{EI\pi^4 a^2}{32L^4}\int_O^L \cos\frac{\pi x}{2L}dx$$

$$= \frac{EI\pi^4 a^2}{64L^3}$$

The vertical movement δ of the elementary load $w \times dx$ is given by Eq. 9.24 as follows:

$$\delta = \frac{1}{2}\int_O^x \left(\frac{dy}{dx}\right)^2 dx = \frac{1}{2}\int_O^x \left(\frac{a\pi}{2L}\sin\frac{\pi x}{2L}\right)^2 dx$$

$$= \frac{a^2\pi^2}{8L^2}\int_O^x \left[\frac{1 - \cos \pi x/L}{2}\right]dx$$

$$= \frac{a^2\pi^2}{16L^2}\left[x - \frac{L}{\pi}\sin\frac{\pi x}{L}\right]$$

Work done by elementary force $w \times dx$,

$$dV = w \times dx \times \delta$$

$$= \frac{wa^2\pi^2}{16L^2}\left[x - \frac{L}{\pi}\sin\frac{\pi x}{L}\right]dx$$

Integrating for total work done,

$$V = \int_O^L dV = \frac{wa^2}{32}(\pi^2 - 4) = 0.1834\ wa^2$$

Equating strain energy stored-work done,

or, $$\frac{EI\pi^4 a^2}{64L^3} = 0.1834\ wa^2$$

or, Critical weight of the column, $wL = 8.3\ EI/L^2$

Example 9.15 Determine the critical load for a cantilever column fixed at the base and free at the top, using Rayleigh–Ritz method to carry a compressive load P at the top.

Solution Assume that the deflected shape is given by,

$$y = a_1\left(\frac{x}{L}\right)^2 + a_2\left(\frac{x}{L}\right)^3$$

It should satisfy the boundary conditions, i.e.,

At $x = 0$, $y = 0$, $\dfrac{dy}{dx} = 0$.

At $x = L$, $\dfrac{d^2y}{dx^2} = 0$.

Strain energy,

$$U = \frac{EI}{2}\int_O^L \left(\frac{d^2y}{dx^2}\right)^2 dx = \frac{EI}{2}\int_O^L \left(\frac{2a_1}{L^2} + \frac{6a_2 x}{L^3}\right)^2 dx$$

or $$U = \frac{2EI}{L^3}(a_1^2 + 3a_1a_2 + 3a_2^2)$$

Loss in potential energy,

$$V = \frac{-P}{2}\int_O^L \left(\frac{dy}{dx}\right)^2 dx = \frac{-P}{2}\int_O^L \left(\frac{2a_1x}{L^2} + \frac{3a_2x^2}{L^3}\right)^2 dx$$

$$= -\frac{P}{30L}(20a_1^2 + 45a_1a_2 + 27a_2^2)$$

$$U + V = \left(\frac{2EI}{L^3} - \frac{2P}{3L}\right)a_1^2 + \left(\frac{6EI}{L^3} - \frac{3P}{2L}\right) a_1a_2 + \left(\frac{6EI}{L^3} - \frac{9P}{10L}\right)a_2^2$$

The values of a_1 and a_2 should be such that,

$$\frac{\partial(U+V)}{\partial a_1} = \left(\frac{4EI}{L^3} - \frac{4P}{3L}\right)a_1 + \left(\frac{6EI}{L^3} - \frac{3P}{2L}\right)a_2 = 0$$

$$\frac{\partial(U+V)}{\partial a_2} = \left(\frac{6EI}{L^3} - \frac{3P}{3L}\right)a_1 + \left(\frac{12EI}{L^3} - \frac{9P}{5L}\right)a_2 = 0$$

which gives,

$$\left(\frac{4EI}{L^3} - \frac{4P}{3L}\right)\left(\frac{12EI}{L^3} - \frac{9P}{5L}\right) = \left(\frac{6EI}{L^3} - \frac{3P}{2L}\right)^2$$

or, $P = 2.49\ EI/L^2$

EXERCISES

PROBLEMS

9.1 Three rigid bars of length L are connected by hinges at both ends. The rotation at the hinges is restrained by springs of stiffness k (N.m/rad). Determine the critical load P corresponding to the antisymmetrical deflected shape shown in Fig. 9.23.

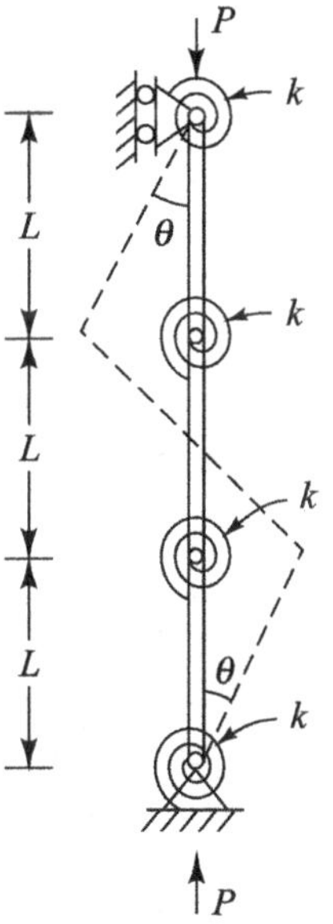

Fig. 9.23

9.2 Determine the critical load P on a rigid bar supported by two horizontal springs of stiffness k (N/mm) as shown in Fig. 9.24.

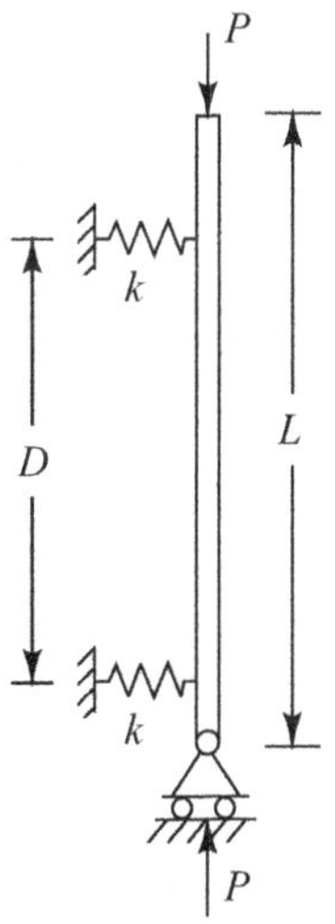

Fig. 9.24

9.3 A hollow circular steel column 5 mm thick has external diameter 150 mm. Determine the Euler's critical load if its length is 3.5 m and, $E = 200$ GPa. The end conditions are as follows.
(a) its both ends are fixed
(b) its one end is free and other end is hinged.

9.4 Solve Problem 9.3 using (i) Rankines formula and (ii) Secant formula if $e = 10$ mm and $\sigma_y = 250$ MPa.

9.5 A hollow circular column has external diameter 50 mm. It is fixed at the base and free to have lateral movement at the top. If the length of the column is 3 m and Euler's buckling load is 3kN, determine the thickness of the column.

9.6 Compare the strength of two circular columns of same cross-section and material. One of the circular column is hollow having wall thickness 1/4th of the external diameter.

9.7 Determine the maximum value of the slenderness ratio for a mild steel column for which Euler's formula is valid if its yield stress in 250 MPa and the modulus of elasticity is 210 GPa.

9.8 A steel column 8 m long consists of two channel sections MC 350 placed back to back. The properties of MC 350 section shown in Fig. 9.8 are as follows
$I_x = 10^8$ mm^4, $I_y = 435 \times 10^4$ mm^4
area $= 5.44 \times 10^3$ mm^2.

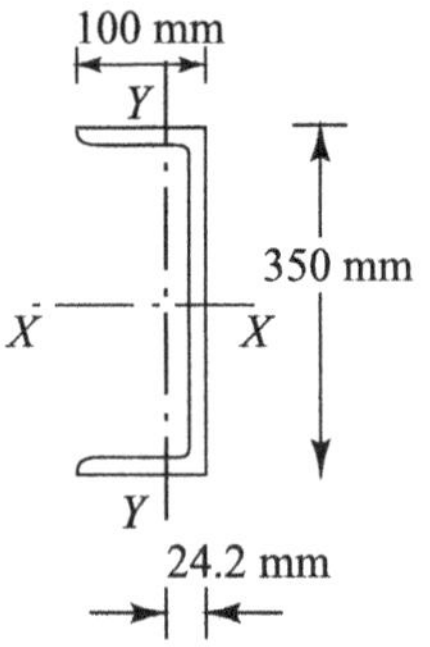

Fig. 9.25

Determine the spacing of the channels so that its strength is equal in X–X and Y–Y direction. Also determine its strength using Rankine's formula if the yield stress is 250 MPa and the Rankine's constant is 1/9000.

9.9 Two channel sections ISLC350 (Fig. 9.26) are placed toe to toe in a built up laced column of length (effective) 4.5m. Determine the distance between webs of the two channel sections if the strength of the column is same about two major axis. For the channel section ISLC350,
$I_x = 9330.9$ cm^4, $I_y = 395.5$ cm^4
area, $A = 49.54$ cm^2. Also determine the permissible load on the column if the eccentricity is 50 mm w.r.t. X–X axis and yield stress is 250 MPa. Take factor of safety equal to 2 and $E = 200$ GPa.

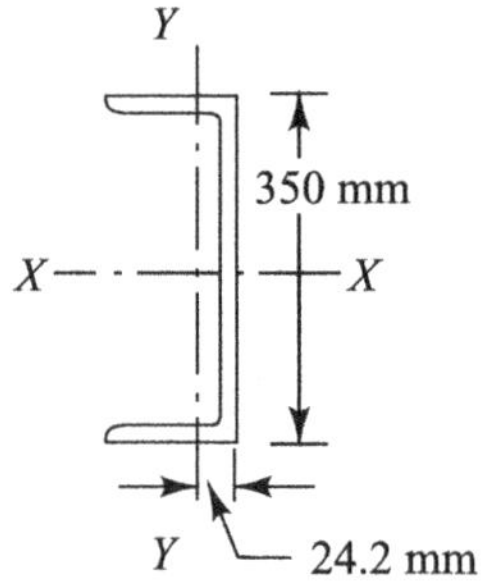

Fig. 9.26

9.10 A simply supported mild steel beam 20 mm × 20 mm is 2 m long. It carries a central load of 80 N and the central deflection is 5 mm. What is the Euler's load for the beam and what is the ratio of maximum stress in both the cases?

9.11 Determine the critical buckling load P for the steel columns shown in Fig. 9.27. The length of the columns is 1.8 m and sectional area is 10 mm × 12 mm. The top beam is very rigid in each case. Take $E = 200$ GPa.

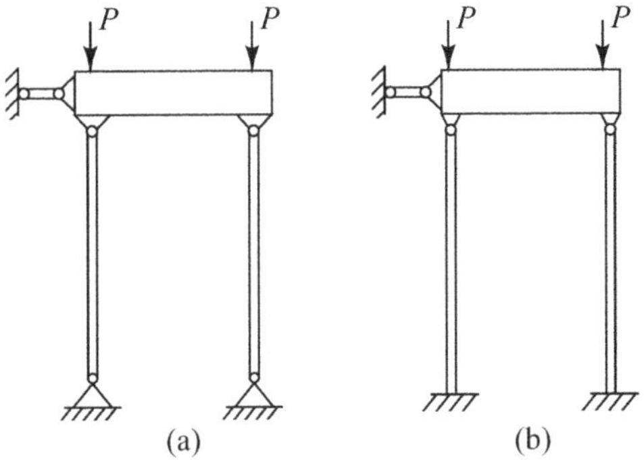

Fig. 9.27(a - b)

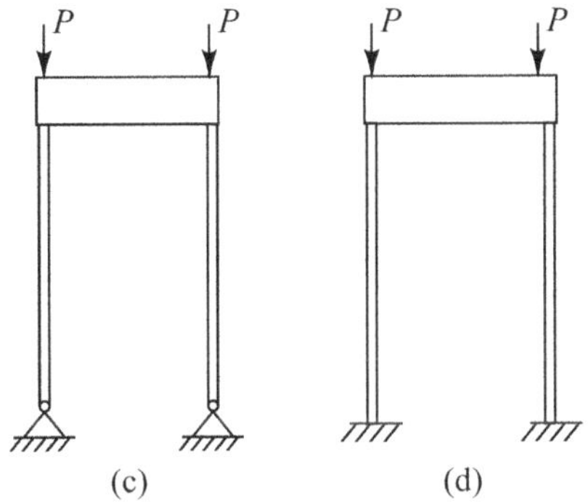

Fig. 9.27(c - d)

9.12 A steel bar of length L, is rigidly fixed at one end and hinged at the other. Determine the maximum value of the slenderness ratio for Euler's instability to occur for a rise in temperature of 40°C. Take $\alpha = 11.7 \times 10^{-6}$ °C.

9.13 A column is rigidly fixed at one end and free to rotate or sway at the other. It carries a vertical load P and a horizontal load $P/4$ at the free end. Determine the lateral displacement of the free end if P is equal to one-fourth of the Euler's load for this column.

9.14 A column of length L is fixed at one end and restrained against sway by a spring of stiffness k (N/mm). Taking displaced position of the column end as the origin, show that the buckling load is given by,

$$\mu \cot \mu L - \frac{k}{\mu^2 EI}(\mu L \cos \mu L - 1) = 0,$$

where $\mu^2 = P/EI$ and EI = flexural rigidity of the column.

9.15 A column has length L and flexural rigidity EI. It is hinged at both ends. Rotation of one end is restrained by a spring of stiffness kEI/L (N m/rad). Show that its critical buckling load P is given by the expression,

$$\tan \mu L = \frac{k \mu L}{k + \mu^2 L^2}$$

where $\quad \mu^2 = P/EI$

9.16 A column has length L and flexural rigidity EI. It is hinged at one end but the rotation is restrained by a spring of stiffness k_1 (N.m/rad). It is free to rotate at the other end but the sway is restrained by a linear spring of stiffness k_2 (N/m). Show that its critical buckling load P is given by the expression,

$$\mu L \cot \mu L + \frac{k_2 L}{P - k_2} - \frac{PL}{k_1} = 0$$

where $\quad \mu^2 = P/EI$

9.17 Determine the safe load P which could be applied on the pin jointed structure shown in Fig. 9.28, if Euler's buckling formula is applicable and factor of safety is 3. Take E = 200 GPa.

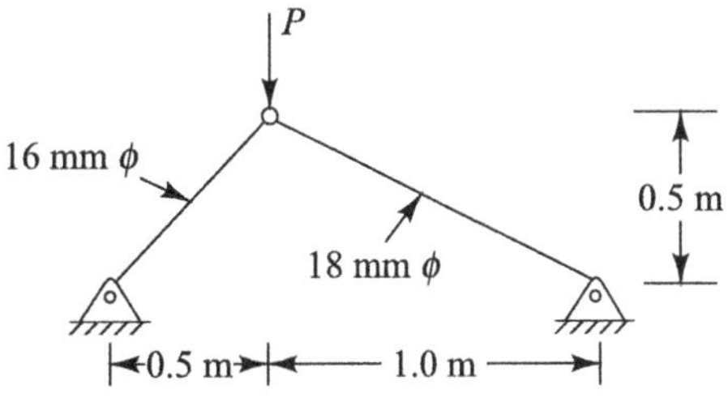

Fig. 9.28

9.18 A hollow circular steel column of external diameter 120 mm and wall thickness 15 mm, has to carry a load of 400 kN at an eccentricity of 20 mm. If the length of the column is 3 m and its both ends are hinged, calculate the maximum and minimum stresses developed in the column. What is the maximum permissible eccentricity so that no tension is induced in the column. Use secant formula and take modulus of elasticity, E = 200 GPa.

9.19 Design a circular steel bar 2 m long, to carry an axial load of 50 kN at an eccentricity of 10 mm if it is free at one end and fixed at the other and the bars are available in even number diameters only. Take E = 200 GPa and allowable compressive stress is 110 MPa.

9.20 Determine the buckling load P for the plane frame ABC shown in Fig. 9.29 if $EI = 1.26 \times 10^{12}$ N mm^2 for both the members AB and BC.

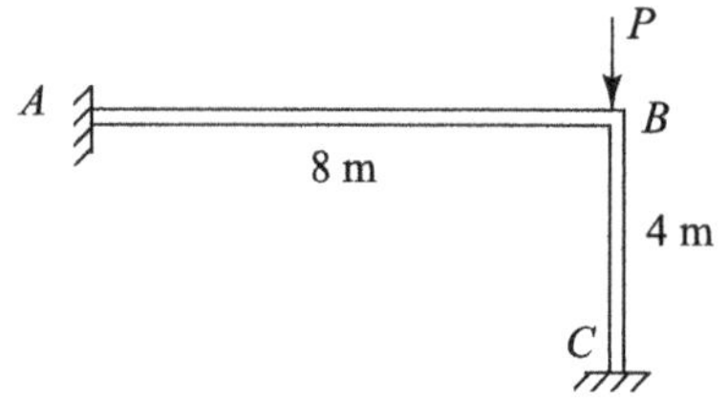

Fig. 9.29

9.21 and 9.22 Determine the critical buckling load P for the plane frame ABC shown in Figs 9.30 and 9.31, if the flexural rigidity EI is same for the members AB and BC.

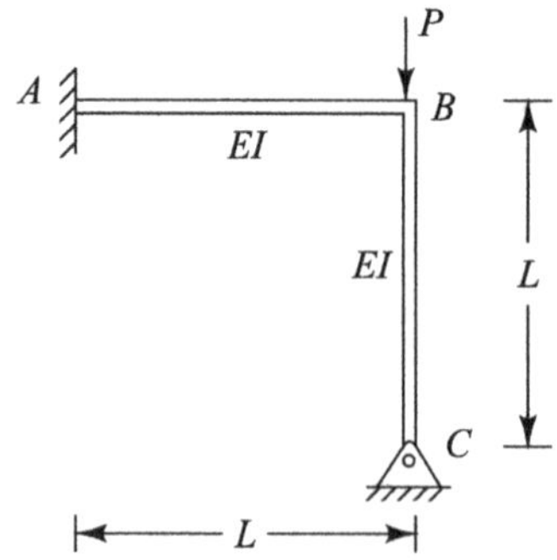

Fig. 9.30

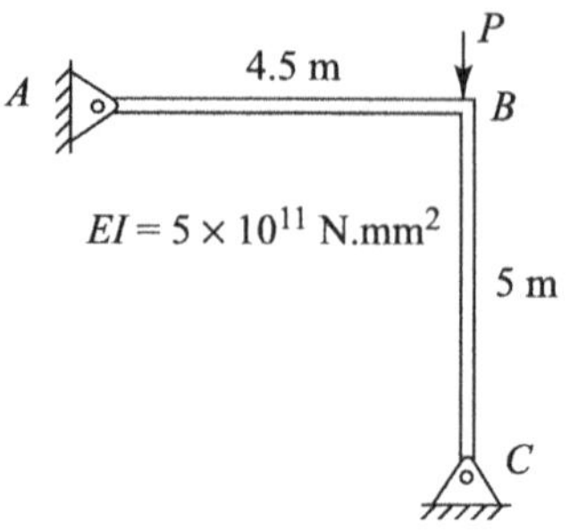

Fig. 9.31

9.23 A horizontal compression member is 7 m long in a steel bridge. It consist of a normal T section ISNT150. Determine the maximum stress when axial loads is 600 kN and self weight is 227 N/m. Take E = 200 GPa and for ISNT150, $I_x = 5.41 \times 10^6$ mm^4 and area = 2890 mm^2.

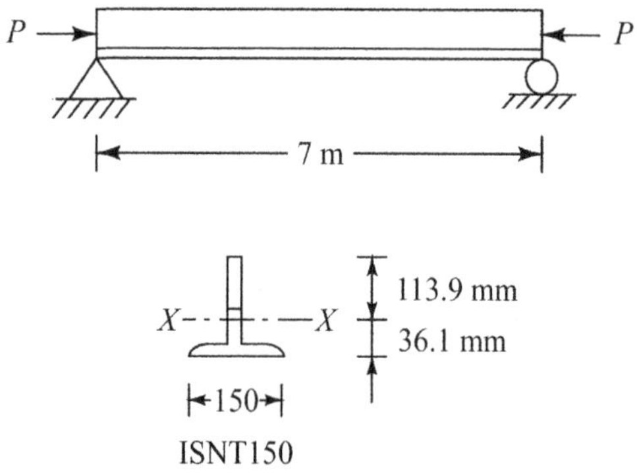

Fig. 9.32

9.24 A steel column I-section SC 250 (depth and width = 250 mm), is 5 m long in a rigidly jointed building frame. It is to carry an axial load of 400 kN and a lateral load of 20 kN at its top. If momemt of inertia of the column about the axis of bending is 12.5×10^7 mm^4, determine the maximum stress in the column. Take E = 200 GPa, A = 109 cm^2 for the column.

9.25 Determine critical buckling load for the column shown in Fig. 9.33 assuming that the deflected shape of the column is approximately given by the expression.
$y = \delta(1 - \cos \pi x/2L)$, where δ is the deflection of the free end.

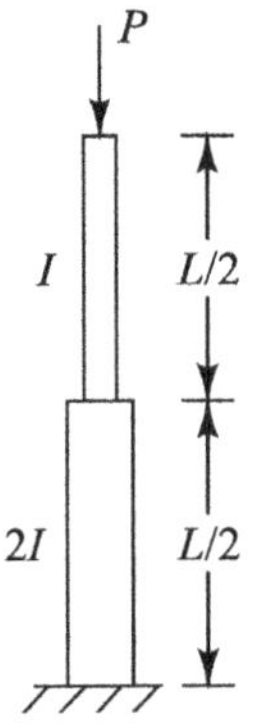

Fig. 9.33

9.26 Determine critical buckling load for the column shown in Fig. 9.34 assuming half sine wave as the approximate deflected shape of the column.

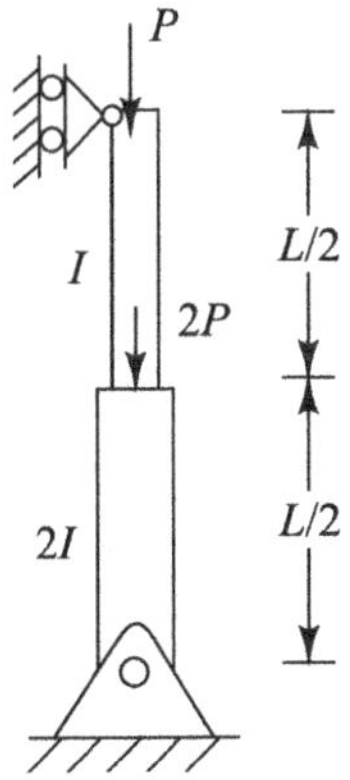

Fig. 9.34

9.27 A column is hinged at both the ends. Determine the critical buckling load for the column if the moment of inertia of the column at any distance x from the fixed end is given by $I_x = I_0(1 + \sin \pi x/L)$. Assume the buckling curve, $y = \sum_{n=1}^{\infty} a_n \sin n\pi x/L$ using Rayleigh–Ritz method.

OBJECTIVE QUESTIONS

9.28 Critical buckling load P_{cr} by Euler's formula is general is given by,
(a) $\pi^2 EA/(KL/r)^2$ (b) $\pi^2 EI/L^2$
(c) $\pi^2 EA/L^2$ (d) $\pi^2 EA/(L/r)^2$

9.29 Euler's hyperbola is the graphical expression between,
(a) stress and strain in the column
(b) critical stress σ_{cr} and flexural rigidity EI
(c) critical stress σ_{cr} and slenderness ratio L_e/r
(d) critical load and minimum radius of gyration

9.30 The effective slenderness ratio of a cantilever column is,
(a) $0.5L/r$ (b) L/r
(c) $\sqrt{2}L/r$ (d) $2L/r$

9.31 A 40 cm diameter circular timber column is 4 m long. The slenderness ratio of the column is,
(a) 10 (b) 20
(c) 40 (d) $20\sqrt{2}$

9.32 A compression member has one end hinged and other end rigidly fixed against rotation and sway. Its Euler's buckling load is 120 kN. What will be its critical buckling load if its both ends are fixed against sway and rotation.
(a) 60 kN (b) 120 kN
(c) 240 kN (d) $120\sqrt{2}$ kN

9.33 The deflected shape of a column fixed at one and free at the other, can be given approximately by,
(a) $y = n(1 - \cos \pi x/L)$
(b) $y = \delta \sin \pi x/L$
(c) $y = \delta(1 - \sin \pi x/L)$
(d) $y = n \sin \pi x/2L$

9.34 The deflected shape of a column hinged at both ends, can be given approximately by,

(a) $y = n(1 - \cos \pi x/2L)/2$
(b) $y = n \sin \pi x/L$
(c) $y = \delta(1 - \sin \pi x/L)$
(d) $y = n \sin \pi x/2L$

9.35 The critical buckling load calculated by the method of strain energy based on an assumed deflected shape is always,

(a) higher than the actual critical load
(b) lower than the actual critical load
(c) higher than or equal to the actual critical load
(d) lower than or equal to the actual critical load

9.36 The allowable stress in a column can be increased by,

(a) increasing the radius of gyration
(b) increasing the eccentricity
(c) increasing the slenderness ratio
(d) increasing the length of the column

9.37 The maximum stress σ_{max} developed in a column with fixed ends and eccentricity e is given by,

(a) $\dfrac{P}{A} + \dfrac{Pe}{Z} \sec \mu L/4$

(b) $\dfrac{P}{A} + \dfrac{Pe}{Z} \sec \mu L/2$

(c) $\dfrac{P}{A} + \dfrac{Pe}{Z} \sec \mu L$

(d) $\dfrac{P}{A} + \dfrac{Pec}{r^2} \tan \mu L/2$

where $\mu = P/EI$, and other terms have usual meaning.

10

Combined Direct and Bending Stress

CONCEPT REVIEW

10.1 STRESSES IN ECCENTRICALLY LOADED MEMBERS

Axial load generally act eccentrically which gives axial stress along with bending stress. If the deflections are small, then the effect of the two can be superpositioned as shown in Fig. 10.1. The total stress can thus be written as,

$$\sigma = \frac{P}{A} \pm \frac{M}{I} \times y \qquad (10.1)$$

where $M = P.e.$

and, e = eccentricity

The stress at extreme fibre can be +ve (compressive) or –ve (tensile).

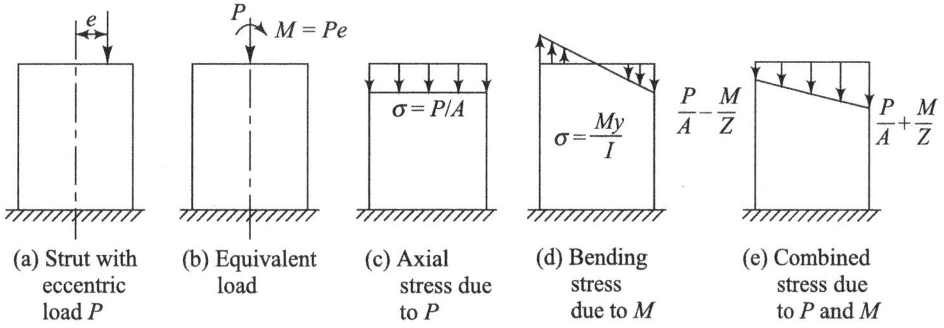

Fig. 10.1

10.2 CORE OR KERN OF A SECTION

No tension is allowed in masonry structures like chimney, retaining walls, etc. Such structures withstand vertical load due to self weight and bending due to lateral thrust.

(1) **Rectangular section:** Consider a rectangular section *EFGH* subjected to compressive load *P* at an eccentricity e_x and e_y as shown in Fig. 10.2. The maximum tensile stress will occur at point *E*, as follows.

$$\sigma_E = \frac{P}{A} - \frac{Pe_x}{Z_y} - \frac{Pe_y}{Z_x} \tag{10.1a}$$

where, $A = bd$

$$Z_x = \frac{bd^2}{6}$$

$$Z_y = \frac{db^2}{6}$$

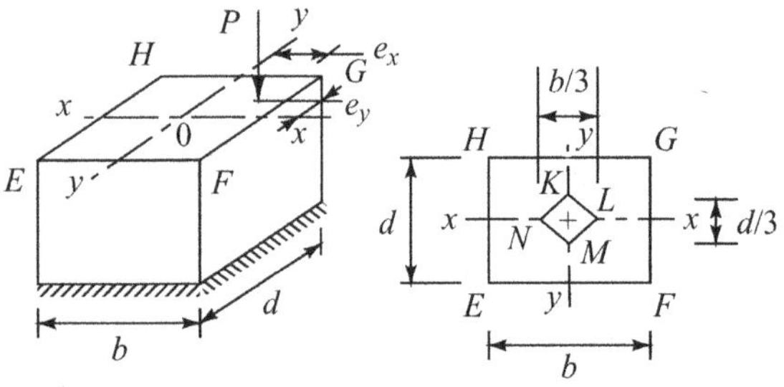

(a) Eccentrically loaded pedestal

(b) Kern of rectangular section

Fig. 10.2

Equating $\sigma_E = 0$, for the condition of no tension.

$$\frac{P}{bd} - \frac{6Pe_x}{db^2} - \frac{6Pe_y}{bd^2} = 0$$

or,

$$\frac{6e_x}{b} + \frac{6e_y}{d} = 1 \tag{10.2}$$

This is an equation of straight line *KL* with limits $e_x \ngtr b/6$ and $e_y \ngtr d/6$. No tension will develop if load *P* is applied at any point within the region *OKL*. Similarly for other quadrants, a point load within a rhombus *KLMN* called kern or core will not cause tensile stress at any point in the section. This is also called as middle third rule.

(2) **Circular section:** Consider Eq. (10.1) for circular section for no tension at extreme fibre, as follows

$$\frac{P}{\pi D^2/4} - \frac{Pe}{\pi D^3/32} = \sigma \nless 0$$

or,

$$e \ngtr D/8 \tag{10.3}$$

The shape of kern in this case is a circle of diameter $2 \times D/8 = D/4$. This is called middle fourth rule.

10.3 INTRACTION FORMULA

Intraction formula is used for elastic design of members subjected to axial load *P* and bending moment *M*.

If σ_a = permissible axial stress

σ_b = permissible bending stress

For an axial load *P*, the area of cross-section required,

$$A_a = \frac{P}{\sigma_a}$$

If A_b is the area required for resisting bending moment *M*, then,

$$\sigma_b = \frac{My}{A_b r^2}$$

or,

$$A_b = \frac{My}{\sigma_b r^2}$$

where, $A_b r^2$ = moment of inertia

and, r^2 = radius of gyration

If *A* is the area of the design section, then,

$$A = A_a + A_b$$

or,

$$A = \frac{P}{\sigma_a} + \frac{My}{\sigma_b r^2}$$

or

$$\frac{P}{A\sigma_a} + \frac{My}{\sigma_b Ar^2} = 1$$

Substituting,

$$\frac{P}{A} = \sigma_{a}\text{, app.} \quad \text{i.e. applied axial stress}$$

and, $$\frac{My}{Ar^2} = \sigma_{b}\text{, app.} \quad \text{i.e applied bending stress}$$

$$\frac{\sigma_{a}\text{, app.}}{\sigma_a} + \frac{\sigma_{b}\text{, app.}}{\sigma_b} = 1 \qquad (10.4)$$

This is the intraction formula shown graphically in Fig. 10.3. Any design section is considered safe which satisfies,

$$\frac{\sigma_{a}\text{, app.}}{\sigma_a} + \frac{\sigma_{b}\text{, app.}}{\sigma_b} \le 1 \qquad (10.5)$$

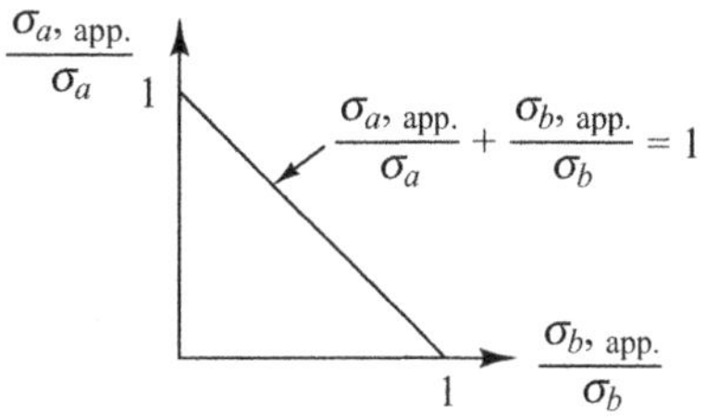

Fig. 10.3

EXAMPLES

Example 10.1 A rectangular strut of size 150 mm by 200 mm carries a point load of 200 kN at an eccentricity of 30 mm from the centre along an axis parallel to longer face as shown in Fig. 10.4. Calculate the stresses.

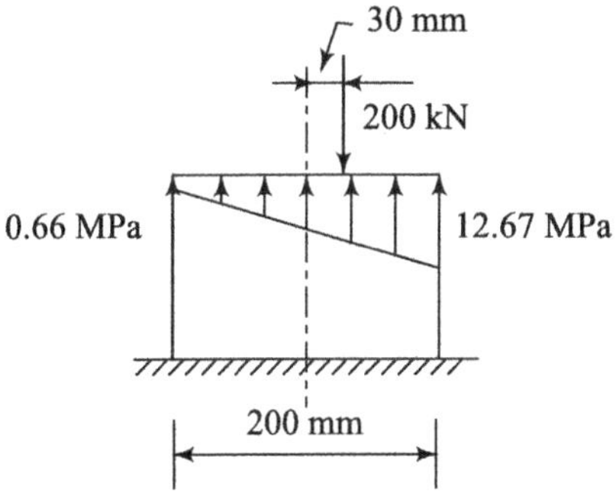

Fig. 10.4

Solution Using Eq. (10.1),

$$\sigma = \frac{P}{A} \pm \frac{M_y}{I}$$

where, $P = 200000$ N

$A = 200 \times 150 = 30000\ \text{mm}^2$

$M = P.\,e = 200000 \times 30 = 6 \times 10^6$ N.mm

$$\frac{I}{y} = Z = \frac{bd^2}{6} = \frac{150 \times 200^2}{6} = 1 \times 10^6\ \text{mm}^3$$

$$\sigma = \frac{200000}{30000} \pm \frac{6 \times 10^6}{1 \times 10^6}$$

$= 12.67$ MPa, 0.66 MPa

Example 10.2 A short column 200 mm by 300 mm carries a load of 200 kN at an eccentricity of 50 mm from centre parallel to shorter axis and 75 mm from centre parallel to longer axis as shown in Fig. 10.5. Calculate stresses at all four corners.

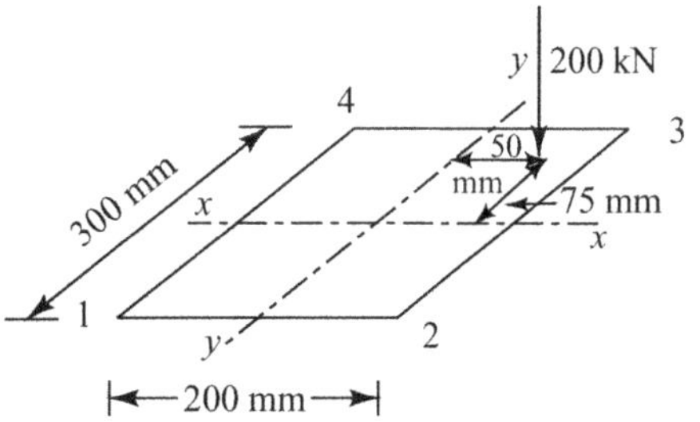

Fig. 10.5

Solution Using Eq. (10.1) for bi-directional eccentricity,

$$\sigma = \frac{P}{A} \pm \frac{Pe_x}{Z_y} \pm \frac{Pe_y}{Z_x}$$

where $P = 200000$ N

$$A = 200 \times 300 = 60000 \text{ mm}^2$$

$$e_x = 50 \text{ mm}$$

$$e_y = 75 \text{ mm}$$

$$Z_x = \frac{200 \times 300^2}{6} = 3 \times 10^6 \text{ mm}^3$$

$$Z_y = \frac{300 \times 200^2}{6} = 2 \times 10^6 \text{ mm}^3$$

At corner 1,

$$\sigma = \frac{200000}{60000} - \frac{200000 \times 50}{2 \times 10^6} - \frac{200000 \times 75}{3 \times 10^6}$$

$$= 3.33 - 5 - 5 = -6.67 \text{ MPa, tension}$$

At corner 2,

$$\sigma = \frac{200000}{60000} + \frac{200000 \times 50}{2 \times 10^6} - \frac{200000 \times 75}{3 \times 10^6}$$

$$= 3.33 + 5 - 5 = 3.33 \text{ MPa}$$

At corner 3,

$$\sigma = \frac{200000}{60000} + \frac{200000 \times 50}{2 \times 10^6} + \frac{200000 \times 75}{3 \times 10^6}$$

$$= 3.33 + 5 + 5 = 13.33 \text{ MPa}$$

At corner 4,

$$\sigma = \frac{200000}{60000} - \frac{200000 \times 50}{2 \times 10^6} + \frac{200000 \times 75}{3 \times 10^6}$$

$$= 3.33 - 5 + 5 = 3.33 \text{ MPa}$$

Example 10.3 Calculate the stresses due to a compressive load of 150 kN applied at the outer edge of a circular hollow pedestal of external and internal diameters 250 mm and 200 mm, respectively.

Solution Using Eq. (10.1)

$$\sigma = \frac{P}{A} \pm \frac{My}{I}$$

where, $P = 150000$ N

$$A = \frac{\pi}{4}(250^2 - 200^2) = 17671.5 \text{ mm}^2$$

$$M = P.e = 150000 \times \frac{250}{2}$$

$$= 18.75 \times 10^6 \text{ N.mm}$$

$$Z = \frac{I}{y} = \frac{\pi(250^4 - 200^4)}{64 \times 125} = 905662 \text{ mm}^3$$

or $$\sigma = \frac{150000}{17671.5} \pm \frac{18.75 \times 10^6}{905662}$$

$$= 29.19 \text{ MPa}, -12.21 \text{ MPa}$$

Example 10.4 A hollow pedestal having outer dimensions 1.5 m square and inner dimensions 1 m square carries vertical point load of 500 kN at an eccentricity of 0.5 m along a diagonal. Determine the stresses at corners (Fig. 10.6).

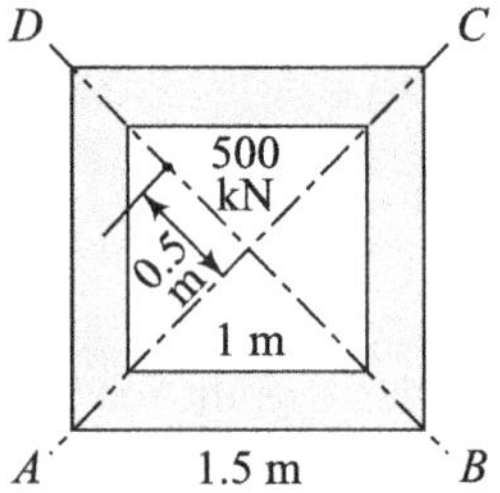

Fig. 10.6

Solution Using Eq. (10.1)

$$\sigma = \frac{P}{A} \pm \frac{M}{I} \times y$$

where, $P = 500$ kN

$$A = 1.5^2 - 1^2 = 1.25 \text{ m}^2$$

$$M = P\,e = 500 \times 0.5 = 250 \text{ kN.m}$$

About diagonal,

$$I = \frac{1.5^4 - 1^4}{12} = 0.33854 \text{ m}^4$$

$$y = 1.5 \sin 45° = \frac{1.5}{\sqrt{2}} \text{ m}$$

$$\therefore \quad \sigma = \frac{500 \times 10^3}{1.25 \times 10^6} \pm \frac{250 \times 10^6 \times 1500}{0.33854 \times 10^{12} \times \sqrt{2}}$$

$$= 1.18 \text{ MPa}, -0.38 \text{ MPa.}$$

Example 10.5 A short column has external diameter 500 mm and internal diameter 300 mm. Determine the maximum eccentricity at which a compressive load can be applied without producing tensile stress on the cross-section.

Solution Using Eq. (10.1) for no tension,

$$\sigma = \frac{P}{A} \pm \frac{M}{Z} = 0$$

or,
$$\frac{P}{\frac{\pi}{4}(D^2 - d^2)} = \frac{Pe}{\frac{\pi}{64}\frac{(D^4 - d^4)}{D}}$$

where D and d are the external and internal diameters, respectively.

or,
$$e \ngtr \frac{D^2 + d^2}{8D} = \frac{500^2 + 300^2}{8 \times 500}$$

or,
$$e \ngtr 85 \text{ mm}$$

Example 10.6 Derive the expression for kern of rectangular hollow section shown in Fig. 10.7.

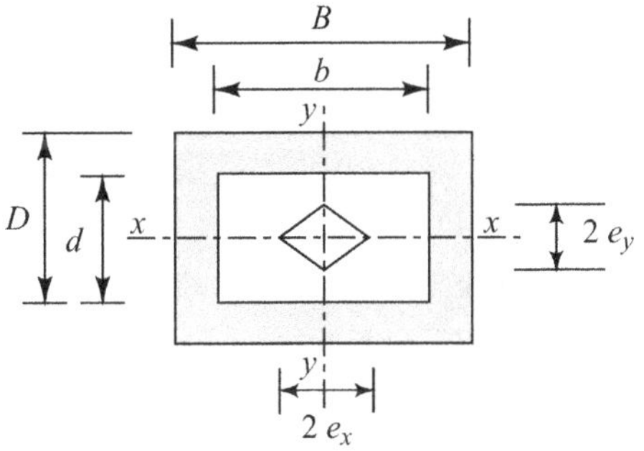

Fig. 10.7

Solution Using Eq. (10.1) for no tension,

$$\frac{P}{A} - \frac{My}{I} \geq 0$$

or,
$$\frac{P}{A} - \frac{Pey}{I} \geq 0$$

or,
$$e \leq \frac{I}{yA}$$

For bending about $x - x$ axis,

$$I = \frac{BD^3}{12} - \frac{bd^3}{12}$$

$$y = \frac{D}{2}$$

$$A = BD - bd$$

or,
$$e_y \geq \frac{BD^3 - bd^3}{6D(BD - bd)}$$

Similarly for bending about $y - y$ axis

$$e_x = \frac{DB^3 - db^3}{6B(BD - bd)}$$

Example 10.7 A free standing masonry wall is 7 m high and has a density of 2000 kg/m^3. It is to resist a horizontal wind pressure of 900 N/m^2 (Fig. 10.8). Calculate the thickness of the wall required if no tension is to develop.

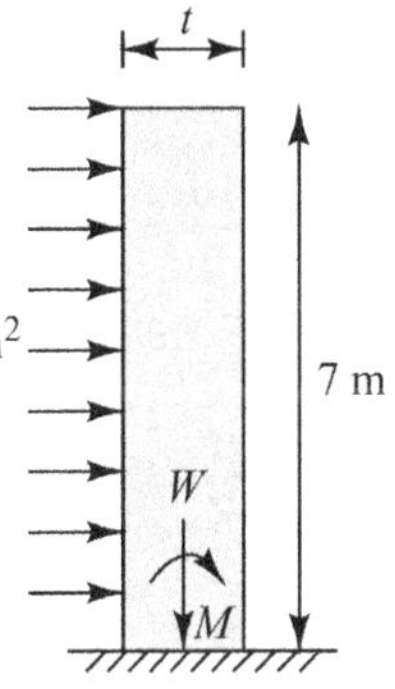

Fig. 10.8

Solution Consider 1 m length of the wall.

Wind load = 900 × 7 × 1 = 6300 N

Moment of wind load about base, (Fig. 10.8)

$$M = 6300 \times \frac{7}{2} = 22050 \text{ N.m}$$

If t is the thickness of the wall, then

Self weight, $W = 2000 \times 9.81 \times (7 \times t \times 1)$
$= 137340t$ N

For no tension at the base, $e = \dfrac{M}{P} \ngtr t/6$

or,
$$\frac{22050}{137340t} \ngtr \frac{t}{6}$$

or,
$$t \nless 0.98 \text{ m}$$

Example 10.8 A 10 m high masonry chimney of external diameter 2 m and internal diameter 1.5 m is subjected to wind pressure of 1000 N/m^2 (Fig. 10.9). Determine the stresses at the base if weight of masonry is 20 kN/m^3.

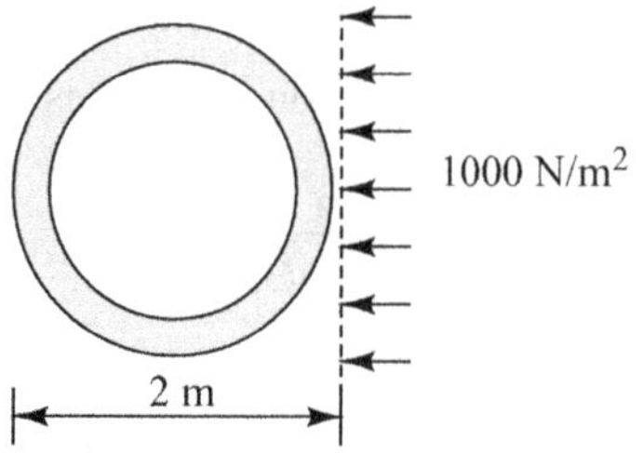

Fig. 10.9

Solution Wind force on the projected area (as shown in Fig. 10.9) = 1000 × 2 × 10 = 20 kN

Overturning moment due to wind,

$$M = 20 \times \frac{10}{2} = 100 \text{ kN.m}$$

Using Eq. (10.1),

$$\sigma = \frac{P}{A} \pm \frac{M}{Z}$$

where, $\frac{P}{A} = 20 \times 10 = 200$ kN/m^2

$$Z = \frac{\pi}{64}\frac{(2^4 - 1.5^4)}{2/2} = 0.5369 \text{ m}^3$$

or, $\sigma = 200 \pm \frac{100}{0.5369} = 386.26$ kN/m^2,

and 13.74 kN/m^2

Example 10.9 A trapezoidal masonry dam is 20 m high. It is 3 m wide at the top and 9 m wide at the base. It retains water to a depth of 18 m against a vertical face. Determine the pressure intensities at the base if the specific weight of masonry is 20 kN/m^3.

Solution Consider 1 m length of the dam. Vertical load due to self weight of the dam is divided in two parts (i) rectangular and (ii) triangular as shown in Fig. 10.10.

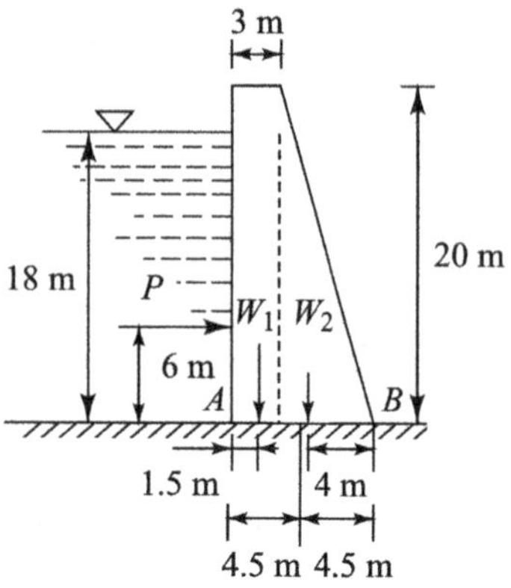

Fig. 10.10

Weight of rectangular section,

$$W_1 = 3 \times 20 \times 1 \times 20 = 1200 \text{ kN}$$

Distance of W_1 from e.g. of the base *AB* of dam,

$$e_1 = 4.5 - 1.5 = 3 \text{ m}$$

Stress at base due to W_1,

$$\sigma_1 = \frac{W}{A} \pm \frac{We}{Z} = \frac{1200}{9 \times 1} \pm \frac{1200 \times 3}{\frac{1 \times 9^2}{6}}$$

$= 133.33 \pm 266.67$

$= 400$ kN/m^2 at edge *A* of base and

-133.33 kN/m^2 at edge *B*

Weight of triangular section,

$$W_2 = \frac{6 \times 20}{2} \times 1 \times 20 = 1200 \text{ kN}$$

W_2 acts at $\frac{2}{3} \times 6 = 4$ m from edge *B* of the base and therefore its eccentricity from c.g. of base,

$$e_2 = 4.5 - 4 = 0.5 \text{ m}$$

Stress at base due to W_2,

$$\sigma_2 = \frac{1200}{9} \pm \frac{1200 \times 0.5}{\frac{1 \times 9^2}{6}}$$

$= 133.33 \pm 44.44$

$= 88.89$ kN/m^2 at edge *A* and

177.78 kN/m^2 at edge *B*

Horizontal water thrust,

$$P = \frac{wh^2}{2} = \frac{9810 \times 18^2}{2}$$

or, $P = 1589220$ N acting at $\frac{18}{3} = 6$ m from the base

Moment due to water thrust,

$M = 1589.22 \times 6 = 9535.32$ kN m

Stress at base due to overturning moment M,

$$\sigma_3 = \frac{M}{Z} = \pm \frac{9535.32}{\frac{1 \times 9^2}{6}} = \pm 706.32 \text{ kN/m}^2$$

$= -706.32$ kN/m^2 at edge A and

$+ 706.32$ kN/m^2 at edge B

Total stress at edge A,

$\sigma = 400 + 88.89 - 706.32$

$= -217.43$ kN/m^2 (uplift)

Total stress at edge B,

$\sigma = -133.33 + 177.78 + 706.32$

$= 750.76$ kN/m^2

Example 10.10 An external column is to withstand loads of 50 kN, 90 kN and 120 kN as shown in Fig. 10.11. Check the safety of I-section if the permissible stresses in axial compression and bending are 100 MPa and 165 MPa, respectively.

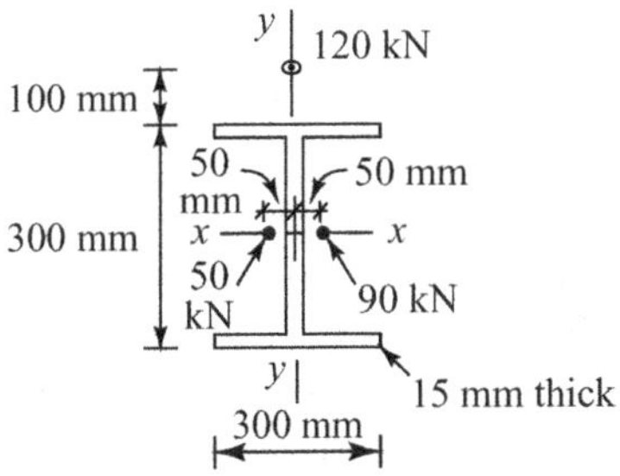

Fig. 10.11

Solution Area of I-section

$= 2 \times 300 \times 15 + 270 \times 15$

$= 13050$ mm^2

$$I_x = \frac{300 \times 300^3}{12} - \frac{285 \times 270^3}{12}$$

$= 207.53 \times 10^6$ mm^4

$$I_y = \frac{2 \times 15 \times 300^3}{12} + \frac{270 \times 15^3}{12}$$

$= 67.57 \times 10^6$ mm^4

Total axial stress applied

$$\sigma_{a\text{, app.}} = \frac{(120 + 90 + 50)\, 10^3}{13050} = 19.92 \text{ MPa}$$

Maximum bending stress at right corner of top flange

$$\sigma_b = \pm \frac{Pey}{I}$$

$$\sigma_{b\text{, app}} = \frac{90000 \times 50 \times 150}{67.57 \times 10^6} + \frac{120000 \times (100 + 150) \times 150}{207.53 \times 10^6} - \frac{50000 \times 50 \times 150}{67.57 \times 10^6}$$

$= 26.12$ MPa (compression)

Using intraction formula,

$$\frac{\sigma_{a\text{, app.}}}{\sigma_a} + \frac{\sigma_{b\text{, app.}}}{\sigma_b} \not> 1$$

or, $$\frac{19.92}{100} + \frac{26.12}{165} = 0.36 \not> 1$$

Hence the section is quite safe.

EXERCISES

PROBLEMS

10.1 Find the maximum stress on the section *AB* of the cramp shown in Fig. 10.12. When a load of 2.5 kN is applied by the screw if the section *AB* is 20 mm by 30 mm.

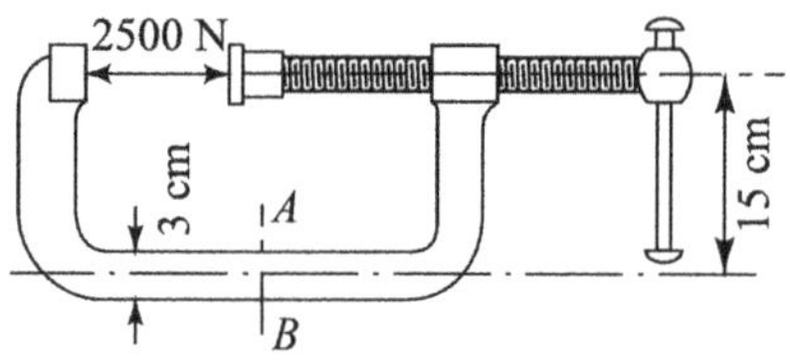

Fig. 10.12

10.2 A rolled steel I-section has 10 mm thickness, 300 mm depth and its flange width is 200 mm. It acts as short column to carry a compressive load *P* at the centroid of one of its flange. Find the value of *P* such that the compressive load does not exceed 100 MPa.

10.3 A masonry pier 1.5 m by 2 m is subjected to a vertical load of 400 kN at an eccentricity of 20 cm from centre parallel to longer axis and 10 cm from centre parallel to shorter axis of the section. Determine the stresses at all four corners.

10.4 A load of 10 kN act on a cast iron bracket as shown in Fig. 10.13. Determine the stresses at extreme fibre of section *AB*.

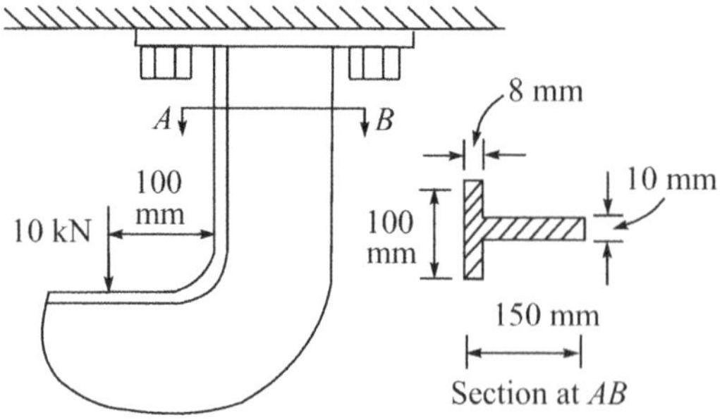

Fig. 10.13

10.5 A masonry chimney of external dimensions 2 m by 2 m has thickness 30 cm. It is 15 m high and is subjected to horizontal wind pressure 1000 N/m^2 normal to one of its diagonals. Determine the stresses at the base if the self weight of masonry is 20 kN/m^3.

10.6 A trapezoidal masonry dam is 8 m high. It is 1.5 m wide at the top and 5 m wide at the base. It retains water to a depth of 7.5 m against a vertical face. Determine the pressure intensity at the base if the specific weight of masonry is 22 kN/m^3.

OBJECTIVE QUESTIONS

10.7 For no tension in a circular section, an axial load can be placed at a maximum eccentricity equal to,

(a) $d/3$ (b) $d/4$ (c) $d/6$ (d) $d/8$

where *d* is the diameter of the circle.

10.8 A short column of external diameter *D* and internal diameter *d* is subjected to load *W* with an eccentricity of *e* causing zero stress at the extreme fibre, then the value of *e* must be,

(a) $\dfrac{D^2+d^2}{8\pi D}$ (b) $\dfrac{D^2+d^2}{8D}$

(c) $\dfrac{D^2-d^2}{8D}$ (d) $\dfrac{D^2-d^2}{8D^2}$

10.9 A short square column of external dimensions *D* and internal dimensions *d* is subjected to laod *W* with an accentricity of *e* along an axis parallel to a face causing zero stress at the extreme fibre, then the value of *e* must be,

(a) $\dfrac{D^2-d^2}{6D}$ (b) $\dfrac{D^2+d^2}{6d}$

(c) $\dfrac{D^2-d^2}{8D}$ (d) $\dfrac{D^2+d^2}{6D}$

10.10 A structural member is to withstand an axial load of 900 kN along with bending moment 160 kN.m. If the allowable stresses in axial compression and bending are 100 MPa and 160 MPa, respectively, then select a suitable section if their area and sectional modulus are given as follows.

(a) Area = 9 cm^2, sectional modulus = 40 cm^3

(b) Area = 90 cm^2, sectional modulus = 400 cm^3

(c) Area = 180 cm^2, sectional modulus = 2000 cm^3

(d) Area = 1800 cm^2, sectional modulus = 4000 cm^3

10.11 A load P is acting on the diagonal of a square column of size D as shown in Fig. 10.14. For no tension, the maximum distance of the load from the centre O is,

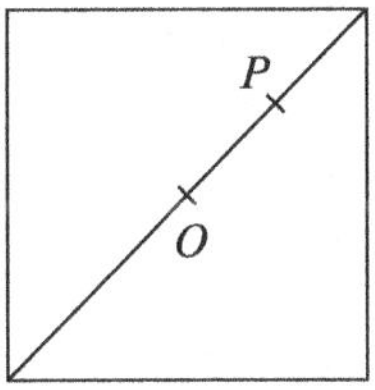

Fig. 10.14

(a) $D/8$ (b) $D/6$

(c) $\sqrt{2}\,D/6$ (d) $\sqrt{2}\,D/12$

10.12 The rectangular column shown in Fig. 10.15 carries a load P having eccentricities e_x and e_y along the x-axis and y-axis respectively, the stress at any point (x, y) is given by,

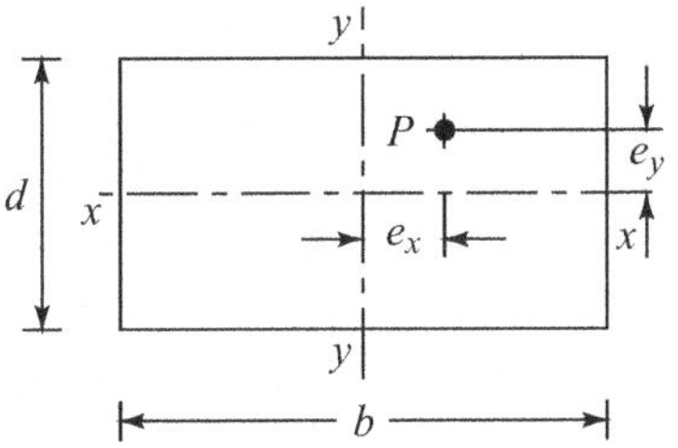

Fig. 10.15

(a) $\frac{P}{bd}\left(1+\frac{12e_y\,y}{d^2}+\frac{12e_x\,x}{b^2}\right)$

(b) $\frac{P}{bd}\left(1+\frac{12e_y\,y}{b^2}+\frac{12e_x\,x}{d^2}\right)$

(c) $\frac{P}{bd}\left(1+\frac{6e_y\,y}{d^2}+\frac{6e_x\,x}{b^2}\right)$

(d) $\frac{P}{bd}\left(1+\frac{6e_y\,y}{b^2}+\frac{6e_x\,x}{d^2}\right)$

10.13 A column of rectangular section $ABCD$ of size $2y_1 \times 2y_2$ carries a load P at a point having coordinates as shown in Fig. 10.16. If the moment of inertia about x and y axis are I_x and I_y, then the stress at the corner C is,

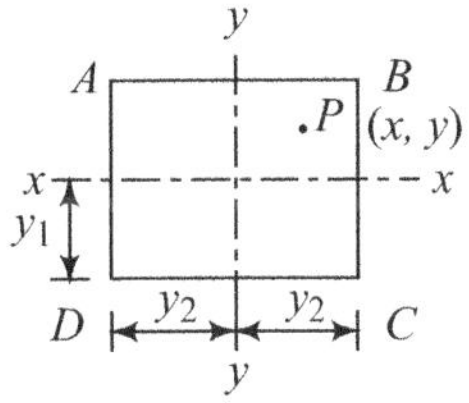

Fig. 10.16

(a) $\frac{P}{A}-\frac{Py}{I_x}y_1+\frac{Px}{I_y}y_2$

(b) $\frac{P}{A}-\frac{Py}{I_x}y_1-\frac{Px}{I_y}y_2$

(c) $\frac{P}{A}-\frac{Py}{I_y}y_1+\frac{Px}{I_x}y_2$

(d) $\frac{P}{A}-\frac{Py}{I_y}y_1-\frac{Px}{I_x}y_2$

10.14 A tie bar 20 mm × 10 mm carries a tensile load of 1 kN as shown in Fig. 10.17. Under this load, the maximum intensity of stress over the mean value will be increased by,

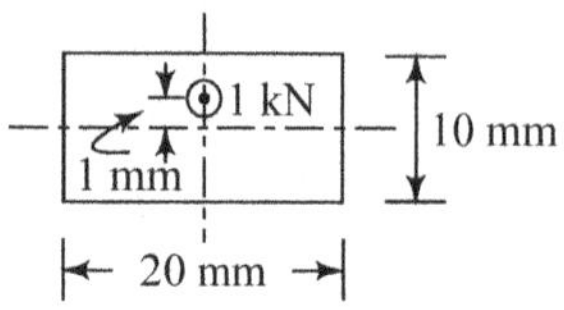

Fig. 10.17

(a) 20% (b) 40% (c) 60% (d) 80%

10.15 A reinforced concrete footing loaded with a concentrated load P as shown in Fig. 10.18 produces maximum bending stresses of 10 kN/m^2 and 15 kN/m^2 due to eccentricities about $x-x$ and $y-y$ axes, respectively. If the direct stress due to load acting at P is 18 kN/m^2 (compressive), then the intensity resultant stress at corner B will be,

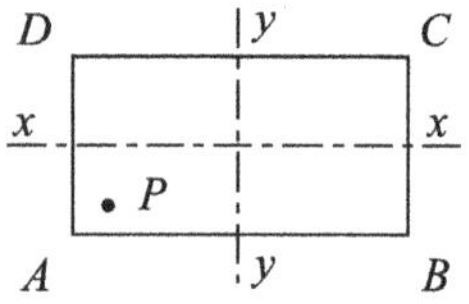

Fig. 10.18

(a) 13 kN/m^2 (tensile)
(b) 13 kN/m^2 (compressive)
(c) 31 kN/m^2 (compressive)
(d) 31 kN/m^2 (tensile)

10.16 *Assertion* (A): In a masonry dam shown in Fig. 10.19, if G is the specific gravity of the masonry and the ratio $H/B = \sqrt{G}$, then the eccentricity is $B/6$, whether the storage in full or nil.

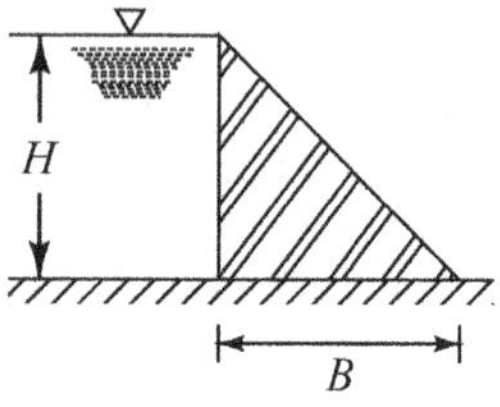

Fig. 10.19

Reason (R): The resultant force will pass through the centroid of the pressure distribution diagram for no tension on the base for nil or full storage if the ratio $H/B = \sqrt{G}$.

(a) A and R are both true and R is the correct explanation of A.
(b) Both A and R are true but R is not the correct explaination of A.
(c) A is true but R is false.
(d) Both A and R are false.

11

Theories of Failure

CONCEPT REVIEW

11.1 YIELD CRITERIA

Yield criteria is most commonly based on yield stress σ_y determined by an axial tensile test. The test conditions in a simple axial tensile test are not sufficient to take into account the varying complexity regarding material behaviour (ductile, brittle), loading pattern (static, impact, fatigue, multiaxial load) Various theories of failure attempt to correlate yield stress σ_y from an axial tensile test and the failure in a multiaxial state of stress to avoid complex testing apparatus for each case.

11.2 MAXIMUM NORMAL STRESS THEORY

The maximum normal stress theory of failure credited to W.J.M. Rankine asserts that yielding will occur when the maximum normal stress exceeds the tensile yield stress σ_y. In a two-dimensional state of stress with principal stresses σ_1 and σ_2 acting in two directions, the failure will occur uniaxially if $\sigma_1 \geq \sigma_y$ or $\sigma_2 \geq \sigma_y$. This is shown graphically as square yield locus *BDFH* in Fig. 11.1. All points outside yield locus indicates failure. For many engineering materials the tensile and compressive stresses are so nearly equal that they are assumed to be equal. The maximum principal stress theory is more reliable for brittle materials.

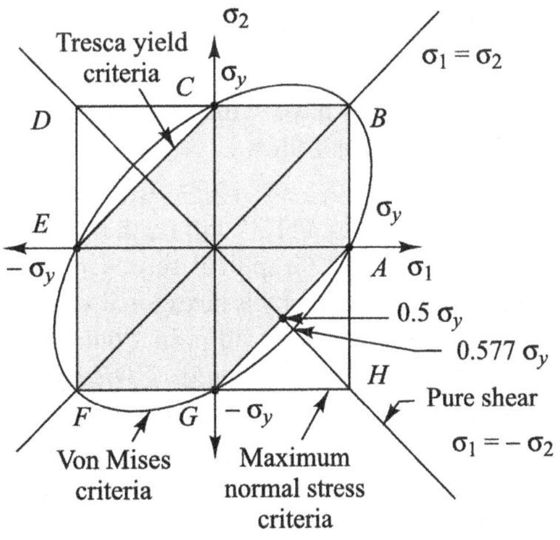

Fig. 11.1 *Yield Locus*

The maximum normal stress has some limitations. Failure of ductile materials is fundamentally a shearing phenomenon. Secondly, this theory completely ignore the principal stresses in other directions. A material can sustain hydrostatic compressive stress much higher than compressive yield stress without failure.

11.3 MAXIMUM SHEAR STRESS THEORY

This theory is also called Tresca or Guest thoery which envisage failure to occur when shear stress is equal to shear stress at the yield point in a simple tensile test. The maximum shear stress and yield stress in an axial tensile test are related as $\tau_{max} = \sigma_y/2$. Consider a body under three-dimensional principal stresses such that $\sigma_1 > \sigma_2 > \sigma_3$. The maximum shear stress at failure will be as follows,

$$\tau_{max} = \frac{\sigma_y}{2} = \frac{\sigma_1 - \sigma_3}{2} \quad (11.1a)$$

or

$$\sigma_y = \sigma_1 - \sigma_3 \quad (11.1b)$$

This criteria thus neglects the effect of intermediate principal stress σ_2 which may not be justified in all situations.

In biaxial stress case where $\sigma_3 = 0$, the tresca yield condition is given as follows,

$$\sigma_y = \sigma_1 - \sigma_2 \quad (11.2)$$

Equation (11.2) will apply when σ_1 and σ_2 are of opposite nature, i.e. Tension/Compression and the limiting τ_{max} will be in the plane of σ_1 and σ_2.

When σ_1 and σ_2 have same sign then the yield condition will be as follows,

$$\sigma_y = \sigma_1 \quad \text{or} \quad \sigma_y = \sigma_2 \quad (11.3)$$

τ_{max} in respect of Eq. (11.3) will be in the plane of $\sigma_1 - \sigma_3$ or $\sigma_2 - \sigma_3$. Graphical representation of Tresca criteria is shown by is hexagonal yield locus *ABCEFG* in Fig. 11.1. As all points outside yield locus represent failure, triangles *CDE* and *AHG* are the areas of disagreement with previous theory. Tresca yield criteria is simple and has good agreement with the experimental test results for ductile materials on a safer side.

11.4 MAXIMUM PRINCIPAL STRAIN THEORY

This is often referred to as St. Venant theory. Failure by yielding as per this criteria cccurs when the maximum principal strain exceeds the tensile or compressive yield strain. For plane strain $\sigma_3 = 0$, the yield criteria is written as follows:

$$|\sigma_1 - \mu\sigma_2| = \sigma_y \quad (11.4a)$$

and

$$|\sigma_2 - \mu\sigma_1| = \sigma_y \quad (11.4b)$$

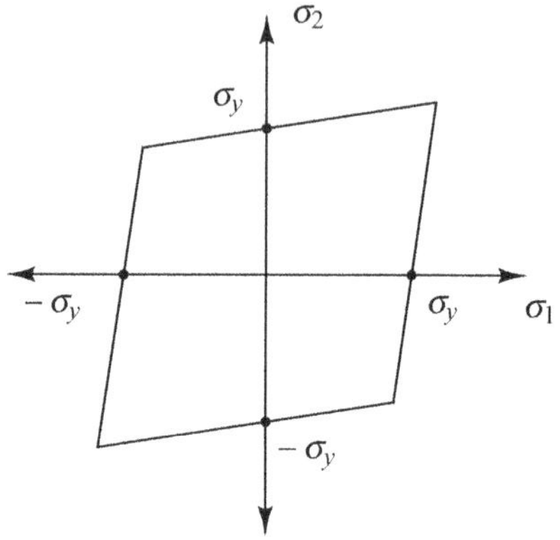

Fig. 11.2 *Yield Locus for Maximum Principal Strain Theory of Failure*

Graphical representation is shown in Fig. 11.2. This theory is not much different than principal stress theory and is therefore used rarely.

11.5 TOTAL STRAIN ENERGY THEORY OF FAILURE

Beltrami and Haigh assumed that yielding commences when the total strain energy stored attains the value of the strain energy for uniaxial yielding. The total strain energy stored by principal stresses σ_1, σ_2, σ_3 and their corresponding principal strain ε_1, ε_2, ε_3 as per Eq. (1.14) is given by,

$$U = \frac{1}{2}(\sigma_1\varepsilon_1 + \sigma_2\varepsilon_2 + \sigma_3\varepsilon_3)$$

$$= \frac{1}{2E}(\sigma_1^2 + \sigma_2^2 + \sigma_3^2)$$

$$- \frac{\mu}{E}(\sigma_1\sigma_2 + \sigma_2\sigma_3 + \sigma_1\sigma_3)$$

At the point yield in simple tension, $\sigma_1 = \sigma_y$, $\sigma_2 = \sigma_3 = 0$ and the strain energy $U = \sigma_y^2/2E$. As per the total strain energy criteria,

$$(\sigma_1^2 + \sigma_2^2 + \sigma_3^2) - 2\mu(\sigma_1\sigma_2 + \sigma_2\sigma_3 + \sigma_1\sigma_3) = \sigma_y^2 \quad (11.5a)$$

For plane stress problem when principal stress $\sigma_3 = 0$, Eq. (11.5a) reduceds to,

$$\sigma_1^2 - 2\mu\sigma_1\sigma_2 + \sigma_2^2 = \sigma_y^2 \quad (11.5b)$$

11.6 MAXIMUM SHEAR STRAIN ENERGY THEORY

This theory of maximum distortion energy is proposed by M.T. Huber and developed by von Mises, Maxwell and Hencky. The theory proposes that failure will occur when the maximum energy of distortion reaches the equivalent value at yielding in a simple tension test. The total strain energy U_t of a body can be regarded comprising of two parts: (i) volumetric strain energy U_v and (ii) distortion energy or shear strain energy U_s. Now consider a body under principal stresses σ_1, σ_2 and σ_3 in three directions. They can be written as,

$$\sigma_1 = \frac{1}{3}(\sigma_1 + \sigma_2 + \sigma_3) + \frac{1}{3}(2\sigma_1 - \sigma_2 - \sigma_3)$$

$$\sigma_2 = \frac{1}{3}(\sigma_1 + \sigma_2 + \sigma_3) + \frac{1}{3}(2\sigma_2 - \sigma_3 - \sigma_1)$$

$$\sigma_3 = \frac{1}{3}(\sigma_1 + \sigma_2 + \sigma_3) + \frac{1}{3}(2\sigma_3 - \sigma_1 - \sigma_2)$$

The first component $\frac{1}{3}(\sigma_1 + \sigma_2 + \sigma_3)$ represent hydrostatic component which causes volumetric changes only while the second part causes distortion in shape. The strains due to these distorting stresses are as follows,

$$\varepsilon_1 = \frac{1}{E}\left(\frac{2\sigma_1}{3} - \frac{\sigma_2}{3} - \frac{\sigma_3}{3} - \mu\left(\frac{2\sigma_2}{3} - \frac{\sigma_3}{3} - \frac{\sigma_1}{3} + \frac{2\sigma_3}{3} - \frac{\sigma_2}{3} - \frac{\sigma_1}{3}\right)\right)$$

or, $$\varepsilon_1 = \frac{1}{3E}(1 + \mu)(2\sigma_1 - \sigma_2 - \sigma_3)$$

or, $$\varepsilon_1 = \frac{1}{6G}(2\sigma_1 - \sigma_2 - \sigma_3)$$

Similarly,

$$\varepsilon_2 = \frac{1}{6G}(2\sigma_2 - \sigma_3 - \sigma_1)$$

$$\varepsilon_3 = \frac{1}{6G}(2\sigma_3 - \sigma_1 - \sigma_2)$$

Strain energy per unit volume is given by Eq. (1.14) as follows:

$$U_s = \frac{1}{2}\sigma\varepsilon = \frac{1}{36G}\left[(2\sigma_1 - \sigma_2 - \sigma_3)^2 + (2\sigma_2 - \sigma_3 - \sigma_1)^2 + (2\sigma_3 - \sigma_1 - \sigma_2)^2\right]$$

or $$U_s = \frac{1}{12G}\left[\begin{matrix}(\sigma_1 - \sigma_2)^2 + (\sigma_2 - \sigma_3)^3 \\ + (\sigma_3 - \sigma_1)^3\end{matrix}\right] \quad (11.6)$$

In case of axial tension test,

$$\sigma_1 = \sigma_y \text{ and } \sigma_2 = \sigma_3 = 0,$$

the shear strain energy, $$U_s = \frac{\sigma_y^2}{6G} \quad (11.7)$$

As per von Mises criteria, failure occurs when U_s given by Eq. (11.6) reaches the value of U_s given by Eq. (11.7) as follows:

$$(\sigma_1 - \sigma_2)^2 + (\sigma_2 - \sigma_3)^2 + (\sigma_3 - \sigma_1)^2 = 2\sigma_y^2 \quad (11.8)$$

For a two-dimensional stress system, $\sigma_3 = 0$ and von Mises criteria of failure is written as follows:

$$\sigma_1^2 + \sigma_2^2 - \sigma_1\sigma_2 = \sigma_y^2 \quad (11.9)$$

The above expression defines the elliptical yield locus shown in Fig. 11.1. The significant difference between Tresca criteria and von Mises criteria: 0.5 σ_y and 0.577σ_y occur under pure shear condition. The test results of most ductile material favour von Mises criteria while Tresca criteria is on safer side and easier for calculations. von Mises criteria is often more convenient as its yield locus is a continuous curve without corners of Tresca's hexagon.

Figure 11.3 shows Tresca and von Mises yield criteria for three-dimensional stress system.

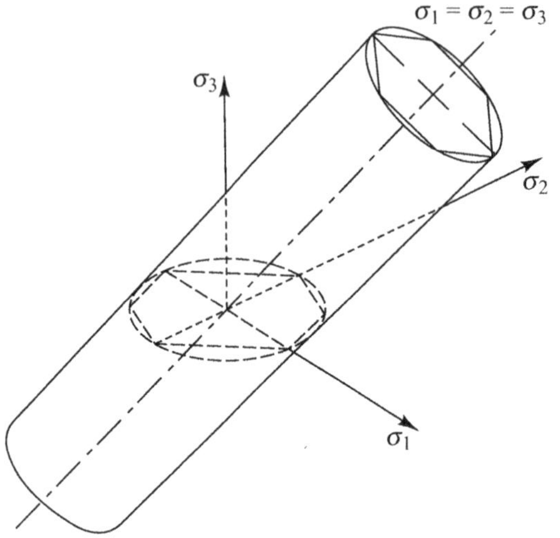

Fig. 11.3 *Tresca and Mises Yield Surfaces.*

Tresca yield envelop is a hexagonal prism with its axis lying along the line $\sigma_1 = \sigma_2 = \sigma_3$ while von Mises yield envelop is a circular cylinder circumscribing Tresca's yield hexagon. Yield never occurs at the axis of these prisms where $\sigma_1 = \sigma_2 = \sigma_3$, i.e. the hydrostatic state of stress.

11.7 OCTAHEDRAL SHEARING STRESS THEORY

Von Mises predicted that yielding will occur in a multiaxial state of stress when the octahedral shearing stress reaches a value equal to the octahedral shearing stress at yield in a simple tension test. The octahedral shear stress in a simple tension test is given by Eq. (2.18) by substituting $\sigma_I = \sigma_y$ an $\sigma_{II} = \sigma_{III} = 0$ as follows:

$$\tau_{ns,oct} = \frac{1}{3}\sqrt{(\sigma_y - 0)^2 + 0 + (0 - \sigma_y)^2}$$

$$= \frac{\sqrt{2}}{3}\sigma_y$$

Equating it to general three directional state of stress,

$$(\sigma_1 - \sigma_2)^2 + (\sigma_2 - \sigma_3)^2 + (\sigma_3 - \sigma_1)^2 = 2\sigma_y^2 \quad (11.10)$$

which is same as Eq. (11.8). This mathematically it is same as previous maximum shear strain theory although it originates from a different concept.

11.8 COULOMB-MOHR FRACTURE THEORY

This criteria takes into consideration the difference in ultimate strength in tension and compression. The yield locus is similar to Tresca's hexagonal yield locus as shown in Fig. 11.4.

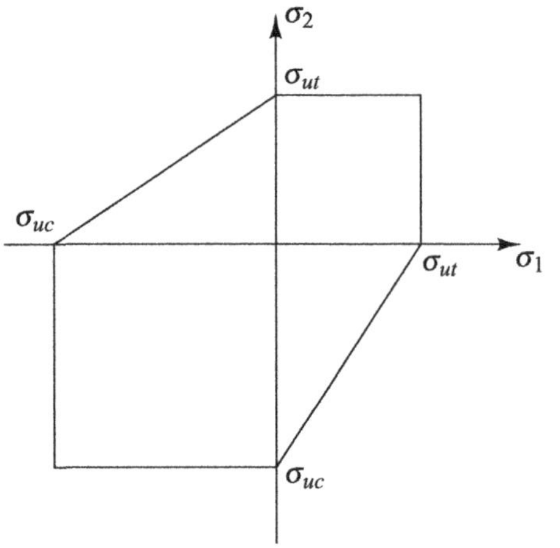

Fig. 11.4

11.9 FATIGUE FAILURE

Structural members can fail under repetitive load much below the ultimate static load. The repetitive load for fatigue test is usually composed of two components: (i) mean stress σ_m and (ii) alternating stress σ_a as shown in Fig. 11.5. The fatigue strength of a material is the alternating stress σ_a that it can withstand for a specified number of cycles without failure. Some materials like mild steel can withstand a stress level for an indefinite number of cycles. This stress is known as endurance limit of the meterial. Endurance limit for mild steel is about half the ultimate stress. No such limit has been

found

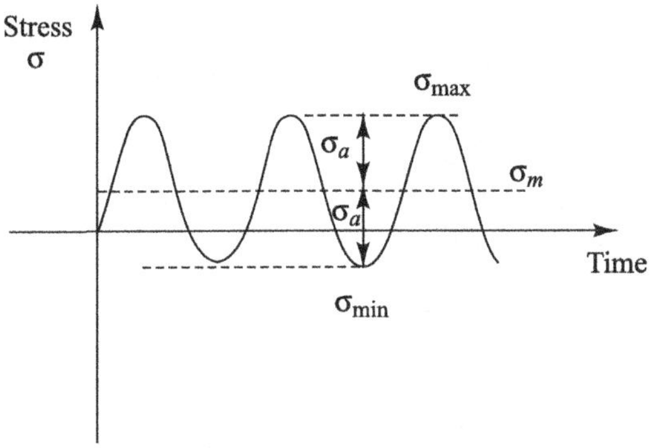

Fig. 11.5

for aluminium and its alloys. The endurance limit is defined arbitrarily for such metal corresponding to 500 million loading cycles.

The fatigue test data are normally presented in the form of S-N curve shown in Fig. 11.6. S-N curve shows that low cycle fatigue corresponds to high stress level but it does not reflects anything about mechanics of failure.

Fatigue failue in brittle in nature and is influenced by minor discontinuities in shape, surface finishing, surface cracks or imperfections and pockets of stress concentration due to welding, etc. Modified Goodman and Soderberg relations are used to predict fatigue failure as follows.

Modified Goodman:

$$\frac{\sigma_a}{\sigma_{cr}} + \frac{\sigma_m}{\sigma_u} = 1 \qquad (11.9)$$

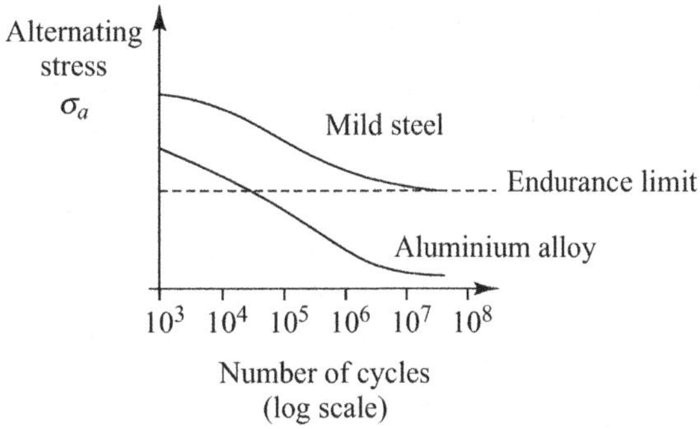

Fig. 11.6

Soderberg:

$$\frac{\sigma_a}{\sigma_{cr}} + \frac{\sigma_m}{\sigma_y} = 1 \qquad (11.10)$$

where,

σ_a = alternating stress

σ_{cr} = alternating stress or completely reversing stress that can be superimposed on the mean stress σ_m and safely cycled N times

σ_u = ultimate tensile strength

σ_y = yield strength

Miner's rule or cumulative fatigue damage formula accounts for loads of varying amplitude and time as follows:

$$\frac{n_1}{N_1} + \frac{n_2}{N_2} + \frac{n_3}{N_3} \le 1 \qquad (11.11)$$

where N is the number of cycles required for fatigue and n is the number of cycles actually applied for a load.

EXAMPLES

Example 11.1 A circular bar 100 mm diameter is carrying a load of 6π kN as shown in Fig. 11.7(a). If the yield stress of the material is 250 MPa, determine the factor of safety for the state of stress at top edge point of the fixed end.

Solution Resolving the applied load in x, y, and z, directions as follows:

$$F_y = 6\pi \sin\theta = 6\pi \times \frac{1}{3} = 2\pi \text{ kN}$$

$$F_x = 6\pi \cos\theta \sin 45^\circ$$

$$= 6\pi \times \frac{\sqrt{8}}{3} \times \frac{1}{\sqrt{2}} = 4\pi \text{ kN}$$

$$F_z = 6\pi \cos\theta_1 \sin 45^\circ$$

$$= 6\pi \times \frac{\sqrt{8}}{3} \times \frac{1}{\sqrt{2}} = 4\pi \text{ kN}$$

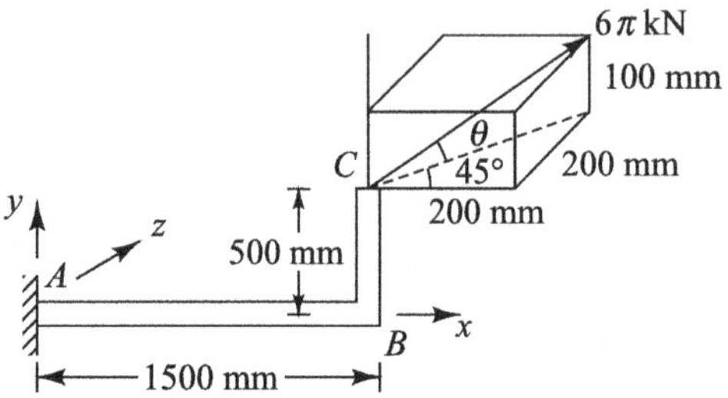

Fig. 11.7(a)

Area of the bar,

$$A = \frac{\pi}{4} \times 100^2 = 2500\ \pi \text{ mm}^2$$

Second moment of area,

$$I = \frac{\pi \times 100^4}{64} = 10^8 \pi/64 \text{ mm}^4$$

Polar moment of inertia $J = 2I = 10^8\ \pi/32$ mm^4

The forces at end B are shown in Fig. 11.7(b) and the normal stress at point A is calculated as follows:

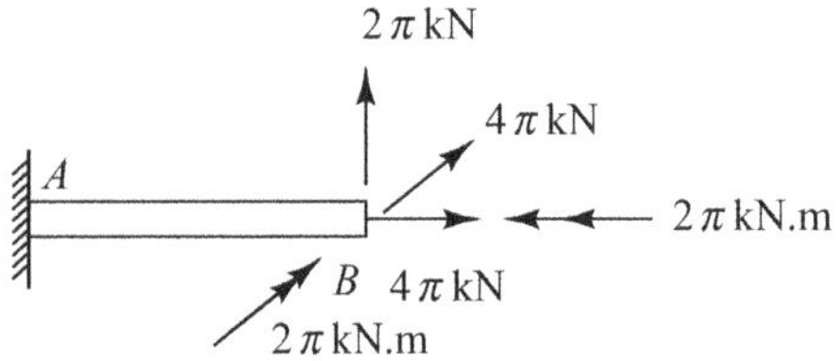

Fig. 11.7(b)

$$\sigma_x = \frac{F}{A} \pm \frac{My}{I}$$

$$\sigma_x = \frac{4000\pi}{2500\pi} + \frac{2\pi \times 10^6 \times 50}{10^8 \pi/64} - \frac{2000\pi \times 1500 \times 50}{10^8 \pi/64}$$

$$= -30.4 \text{ MPa} \quad \text{(compression)}$$

Shear stress due to torsion 2π kN.m and 4π kN force will be additive as shown in Fig. 11.7(c).

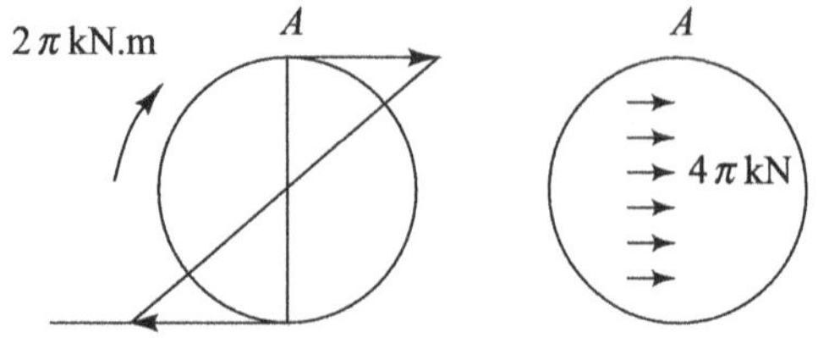

Fig. 11.7(c)

$$\tau = \frac{Tr}{J} + \frac{3}{2}\frac{V}{A}$$

or $$\tau = \frac{2\pi \times 10^6 \times 50}{10^8 \pi/32} + \frac{3}{2} \times \frac{4\pi \times 10^3}{2500\pi}$$

$$= 34.4 \text{ MPa}$$

Principal stresses at A,

$$\sigma = \frac{\sigma_x + \sigma_y}{2} \pm \sqrt{\left(\frac{\sigma_x - \sigma_y}{2}\right)^2 + \tau_{xy}^2}$$

$$= \frac{-30.4}{2} \pm \sqrt{\left(\frac{30.4}{2}\right)^2 + 34.4^2}$$

$$= -52.8 \text{ MPa, } 7.2 \text{ MPa}$$

$$\text{Factor of safety} = \frac{250}{52.8} = 4.73$$

Example 11.2 A circular bar is to withstand a bending moment of 10 kN.m and a torque of 30 kN.m. Determine the diameter of the bar if the yield stress of the material is 250 MPa and the factor of safety is 2. Use maximum normal strain theory of failure. Take Poisson's ratio $\mu = 0.3$.

Solution Normal stress and shear stress in the bar are as follows,

$$\sigma = \frac{My}{I} = \frac{32M}{\pi d^3}$$

$$\tau = \frac{Tr}{J} = \frac{16T}{\pi d^3}$$

where, $y = r = d/2$

$J = 2I = \pi d^4/32$

The principal stresses σ_1 and σ_2 are as follows:

$$\sigma_{1,2} = \frac{16M}{\pi d^3} \pm \sqrt{\left(\frac{16M}{\pi d^3}\right)^2 + \left(\frac{16T}{\pi d^3}\right)^2}$$

$$= \frac{16M}{\pi d^3}\left(1 \pm \sqrt{1 + \left(\frac{T}{M}\right)^2}\right)$$

$$= \frac{16M}{\pi d^3}\left(1 \pm \sqrt{10}\right) = \frac{21.2M}{d^3}, \frac{-11.01M}{d^3}$$

The maximum strain corresponding to principal stresses is,

$$\varepsilon = \frac{1}{E}(\sigma_1 - \mu\sigma_2) = \frac{M}{Ed^3}(21.2 + 0.3 \times 11.01)$$
$$= 24.503\ M/Ed^3$$

If failure is to occur as per the maximum normal strain failure theory,

$$\varepsilon_1 = \frac{\varepsilon_y}{F.S.} = \frac{\sigma_y}{E \times F.S.}$$

or, $$\frac{24.503M}{Ed^3} = \frac{250}{E \times 2}$$

or, $$d = \sqrt[3]{\frac{24.503 \times 10^7}{125}}$$
$$= 125.15 \text{ mm}$$

Example 11.3 Solve Example 11.2 using maximum shear stress theory.

Solution Maximum shear stress,

$$\tau_{max} = \sqrt{\left(\frac{\sigma_x}{2}\right)^2 + \tau_{xy}^2}$$

$$= \sqrt{\left(\frac{16M}{\pi d^3}\right)^2 + \left(\frac{16T}{\pi d^3}\right)^2}$$

$$= \frac{16M}{\pi d^3}\sqrt{1 + \left(\frac{T}{M}\right)^2} = 16.1\frac{M}{d^3}$$

The Maximum shearing stress, τ_{max} will be within safe limit as per the maximum shear stress theory of failure when,

$$\tau_{max} = \frac{\sigma_y}{2 \times F.S.}$$

or, $$\frac{16.1M}{d^3} = \frac{250}{2 \times 2}$$

or, $$d = \sqrt[3]{\frac{16.1 \times 10 \times 10^6 \times 4}{250}}$$
$$= 137 \text{ mm}$$

Example 11.4 Figure 11.8 shows the state of stress in a material whose yield stress is 250 MPa. Determine whether it is safe as per the maximum energy of distortion theory of failure if the factor of safety is 2.

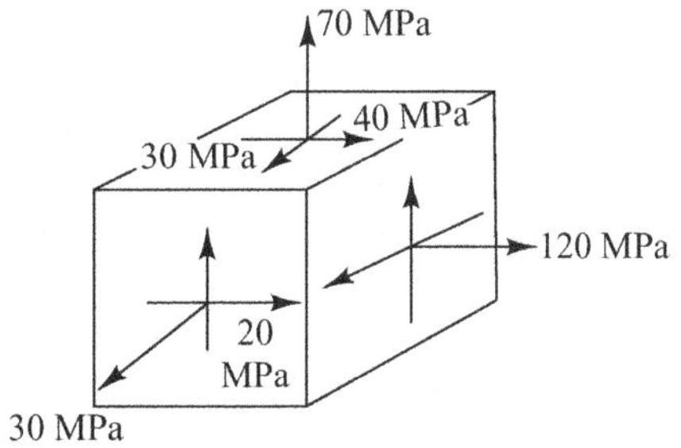

Fig. 11.8

Solution The principal stresses in three-dimensional stress condition is given by stress the roots of Eq. (2.13) as follows:

$$\tau_{nn}^3 - I_1\,\tau_{nn}^2 + I_2\,\tau_{nn} - I_3 = 0$$

where,

$$I_1 = \tau_{xx} + \tau_{yy} + \tau_{zz}$$
$$= 120 + 70 + 30 = 220 \text{ MPa}$$

$$I_2 = \tau_{xx}\,\tau_{yy} + \tau_{yy}\,\tau_{zz} + \tau_{zz}\tau_{xx} - \tau_{xy}^2 - \tau_{yz}^2 - \tau_{zx}^2$$
$$= 120 \times 70 + 70 \times 30 + 30 \times 120$$
$$-40^2 - 20^2 - 30^2$$
$$= 11200$$

$$I_3 = \begin{vmatrix} \tau_{xx} & \tau_{yx} & \tau_{zx} \\ \tau_{xy} & \tau_{yy} & \tau_{zy} \\ \tau_{xz} & \tau_{yz} & \tau_{zz} \end{vmatrix} = \begin{vmatrix} 120 & 40 & 20 \\ 40 & 70 & 30 \\ 20 & 30 & 30 \end{vmatrix}$$
$$= 116000$$

Substituting in Eq. (2.13),

$$\tau_{nn}^3 - 220\,\tau_{nn}^2 + 11200\,\tau_{nn} - 116000 = 0$$

The solution of this equation gives three values of principal stress, $\tau_{nn} = 150.85, 55, 13.92$. Substituting three values of principal stresses in Eq. (11.8),

$$(\sigma_1 - \sigma_2)^2 + (\sigma_2 - \sigma_3)^2 + (\sigma_3 - \sigma_1)^2 \le 2\left(\frac{\sigma_y}{\text{F.S.}}\right)^2$$

or
$$(150.85 - 55)^2 + (55 - 13.92)^2 + (13.92 - 150.85)^2 \le 2\left(\frac{250}{2}\right)^2$$

or $29624.6 \le 31250$, which is safe.

Example 11.5 A steel cantilever 50 mm wide, 100 mm deep is 1 m long. It is to carry a concentrated load of varying magnitude at its free end. The static tensile yield stress and fatigue strength for completely reversed stress are 300 MPa and 200 MPa, respectively. If the minimum value of concentrated load of 10 kN, determine the maximum value of varying load which will cause failure as per Soderberg criteria.

Solution Let P = maximum value of varying load

$$\text{Average load} = \frac{P+10}{2} = \frac{P}{2} + 5 \text{ kN}$$

$$\text{Alternating load} = \frac{P-10}{2} = \frac{P}{2} - 5 \text{ kN}$$

Mean stress,

$$\sigma_m = \frac{My}{I} = \frac{(P/2+5)\times 10^3 \times 1000 \times 50}{50 \times 100^3/12}$$

$$= 6P + 60 \text{ MPa}$$

Alternating stress,

$$\sigma_a = \frac{(P/2-5)\times 10^3 \times 1000 \times 50}{50 \times 100^3/12}$$

$$= 6P - 60 \text{ MPa}$$

Soderberg criteria,
$$\frac{\sigma_m}{\sigma_y} + \frac{\sigma_a}{\sigma_{cr}} = 1$$

or
$$\frac{6P+60}{300} + \frac{6P-60}{200} = 1$$

or,
$$P = 22 \text{ kN}$$

EXERCISES

PROBLEMS

11.1 A circular cantilever of diameter 40 mm is carrying 10π kN and 5π kN load as shown in Fig. 11.9. Determine the factor of safety at mid point A at the fixed end if yield stress is 250 MPa.

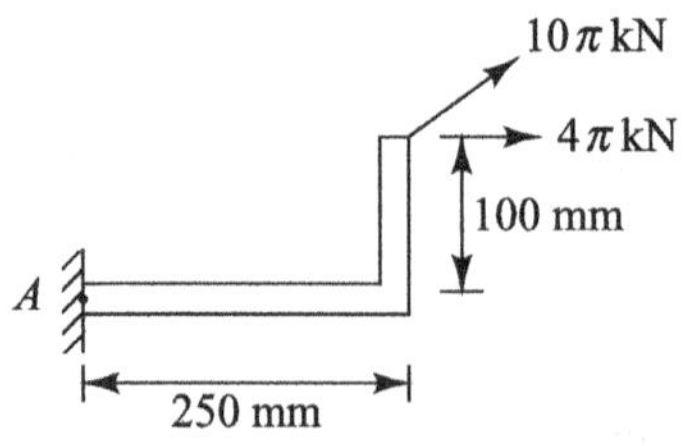

Fig. 11.9

11.2 A circular pipe of cast iron has inner and outer diameters 200 and 250 mm, respectively. It is subjected to a torque of 50 kN.m and axial compressive load of P kN. If the ultimate strength of the material is 250 MPa in tension and 850 MPa in compression, determine the maximum value of P such that the material does not fail as per maximum principal strain theory. Take Poisson's ratio equal to 0.26.

11.3 and 11.4 Determine whether the material is safe as per von Mises yield criteria under the state of stress shown in Figs 11.10 and 11.11 if the yield stress of the material is 540 MPa and factor of safety is 1.5.

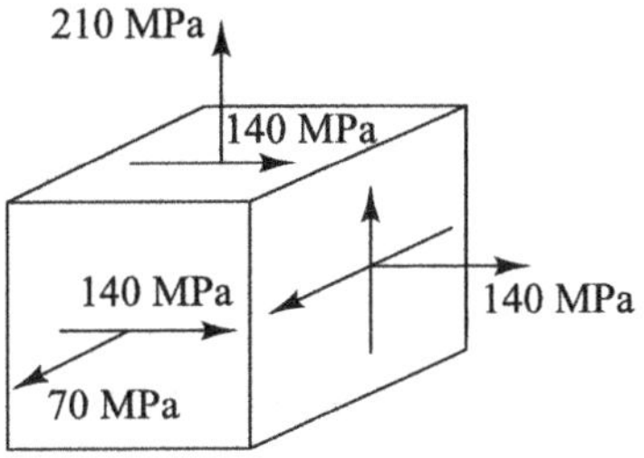

Fig. 11.10

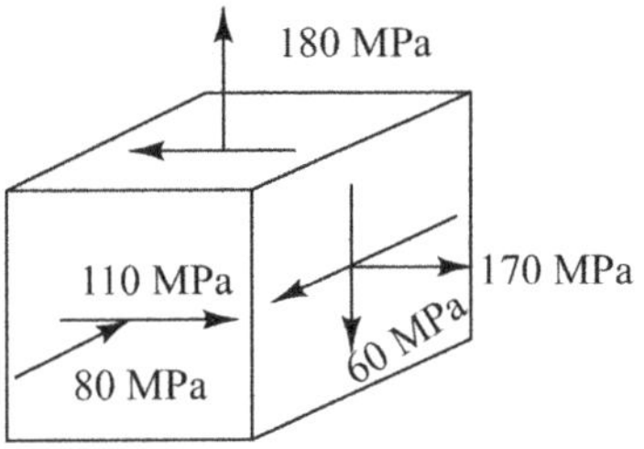

Fig. 11.11

11.5 The state of stress at a point in a material at which yielding was initiated is shown in Fig. 11.12. Determine the value of uniaxial tensile yield strength of the material predicted by each of the following theories of failure; (a) maximum normal stress,(b) maximum normal strain, (c) maximum shearing stress and (d) octahedral shearing stress. The Poisson's ratio is 0.3.

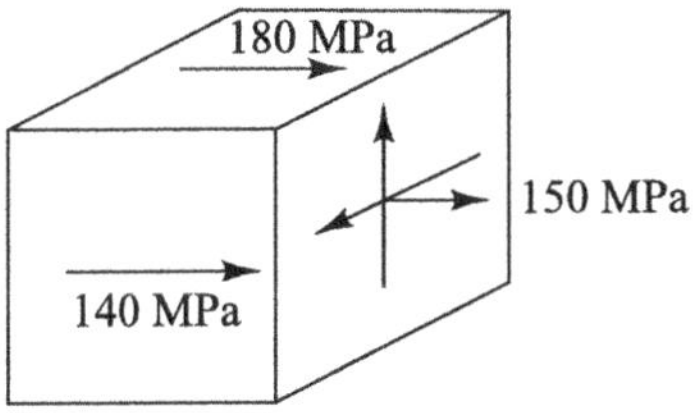

Fig. 11.12

11.6 A circular bar of diameter 54 mm is subjected to an axial tension of 40 MPa. The fatigue strength for complete reversal of stress is 200 MPa and the static yield strength is 300 MPa. Determine the limiting value of completely reversed axial load that can be applied at an eccentricity of 20 mm without causing fatigue failure as per Soderberg criteria.

OBJECTIVE QUESTIONS

11.7 At a point in a steel member, the major principal stress is 200 MPa (tensile) and the minor principal stress is compressive. If the uniaxial tensile yield stress is 250 MPa, then according to the maximum shear stress theory, the magnitude of the minor principal stress (compressive) at which yielding will commence is,

(a) 200 MPa (b) 100 MPa
(c) 50 MPa (d) 25 MPa

11.8 According to maximum shear stress failure criteria, yielding in material occur when the maximum shear stress is equal to

(a) $1/2 \times$ yield stress (b) $\sqrt{2} \times$ yield stress

(c) $\sqrt{\frac{2}{3}} \times$ yield stress (d) $2 \times$ yield stress

11.9 The limit of proportionality of a certain steel specimen is 300 MPa in simple tension. It is subjected to principal tensile stresses of 150 MPa, 60 MPa and 30 MPa. According to the maximum principal stress theory, the factor of safety in this case would be,

(a) 10 (b) 5
(c) 4 (d) 2

11.10 A cube is subjected to equal tensile stress on all the three faces. If the yield stress of material is σ_y, then, based on the strain energy theory, the maximum tensile stress will be,

(a) $\dfrac{\sigma_y}{\sqrt{3(1-2\mu)}}$ (b) $\dfrac{\sigma_y}{\sqrt{3(2-\mu)}}$

(c) $\dfrac{\sigma_y}{\sqrt{3(1-\mu)}}$ (d) $\dfrac{\sigma_y}{\sqrt{3-2\mu}}$

11.11 As per the maximum principal stress theory, when a shaft is subjected to a bending moment M and torque T, and if σ is the allowable stress in axial tension then the diameter d of the shaft is given by,

(a) $d^3 = \frac{16}{\pi\sigma}\left[M + \sqrt{M^2 + T^2}\right]$

(b) $d^2 = \frac{32}{\pi\sigma}\left[M + \sqrt{M^2 + T^2}\right]$

(c) $d^3 = \frac{4}{\pi\sigma}\left[M + \sqrt{M^2 + T^2}\right]$

(d) $d^2 = \frac{8}{\pi\sigma}\left[M + \sqrt{M^2 + T^2}\right]$

11.12 The stress at which a material fracture under large number of reversal of stress is called,
(a) endurance limit (b) creep
(c) ultimate strength (d) residual stress

11.13 At a point is a structure, there are two mutually perpendicular tensile stresses of 800 MPa and 400 MPa at failure. If Poisson's ratio is 0.25, what would be the equivalent stress in simple tension according to maximum principal strain theory?
(a) 1200 MPa (b) 900 MPa
(c) 700 MPa (d) 400 MPa

11.14 A machine part is subjected to a series of four cyclic loads. The fatigue life cycles at these four stress levels are 3×10^4, 5×10^5, 12×10^6 and 7×10^7, respectively. If the number of cycles for first three loads are 10^4, 2×10^5 and 10^6, then the number of cycles of fourth load which will cause failure is,
(a) 12×10^6 (b) 2×10^7
(c) 5×10^6 (d) 10^7

12

Deflection–Geometrical Methods

CONCEPT REVIEW

12.1 DEFLECTION DIAGRAMS

For statically indeterminate structures, displacement conditions are used in addition to equilibrium conditions for determination of the unknown forces and reactions. Deflection calculations are also necessary where the excessive deflections can cause instability of a structure or may generate excessive secondary stresses. Deflections are also checked to remain within limits since it may damage the brittle finishings of floors, etc. The deflection pattern also indicates the nature of bending stresses and the location of points of contraflexure or zero moment as shown in Fig. 12.1.

Deflections in structures within the elastic range are usually very small and the deflected shapes shown in this book are exaggerated. The commonly used methods for the determination of deflections are as follows.

- Moment area method
- Conjugate beam method
- Strain energy method
- Double integration method
- Singularity function method
- Graphical method

Some of the result derived by them are used in the advanced stiffness and flexibility methods for the analysis of very complex structures.

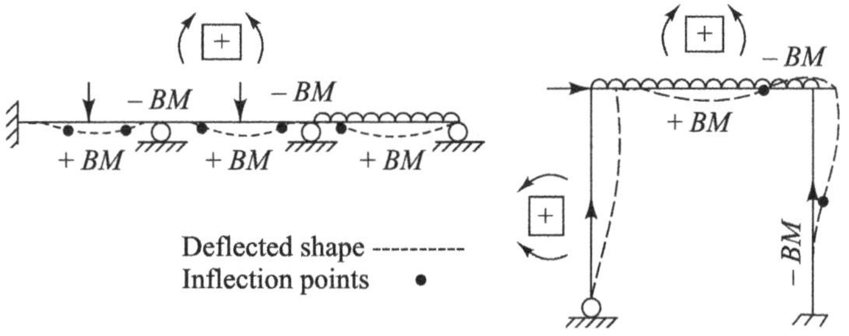

Fig. 12.1

12.2 MOMENT AREA METHOD

Figure 12.2 shows the deflected shape and bending moment of a beam subjected to arbitrary load. Consider two points C and D at a small distance dx on the beam. The radius of curvature and bending moment for any such section is given by the usual bending equation.

$$\frac{M}{I} = \frac{E}{R} \qquad (12.1)$$

The elementary lengh dx and the differential $d\theta$ between the radii of curvature at points C and D are related as,

$$dx = -\,d\theta \times R \qquad (12.2)$$

The negative sign is used with $d\theta$ since the slope θ decreases with increasing x from a positive maximum value at $x = 0$ to a negative minimum value at $x = L$ as per $x - y$ axes shown in Fig. 12.4(d). Eq.(12.1) and (12.2) give,

$$\int_A^B d\theta = \int_A^B \frac{-Mdx}{EI} = \int_A^B \frac{-dA}{EI} \qquad (12.3)$$

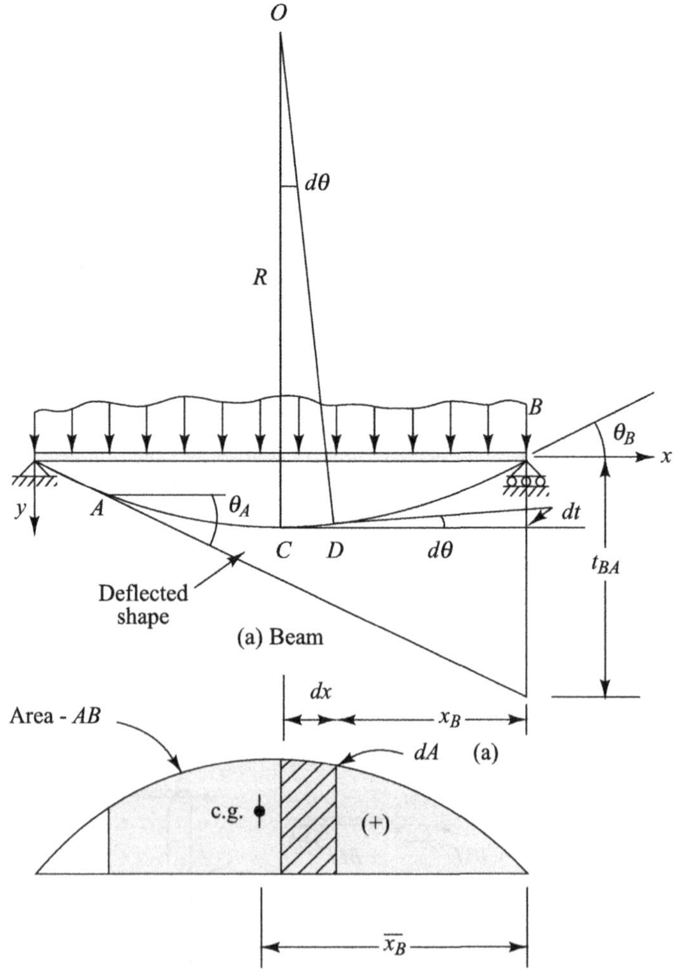

Fig. 12.2 (b) B.M. Diagram

where dA is the area of bending moment diagram between points C and D. Integration of Eq. (12.3) gives,

$$\theta_B - \theta_A = \frac{-1}{EI} \times (\text{area} - AB) \qquad (12.4)$$

This is the first moment area theorem.

12.2.1 First Moment Area Theorem

The difference of deflection angles at two points B and A on the deflected curve of a beam is equal to the area under the bending moment diagram between two points A and B divided by EI.

12.2.2 Second Moment Area Theorem

In Fig.12.2 the distance dt along the vertical line through point B is nearly equal to,

$$dt \approx -x_B \times d\theta \qquad (12.5)$$

Negative sign is again used with $d\theta$ for the reason similar to Eq. (12.2) as θ also decreases with increasing t in the positive x-direction. Integrating dt yield the vertical distance t_{BA} between the point B and the tangent from point A on the deflected beam.

$$t_{BA} = \int_A^B dt = -\int_A^B x_B d\theta = \frac{1}{EI}\int_A^B x_B dA$$

$$= \frac{1}{EI}(\text{area} - AB) \times \bar{x}_B \qquad (12.6)$$

This is the second moment area theorem, where, $\bar{x}_B$ = distance of the centroid of moment diagram (area – AB) from point B.

If A and B are two points on the deflected shape of a beam, the vertical distance of point B from the tangent drawn to the elastic curve at point A is equal to the moment of bending moment diagram area, between the points A and B about the vertical line from point B, divided by EI.

12.2.3 Sign Convention for Moment Area Theorem

Any sign convention can be chosen but it should be consistent throughout the calculations. Sagging moment has the positive B.M. diagram. Clockwise rotation is taken as positive. Positive deviation t_{BA} indicates that the point B lies above the tangent from point A.

12.2.4 Moment Diagram by Parts

Since Moment Area theorems require computation of area and their centroids, it is simpler to consider each load or reaction separate and then add to get the total moment area or its moment. Figure 12.3 gives some basic result required for the calculations. In case of beams with overhangs as in Figs. 12.13 and 12.14, moment diagram can be drawn in parts from both sides of the middle support.

12.2.5 Statically Indeterminate Beam

Moment Area theorem can solve the indeterminate beam when used alongwith geometrical constrains as explained in Examples 12.8 through 12.13.

12.3 CONJUGATE BEAM METHOD

It is a simple method presented by Otto Mohr in 1860 to compute displacements in beams and frames. It is based on similarity between loading equations.

Applied load w and shear force V are related by,

$$w = -dV/dx$$

Shear force V and bending moment M are related by,

$$V = dM/dx$$

Bending moment M and vertical displacements y are related by,

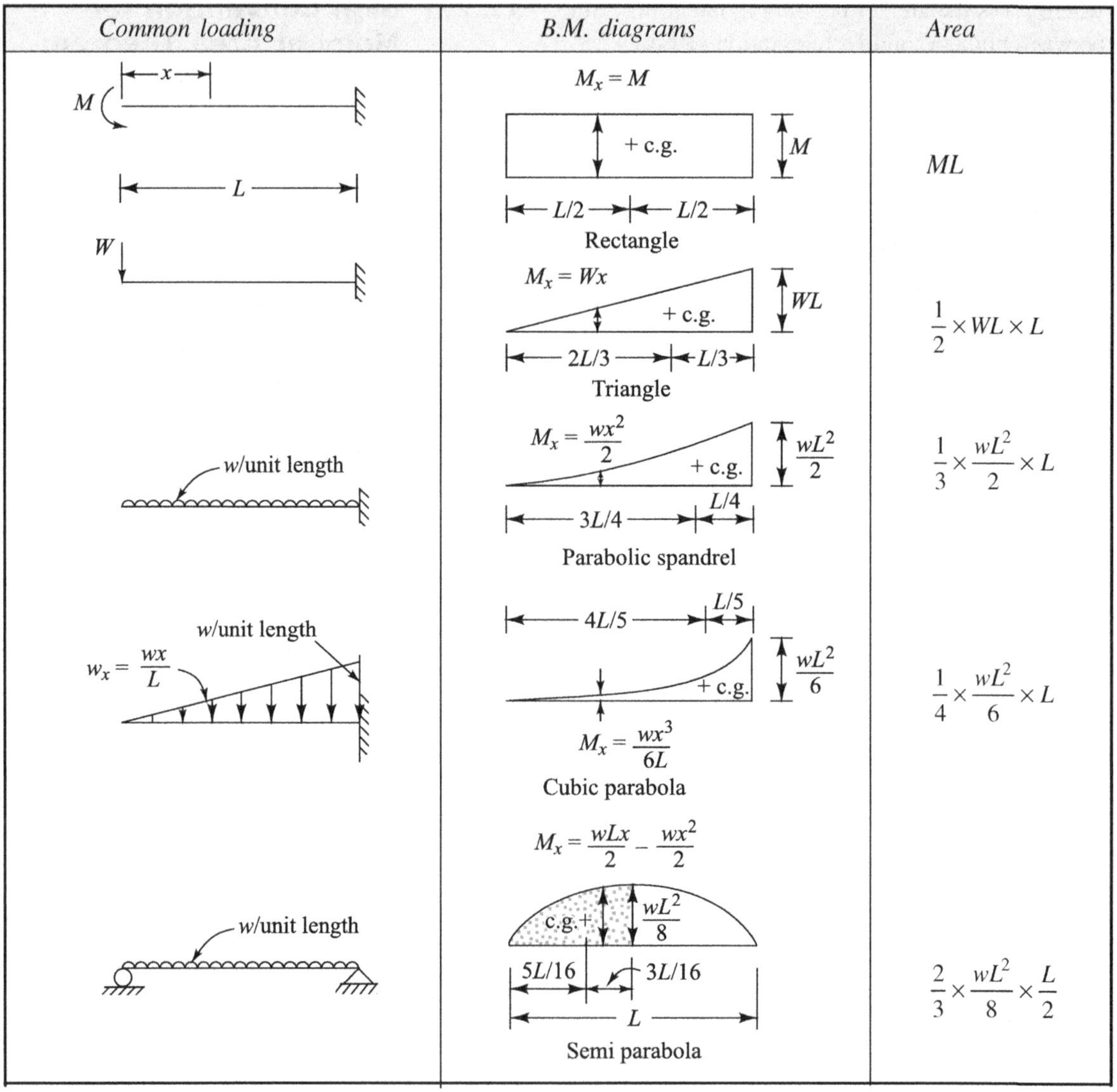

Fig. 12.3

$$\frac{M}{EI} = \frac{1}{R} = \frac{\pm\, d^2y/dx^2}{\left[1 + (dx/dx)^2\right]^{3/2}} \approx -\frac{d^2y}{dx^2}$$

The negative sign is used here with d^2y/dx^2, since slope $dy/dx = \theta$, decreases with increasing x for positive B.M. Now consider the similarity between these equations as follows.

$$\frac{dV}{dx} = -w \Leftrightarrow \frac{d\theta}{dx} = \frac{-M}{EI} \qquad \text{(i)}$$

and

$$\frac{d^2M}{dx^2} = -w \Leftrightarrow \frac{d^2y}{dx^2} = \frac{-M}{EI} \qquad \text{(ii)}$$

Here, shear V is analogous to the slope θ, M/EI is analogous to the external load w, and moment M is analogous to the displacement y.

If an actual structure is replaced by an imaginary structure called conjugate beam, which is loaded by M/EI, then its shear diagram will give a slope θ and the moment diagram will give a deflection y of the real structure as shown in Fig. 12.4.

Theorem 1: The slope θ_x at a point x in a beam is equal to the shear in its conjugate beam at the corresponding point.

Theorem 2: The deflection y_x at a point x in a beam is equal to the moment in its conjugate beam at the corresponding point.

Conjugate Beam Supports: The supports of the conjugate beam should be selected such that its shear V corresponds to the slope θ and the moment M corresponds to the deflection y of the real beam as shown in Fig. 12.5.

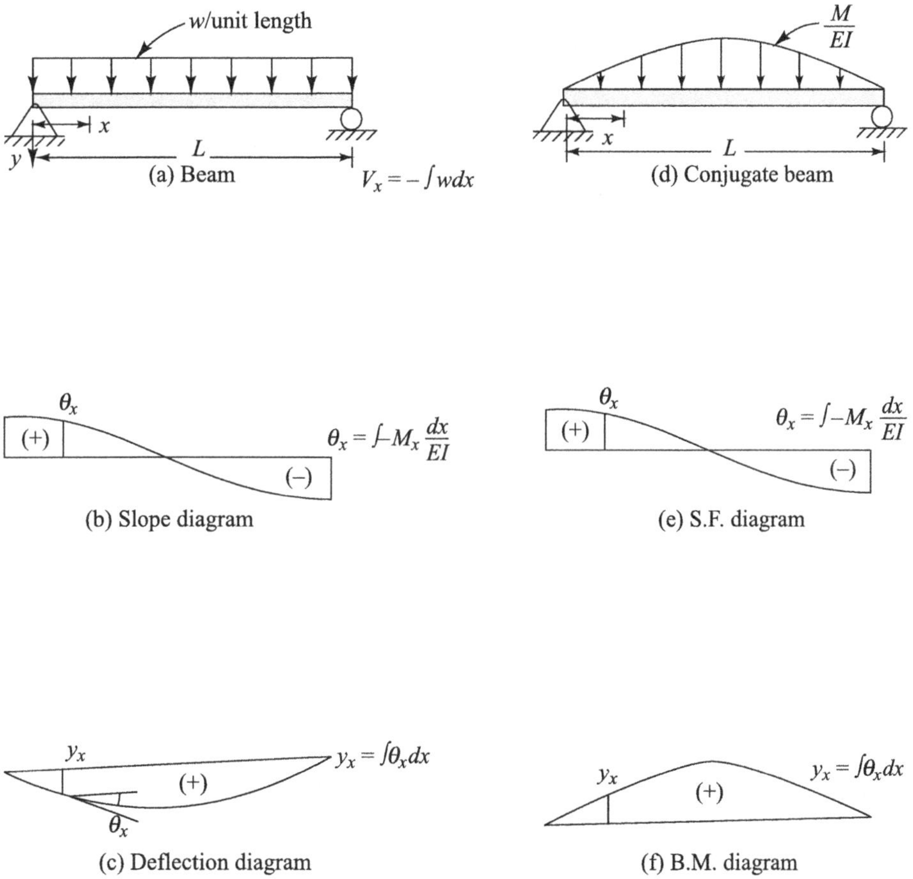

Fig. 12.4

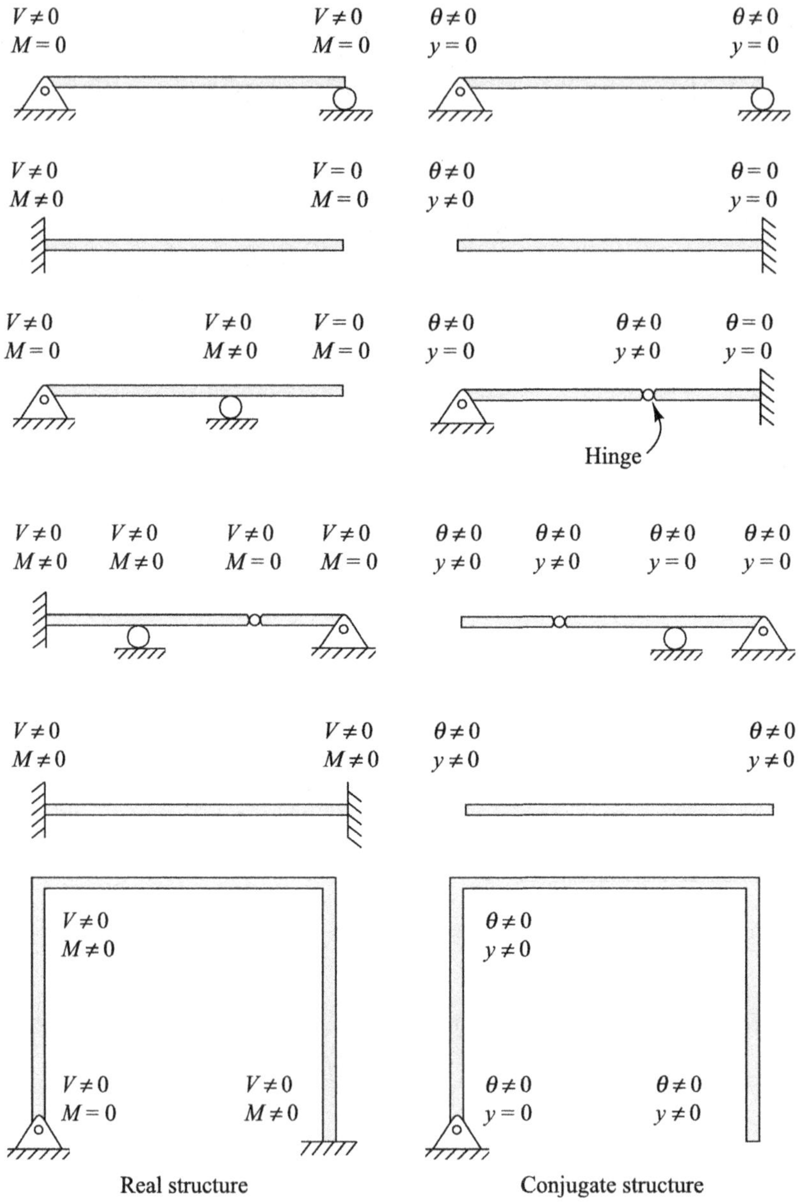

Fig. 12.5

Sign Conventions: Positive *M/EI* will be applied downwards on the conjugate beam. Sign convention for the S.F. and B.M. diagram will be same as for the real beam. If the shear force and the bending moment of the conjugate beam are positive, the corresponding rotation θ and deflection y of the real beam will be clockwise and downward, respectively.

12.4 DOUBLE INTEGRATION METHOD

The deflected shape can be obtained by integration the bending equation twice.

$$\iint \frac{d^2y}{dx^2} = \iint \frac{-M}{EI},$$

or, $$y = \iint \left(\frac{-M}{EI}\right) dx^2$$

Sometimes the integration process is more difficult but it is convenient if the deflected shape is to be evaluated at all points.

12.5 SINGULARITY FUNCTION METHOD

Consider a case of simply supported beam carrying a concentrated point load as shown in Fig. 12.6. Bending moment at any section at a distance x from left end is given by,

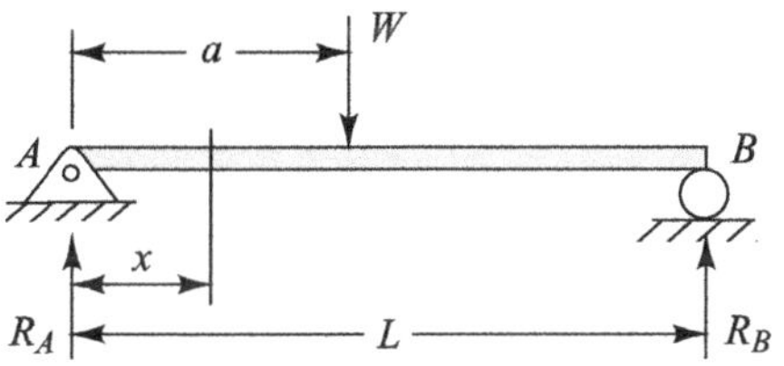

Fig. 12.6

when, $0 < x < a$, $M(x) = R_A x$ (12.7)

when, $a < x < L$, $M(x) = R_A x - W(x - a)$ (12.8)

$M(x)$ can be expressed by a single equation by introducing a Singularity function or Half range function $<x - a>$ such that,

$$<x - a> = (x - a), \quad \text{when } x > a$$
$$= 0, \quad \text{when } x < a$$

Then Eqs (12.7) and (12.8) can be expressed by Eq. (12.9) as follows.

$$M(x) = R \times x - W <x - a> \qquad (12.9)$$

Singularity functions are defined to follow the laws of integration as follows.

$$\int_0^x <c - a>^{-1} dc = <x - a>^0 = 0, \quad \text{when } x < a$$
$$= 1, \quad \text{when } x > a \qquad (12.10)$$

$$\int_0^x <c - a>^{-2} dc = <x - a>^{-1} \qquad (12.11)$$

$$\int_0^x <c - a>^n dc = \frac{<x - a>^{n+1}}{n+1} \quad \text{when } x \geq a \qquad (12.12)$$

Table 12.1 gives the singularity functions for the different types of loads. Singularity functions for load intensity of concentrated load and moment, $<x - a>^{-1}$ and $<x - a>^{-2}$ are zero everywhere except at $x = a$ where they are infinite in such a way that Eqs (12.10) and (12.11) are valid.

12.5.1 Sign Convention

In continuation to the sign convention used for B.M. and S.F. as shown in Fig. 12.4, the downward load and anticlockwise moment is taken positive for the load function $w(x)$ as shown in Table 12.1. As earlier for the moment area and conjugate beam methods, clockwise rotation is positive.

12.6 GRAPHICAL METHOD

The deflection of a truss can be found by plotting it on an enlarged scale, the displacement of the joints due to extension/shortening of each member in the frame. Graphical method is specially useful if it is desired to plot the deflected shape of the truss to fix camber or other properties.

12.6.1 Williot Diagram

Williot diagram is a truss deflection diagram excluding the length of the members and thus a larger scale can be used for the plot. Consider a frame in Fig. 12.7 carrying some load at B. If the length of the member AB is extended by Ba and the length of the member CB is shortened by Bc, then to locate the new position of joint B, draw arcs from A and C of length Aa and Cc, respectively, to intersect as B'.

Since the extension/shortening of the members is very small in working load range, arcs aB' and cB' can be replaced by perpendicular lines from a and c, respectively, ΔBx and ΔBy are the horizontal and vertical displacements of the joint B.

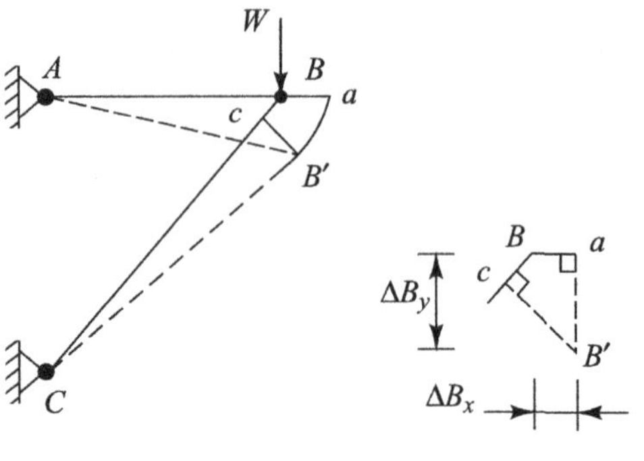

(a) Frame and deformed shape (b) Williot diagram

Fig. 12.7

Table 12.1

S. No.	Type of Loading	Singularity function	Bending moment $M_x = \int\int w(x)dx$
(1)	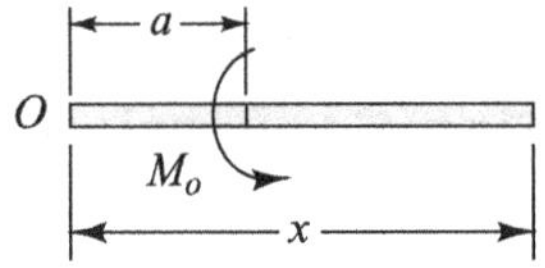 Concentrated moment	$w(x) = M_o < x - a >^{-2}$ (Doublet function)	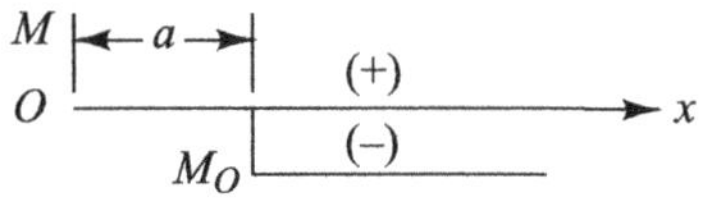 $M_x = -M_o <x - a>^0$
(2)	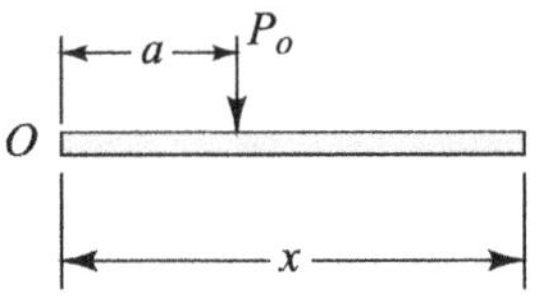 Point load	$w(x) = P_o < x - a >^{-1}$ (Dirac delta or impulse function)	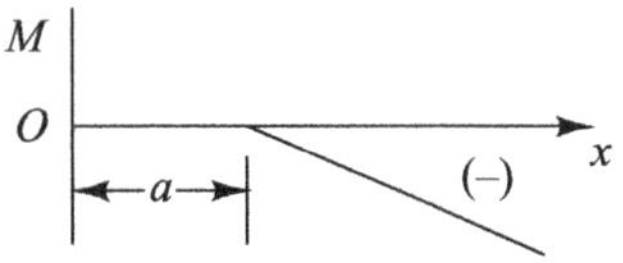 $M_x = -P_o <x - a>^1$
(3)	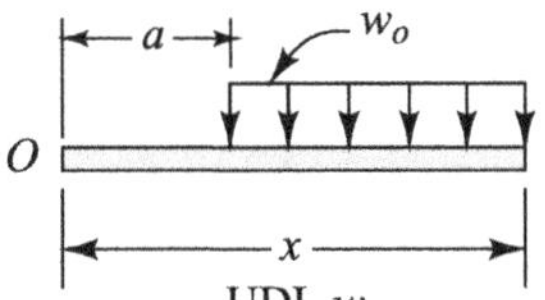 UDL w_o	$w(x) = w_o < x - a >^0$ (Step function)	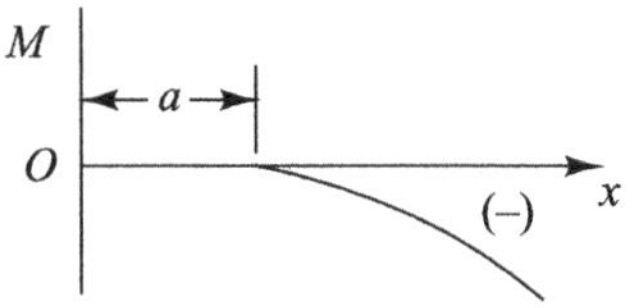 $M_x = \frac{-1}{2} w_o <x - a>^2$
(4)	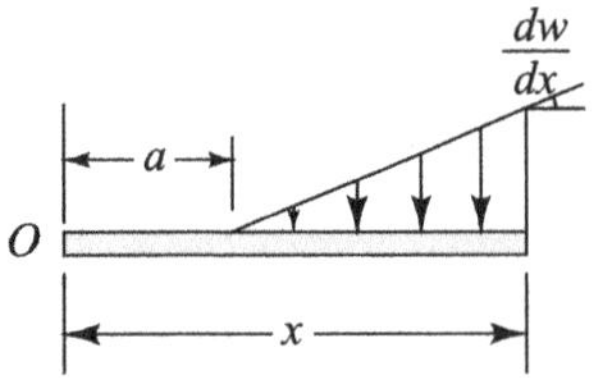 Linearly varying load	$w(x) = \frac{dw}{dx} <x - a>^1$	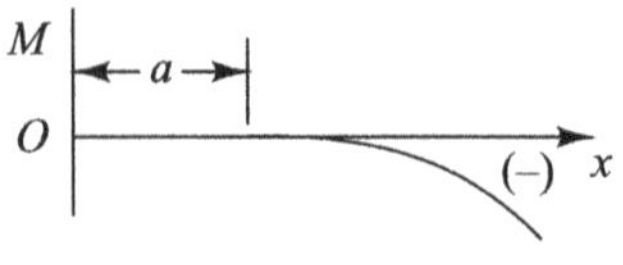 $M_x = \frac{-1}{2 \times 3} \frac{dw}{dx} <x - a>^3$

12.6.2 Williot–Mohr Diagram

For the Williot diagram, the solution requires a reference line passing through a joint, which does not rotate and it may or may not be a member of frame, with which the displacements of the joints can be related. It is however not possible in case of asymmetric loading/structure. Mohr's correction is used with Williot diagram in *such cases and the deflection diagram is then known as Williot-Mohr diagram.*

12.6.3 Sign Convention

Move on Williot diagram from one joint to the other in the opposite direction of the actual structure in case of shortening or negative strain. If the member forces are marked on the truss, then move in the opposite direction of the member force marked at the other end of the member.

EXAMPLES

Moment Area Method

Example 12.1 Determine the slope and deflection at the free end of the cantilever beam carrying a concentrated load W at the free end. Neglect weight of the beam.

Solution Figure 12.8 shows the deflected shape of the beam. Figure 12.3(ii) shows the bending moment diagram and its e.g. for this type of loading. From Eq. (12.4), (B.M. is –ve)

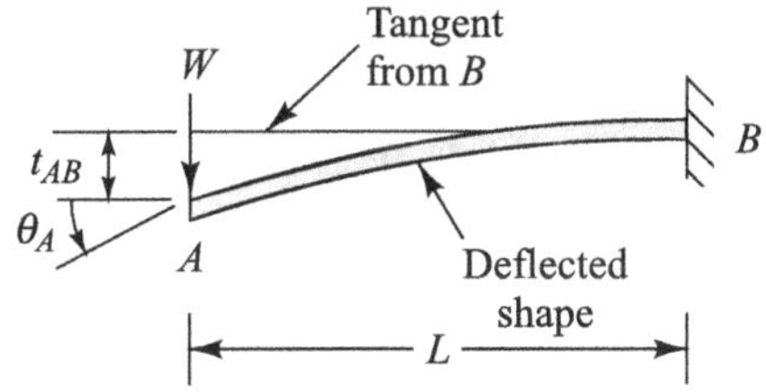

Fig. 12.8

$$\theta_B - \theta_A = \frac{-1}{EI}(\text{area} - AB) = \frac{WL^2}{2EI}$$

Since $\theta_B = 0$, therefore $\theta_A = -WL^2/2EI$ Anticlockwise

From Eq. (3.6)

$$t_{AB} = \frac{1}{EI}(\text{area} - AB) \times \bar{x}_A = \frac{1}{EI} \times \frac{-WL^2}{2} \times \frac{2L}{3}$$

or $t_{AB} = -WL^3/6EI$

The negative sign shows that pont A lies below point B on the deflected shape.

Example 12.2 Derive an expression for the end slope and maximum deflection of a simply supported beam loaded with uniformly distributed load w per unit length.

Solution Bending moment diagram in Fig. 12.9 (b) is drawn in parts by considering the moment of reaction $R_A(= wL/2)$ and udl w about end B. Taking moment of B.M. diagram (area AB) about point B,

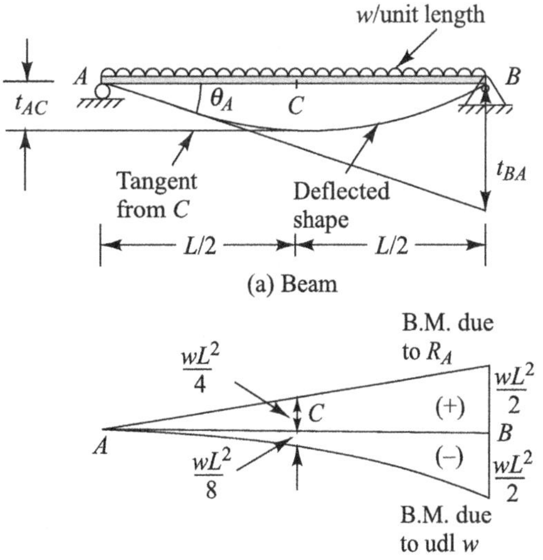

(a) Beam

(b) B.M. diagram by parts

Fig. 12.9

$$t_{BA} = \frac{1}{EI}\left[\left(\frac{1}{2}\times\frac{wL^2}{2}\times L\right)\times\frac{L}{3}\right.$$
$$\left.-\left(\frac{1}{3}\times\frac{wL^2}{2}\times L\right)\times\frac{L}{4}\right] = \frac{wL^4}{24E}$$

therefore,

$$\theta_A = \frac{t_{BA}}{L} = \frac{wL^3}{24EI} = -\,\theta_B$$

Due to symmetry, end slopes will be equal and deflection will be moment maximum at the mid span. Taking of B.M. diagram (area – AC) about point A, mid span deflection,

$$t_{AC} = \frac{1}{EI}\left[\frac{1}{2}\times\frac{wL^2}{4}\times\frac{L}{2}\times\frac{2}{3}\times\frac{L}{2}-\frac{1}{3}\right.$$
$$\left.\times\frac{wL^2}{8}\times\frac{L}{2}\times\frac{3}{4}\times\frac{L}{2}\right] = \frac{5}{384}\times\frac{wL^4}{EI}$$

Example 12.3 Determine the deflection at points B and C of the beam shown in Fig. 12.10. Take $E = 210$ GPa and $I = 8 \times 10^6$ mm^4.

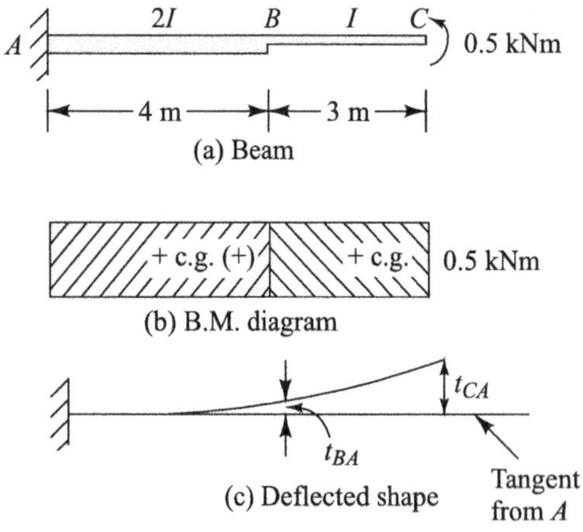

Fig. 12.10

Solution Deflection at B, by taking moment of B.M. diagram area A to B about B.

$$t_{BA} = \frac{1}{2EI}(0.5\times10^6\times4000)\times2000 = 1.19 \text{ mm}$$

Deflection at C, by taking moment of B.M: diagram about C,

$$t_{CA} = \frac{(0.5\times10^6\times4000)\times5000}{16\times10^6\times210\times10^3}$$
$$+\frac{(0.5\times10^6\times3000)\times1500}{8\times10^6\times210\times10^3} = 4.315 \text{ mm}$$

Positive t_{BA} and t_{CA} indicates that the points B and C are higher than the tangent from point A.

Example 12.4 Determine the end displacements of a cantilever beam carrying u.d.l.w per unit length over half span as shown in Fig. 12.11(a).

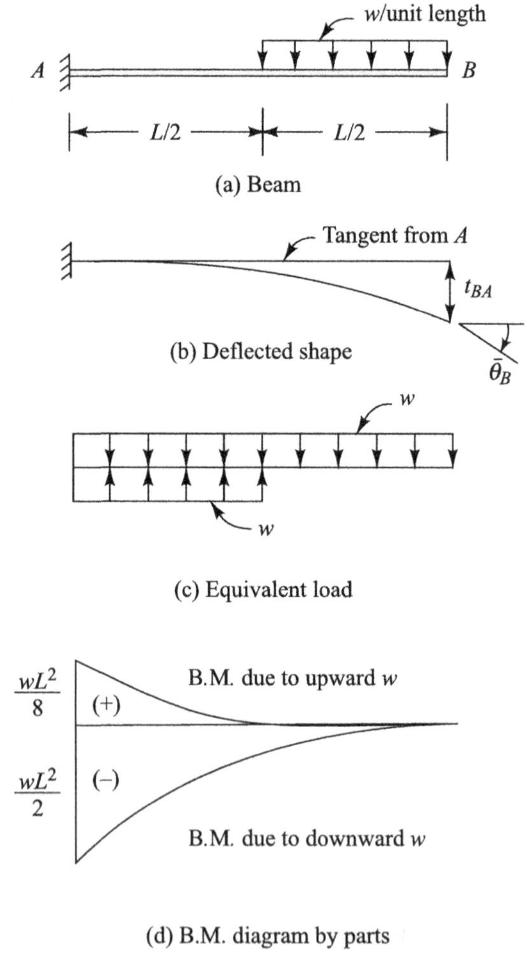

Fig. 12.11

Solution The applied load is replaced by an equivalent load as shown in Fig.12.11(c).

$$\theta_B - \theta_A = \theta_B - 0$$

$$= \frac{-1}{EI}\left[\frac{1}{3}\times\frac{wL^2}{8}\times\frac{L}{2} - \frac{1}{3}\times\frac{wL^2}{2}\times L\right]$$

$$= \frac{7}{48}\times\frac{wL^3}{EI}$$

Taking moment of B.M. diagram area about B,

$$t_{BA} = \frac{1}{EI}\left[\frac{1}{3}\times\frac{wL^2}{8}\times\frac{L}{2}\times\left(\frac{3}{4}\times\frac{L}{2}+\frac{L}{2}\right)\right.$$

$$\left. -\frac{1}{3}\times\frac{wL^2}{2}\times L\times\frac{3L}{4}\right] = \frac{-41}{384}\times\frac{wL^4}{EI}$$

Negative t_{BA} indicates that the point B is lower than the tangent from point A.

Example 12.5 A timber beam 100 mm × 200 mm carries an u.d.l. 6 kN/m as shown in Fig.12.12. Determine the deflection at every meter interval if E = 10 GPa for the timber.

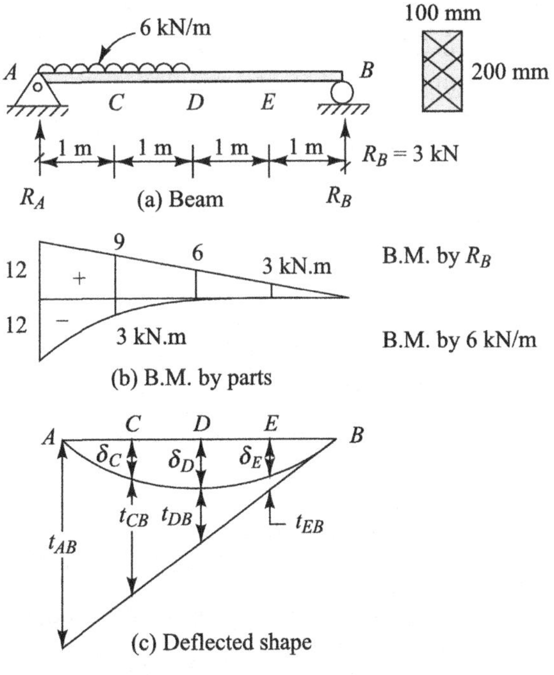

Fig. 12.12

Solution

$$EI = 10\times10^3\times\frac{100\times200^3}{12} = 6.67\times10^{11}\ \text{N mm}^2$$

Taking the moment of forces about A,

$$R_B = \frac{1}{4}\times\left(\frac{6\times2\times2}{2}\right) = 3\ \text{kN}$$

Figure 12.12 shows the B.M. diagram by parts. Taking the moment of B.M. diagram about points A, C, D and E,

$$t_{AB} = \frac{1}{EI}\left(\frac{12\times10^6\times4000}{2}\times\frac{4000}{3}\right.$$

$$\left. -\frac{12\times10^6\times2000}{3}\times\frac{2000}{4}\right) = 41.98\ \text{mm}$$

$$t_{CB} = \frac{1}{EI}\left(\frac{9\times10^6\times3000}{2}\times\frac{3000}{3}\right.$$

$$\left. -\frac{3\times10^6\times1000}{3}\times\frac{1000}{4}\right) = 19.86\ \text{mm}$$

$$t_{DB} = \frac{1}{EI}\left(\frac{6\times10^6\times2000}{2}\times\frac{2000}{3}\right) = 6.0\ \text{mm}$$

$$t_{EB} = \frac{E}{EI}\left(\frac{3\times10^6\times1000}{2}\times\frac{1000}{3}\right) = 0.75\ \text{mm}$$

$$\delta_C = \frac{3}{4}\times t_{AB} - t_{CB} = 11.625\ \text{mm}$$

$$\delta_D = \frac{1}{2}\times t_{AB} - t_{DB} = 14.99\ \text{mm}$$

$$\delta_E = \frac{1}{4}\times t_{AB} - t_{EB} = 9.75\ \text{mm}$$

Example 12.6 Determine the deflected shape of a simply supported beam under cantilever load W as shown in Fig. 12.13(a).

Solution End reactions,

$$R_A = \frac{WL_2}{L_1} \quad \text{and} \quad R_B = \frac{W(L_1 + L_2)}{L_1}$$

Figure 12.13(c) shows the B.M. Diagram by parts. Taking the moment of B.M diagram (area–AB) about B,

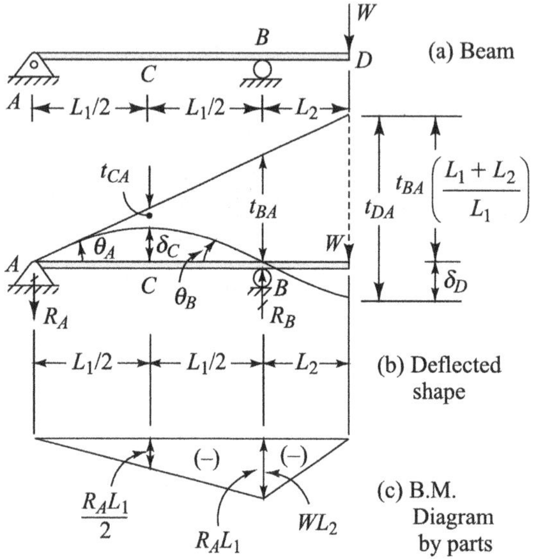

Fig. 12.13

$$t_{BA} = \frac{-1}{EI} \times \frac{R_A L_1}{2} \times L_1 \times \frac{L_1}{3} = \frac{-W L_2 L_1^2}{6EI}$$

Negative sign indicates that B is below the tangent from A.

$$\theta_A = \frac{t_{BA}}{L_1} = \frac{-W L_2 L_1}{6EI} \quad \text{(anticlockwise)}$$

$$\theta_B - \theta_A = \frac{-1}{EI} \times (\text{area} - AB)$$

$$= \frac{-1}{EI} \times \frac{(-R_A L_1 \times L_1)}{2} = \frac{W L_1 L_2}{2EI}$$

$$\theta_B = \frac{W L_1 L_2}{EI}\left(\frac{1}{2} - \frac{1}{6}\right) = \frac{W L_1 L_2}{3EI} \quad \text{(clockwise)}$$

Taking the moment of B.M. diagram (area – AC) about C,

$$t_{CA} = \frac{-1}{EI} \times \left[\frac{1}{2} \times \frac{R_A L_1}{2} \times \frac{L_1}{2} \times \left(\frac{1}{3} \times \frac{L_1}{2}\right)\right]$$

$$= \frac{-W L_2 \times L_1^2}{48EI}$$

$$\delta_C = \frac{1}{2} \times t_{BA} - t_{CA} = \frac{-W L_2 L_1^2}{EI}\left(\frac{1}{12} - \frac{1}{48}\right)$$

$$= \frac{-W L_2 L_1^2}{16EI} \quad \text{(upward)}$$

Taking the moment of B.M. diagram (area – AD) about D,

$$t_{DA} = \frac{-1}{EI}\left[\frac{R_A L_1 \times L_1}{2} \times \left(\frac{L_1}{3} + L_2\right) + \frac{W L_2 \times L_2}{2} \times \frac{2L_2}{3}\right]$$

$$= \frac{-W L_2}{EI}\left(\frac{L_1^2}{6} + \frac{L_1 L_2}{2} + \frac{L_2^2}{3}\right)$$

Negative sign indicates that D is below the tangent from A.

$$\delta_D = t_{DA} - t_{BA} \times \frac{L_1 + L_2}{L_1}$$

$$= \frac{-W L_2^2}{3EI}(L_1 + L_2) \quad \text{(downward)}$$

Example 12.7 A steel beam *MB* 225 (I_x = 3442 cm^4) is loaded as shown in Fig. 12.14(a). Determine the end slopes and deflection under both the concentrated loads. Take E = 210 GPa.

Solution Taking the moment of forces about B,

$$R_A = \frac{1}{4}(30 \times 2 + 10 \times 4 \times 2 - 20 \times 2) = 25 \text{ kN}$$

$$EI = 3442 \times 10^4 \times 210 \times 10^3$$

$$= 7.23 \times 10^{12} \text{ Nmm}^2$$

Figure 12.14 shows the B.M. diagram by parts. Taking moment of B.M. diagram area-AC about C,

$$t_{CA} = \frac{1}{EI}\left(\frac{50 \times 10^6 \times 2000}{2} \times \frac{2000}{3} - \frac{20 \times 10^6 \times 2000}{3} \times \frac{2000}{4}\right) = 3.7 \text{ mm}$$

Taking moment of B.M. diagram area-AB about B,

$$t_{BA} = \frac{1}{EI}\left(\frac{100 \times 10^6 \times 4000}{2} \times \frac{4000}{3}\right.$$

$$- \frac{60\times10^6\times2000}{2}\times\frac{2000}{3}$$

$$\left.- \frac{80\times10^6\times4000}{3}\times\frac{4000}{4}\right) = 16.6 \text{ mm}$$

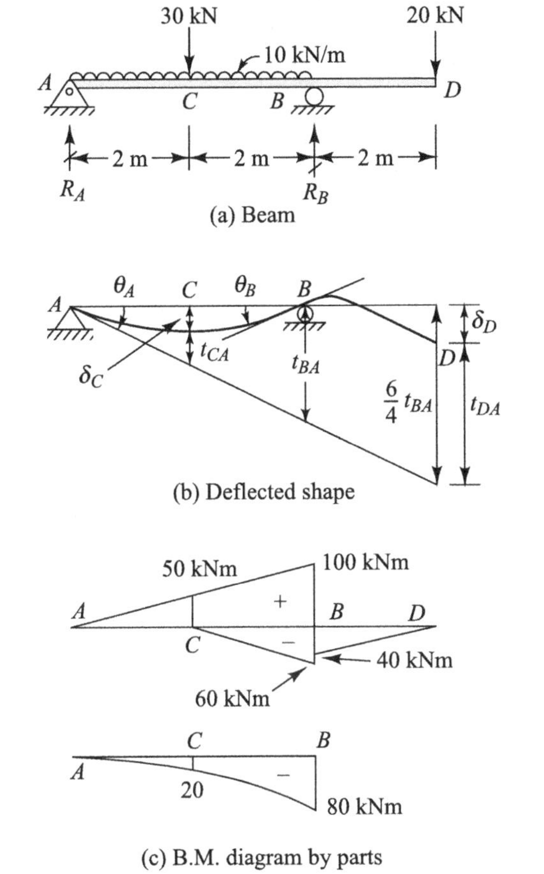

(a) Beam

(b) Deflected shape

(c) B.M. diagram by parts

Fig. 12.14

$$\delta_C = \frac{1}{2}t_{BA} - t_{CA} = \frac{16.6}{2} - 3.7 = 4.6 \text{ mm}$$

$$\theta_A = t_{BA}/4000 = 16.6/4000$$

$$= 4.15\times10^{-3} \text{ rad. (clockwise)}$$

$$\theta_B - \theta_A = \frac{-1}{EI}(\text{area} - AB)$$

$$= \frac{-1}{EI}\left(\frac{100\times10^6\times4000}{2} - \frac{60\times10^6\times2000}{2}\right.$$

$$\left.- \frac{80\times10^6\times4000}{3}\right) = -4.61\times10^{-3} \text{ rad.}$$

$$\theta_B = -4.61\times10^{-3} + 4.15\times10^{-3}$$

$$= -4.6\times10^{-4} \text{ rad. (anticlockwise)}$$

Taking the moment of B.M. diagram Area-*A* to *D* about *D*,

$$t_{DA} = \frac{1}{EI}\left[\frac{100\times10^6\times4000}{2}\times\left(\frac{4000}{3}+2000\right)\right.$$

$$- \frac{60\times10^6\times2000}{2}\times\left(\frac{2000}{3}+2000\right)$$

$$- \frac{80\times10^6\times4000}{3}\times\left(\frac{4000}{4}+2000\right)$$

$$\left.- \frac{40\times10^6\times2000}{2}\times\frac{2}{3}\times2000\right]$$

$$= 18.44 \text{ mm}$$

$$\delta_D = \frac{6}{4}t_{BA} - t_{DA} = 6\times16.6/4 - 18.44$$

$$= 6.46 \text{ mm}$$

Alternate method This problem can be solved by the superposition of displacement of the beam under each load using results of other solved examples.

(1) If 30 kN is acting alone.

$$\theta_A = \frac{WL^2}{16EI} = \frac{30\times10^3\times4000^2}{16\times7.23\times10^{12}}$$

$$= 4.15\times10^{-3} \text{ rad (clockwise)}$$

$$\theta_B = \theta_A = 4.15\times10^{-3} \text{ rad (anticlockwise)}$$

$$\delta_C = \frac{WL^3}{48EI} = \frac{30\times10^3\times4000^3}{48\times7.23\times10^{12}}$$

$$= 5.53 \text{ mm (downward)}$$

$$\delta_D = 2000\times\theta_B = 8.3 \text{ mm}$$

(2) If udl 10 kN/m = 10 N/mm is acting alone.

$$\theta_A = \frac{WL^3}{24EI} = \frac{10\times10^3\times4000^3}{24\times7.23\times10^{12}}$$

$$= 3.688\times10^{-3} \text{ rad. (clockwise)}$$

$\theta_B = \theta_A = 3.688 \times 10^{-3}$ rad. (anticlockwise)

$$\delta_C = \frac{5wL^4}{384EI} = \frac{5 \times 10 \times 4000^4}{384 \times 7.23 \times 10^{12}}$$

$= 4.61$ (downward)

$\delta_D = 2000 \times \theta_B = 7.38$ mm (upward)

(3) If 20 kN is acting alone,

$$\theta_A = \frac{WL_1L_2}{6EI} = \frac{20 \times 10^3 \times 4000 \times 2000}{6 \times 7.23 \times 10^{12}}$$

$= 3.688 \times 10^{-3}$ rad. (anticlockwise)

$$\theta_B = \frac{WL_1L_2}{3EI} = \frac{20 \times 10^3 \times 4000 \times 2000}{3 \times 7.23 \times 10^{12}}$$

$= 7.377 \times 10^{-3}$ rad. (clockwise)

$$\delta_C = \frac{WL_2L_1^2}{16EI} = \frac{20 \times 10^3 \times 4000^2 \times 2000}{16 \times 7.23 \times 10^{12}}$$

$= 5.53$ mm (upward)

$$\delta_D = \frac{WL_2^2}{3EI} \times (L_1 + L_2)$$

$$= \frac{20 \times 10^3 \times 2000^2 \times 6000}{3 \times 7.23 \times 10^{12}}$$

$= 22.13$ mm (downward)

Final displacements

$\theta_A = (4.15 + 3.688 - 3.688) \times 10^{-3}$

$= 4.15 \times 10^{-3}$ rad. (clockwise)

$\theta_B = (4.15 + 3.688 - 7.377) \times 10^{-3}$

$= 4.61 \times 10^{-4}$ rad. (anticlockwise)

$\delta_C = 5.53 + 4.61 - 5.53$

$= 4.61$ mm (downward)

$\delta_D = 22.13 - 8.33 - 7.38$

$= 6.42$ mm (downward)

Example 12.8 Determine the forces acting at the ends of a beam which will cause clockwise rotation of the end A through angle θ_A as shown in Fig. 12.15. EI = constant for the beam.

Solution $\theta_B - \theta_A = 0 - \theta_A = \frac{-1}{EI}(\text{area} - AB)$

$$= \frac{-1}{EI}\left(M_A \times L - \frac{R_A \times L \times L}{2}\right) \quad \text{(i)}$$

Taking the moment of B.M. diagram area-AB about A,

$$t_{AB} = 0 = \frac{1}{EI}\left(M_A \times L \times \frac{L}{2} - \frac{R_A \times L \times L}{2} \times \frac{2L}{3}\right) \quad \text{(ii)}$$

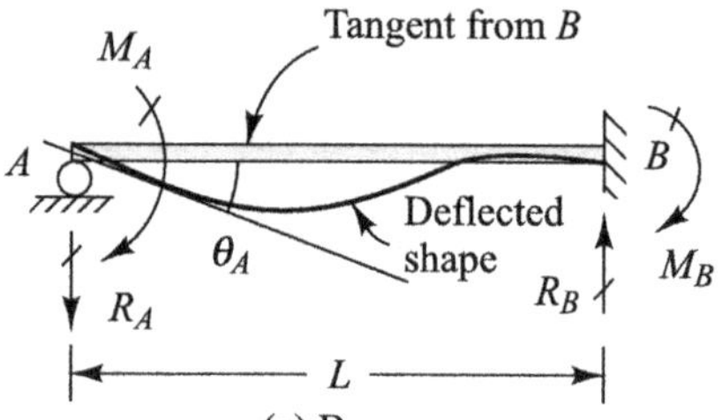

(a) Beam

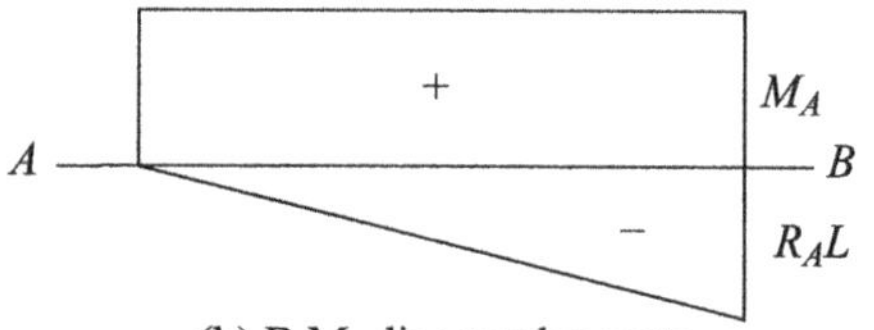

(b) B.M. diagram by parts

Fig. 12.15

or $M_A = 2R_A \times L/3$

Substituting in Eq.(i),

$R_A = \frac{6EI}{L^2}\theta_A$ (downward) and $M_A = \frac{4EI}{L}\theta_A$

Reactions at the end B using static equilibrium equations,

$\Sigma F = 0, \quad R_B = R_A = \frac{6EI}{L^2}\theta_A$ (upward)

$\Sigma M = 0, \quad M_B + M_A - R_AL = 0$

or, $M_B = \frac{2EI}{L}\theta_A$

Example 12.9 Determine the reaction of spring at the free end of a cantlever beam loaded with u.d.l. w per unit length as shown in Fig. 12.16(a). Stiffness of the spring is equal to k and EI = constant for the beam.

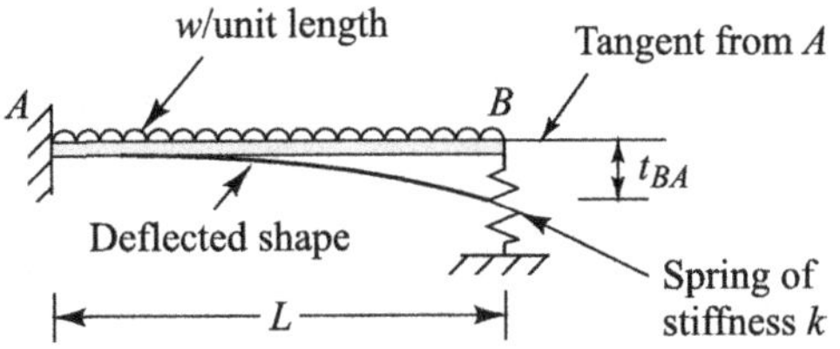

Fig. 12.16(a) *Beam and Deflected Shape*

Solution Let the force in the spring = $R = k \times t_{BA}$ where, t_{BA} = deflection at the spring end B.
Figure 12.16(b) gives the B.M. diagram for u.d.l. and point load in parts. Taking the moment of B.M. diagram about B,

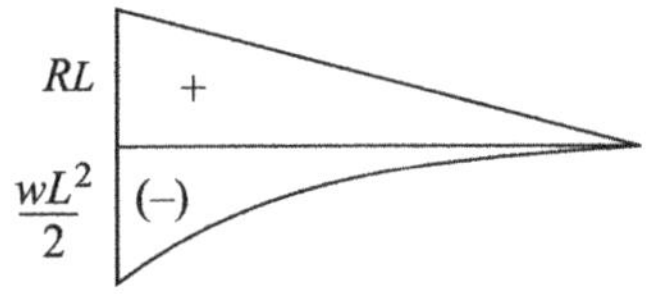

Fig. 12.16(b) *B.M Diagram in Parts*

$$-t_{BA} = \frac{1}{EI}\left[\frac{RL \times L}{2} \times \frac{2L}{3} - \frac{wL^2}{2} \times \frac{L}{3} \times \frac{3L}{4}\right]$$

$$= \frac{1}{EI}\left[\frac{RL^3}{3} - \frac{wL^4}{8}\right]$$

t_{BA} is negative since the end B deflects lower than the tangent from end A.

or, $$\frac{-R}{k} = \frac{1}{EI}\left[\frac{RL^3}{3} - \frac{wL^4}{8}\right]$$

or, $$R\left(\frac{1}{k} + \frac{L^3}{3EI}\right) = \frac{wL^4}{8EI}$$

or, $$R = \frac{3wL/8}{1 + \dfrac{3EI}{kL^3}}$$

Example 12.10 A beam AB is rigidly fixed at both the ends. Determine the end reactions if the end B sinks by a distance Δ.

Solution End reactions are assumed in the direction shown in Fig. 12.17(a). Taking moment of B.M. diagram about B,

$$-t_{BA} = -\Delta$$

$$= \frac{1}{EI}\left(\frac{R_A \times L \times L}{2} \times \frac{L}{3} - M_A \times L \times \frac{L}{2}\right)$$

t_{BA} is negative since B deflects lower than the tangent from A.

or, $$-\Delta EI = \frac{R_A L^3}{6} - \frac{M_A L^2}{2} \qquad \text{(i)}$$

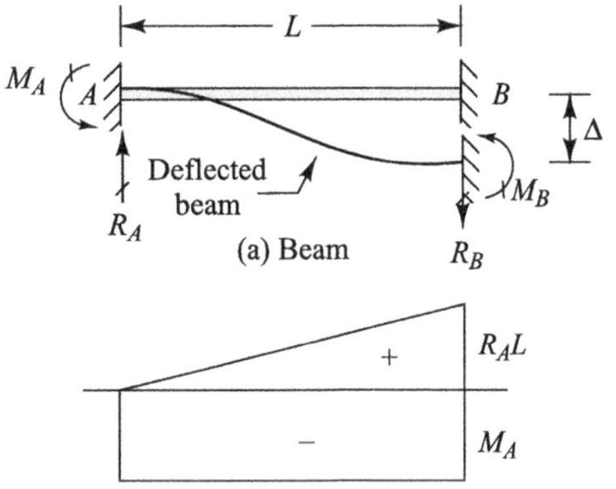

(b) B.M. diagram by parts

Fig. 12.17

Taking the moment of B.M. diagram about A,

$$t_{AB} = \Delta = \frac{1}{EI}\left(\frac{R_A \times L \times L}{2} \times \frac{2L}{3} - \frac{M_A \times L \times L}{2}\right)$$

or, $$\Delta EI = \frac{R_A L^3}{3} - \frac{M_A L^2}{2} \qquad \text{(ii)}$$

From Eqs (i) and (ii),

$$M_A = \frac{6EI}{L^2}\Delta \quad \text{and} \quad R_A = \frac{12EI}{L^3}\Delta$$

From the static equilibrium equations,

$$\Sigma F = 0, \qquad R_B = R_A = \frac{12EI}{L^3}\Delta$$

$$\Sigma M = 0, \qquad R_A L = M_A + M_B$$

or, $$M_B = \frac{6EI}{L^2}\Delta$$

Example 12.11 A fixed beam is loaded with a moment M as shown in Fig. 12.18. Determine the end reactions.

Solution The directions of the end reactions are assumed as shown in Fig. 12.18.

$$\theta_B - \theta_A = 0 = \frac{-1}{EI}(\text{area} - AB)$$

$$= \frac{-1}{EI}\left(\frac{R_A \times L \times L}{2} - M_A \times L - M \times b\right) \quad \text{(i)}$$

Similarly taking the moment of B.M. diagram about B,

$$t_{BA} = 0 = \frac{1}{EI}\left(\frac{R_A \times L \times L}{2} \times \frac{L}{3} - M_A \times L \times \frac{L}{2} - M \times b \times \frac{b}{2}\right) \quad \text{(ii)}$$

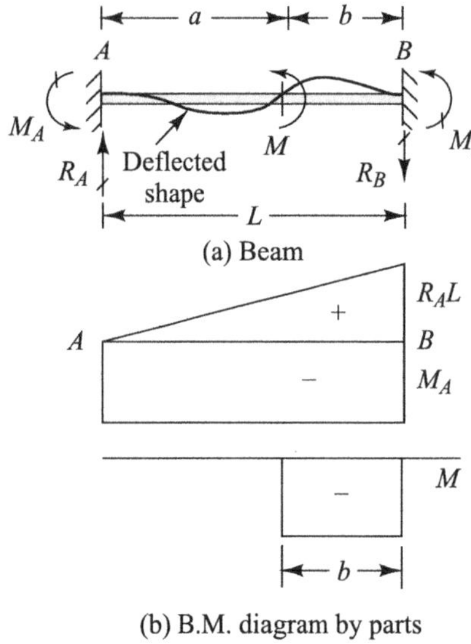

Fig. 12.18

From Eqs. (i) and (ii),

$$M_A = \frac{(ab - b^2)}{L^2}M \quad \text{and} \quad R_A = \frac{6ab}{L^3}M$$

From the static equilibrium equations,

$$\Sigma F = 0, \qquad R_B = R_A = \frac{6ab}{L^3}M$$

Taking the moment about B,

$$\Sigma M = 0, \qquad R_AL = M_A + M_B,$$

or,

$$M_B = \frac{(2ab - a^2)}{L^2}M$$

It should be seen that the direction of M_A or M_B is reversed if $2ab < a^2$ or b^2.

Example 12.12 Determine the reactions of the two span beam shown in Fig. 12.19. EI = constant for the beam.

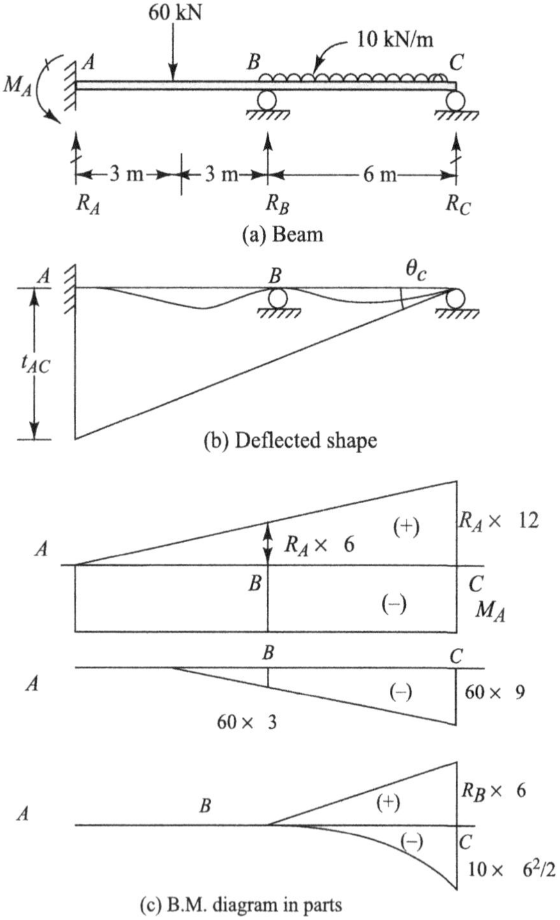

Fig. 12.19

Soluiton Three are four unknown reactions R_A, R_B, R_C and M_A.

Static equilibrium conditions give,

$\Sigma F = 0, \quad R_A + R_B + R_C = 60 + 10 \times 6 = 120 \quad$ (i)

$\Sigma M = 0, \quad$ Taking the moment about C,

$$R_A \times 12 - M_A - 60 \times 9 + R_B \times 6 - \frac{10 \times 6^2}{2} = 0$$

or, $$12R_A - M_A + 6R_B = 720 \qquad \text{(ii)}$$

Taking the moment of B.M. diagram area-AB about B,

$$t_{BA} = 0 = \frac{1}{EI}\left[\frac{R_A \times 6 \times 6}{2} \times \frac{6}{3} - M_A \times 6 \times \frac{6}{2} - \frac{60 \times 3 \times 3}{2} \times \frac{3}{3}\right]$$

or, $$2R_A - M_A = 15 \qquad \text{(iii)}$$

$$\theta_B \ \theta_A = \theta_B = \frac{-1}{EI}(\text{area} - AB)$$

$$\theta_B = \frac{-1}{EI}\left[\frac{R_A \times 6 \times 6}{2} - M_A \times 6 - \frac{60 \times 3 \times 3}{2}\right]$$

or, $$EI\theta_B = -18R_A + 6M_A + 270 \qquad \text{(iv)}$$

$$-\theta_C = \frac{t_{AC}}{12} = \frac{1}{12EI}\left[\frac{R_A \times 12 \times 12}{2}\left(\frac{2}{3} \times 12\right) - M_A \times 12 \times \frac{12}{2} - \frac{60 \times 9 \times 9}{2}\left(12 - \frac{9}{3}\right) + \frac{R_B \times 6 \times 6}{2}\left(12 - \frac{6}{3}\right) - \frac{10 \times 6^2}{2} \times \frac{6}{2}\left(12 - \frac{6}{4}\right)\right]$$

or, $$EI\theta_C = -48R_A + 6M_A - 15R_B + 2137.5 \qquad \text{(v)}$$

Note: The negative sign is used with θ_C as $+ t_{AC}/12$ gives an anticlockwise θ_C which is negative.

$$\theta_C - \theta_B = \frac{-1}{EI}(\text{area} - BC)$$

$$= \frac{-1}{EI}\left[\left(\frac{R_A \times 12 \times 12}{2} - \frac{R_A \times 6 \times 6)}{2}\right) - M_A \times 6 - \left(\frac{60 \times 9 \times 9}{2} - \frac{60 \times 3 \times 3}{2}\right) + \frac{R_B \times 6 \times 6}{2} - \frac{10 \times 6^2}{2} \times \frac{6}{3}\right]$$

or, $$EI\theta_B - EI\theta_C = 54R_A - 6M_A + 18R_B - 2520 \qquad \text{(vi)}$$

Substituting Eqs (iv) and (v) in Eq.(vi),

$$24R_A - 6M_A + 3R_B = 652.5 \qquad \text{(vii)}$$

Eqs (i), (ii), (iii) and (vii) yield,

$R_A = 30$ kN, $M_A = 45$ kN m,

$R_B = 67.5$ kN, and $R_C = 22.5$ kN

Example 12.13 Determine end reactions of the beam shown in Fig. 12.20 if the support B sinks by an amount Δ. Take EI = constant.

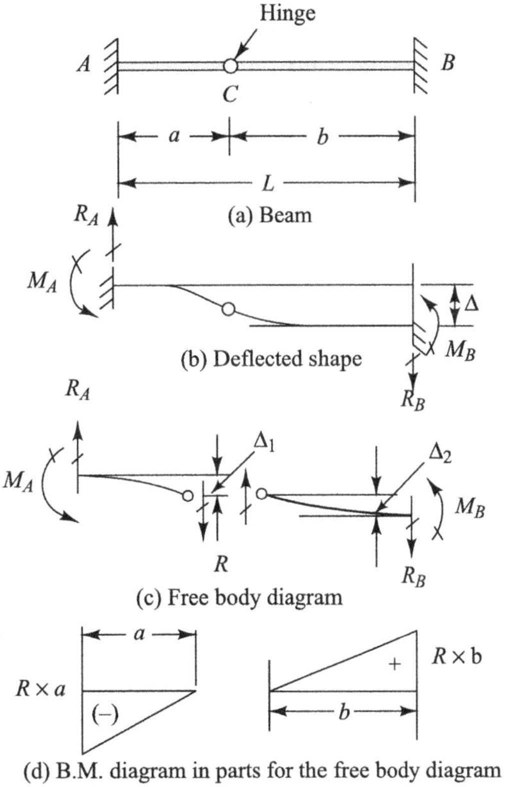

Fig. 12.20

Solution The directions of end reactions are

assumed as in Fig. 12.20(b). The deflected shape is analyzed by considering the free body diagram in Fig. 12.20(c). Figure 12.20(d) shows the bending moment diagram in parts for the free body diagram. Taking the moment of the bending moment diagram about the hinge.

$$t_{CA} = \frac{-1}{EI} \times \frac{R \times a \times a}{2} \times \frac{2a}{3} = \Delta_1$$

(C is lower than A)

or, $$-Ra^3/3EI = \Delta_1 \qquad \text{(i)}$$

$$t_{CB} = \frac{1}{EI} \times \frac{R \times b \times b}{2} \times \frac{2b}{3} = \Delta_2$$

(C is higher than B)

or, $$Rb^3/3EI = \Delta_2 \qquad \text{(ii)}$$

Total deflection,

$$\Delta = \Delta_1 + \Delta_2 = \frac{Rb^3}{3EI} + \frac{Ra^3}{3EI} \quad \text{or,} \quad R = \frac{3EI\Delta}{a^3 + b^3}$$

and from equilibrium of the free body diagrams,

$$M_a = -\frac{3EIa\Delta}{a^3 + b^3} \text{ and } M_B = \frac{3EIb\Delta}{a^3 + b^3}$$

Example 12.14 Determine the joint displacements of the frame shown in Fig. 12.21.

Solution Taking the moment about A,

$$R_c L = wL^2/2 \quad \text{or} \quad R_c = wL/2$$

Figure 12.21(c) shows the bending moment diagram.

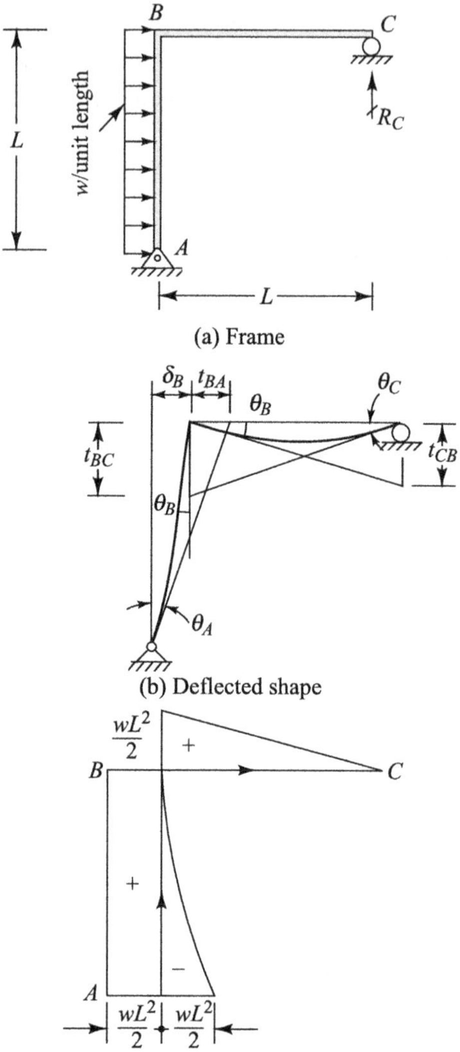

(a) Frame

(b) Deflected shape

(c) B.M. diagram by parts

Fig. 12.21

$$\theta_B - \theta_A = \frac{-1}{EI}(\text{area} - AB)$$

$$= \frac{-1}{EI}\left(\frac{wL^2}{2} \times L - \frac{wL^2}{2} \times \frac{L}{3}\right)$$

$$\theta_B - \theta_A = -wL^3/3EI \qquad \text{(i)}$$

From the deflected shape Fig. 12.21(b)

$$\theta_A \times L = \delta_B + t_{BA}$$

$$= \delta_B + \frac{1}{EI}\left(\frac{wL^2}{2} \times L \times \frac{L}{2} - \frac{wL^2}{2} \times \frac{L}{3} \times \frac{3L}{4}\right)$$

or, $$\theta_A \times L = \delta_B + wL^4/8EI \qquad \text{(ii)}$$

$$\theta_B \times L = t_{CB} = \frac{wL^2}{2EI} \times \frac{L}{2} \times \frac{2L}{3}$$

or, $$\theta_B = wL^3/6EI \qquad \text{(iii)}$$

Substituting θ_B in Eq. (i), $\theta_A = \dfrac{wL^3}{2EI}$

Substituting θ_A in Eq. (ii), $\delta_B = \dfrac{wL^4}{8EI}$

$$-\theta_C \times L = t_{BC} = \frac{wL^2}{2EI} \times \frac{L}{2} \times \frac{L}{3}$$

or $$\theta_C = -\frac{wL^3}{12EI}$$

The negative sign is used with θ_C as t_{BC}/L is anticlockwise.

Conjugate Beam Method

Example 12.15 Determine the mid span deflection and end slope of a simply supported beam carrying a concentrated point load at its mid span.

Solution Figure 12.22(a) shows the real beam and Fig. 12.22(b) shows its conjugate beam.

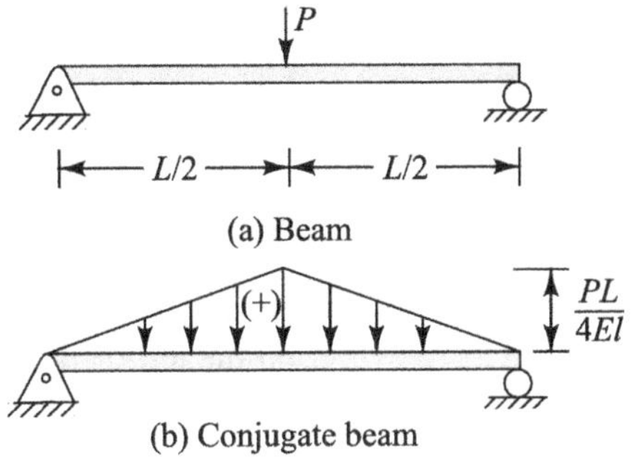

Fig. 12.22

End slope of the real beam

= end shear of the conjugate beam

$$= \frac{1}{2} \times \frac{PL}{4EI} \times \frac{L}{2} = \frac{PL^2}{16EI}$$

Mid span deflection of the real beam

= mid span moment of the conjugate beam

$$= \frac{PL^2}{16EI} \times \frac{L}{2} - \frac{1}{2} \times \frac{PL}{4} \times \frac{L}{2} \times \frac{1}{3} \times \left(\frac{L}{2}\right) = \frac{PL^3}{48EI}$$

Example 12.16 Determine the end slopes and mid span deflection for a simply supported beam carrying u.d.l w per unit length.

Solution Figure 12.23 shows the simply supported beam and its conjugate beam. The load on conjugate beam is drawn in parts, (1) Positive B.M. due to the support rection R_B, (2) Negative B.M. due to u.d.l. w.

For conjugate beam support reaction P_A in Fig. 12.23(b), take the moment of the forces about B,

$$P_A = \frac{1}{EI\,L} \times \left[\frac{wL^2}{2} \times \frac{L}{2} \times \left(\frac{2}{3}L\right) - \frac{wL^2}{2} \times \frac{L}{3} \times \left(\frac{3}{4}L\right)\right] = \frac{wL^3}{24EI}$$

Slope at A in real beam

θ_A = S.F. of conjugate beam at A

$= P_A = wL^3/24EI$ (clockwise)

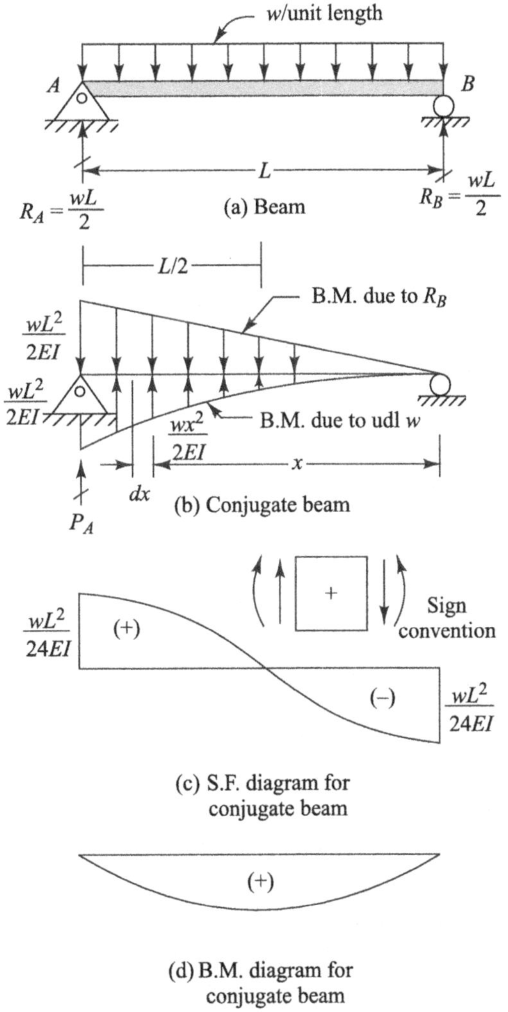

Fig. 12.23

Mid span slope

$$= P_A + \int_{L/2}^{L} \left(\frac{wx^2}{2EI}\right) dx - \frac{1}{2}\left(\frac{wL^2}{2EI} + \frac{wL^2}{4EI}\right) \times \frac{L}{2}$$

$$= \frac{wL^3}{24EI} + \frac{7wL^3}{48EI} - \frac{3wL^3}{16EI} = 0$$

Slope at B,

$$\theta_B = P_A + \int_0^L \left(\frac{wx^2}{2EI}\right) dx - \frac{wL^2}{2EI} \times \frac{L}{2}$$

$$= \frac{wL^3}{24EI} + \frac{wL^3}{6EI} - \frac{wL^3}{4EI} = -\frac{wL^3}{24EI}$$

Mid span deflection = Mid span B.M. of the conjugate beam

$$= \frac{P_A \times L}{2} + \int_{L/2}^{L} \frac{(wx^2/2) \times x \times dx}{EI} - \frac{wL^2}{4EI} \times \frac{L}{2} \times \frac{L}{4}$$

$$- \frac{wL^2}{2 \times 4EI} \times \frac{L}{2} \times \left(\frac{2}{3} \times \frac{L}{2}\right) = \frac{5wL^4}{384EI}$$

(Moment of trapezoidal portion has been taken by dividing it in triangular and rectangular portions of $wL^2/4$.)

Example 12.17 For the beam shown in Fig. 12.24, plot the deflected shape. Take $E = 210$ GPa and $I = 15000$ cm^4.

Solution $EI = 210000 \times 15000 \times 10^4$

$= 3.15 \times 10^{13}$ N.mm

$= 3.15 \times 10^4$ kNm

For, real beam $R_C = \frac{1}{3} \times \left(30 \times 4 \times \frac{4}{2}\right) = 80$ kN

Positive moment by R_C at A

$= 80 \times 6 = 480$ kNm

Negative moment by u.d.l at A

$= 30 \times 7 \times 7/2 = 735$ kN m

Negative moment by u.d.l at B

$= 30 \times 4 \times 4/2 = 240$ kN m

Negative moment by u.d.l at C

$= 30 \times 1 \times 1/2 = 15$ kNm

For conjugate beam reactions,

$$P_B + P_D = \frac{1}{EI}\left(\frac{480 \times 6}{2} - \frac{735 \times 5}{3}\right) = -\frac{275}{EI}$$

The negative sign shows that $P_B + P_D$ is acting downward.

Taking moment about the hinge at C in the conjugate beam,

$$P_B = \frac{1}{3EI}\left[\frac{480}{2} \times 6 \times \frac{2}{3} \times 6 - \int_1^7 \frac{30x^2}{2}(x-1)dx\right]$$

$$= -\frac{510}{EI}$$

The negative sign show P_B is is acting downward, i.e. opposite of the, i.e. opposite of the sign assumed in Fig. 12.24(b).

$$P_D = -\frac{275}{EI} + \frac{510}{EI} = \frac{235}{EI} \text{ upward}$$

Slope at A in real beam, θ_A = Shear in conjugate beam at $A = 0$

Slope on left of pont B = Shear in conjugate beam on the left of B

$$\theta_{BA} = \frac{1}{EI}\left[\left(\frac{735 \times 7}{3} - \frac{240 \times 4}{3}\right) - \frac{(480 + 240) \times 3}{2}\right]$$

$$= \frac{315}{EI} = 0.01 \text{ rad. (clockwise)}$$

Slope on right of point B,

$$\theta_{BC} = \frac{315}{EI} - \frac{510}{EI} = -\frac{195}{EI}$$

$= -0.0062$ rad. (anticlockwise)

Slope at C in real beam,

$$\theta_C = \frac{1}{EI}\left[\frac{735 \times 7}{3} - \frac{480 \times 6}{2} - 510 - \frac{15 \times 1}{3}\right]$$

$$= -\frac{240}{EI} = -0.0076 \text{ rad. (anticlockwise)}$$

Slope at D in real beam,

$$\theta_D = -\frac{240}{EI} + \frac{15 \times 1}{3} = -\frac{235}{EI}$$

$= -0.0074$ rad. (anticlockwise)

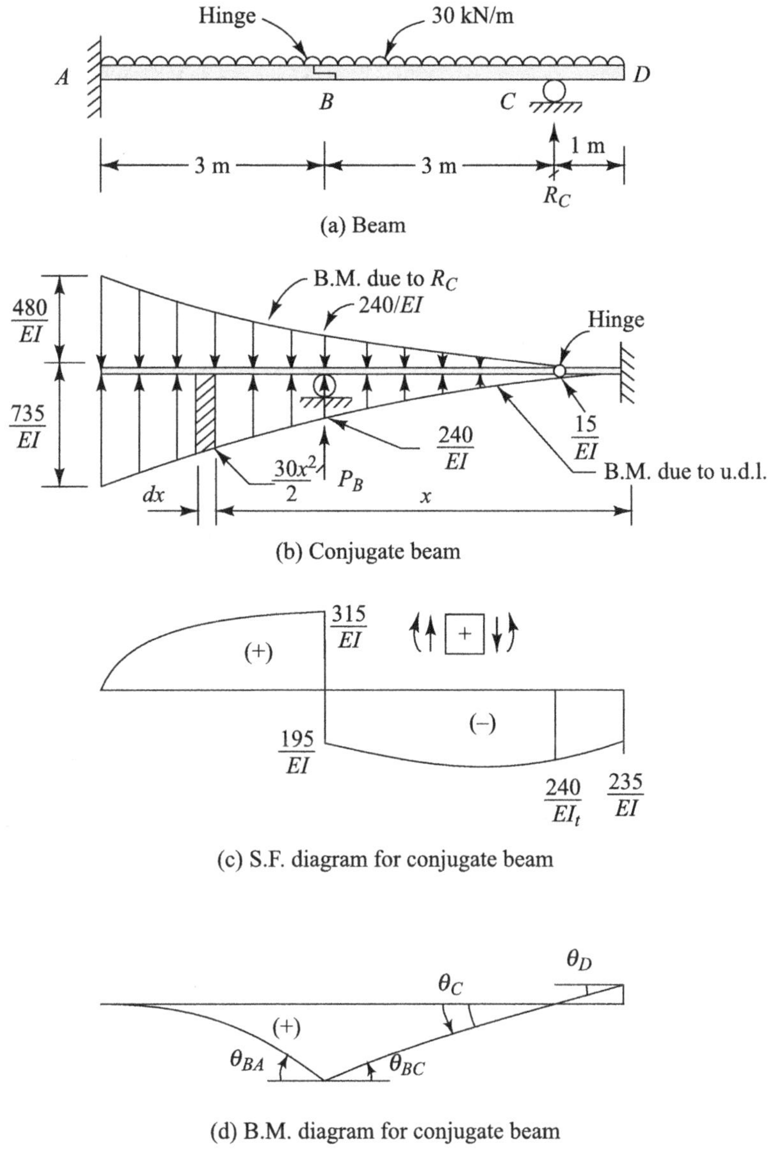

Fig. 12.24

Deflection in real beam at A = B.M. of conjugate beam at $A = 0$

Deflection at B = B.M. of conjugate Beam at B

$$= \frac{1}{EI}\left[\int_4^7 \frac{30x^2}{2}(x-4)dx - 240\times\frac{3}{2}\times\left(3\times\frac{2}{3}\right) - 240\times 3\times\frac{3}{2}\right]$$

$$= \frac{15}{EI}\left[\int_4^7 x^3dx - \int_4^7 4x^2dx\right] - \frac{1800}{EI}$$

$$= \frac{663.75}{EI} = 0.021 \quad \text{(down)}$$

Deflection at $C = 0$ (being hinge in the conjugate beam)

Deflection at D

$$= \frac{1}{EI}\left[\frac{15\times 1}{3}\times\frac{3}{4} + \text{shear transferred at } C\times 1\right]$$

$$= 15/4EI - 240/EI = -\,236.25/EI$$

$$= -\,0.0075 \text{ m(up)}.$$

Example 12.18 The fixed end of a propped cantilever sinks by Δ. Determine the end reaction due to the same.

Solution End reaction due to settlement will be R_A, R_B and M_A. For equilibrium,

$$R_A = -R_B, \qquad \text{(i)}$$

and $$M_A = -R_A \times L \qquad \text{(ii)}$$

Figure 12.25(b) shows the conjugate beam.

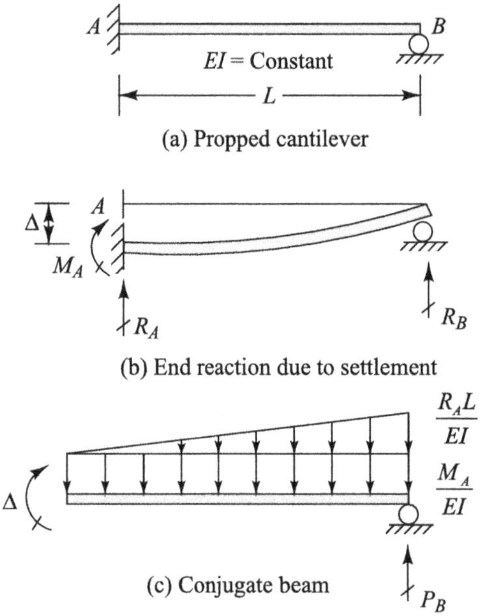

(a) Propped cantilever

(b) End reaction due to settlement

(c) Conjugate beam

Fig. 12.25

B.M. of the conjugate beam at B

= Displacement of the real beam at $B = 0$

or, $$\frac{1}{2}\times\frac{R_A L}{EI}\times L\times\frac{L}{3} + \frac{M_A L}{EI}\times\frac{L}{2} - \Delta = 0 \qquad \text{(iii)}$$

Eqs (i) through (iii) give,

$$M_A = \frac{3EI}{L^2}\Delta \quad \text{and} \quad R_A = -R_B = \frac{-3EI}{L^3}\Delta$$

Example 12.19 Determine the mid span deflection of a simply supported beam of span 4 m carrying udl of 10 kN/m if the moment of inertia of middle 2 m length is twice the value of the rest.

Soution End reactions of the beam by symmetry

$$= 10 \times 2 = 20 \text{ kN}$$

B.M. at mid span

$$= wL^3/8 = 10 \times 16/8 = 20 \text{ kNm}$$

$$\text{B.M. at } B = 20 \times 1 - 10 \times 1 \times 1/2 = 15 \text{ kNm}$$

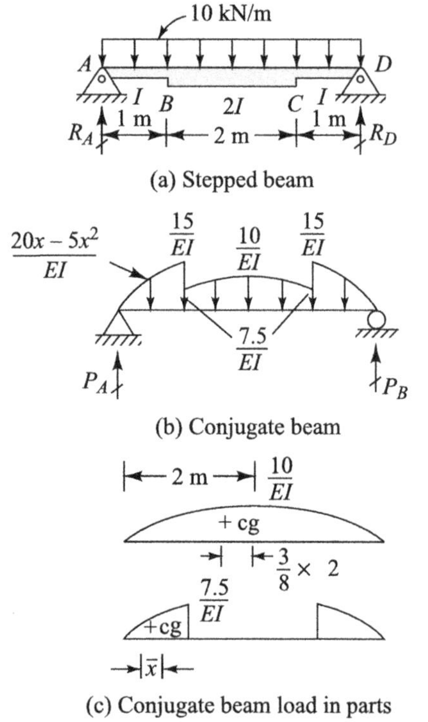

(a) Stepped beam

(b) Conjugate beam

(c) Conjugate beam load in parts

Fig. 12.26

Figure 12.26(b) shows the conjugate beam with M/EI diagram load. M/EI load is divided in two parts as in Fig. 12.26(c) for the ease of calculation of its area and moment. For the 1st part, Fig. 12.3 is used and integration is done for the 2nd part.

Conjugate beam end reaction,

$$P_A = \frac{2}{3\,EI} \times 10 \times 2 + \frac{1}{2\,EI}\int_0^1 (20x - 5x^2)\,dx$$

$$= 40/3\,EI + 25/6EI = 17.5/EI$$

Mid span deflection = Mid span B.M. of the conjugate beam

$$= P_A \times 2 - \frac{40}{3\,EI} \times \left(\frac{3}{2} \times 2\right) - \frac{25}{6\,EI} \times (2 - \bar{x})$$

$$= \frac{35}{EI} - \frac{10}{EI} - \frac{25}{6\,EI}\left(2 - \frac{\int_0^1 (20x - 5x^2)\,x\,dx}{\int_0^1 (20x - 5x^2)\,dx}\right)$$

$$= \frac{19.375}{EI}$$

Double Integration Method

Example 12.20 Determine the end slope and mid span deflection of a simply supported beam carrying udl w over span L.

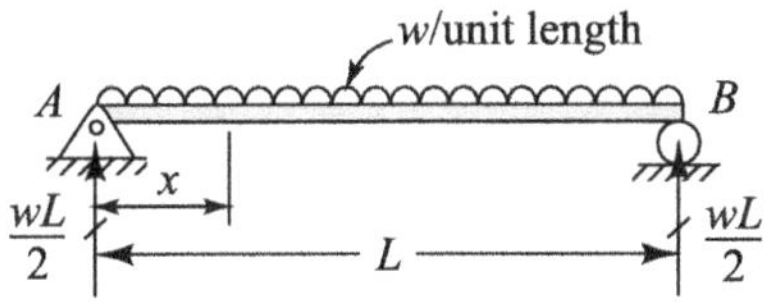

Fig. 12.27

Solution B.M. at x from A (Fig. 12.27)

$$M_x = \frac{wLx}{2} - \frac{wx^2}{2}$$

or $$-EI\frac{d^2y}{dx^2} = \frac{wLx}{2} - \frac{wx^2}{2}$$

On integration,

$$EI\,\frac{dy}{dx} = \frac{-wLx^2}{4} + \frac{wx^3}{6} + C_1 \qquad \text{(i)}$$

Again integrating,

$$Ely = \frac{-wLx^2}{12} + \frac{wx^4}{24} + C_1x + C_2 \qquad \text{(ii)}$$

where C_1 and C_2 are the constants of integration.

Boundary condtions:

At $x = 0,\ \ y = 0 \ \Rightarrow\ C_2 = 0,$

At $x = L,\ \ y = 0 \ \Rightarrow\ C_1 = \dfrac{wL^3}{24},$

Substituting $x = 0$ and $C_1 = wL^3/24$ in Eq. (i)

$$\frac{dy}{dx} = \frac{wL^3}{24EI} = \text{end slope at } A$$

Similarly substituting $x = L$ in Eq. (i)

$$\frac{dy}{dx} = \frac{-wL^3}{24EI} = \text{end slope at } B$$

For mid span deflection substiture $x = L/2$ in Eq. (ii)

$$y = \frac{5}{384} \times \frac{wL^4}{EI}$$

Example 12.21 Determine end slope and deflection for a cantilever of length L carrying a concentrated load of W at a distance of a from the fixed end.

Solution Figure 12.28 shows the cantilever beam, its deflected shape and B.M. diagram for the same.

Bending moment $0 < x < a$, $M(x) = -W(a - x)$

or, $$EI\,\frac{d^2y}{dx} = -M(x) = -W(a - x)$$

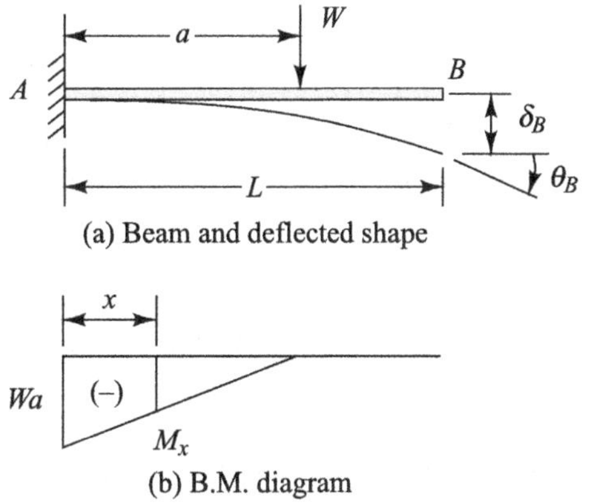

Fig. 12.28

Integrating twice,

$$EI\,\frac{dy}{dx} = W\left(ax - \frac{x^2}{2}\right) + C_1 \qquad \text{(i)}$$

$$EIy = W\left(\frac{ax^2}{2} - \frac{x^3}{6}\right) + C_1x + C_2 \qquad \text{(ii)}$$

Bending moment for $a < x < L$,

$$M(x) = -\,EI\,(d^2y/dx^2) = 0$$

Integrating twice, $EI\,(dy/dx) = C_3$ (iii)

and, $EI(y) = C_3x + C_4$ (iv)

where C_1, C_2, C_3 and C_4 are the constants of integration.

Boundary condition :

At $x = 0,\ dy/dx = 0 \quad \Rightarrow \quad C_1 = 0$

At $x = 0,\ y = 0 \quad \Rightarrow \quad C_2 = 0$

At $x = a$, from Eqs (i) and (iii),

$$EI\ dy/dx = Wa^2/2 = C_3$$

At $x = a$, from Eqs (ii) and (iv),

$$C_4 = -Wa^3/6$$

Therefore, slope for $a < x < L$,

$$dy/dx = Wa^2/2\,EI \qquad \text{(constant)}$$

and, displacement for $a < x < L$,

$$y = Wa^2x/2EI - Wa^3/6EI$$

At $x = L$, $\quad y = \dfrac{Wa^2}{6\,EI}\,(3L - a)$

Example 12.22 Determine the equation of the deflected shape for a long beam resting on an elastic foundation carrying a single point load at its middle.

Solution For any elementary length dx of the beam at a distance x from P, the foundation reaction is $k \times y \times dx$, where k is the stiffness of the foundation (force/unit length/unit area). Considering the equilibrium of the free body diagram of the element dx in Fig. 12.29.

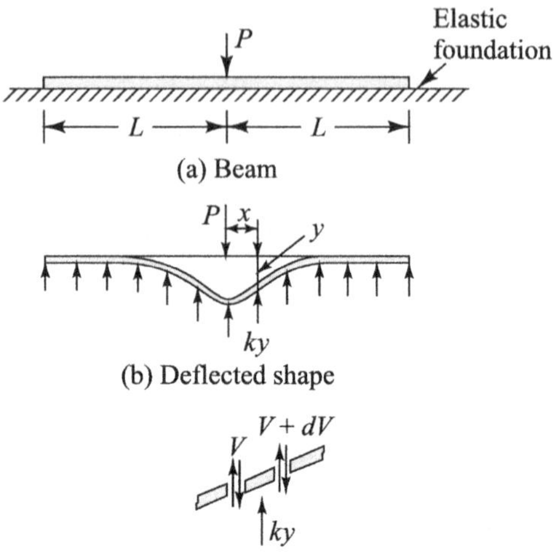

(a) Beam
(b) Deflected shape
(c) Free body diagram

Fig. 12.29

$$-(V + dV) + V - k \times y \times dx = 0$$

or,

$$-\,dV/dx - ky = 0$$

Substituting,

$$V = \frac{dM}{dx} = \frac{d}{dx}\left(-\,EI\,\frac{d^2y}{dx^2}\right) = -\,EI\,\frac{d^3y}{dx^3}$$

or, $$EI\,\frac{d^4y}{dx^4} - ky = 0 \qquad \text{(i)}$$

or, $$\frac{d^4y}{dx^4} - K^4y = 0 \qquad \text{(ii)}$$

where, $K^4 = k/EI$

Assuming the solution of Eq. (i) in the form of

$$y = e^{mx} \qquad \text{(iii)}$$

and substituting in Eq. (ii),

$$m^4 - K^4 = 0$$

or, $$(m^2 + K^2)\,(m^2 - K^2) = 0 \qquad \text{(iv)}$$

The four roots of Eq. (iv) for which Eq. (iii) satisfies Eq. (i) are,

$$m = \pm\,K$$

$$m = \pm\,jK \qquad (\,j = \sqrt{-1}\,)$$

The solution Eq. (iii) of Eq. (i) can then be written as the sum of solutions,

$$y = Ae^{Kx} + Be^{-Kx} + Ce^{jKx} + De^{-jKx} \quad \text{(v)}$$

Using the identities,

$$e^{Kx} = \cosh Kx + \sinh Kx$$
$$e^{-Kx} = \cosh Kx - \sinh Kx$$
$$e^{jKx} = \cos Kx + j \sin Kx$$
$$e^{-jKx} = \cos Kx - j \sin Kx$$

in Eq. (v), $y = C_1 \cosh Kx + C_2 \sinh Kx + C_3 \cos Kx + C_4 \sin Kx$

in which C_1, C_2, C_3 and C_4 are the real constants as follows.

$$C_1 = A + B, \qquad C_2 = A - B$$
$$C_3 = C + D, \qquad C_4 = j\,(C - D)$$

They require four boundary conditions for their solution.

Boundary conditions: At $x = 0$, slope $dy/dx = 0$

or, $C_1K \sinh Kx + C_2K \cosh Kx - C_3K \sin Kx + C_4K \cos Kx = 0$

or, $$C_2 + C_4 = 0 \quad \text{(vi)}$$

At $x \approx 0$, shear $V = -EId^3y/dx^3 = P/2$

or, $$C_2 - C_4 = -\frac{P}{2\,EIK^3} \quad \text{(vii)}$$

Eqs (vi) and (vii) give, $C_2 = C_4 = -\dfrac{P}{4\,EIK^3}$

At $x = L$, deflection $y = 0$, which gives,

$$C_1 \cosh KL + C_2 \sinh KL + C_3 \cos KL + C_4 \sin KL = 0 \quad \text{(viii)}$$

At $x = L$, bending moment $M = -EId^2y/dx^2 = 0$, which gives

$$C_1 \cosh KL + C_2 \sinh KL - C_3 \cos KL - C_4 \sin KL = 0 \quad \text{(ix)}$$

Eqs (viii) and (ix) give,

$$C_1 \cosh KL + C_2 \sinh KL = 0$$

or, $C_1 = -C_2 \tanh KL$

Substituting in Eq. (ix),

$$-C_2 \tanh KL \times \cosh KL + C_2 \sinh KL - C_3 \cos KL - C_4 \sin KL = 0$$

or, $$C_3 = -C_4 \tanh KL,$$

Substituting values of C_1 to C_4 in y,

$$y = -\frac{P}{4\,E\,I\,K^3}\,[-\tanh KL \times \cosh Kx + \sinh Kx - \tan KL \times \cos Kx + \sin Kx]$$

Example 12.23 Determine the variation of S.F., B.M., slope and deflection for a simply supported beam of span L using Fourier series, if it is to carry (i) udl w over its entire span and (ii) point load W at the mid span.

Solution Load itensity in Fourier series is given by,

$$w(x) = \frac{b_0}{2} + \sum_{n=1}^{\infty} b_n \cos\frac{n\pi x}{L} + \sum_{n=1}^{\infty} a_n \sin\frac{n\pi x}{L}$$

Simply supported beams have maximum shear and zero moment at the end supports and therefore, half range Fourier series of sine function is used for the load intensity function as follows.

$$w(x) = \sum_{n=1}^{\infty} a_n \sin\frac{n\pi x}{L}$$

where, $$a_n = \frac{2}{L}\int_0^L w(x)\sin\frac{n\pi x}{L}\,dx, \quad n = 0,1,2,3$$

For a general case of loading as shown in Fig. 12.30,

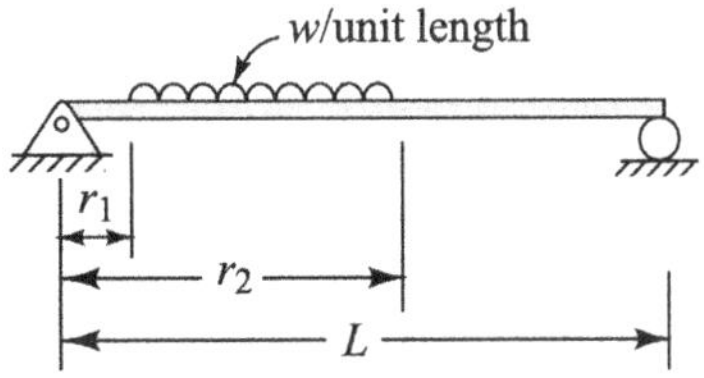

Fig. 12.30

Load function, $w(x) = 0$ when $0 \le x \le r_1$

$= w$ when $r_1 \le x \le r_2$

$= 0$ when $r_2 \le x \le L$

$$\therefore \quad a_n = \frac{2}{L}\int_{r_1}^{r_2} w \sin\frac{n\pi x}{L}\,dx$$

$$= \frac{2wL}{Ln\pi}\left(\cos\frac{n\pi r_1}{L} - \cos\frac{n\pi r_2}{L}\right) \qquad \text{(i)}$$

Case (i): When udl w is on the entire span, $r_1 = 0$ and $r_2 = L$

$$\text{and, } a_n = \frac{2w}{n\pi}(1 - \cos n\pi)$$

$$= 4w/n\pi \qquad \text{when } n = 1, 3, 5, \ldots$$

$$= 0 \qquad \text{when } n = 0, 2, 4, \ldots$$

Substituting a_n in Eq. (i), load intensity function is given by,

$$w(x) = \frac{4w}{\pi}\left(\sin\frac{\pi x}{L} + \frac{1}{3}\sin\frac{3\pi x}{L} + \frac{1}{5}\sin\frac{5\pi x}{L} + \cdots\right)$$

Shear force, $V(x) = -\int w(x)dx$

$$\text{or, } V(x) = \frac{4wL}{\pi^2}\left(\cos\frac{\pi x}{L} + \frac{1}{9}\cos\frac{3\pi x}{L} + \frac{1}{25}\cos\frac{5\pi x}{L} + \cdots\right) + C_1$$

Boundary condition:

At $x = L/2$, $V(x) = 0 \Rightarrow C_1 = 0$

Bending moment $M(x) = \int V(x)\,dx$

$$\text{or, } M(x) = \frac{4wL^2}{\pi^3}\left(\sin\frac{\pi x}{L} + \frac{1}{27}\sin\frac{3\pi x}{L} + \frac{1}{125}\sin\frac{5\pi x}{L} + \cdots\right) + C_2$$

Boundary condition:

At $x = 0$, $M(x) = 0 \Rightarrow C_2 = 0$

Slope $dy/dx = \theta(x) = -\int M(x)\frac{dx}{EI}$

$$\text{or, } EI\theta(x) = \frac{4wL^3}{\pi^4}\left(\cos\frac{\pi x}{L} + \frac{1}{81}\cos\frac{3\pi x}{L} + \frac{1}{625}\cos\frac{5\pi x}{L} + \cdots\right) + C_3$$

Deflection $y(x) = \int \theta(x)dx$

$$\text{or, } EIy(x) = \frac{4wL^4}{\pi^5}\left(\sin\frac{\pi x}{L} + \frac{1}{243}\sin\frac{3\pi x}{L} + \cdots\right) + C_3x + C_4$$

Boundary condition: At $x = 0$ and L, $y(x) = 0$

$$\Rightarrow \qquad C_3 = C_4 = 0$$

Maximum deflection occurs at $x = L/2$,

$$y\left(\frac{L}{2}\right) = \frac{4wL^4}{EI\pi^5}\left(\sin\frac{\pi L}{2L} + \frac{1}{243}\sin\frac{3\pi L}{2L} + \frac{1}{3125}\sin\frac{3\pi L}{2L} + \cdots\right)$$

$$= \frac{4wL^4}{EI\pi^5}\left(1 - \frac{1}{243} + \frac{1}{3125} + \cdots\right)$$

$$= \frac{5wL^4}{384EI} \quad \text{(for the first three terms)}$$

End slope at $x = 0$,

$$\theta(x) = \frac{4wL^3}{EI\pi^4}\left(\cos\frac{\pi x}{L} + \frac{1}{81}\cos\frac{3\pi x}{L} + \frac{1}{625}\cos\frac{5\pi x}{L} + \cdots\right)$$

$$= \frac{4wL^3}{EI\pi^4}\left(1 + \frac{1}{81} + \frac{1}{625} + \cdots\right)$$

$$= \frac{1}{24.017} \times \frac{wL^3}{EI} \quad \text{(for the first three terms)}$$

The approximate results by Fourier series are quite accurate and are accepted for practical purposes.

Case (ii): When concentrated load W is at mid span, substitute $r_1 = r_2 = L/2$ in Eq. (i). Rest of the analysis is similar as in Case (i).

Singularity Function Method

Example 12.24 Determine the end slope and deflection of a cantilever beam carrying a point load W at its free end.

Solution Figure 12.31 shows the cantilever beam, S.F., B.M., slope and deflection diagrams for the cantilever beam.

Load function,

$$w(x) = + W< x - 0 >^{-1}$$

Shear force,

$$V(x) = - \int w(x)dx$$
$$= -W< x - 0 >^0 + C_1$$

Boundary condition:

At $x > 0$, $V(x) = -W \Rightarrow C_1 = 0$

Bending moment:

$$M(x) = \int V(x) \quad dx = -W < x - 0 >^1 + C_2$$

Boundary condition:

At $x = 0$, $M(x) = 0 \Rightarrow C_2 = 0$

Slope

$$dy/dx = \theta(x) = - \int M(x)dx/EI$$
$$= W < x - 0 >^2/2EI + C_3$$

Boundary condition:

At $x = L$, $\theta(x) = 0 \Rightarrow C_3 = -WL^2/2EI$

Slope

$$dy/dx = \theta(x) = (-WL^2+W< x - 0 >^2)/2EI$$

Deflection

$$y(x) = \int \theta(x)\ dx$$
$$= (-WL^2x + W < x - 0>^3/3)/2EI + C_4$$

Boundary condition:

At $x = L$, $y(x) = 0 \Rightarrow C_4 = WL^3/3EI$

where C_1 to C_4 are the constants of integration. So finally w, V, M, θ and y are as follows and as shown in Fig. 12.31 with proper signs.

$$w(x) = +W < x - 0 >^{-1},\ V(x) = -W < x - 0 >^0,$$
$$M(x) = -W < x - 0 >^1$$
$$\theta(x) = (-WL^2 + W < x - 0 >^2)/2EI$$
$$= -WL^2/2EI \quad \text{at } x = 0$$
$$y(x) = (-WL^2x + W < x - 0 >^3/3)2EI +$$
$$WL^3/3EI = WL^3/3EI \qquad \text{at } x = 0$$

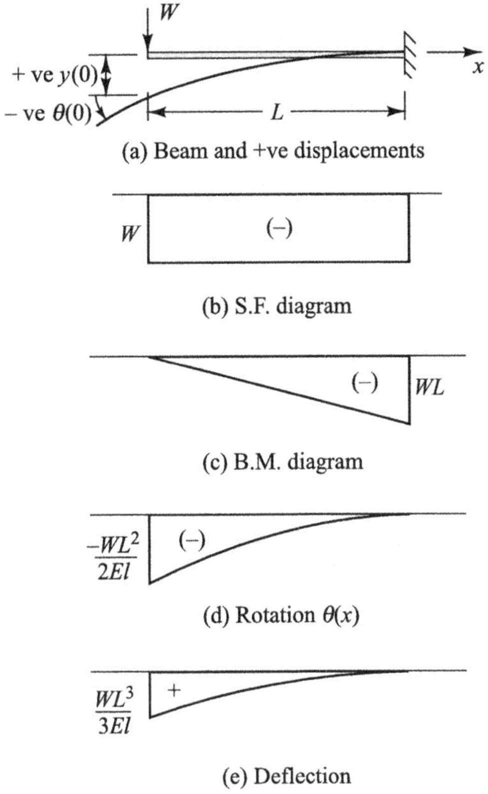

Fig. 12.31

Example 12.25 Determine the shear force, bending moment and displacements of a simply supported beam of span 4 m, carrying a point load of 15 kN at mid span (Fig. 12.32).

Solution Load intensity function,

$$w(x) = 15 < x - 2 >^{-1} \text{ kN}$$

Integrating load intensity function twice,

$$V(x) = - \int w(x)dx = -15 < x - 2 >^0 + C_1$$

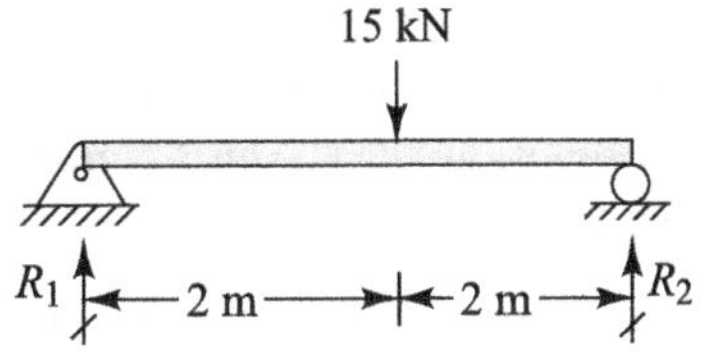

Fig. 12.32

$$M(x) = \int V(x)dx = -15 < x - 2 >^1 + C_1x + C_2$$

Boundary conditions:

At $x = 0$, $M(x) = 0 \Rightarrow C_2 = 0$

At $x = 4$ m, $M(x) = 0 \Rightarrow C_1 = 7.5$ kN

S.F. variation is given by,

$$V(x) = 7.5 - 15 < x - 2 >^0$$

and B.M. variation is given by,

$$M(x) = 7.5 < x - 0 >^1 - 15 < x - 2 >^1$$

Integrating moment function twice,

$$EI\theta(x) = -\int M(x)dx$$

$$= \frac{-7.5 < x - 0 >^2}{2} + \frac{15 < x - 2 >^2}{2} + C_3$$

$$EIy(x) = EI \int \theta(x)dx$$

$$= \frac{-7.5 < x - 0 >^3}{6} + \frac{15 < x - 2 >^3}{6} + C_3x + C_4$$

Boundary conditions:

At $x = 0$, $y(x) = 0 \Rightarrow C_4 = 0$

At $x = 4$ m, $y(x) = 0 \Rightarrow C_3 = 15$ kN m^2

Therefore variation of θ and y is given by,

$$EI\theta(x) = \frac{-7.5 < x - 0 >^2}{2} + \frac{15 < x - 2 >^2}{2} + 15$$

$$= 15 \text{ at } x = 0$$

$$EIy(x) = \frac{-7.5 < x - 0 >^3}{6} + \frac{15 < x - 2 >^3}{6} + 15x$$

$$= 20 \text{ at mid span } x = 2 \text{ m}$$

Note: Support reactions may or may not be included in the load function $w(x)$. They can be evaluated in both ways. In Example 12.25, $R_1 = V(x) = 7.5$ at $x = 0$. If R_1 and R_2 are both included in the load function $w(x)$, then C_1 and C_2 are not included in $V(x)$ and $M(x)$.

Example 12.26 Determine the shear force, bending moment and displacement of the beam shown in Fig. 12.33.

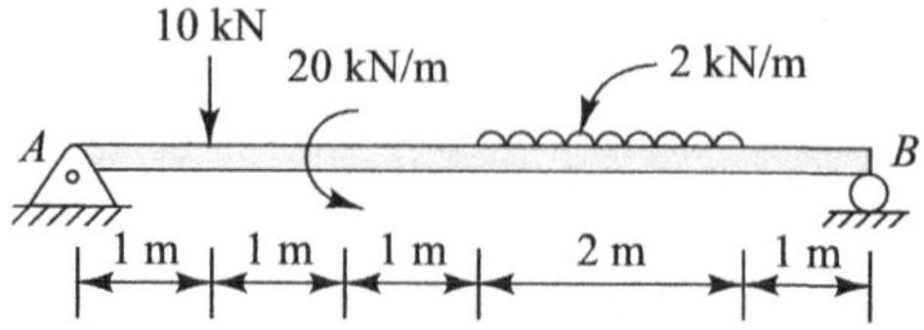

Fig. 12.33

Solution Load intensity function: udl 2 kN/m is first considered up to roller end B and upward -2 kN/m is applied for the last 1 m span.

$$w(x) = 10 < x - 1 >^{-1} + 20 < x - 2 >^{-2} + 2 < x - 3 >^0 - 2 < x - 5 >^0 \text{ kN}$$

Integrating load intensity function twice,

$$V(x) = -10 < x - 1 >^0 - 20 < x - 2 >^{-1} - 2 < x - 3 >^1 + 2 < x - 5 >^1 + C_1$$

$$M(x) = -10 < x - 1 >^1 - 20 < x - 2 >^0 - \frac{2 < x - 3 >^2}{2} + \frac{2 < x - 5 >^2}{2} + C_1x + C_2$$

Boundary conditions:

At $x = 0$, $M(x) = 0 \Rightarrow C_2 = 0$

At $x = 6$ m, $M(x) = 0 \Rightarrow C_1 = 13$ kN

$$V(x) = 13 - 10 < x - 1 >^0 - 20 < x - 2 >^{-1} - 2 < x - 3 >^1 + 2 < x - 5 >^1$$

$$M(x) = 13x - 10 < x - 1 >^1 - 20 < x - 2 >^0 - \frac{2 < x - 3 >^2}{2} + \frac{2 < x - 5 >^2}{2}$$

Integrating moment function twice,

$$EI\theta(x) = \frac{-13x^2}{2} + \frac{10 < x - 1 >^2}{2} + 20 < x - 2 >^1 + \frac{< x - 3 >^3}{3} - \frac{< x - 5 >^3}{3} + C_3$$

$$EIy(x) = \frac{-13x^3}{6} + \frac{10 < x - 1 >^3}{6} + \frac{20 < x - 2 >^2}{2} + \frac{< x - 3 >^4}{12} - \frac{< x - 5 >^4}{12} + C_3x + C_4$$

Boundary conditions:

At $x = 0$, $y(x) = 0 \Rightarrow C_4 = 0$

At $x = 6$ m, $y(x) = 0 \Rightarrow C_3 = -15.5$ kNm^2

Therefore variation of θ and y is given by,

$$EI\theta(x) = \frac{-13x^2}{2} + \frac{10 < x-1 >^2}{2} + 20 < x-2 >^1 + \frac{< x-3 >^3}{3} - \frac{< x-5 >^3}{3} - 15.5$$

$$EIy(x) = \frac{-13x^3}{6} + \frac{10 < x-1 >^3}{6} + \frac{20 < x-2 >^2}{2} + \frac{< x-3 >^4}{12} - \frac{< x-5 >^4}{12} - 15.5x$$

Example 12.27 Determine the mid span deflection of the beam shown in Fig. 12.34(a) if $EI = 10{,}000$ kN m^2 for the beam.

Solution Load intensity function: The applied load is replaced by an equivalent uniformly increasing upward and downward load as shown in Fig. 12.34(b) for which,

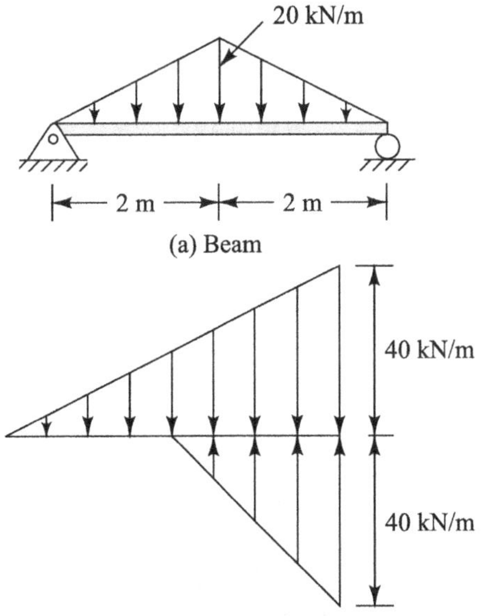

(a) Beam

(b) Equivalent load

Fig. 12.34

$$w(x) = \frac{dw}{dx} < x-a >^1$$

$$= \frac{40}{4} < x-0 >^1 - \frac{40}{2} < x-2 >^1$$

Integrating load intensity function twice,

$$V(x) = -\int w(x)\,dx$$

$$= -\frac{10 < x-0 >^2}{2} + \frac{20 < x-2 >^2}{2} + C_1$$

$$M(x) = \int V(x)\,dx$$

$$= -\frac{5 < x-0 >^3}{3} + \frac{10 < x-2 >^3}{3} + C_1x + C_2$$

Boundary conditions:

At $x = 0$, $M(x) = 0 \Rightarrow C_2 = 0$

At $x = 4$ m, $M(x) = 0 \Rightarrow C_2 = 20$ kN,

$$M(x) = 20x - \frac{5 < x-0 >^3}{3} + \frac{10 < x-2 >^3}{3}$$

Integrating moment function twice,

$$EI\theta(x) = -\int M(x)\,dx = -\frac{20x^2}{2} + \frac{5}{3} \times \frac{< x-0 >^4}{4} - \frac{10}{3} \times \frac{< x-2 >^4}{4} + C_3$$

$$EIy(x) = EI\int \theta(x)\,dx$$

$$= -\frac{20x^3}{6} + \frac{5}{12} \times \frac{< x-0 >^5}{5} - \frac{10}{12} \times \frac{< x-2 >^5}{5} + C_3x + C_4$$

Boundary conditions:

At $x = 0$, $y(x) = 0 \Rightarrow C_4 = 0$

At $x = 4$ m, $y(x) = 0 \Rightarrow C_3 = 33.33$ kN m

Therefore,

$$EIy(x) = 33.33x - \frac{20x^3}{6} + \frac{< x-0 >^5}{12} - \frac{< x-2 >^5}{6}$$

At $x = 2$ m, $y(2) = 42.66/EI = 4.266$ mm

Example 12.28 Determine the roller reaction for, the beam shown in Fig. 12.35.

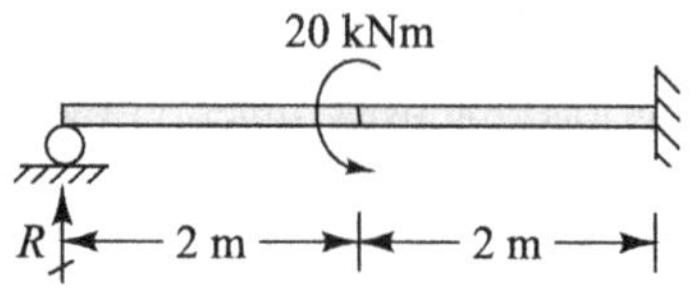

Fig. 12.35

Solution Load function,

$$w(x) = 20 < x - 2 >^{-2}$$

Shear force,

$$V(x) = -\int w(x)dx = -20 < x - 2 >^{-1} + C_1$$

Bending moment

$$M(x) = \int V(x)dx = -20 < x - 2 >^{0} + C_1 x + C_2$$

Boundary condition:

At $x = 0$, $M(0) = 0 \Rightarrow C_2 = 0$

Slope

$$EI\theta(x) = -\int M(x)dx$$

$$= 20 < x - 2 >^{1} - \frac{x^2}{2} C_1 + C_3$$

Boundary condition:

At $x = 4$ m, $\theta(4) = 0$

which gives, $8C_1 - C_3 = 40$ (i)

Deflection,

$$EIy(x) = EI \int \theta(x)dx$$

$$= \frac{20 < x - 2 >^2}{2} - \frac{x^3}{6} C_1 + C_3 x + C_4$$

C_1 to C_4 are the constants of integration.

Boundary condition:

At $x = 0$, $y(0) = 0 \Rightarrow C_4 = 0$

At $x = 4$ m, $y(4) = 0$

which gives $64/6C_1 - 4C_3 = 40$ (ii)

From Eqs (i) and (ii), $C_1 = 45/8$ and $C_3 = 5$

Shear force at $x = 0$,

$$V(0) = -20 < x - 2 >^{-1} + C_1$$

$$= 45/8 \text{ kN} = \text{reaction at roller}$$

Graphical Method

Example 12.29 Determine the displacement of joints for the truss shown in Fig. 12.36(a). Take $EA = 10000$ kN m^2.

Solution Truss members are marked 1 through 5 in Fig. 12.36(a). Member forces are also marked as follows.

Member	*Length (m)*	*Force (kN)*	*Change in length (m)*
1, 4	5	12.5(c)	5 × 12.5/10000 = 0.00625
2, 5	4	10.0(t)	4 × 10/10000 = 0.004
3	3	15.0(t)	3 × 15/10000 = 0.0045

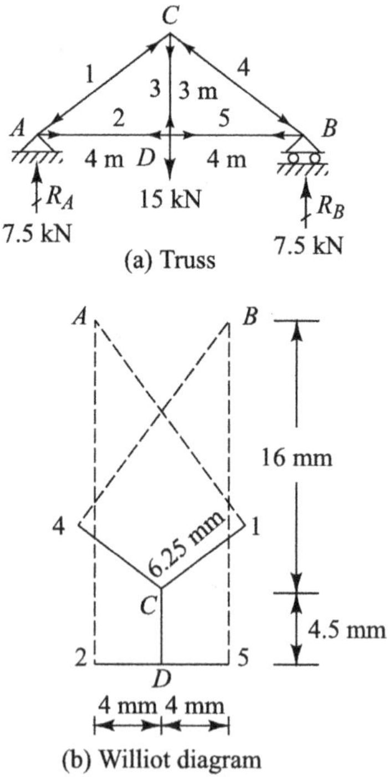

Fig. 12.36

Assume member CD is the reference line as it will not rotate due to symmetric loading. Also assume joint D as stationary.

Starting from the point D on Williot diagram, extension of member 2 is drawn as $D2$ on the left of D which is opposite of the direction of arrow at joint A. Again from the joint D, extension of member 3 is drawn as CD on top of D as the arrow at C is downward. Similarly $C1$ is drawn on the Williot diagram. Draw perpendiculars from 1 and 2 intersecting at A. Joints A, B, C and D are thus located on the Williot diagram. Their position relative to A gives their displacement in the x–y direction, since it is actually the joint A which is stationary.

Example 12.30 Figure 12.37 shows the change in length in mm for the members of a truss due to certain loading. Determine the displacement of the joints for the truss.

Solution Williot diagram is drawn in the usual way for the truss by assuming member AC to be fixed. Starting from joint A, draw AC on the Williot diagram = 3.18 mm to some convenient scale. Since the deformation of member AC is positive, point C on the Williot diagram will be on the same side of A as in the real structure (see Sec. 12.6.3). From point A, draw a line = – 4.76 mm parallel to BA in the truss (opposite side). From point C, draw a line = + 1.59 mm parallel to BC in the truss (same side). Intersection of the perpendicular to these two lines will give point B on the Williot diagram. In this way complete the Williot diagram at the point F. Note that the points F and A, should lie on the same horizontal line in the displacement diagram but it is not so due to the rotation of the reference member AC in the actual structure. Mohr's correction is now applied for this error due to a wrong assumption. Draw a vertical line AF' from A and a horizontal line FF' from F on Williot diagram to intersect AF' at F'. The displacement of joint F is now measured with respect to F'. The reference for other joints is also shifted similarly by drawing the truss using scale: AF' in Williot–Mohr diagram = AF in space diagram. The displacements of all joints are now measured with reference to the truss joints in Williot–Mohr diagram as marked for D in Fig. 12.37(b).

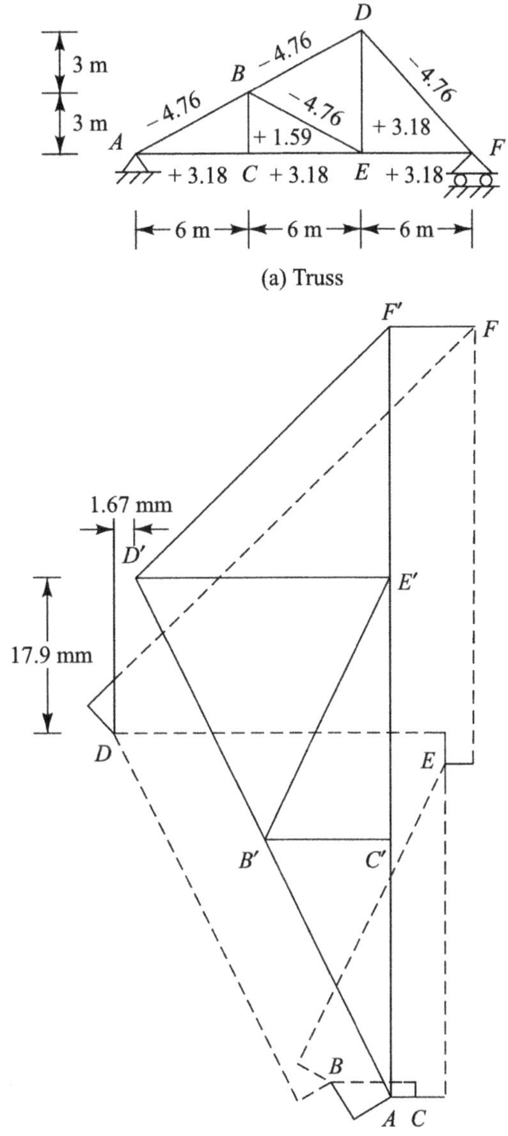

(a) Truss

(b) Williot–Mohr diagram

Fig. 12.37

Joint	Displacements (mm)	
	Horizontal	Vertical
B	8.5	27.6
C	3.18	29.1
D	– 1.67	17.9
E	6.4	21.1

EXERCISES

PROBLEMS

12.1 Draw the deflected shapes of the structures shown in Fig. 12.38 indicating location of points of contraflexure and bending tension. Also show the alternate deflected shapes if the relative magnitude of the applied loads is changed.

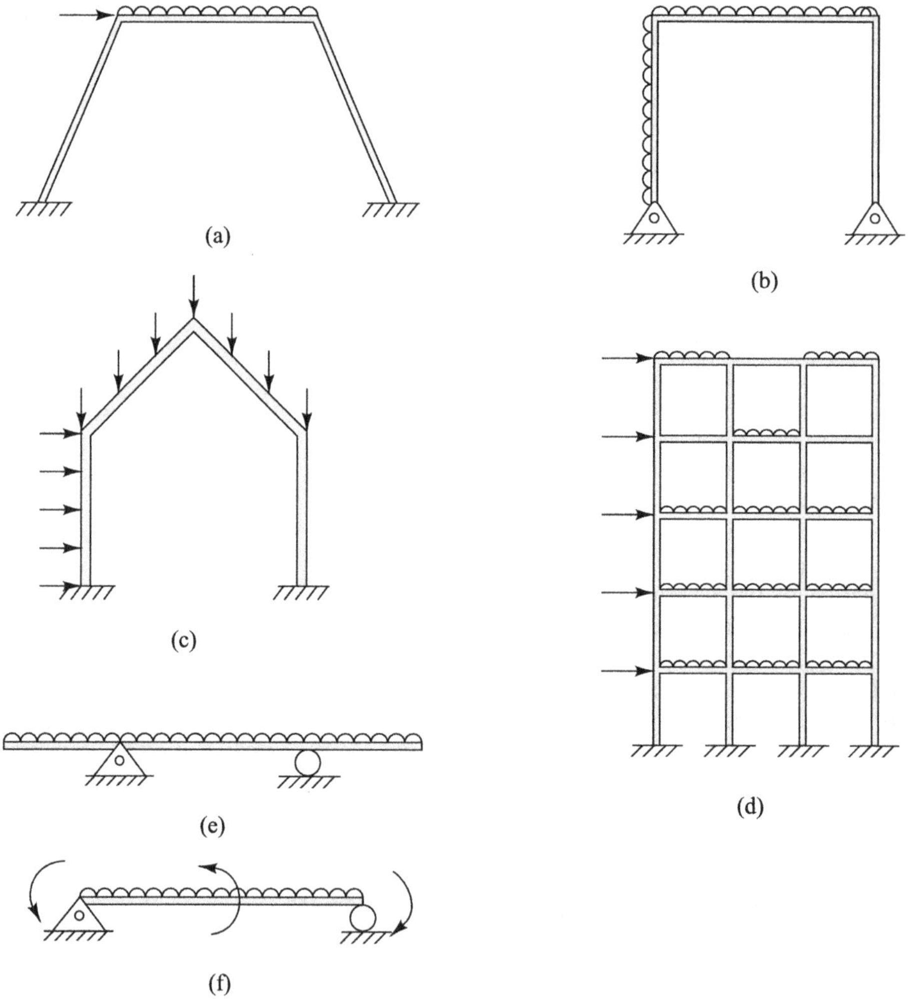

Fig. 12.38

12.2 Using moment area theorem determinate the ratio P/Q for the beam shown in Fig. 12.39 which is loaded with P and Q if (i) slope at C is zero, and (ii) deflection at C is zero.

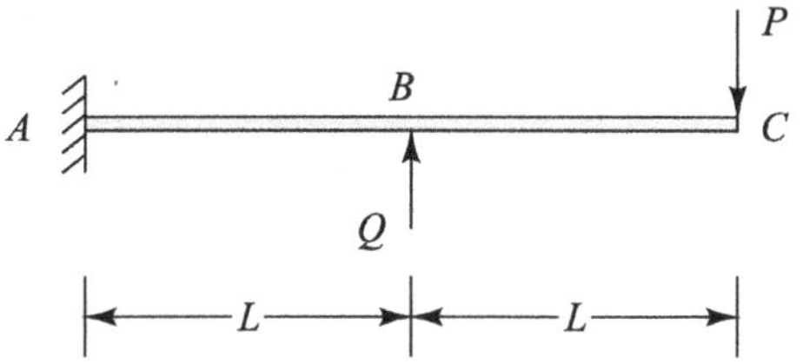

Fig. 12.39

12.3 and 12.4 Solve Examples 12.18 and 12.19 by the moment-area method.

12.5 Determine the slope and deflection of the joint C in the frame shown in Fig. 12.40 using moment-area method.

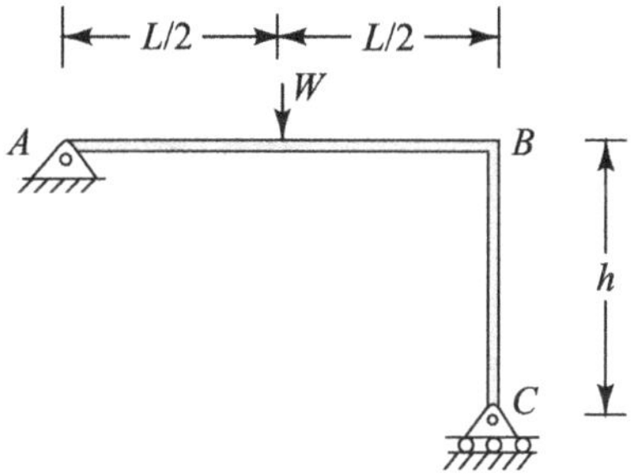

Fig. 12.40

12.6 Determine the deflection of the ends of the beam shown in Fig. 12.41.

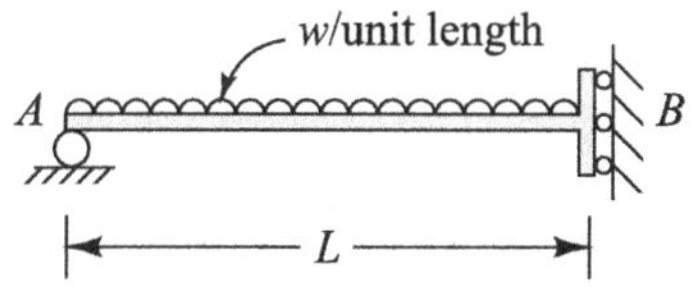

Fig. 12.41

12.7 Determine the end reactions and deflection at the hinge of the beam shown in Fig. 12.42.

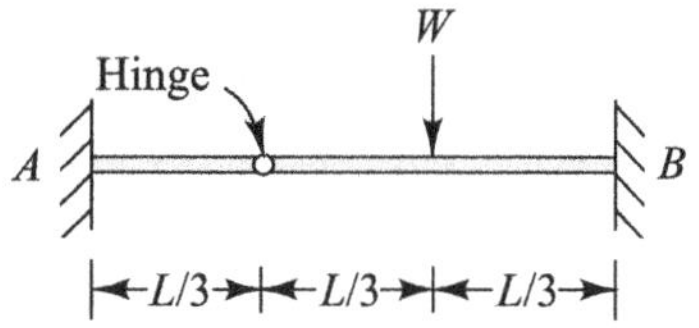

Fig. 12.42

12.8 Determine the mid span deflection for the beam shown in Fig. 12.43 if its flexural rigidity $EI = 10{,}000$ kN m^2.

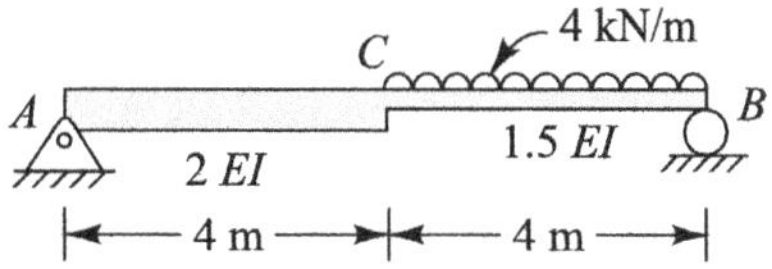

Fig. 12.43

12.9 Two cantilever beams are connected by a tie bar of diameter d and the lower beam carries a point load W at its middle as shown in Fig. 12.44. Show that the tension in the tie bar is given by,

$$P = \frac{5\,WL^3}{32\left(L^3 + 6aI/\pi d^2\right)}$$

if EI = constant for both the beams

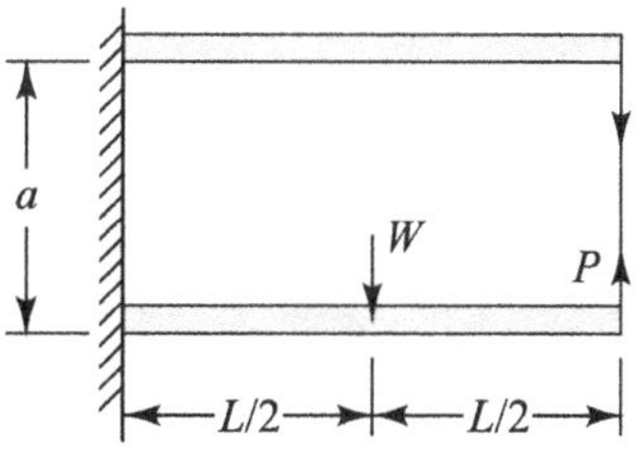

Fig. 12.44

12.10 A cantilever beam of length L and depth d, has a width b at its fixed end which decreases uniformly to zero at its free end. Calculate the deflection of the free end under a vertical point load P.

12.11 Figure 12.45 shows a cantilever beam supported by a 10 mm dia bar at the free end. Determine the deflection of the free end if $I = 300\ \text{cm}^4$ and $E = 200$ GPa.

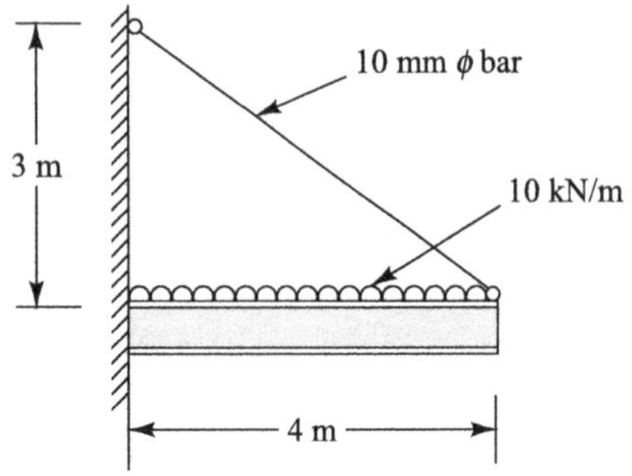

Fig. 12.45

12.12 and 12.13 A small gap of 3 mm exists between the beam and the roller support before application of the load as shown in Figs. 12.46 and 12.47. Determine the roller support reaction. Take $EI = 5 \times 10^{11}$ N mm^2.

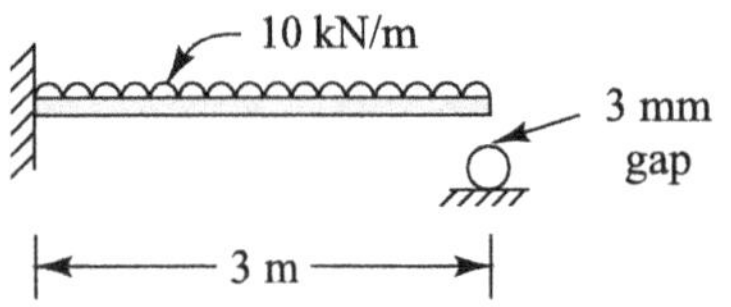

Fig. 12.46

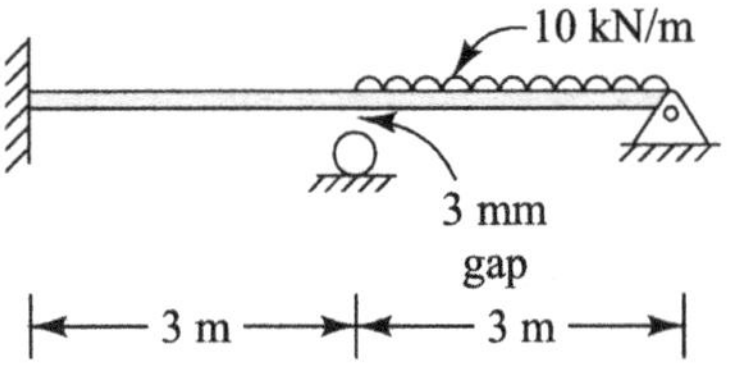

Fig. 12.47

12.14 Determine the deflection of the cantilever beam shown in Fig. 12.48 at beam support and at the free end. Take EI = constant for both beams.

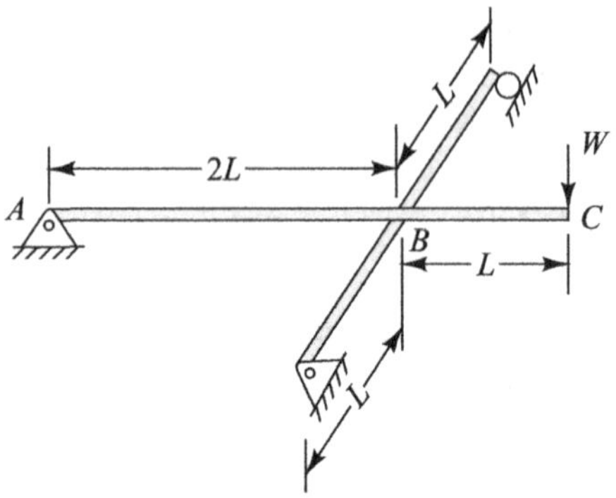

Fig. 12.48

12.15 Determine the uniformly distributed load w required to cause a deflection of 20 mm in the lower beam shown in Fig. 12.49. Take $EI = 5 \times 10^{11}$ N mm^2 for both beams.

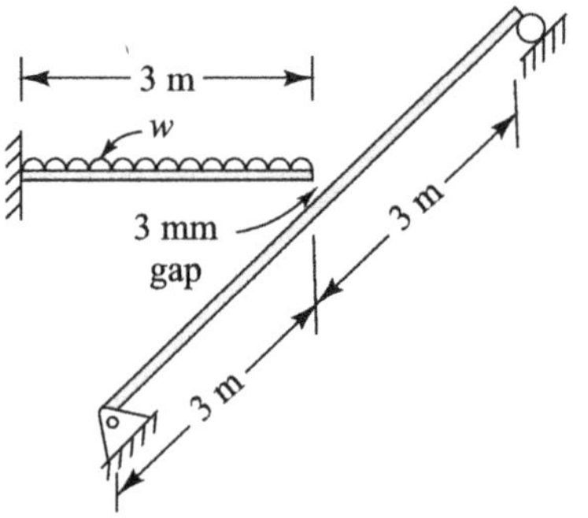

Fig. 12.49

12.16 through 12.19 Determine the equation of the elastic curve for the beams shown in Figs 12.50 through 12.53.

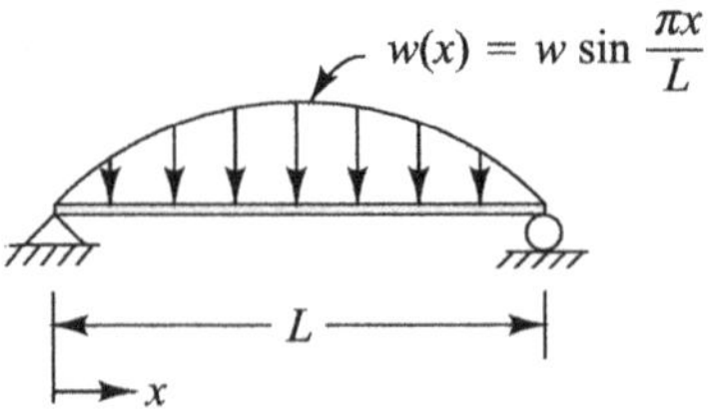

Fig. 12.50

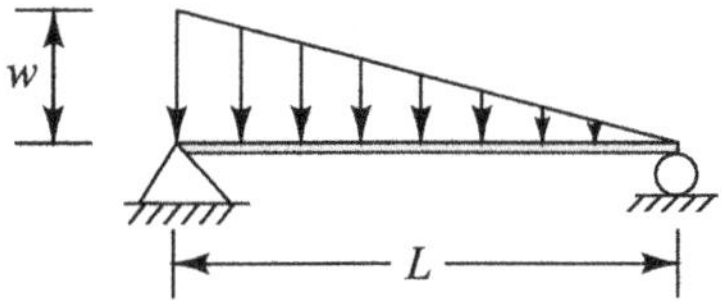

Fig. 12.51

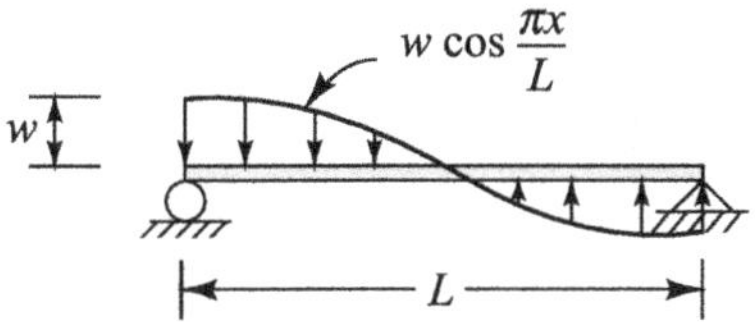

Fig. 12.52

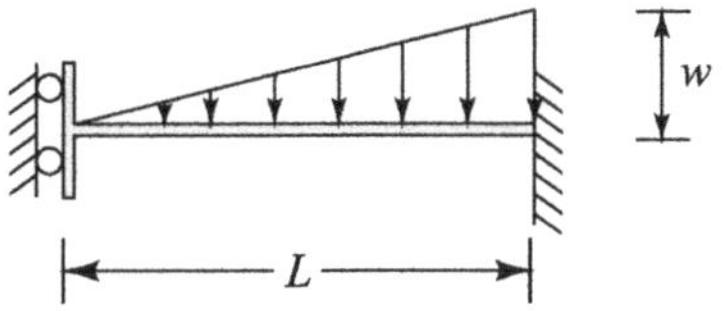

Fig. 12.53

12.20 through 12.22 Determine the maximum deflection in the beams carrying total load W over simply supported span L as shown in Figs. 12.54 through 12.56.

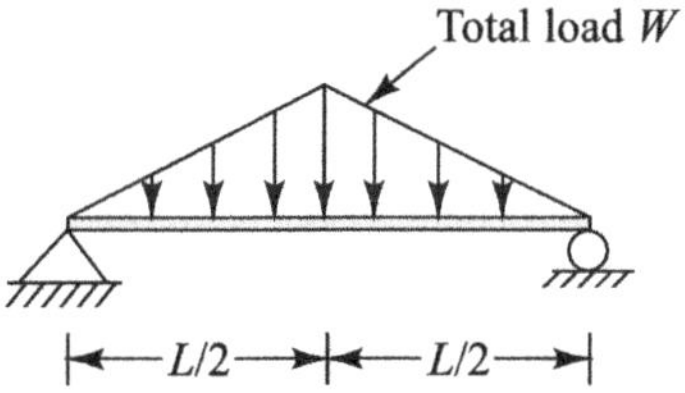

Fig. 12.54 *Prob 12.20*

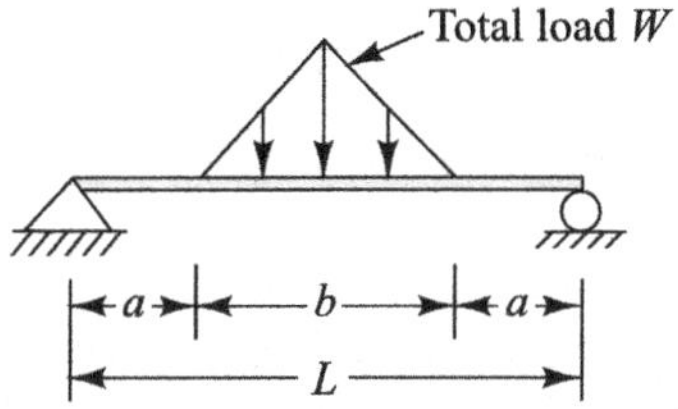

Fig. 12.55 *Prob 12.21*

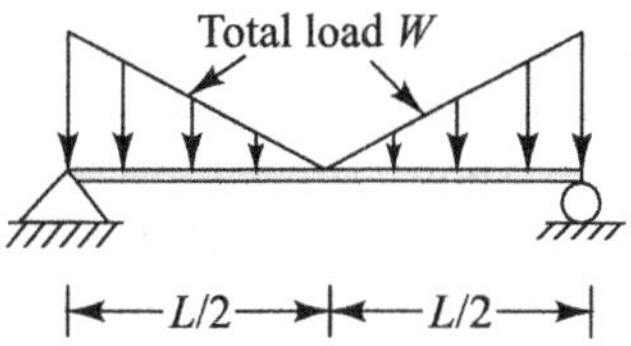

Fig. 12.56 *Prob 12.22*

12.23 through 12.26 Determine the propped end reactions for the cantilever beams shown in Figs. 12.57 through 12.60.

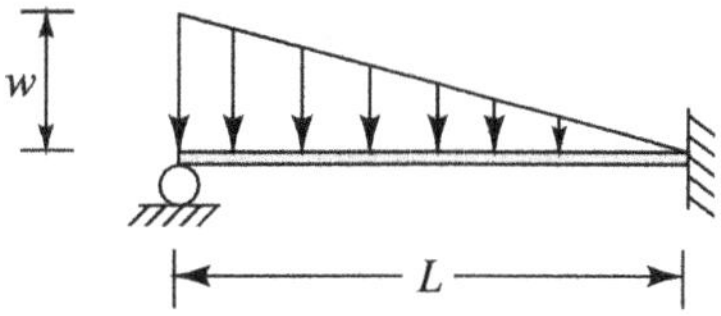

Fig. 12.57 *Prob 12.23*

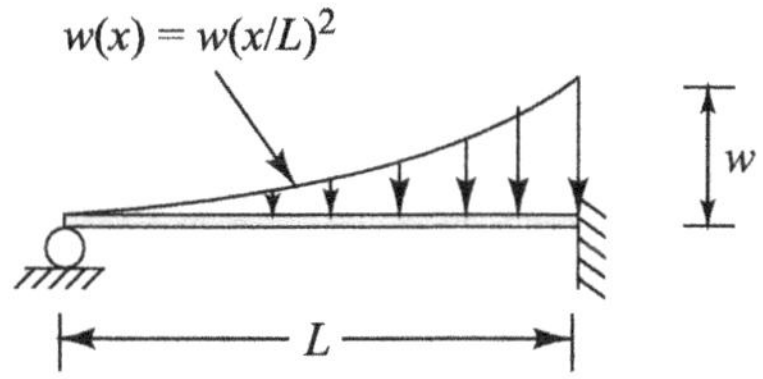

Fig. 12.58 *Prob 12.24*

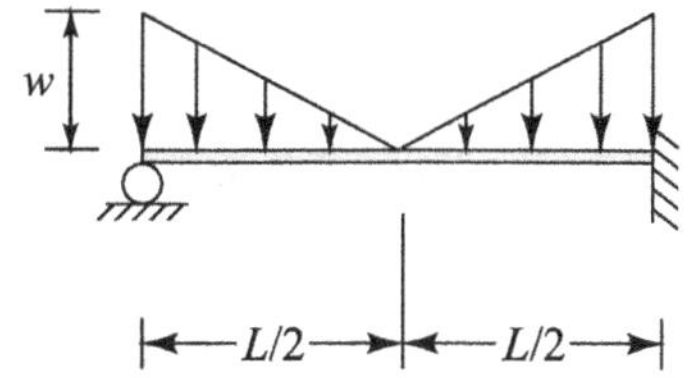

Fig. 12.59 *Prob 12.25*

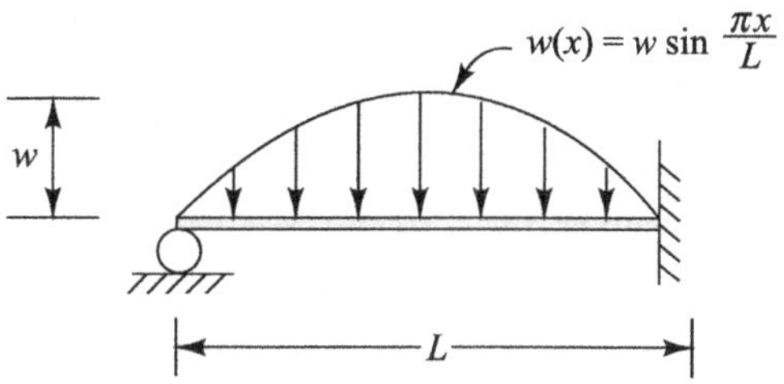

Fig. 12.60 *Prob 12.26*

12.27 and 12.28 Determine horizontal and vertical displacement of joint B for the trusses shown in Figs. 12.61 and 12.62 if area of cross-section is 500 mm^2 and $E = 200$ GPa for all members.

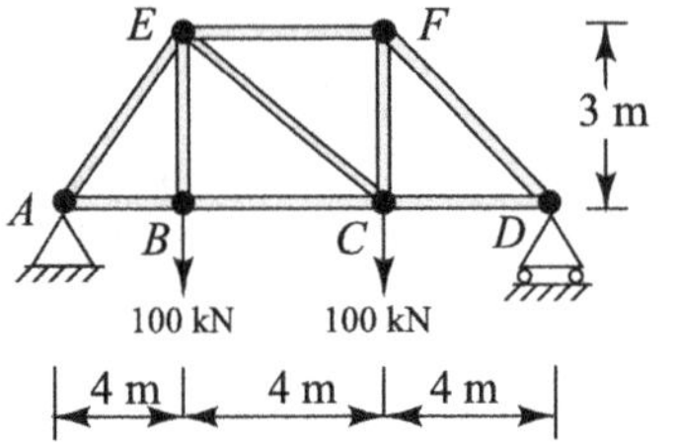

Fig. 12.61 *Prob 12.27*

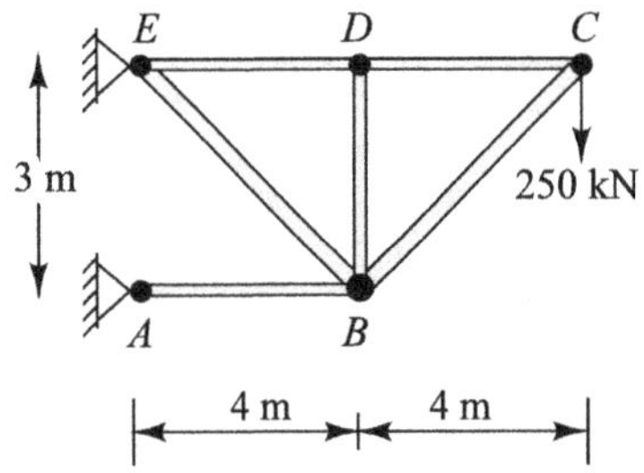

Fig. 12.62 *Prob 12.28*

OBJECTIVE QUESTIONS

12.29 The principle of superposition states that the total deflection of a structure under different sets of loads is equal to the sum of deflections under each set of loads acting separately on the structure if the loads are within
(a) Limit state
(b) Elastic limit
(c) Proportionality limit without buckling
(d) Elastic limit including buckling

12.30 The curvature at any point along the curve representing the deformed shape of a beam is given by,

(a) $$\frac{1}{R} = \frac{\pm \frac{dy}{dx}}{\left[1 + \frac{d^2y}{dx^2}\right]^{1/2}}$$

(b) $$\frac{1}{R} = \frac{\pm \frac{d^2y}{dx^2}}{\left[1 + \left(\frac{dy}{dx}\right)^2\right]^{3/2}}$$

(c) $$\frac{1}{R} = \frac{\pm \frac{d^2y}{dx^2}}{\left[1 + \frac{d^2y}{dx^2}\right]^{1/2}}$$

(d) $$\frac{1}{R} = \frac{\pm \frac{dy}{dx}}{\left[1 + \frac{d^2y}{dx^2}\right]^{2}}$$

12.31 For a cantilever beam of length L carrying a triangular load of intensity w at the support and zero at the free end, the slope at the free end is given by,

(a) $\frac{wL^3}{8EI}$ (b) $\frac{wL^3}{12EI}$ (c) $\frac{wL^3}{24EI}$ (d) $\frac{wL^3}{48EI}$

12.32 If a cantilever beam of span L and flexural rigidity EI carries a moment M concentrated at the free end, the deflection at the end will be,

(a) $\frac{ML}{24EI}$ (b) $\frac{ML}{12EI}$ (c) $\frac{ML}{6EI}$ (d) $\frac{ML^2}{2EI}$

12.33 For a cantilever beam of length L carrying a udl w per unit length, the deflection at the free end is given by,

(a) $\frac{wL^3}{8EI}$ (b) $\frac{wL^2}{24EI}$ (c) $\frac{wL^4}{8EI}$ (d) $\frac{wL^3}{12EI}$

12.34 A cantilever of length L has a flexural rigidity EI up to length $L/2$ from the fixed end and $EI/2$ for the rest. It carries a moment M at the free end. The slope at the free end is given by,

(a) $\frac{ML^2}{2EI}$ (b) $\frac{ML}{2EI}$ (c) $\frac{3ML}{2EI}$ (d) $\frac{2ML}{3EI}$

13

Thin Shells

CONCEPT REVIEW

13.1 THIN WALLED PRESSURE VESSEL

If the wall of a pressure vessel is thin, it will not sustain any bending. As per the relation between shear force and bending moment, if a body is subjected to a constant bending moment (including zero), the shear force is zero. According to this, thin walled pressure vessel is analyzed by membrane theory wherein all the stresses lie in the plane of the shell. For example consider an elementary section *ABCD* of a general shell of revolution shown in Fig. 13.1. The radius of

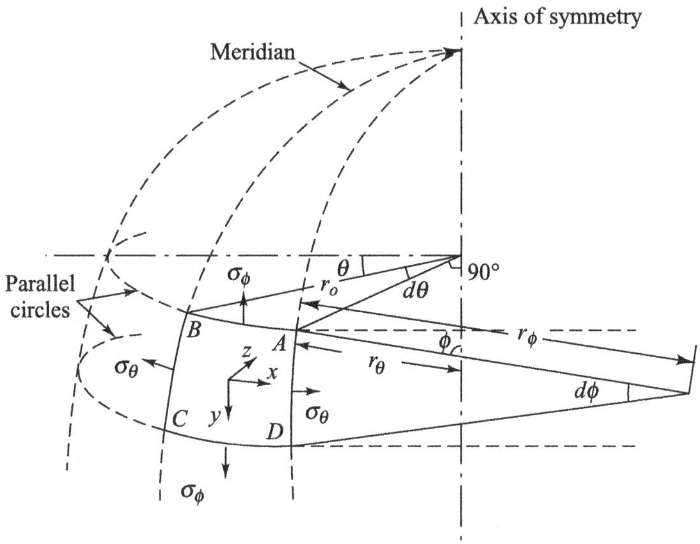

Fig. 13.1 *Shell of Revolution*

curvature of meridian is denoted by r_ϕ and this may vary along the length of the meridian. Radius of curvature of the shell surface in x-direction (perpendicular to meridian) is denoted by r_θ and its centre lies on the axis of symmetry. r_ϕ is perpendicular to the shell through points A to D while r_θ is perpendicular to points A to B. An internal pressure p gives rise to membrane stresses σ_ϕ and σ_θ. No shear force will act in the plane $ABCD$ and there will be no variation in meridional stress σ_ϕ and hoop stress σ_θ as there is no variation in internal pressure p. Outward force due to internal pressure =

$$pr_\phi d\phi \times r_\theta d\theta.$$

The meridional forces on edges AB or CD as shown in Fig. 13.2 = $\sigma_\phi (r_0\, d\theta)t$, where t = thickness of the shell.

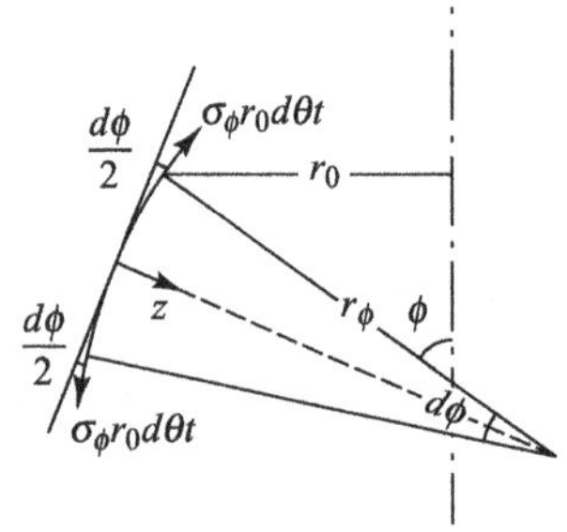

Fig. 13.2

The z-directional component (normal to shell) of meridional forces

$$= 2\sigma_\phi \left(r_0 d\theta\right) t \sin d\phi/2 \approx \sigma_\phi r_0 t d\theta d\phi$$

Horizontal component of hoop stress σ_θ on edges BC and AD shown in Fig. 13.3, in the direction of r_0,

$$= 2\sigma_\theta \left(r_\phi d\phi\right) t \sin d\theta/2 \approx \sigma_\theta r_\phi t d\theta d\phi$$

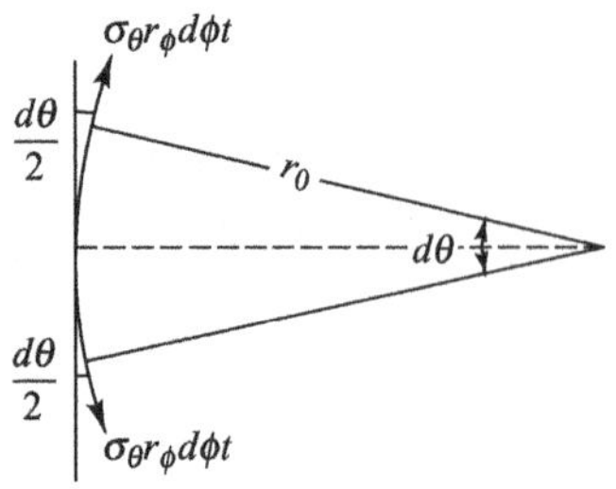

Fig. 13.3

The z-directional component (normal to shell) of these forces

$$= \sigma_\theta r_\phi t d\theta d\phi \sin\phi$$

Equating, Outward force = Inward force,

or, $\quad pr_\phi d\phi \times r_0 d\theta$

$$= \sigma_\phi r_0 t d\theta d\phi + \sigma_\theta r_\phi t d\theta d\phi \sin\phi$$

Substituting $r_0 = r_\theta \sin\phi$, the above equation yields,

or, $$\frac{p}{t} = \frac{\sigma_\phi}{r_\phi} + \frac{\sigma_\theta}{r_\theta} \quad (13.1)$$

This fundamental equation for axisymmetric deformation of this shells of revolution is also valid in case of plastic deformations as no displacement condition was used to determine the stresses. This equation has two unknowns σ_ϕ and σ_θ and another equation required for solution is obtained by considering vertical equilibrium of the entire shell above some convenient horizontal parallel circle as explained in Example 13.2.

In case of sphere, $r_\phi = r_\theta = r$ and $\sigma_\theta = \sigma_\phi = \sigma$, Eq. (13.1) reduces to,

$$\sigma = \frac{pr}{2t} \quad (13.2)$$

In case of cylinders and cones, $r_\phi = \infty$ and Eq. (13.1) reduces to,

$$\sigma_\theta = \frac{pr_\theta}{t} \quad (13.3)$$

13.2 ELASTIC DEFORMATION OF THIN CYLINDERS

The strains corresponding to circumferential and longitudinal stresses (see Example 13.1 for σ_ϕ) are given by Eq. (1.24) (h) and (i).

$$\varepsilon_\theta = \frac{1}{E}\left(\sigma_\theta - \mu\sigma_\phi\right) = \frac{pr}{Et}(1 - \mu/2) \quad (13.4)$$

$$\varepsilon_\phi = \frac{1}{E}\left(\sigma_\phi - \mu\sigma_\theta\right) = \frac{pr}{Et}\left(\frac{1}{2} - \mu\right) \quad (13.5)$$

Increase in circumference = $2\pi r \varepsilon_\theta$

Increase in radius = $r\varepsilon_\theta$

Increase in length per unit length = ε_ϕ

Increase in volume per unit length,

$$\Delta V = \pi(r + r\varepsilon_\theta)^2 (1 + \varepsilon_\phi) - \pi r^2$$

Volumetric strain, $\varepsilon_v = \dfrac{\Delta V}{V}$

$$= \frac{\pi(r + r\varepsilon_\theta)^2 (1 + \varepsilon_\phi) - \pi r^2}{\pi r^2}$$

or, $$\varepsilon_v \approx 2\varepsilon_\theta + \varepsilon_\phi = \frac{pr}{2Et}(5 - 4\mu)$$

Change in volume, $\Delta V = \varepsilon_v \times V$

$$= \frac{pr}{2Et}(5 - 4\mu) \times \pi r^2 L \qquad (13.6)$$

where, L = length of the cylinder.

13.3 ELASTIC DEFORMATION OF THIN SPHERES

The circumferential strain for spheres is given by Eq. (1.24)(h).

$$\varepsilon = \frac{1}{E}(\sigma - \mu\sigma) = \frac{pr}{2Et}(1 - \mu) \qquad (13.7)$$

$$\text{Volumetric strain, } \varepsilon_v = 3\varepsilon = \frac{3pr}{2Et}(1 - \mu) \qquad (13.8)$$

Change in volume, $\Delta V = \varepsilon_v \times V$

$$= \frac{3pr}{2Et}(1 - \mu) \times 4\pi r^3 / 3$$

13.4 CYLINDRICAL SHELLS WITH HEMISPHERICAL HEADS

Bending stresses near ends of a cylindrical pressure vessel can be reduced by hemispherical ends. The thicknesses t_1 and t_2 of cylindrical and hemispherical ends respectively are proportioned in such a way that the radial expansion of the two is same at the junction.

Circumferential strain of cylinder [Eq. (13.4)]

$$\varepsilon_\theta = \frac{pr}{Et_1}(1 - \mu/2)$$

Circumferential strain of sphere [(Eq. 13.7)]

$$\varepsilon = \frac{pr}{2Et_2}(1 - \mu)$$

These strains should be equal for compatibility, which gives,

$$\frac{pr}{Et_1}(1 - \mu/2) = \frac{pr}{2Et_2}(1 - \mu)$$

or, $$\frac{t_1}{t_2} = \frac{2 - \mu}{1 - \mu} \qquad (13.9)$$

For most metals $\mu \approx 0.3$, and $t_1/t_2 \approx 2.4$. The hemispherical end is therefore thinner than the cylindrical portion.

13.5 SPHERICAL DOME UNDER SELF WEIGHT

Figure 13.4 shows a spherical dome of radius r and thickness t. Consider a ring shaped element of horizontal radius $r \sin \alpha$. If self weight of the dome is w per unit volume then the weight of the annular ring = $w\,(2\pi r \sin \alpha)\,(r\,d\alpha)\,t$.

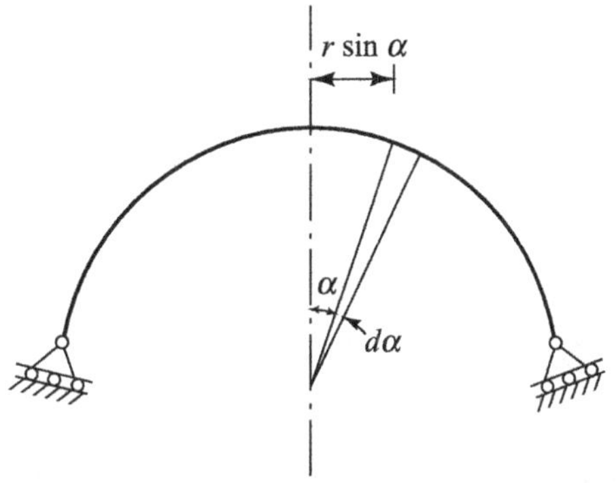

Fig. 13.4(a)

For meridional force σ_ϕ at any angle ϕ, consider equilibrium of upward force and downward force in Fig. 13.4(b).

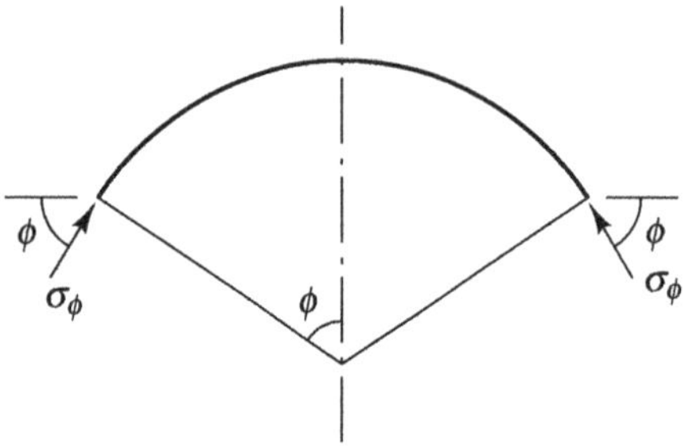

Fig. 13.4(b)

$$(2\pi r \sin \phi)\, t\sigma_\phi \sin \phi = \int_0^\phi wt(2\pi r \sin \alpha) r\, d\alpha$$

or, $$\sigma_\phi = \frac{-wr}{1+\cos\phi} \quad (13.10)$$

Minus sign has been added since σ_ϕ is compressive in nature. Substituting σ_ϕ in Eq. (13.1),

$$\frac{\sigma_\theta}{r_\theta} + \frac{\sigma_\phi}{r_\phi} = \frac{p}{t}$$

where, $$r_\theta = r_\phi = r$$

and, $$p = -\,wt \cos \phi$$

$$\frac{\sigma_\theta}{r} - \frac{wr}{1+\cos\phi} = \frac{-wt\cos\phi}{t}$$

or, $$\sigma_\theta = wr\left(\frac{1}{1+\cos\phi} - \cos\phi\right) \quad (13.11)$$

13.6 TOROIDAL SHELL

Figure 13.5 shows a Toroidal shell of mean radius b. The radius of its cross section is R. If the internal pressure is p, then for longitudinal stress σ_θ, consider the equilibrium of forces at a vertical section as follows.

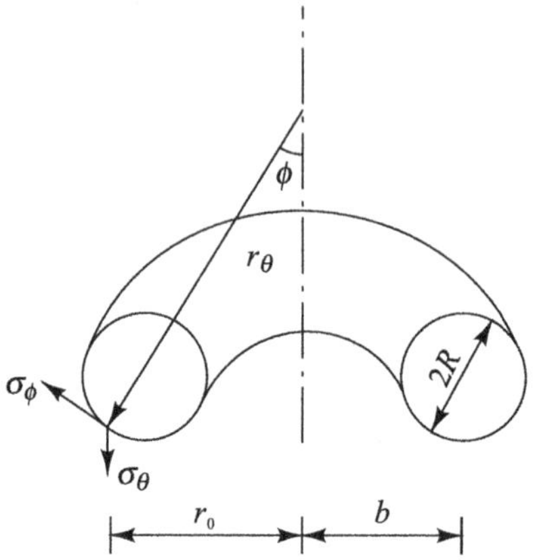

Fig. 13.5

$$2\pi R \times t \times \sigma_\theta = \pi R^2 p$$

or, $$\sigma_\theta = \frac{pR}{2t} \quad (13.12)$$

Substituting in Eq. (13.1),

$$\frac{\sigma_\theta}{r_\theta} + \frac{\sigma_\phi}{r_\phi} = \frac{p}{t}$$

where, $$r_\phi = R,$$

$$r_\theta = \frac{b + R\sin\phi}{\sin\phi}$$

or, $$\sigma_\phi = \frac{pR}{t}\left(1 - \frac{R\sin\phi}{2(b + R\sin\phi)}\right)$$

Substituting $$\sin\phi = \frac{r_0 - b}{R}$$

$$\sigma_\theta = \frac{pR}{tr_0}\left(\frac{r_0 + b}{2}\right) \quad (13.13)$$

Equation 13.12 and Eq. 13.13 reduce to Eq. (13.2) for sphere when $b = 0$.

EXAMPLES

Example 13.1 A compressed air cylinder of internal diameter 600 mm has a wall thickness of 3 mm. It contains air at a pressure of 1.5 MPa. Calculate the stresses in the cyclinder wall. Also determine the shear stress in the cylinder.

Solution From Eq. (13.3),

$$\text{Hoop stress, } \sigma_\theta = \frac{pr_\theta}{t} = \frac{1.5 \times 300}{3} = 150 \text{ MPa}$$

For longitudinal stress, consider equilibrium of forces at a vertical section shown in Fig. 13.6(a)

$$\pi r^2 p = 2\pi r t \sigma_\phi$$

or,

$$\sigma_\phi = \frac{pr}{2t} = \frac{1.5 \times 300}{2 \times 3} = 75 \text{ MPa}$$

It should be noted that hoop stress σ_θ is double than longitudinal stress σ_ϕ in case of cylinders.

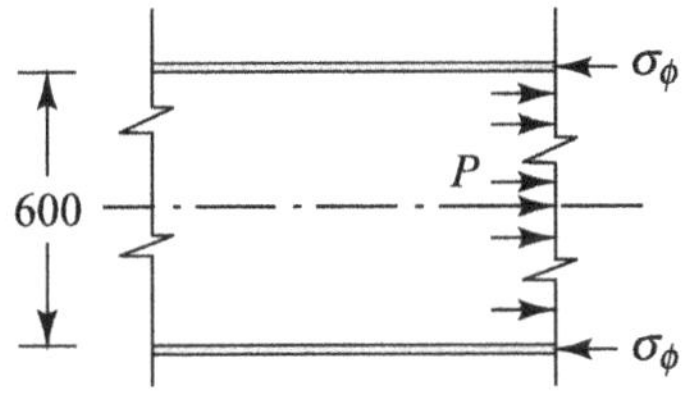

Fig. 13.6(a)

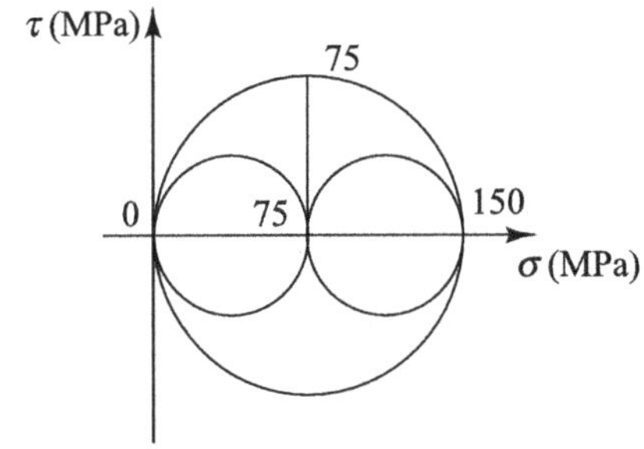

Fig. 13.6(b)

Figure 13.6(b) shows three dimensional Mohr's circle for the surface of the cylinder. The maximum shear occurs in the plane containing σ_θ and the perpendicular to the surface of the cylinder and is equal to $\sigma_\theta/2$ = 75 MPa. The shear stress in the plane of surface of the cylinder = $(\sigma_\theta - \sigma_\phi)/2$ = 37.5 MPa.

Example 13.2 A mild steel cylindrical cone of thickness 2 mm, base diameter 2 m and height 1.5 m, is suspended from the upper rim as shown in Fig. 13.7(a). It contains water upto depth 1 m. Determine the stresses in the wall of the cone at the surface and at 0.5 m depth of water.

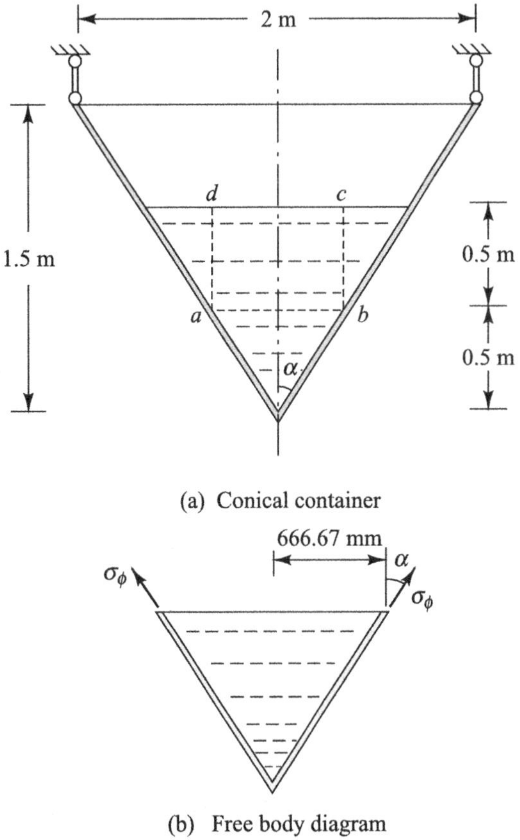

Fig. 13.7

Solution Substituting $r_\phi = \infty$ in Eq. 13.1,

$$\sigma_\theta = \frac{pr_\theta}{t}.$$

1. At the surface of water, $\sigma_\theta = 0$ as $p = 0$. Meridional stress σ_ϕ is determined by equating its upward component with downward weight of water as shown in Fig. 13.7(b).
   Upward force $= 2\pi rt\sigma_\phi \cos\alpha$

   $$\text{Downward force} = \frac{1}{3}\pi r^2 hw$$

where, r = radius at free water surface = 666.67 mm
t = thickness of cone = 2 mm
α = apex angle of cone
$= \tan^{-1}(1/1.5) = 33.69°$
h = depth of water = 1000 mm
w = specific weight of water = 9810 N/m^3
$= 9810 \times 10^{-9}$ N/mm^3

Equating upward force = Downward force

$$\sigma_\phi = \frac{rhw}{6t\cos\alpha}$$

$$= \frac{666.67 \times 1000 \times 9810 \times 10^{-9}}{6 \times 2\cos 33.69°}$$

$= 0.655$ N/mm^2

2. At 0.5 m depth of water, meridional stress σ_ϕ is determined by equating its upward component with downward weight of water in the lower half of the cone along with its vertical projection in the cylindrical portion *abcd* shown in Fig. 13.7(a).

   $$\text{Upward force} = 2\pi rt\sigma_\phi \cos\alpha$$

   $$\text{Downward force} = \frac{1}{3}\pi r^2 hw + \pi r^2 hw$$

where, r = radius at 0.5 m depth = 333.33 mm
h = depth of water in the cone
= depth of water in the cylinder *abcd*
= 500 mm

Equating upward force = Downward force

$$\sigma_\phi = \frac{rhw}{6t\cos\alpha} + \frac{rhw}{2t\cos\alpha}$$

$$= \frac{333.33 \times 500 \times 9810 \times 10^{-9}}{6 \times 2\cos 33.69°} + \frac{333.33 \times 500 \times 9810 \times 10^{-9}}{2 \times 2\cos 33.69°}$$

$= 0.655$ N/mm^2

Hoop stress, $\sigma_\theta = \dfrac{pr_\theta}{t}$

where, $r_\theta = \dfrac{r_0}{\cos\alpha} = \dfrac{333.33}{\cos 33.69°} = 400.61$ mm

$p = wh = 9810 \times 10^{-9} \times 500 = 0.004905$ N/mm^2

or, $\sigma_\theta = \dfrac{0.004905 \times 400.61}{2} = 0.982$ N/mm^2

Example 13.3 The ends of a pressure vessel are of the shape shown in Fig. 13.8(a). If the thickness of the pressure vessel is 5 mm and internal pressure is 5 MPa. Determine the stresses at points *X* and *Y* on the shell as shown in Fig. 13.8(a).

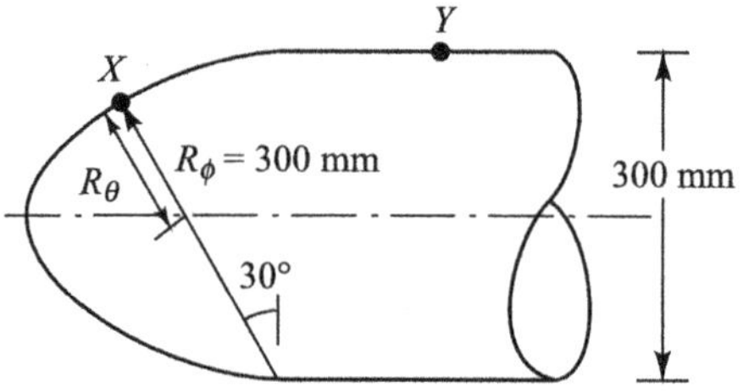

(a)

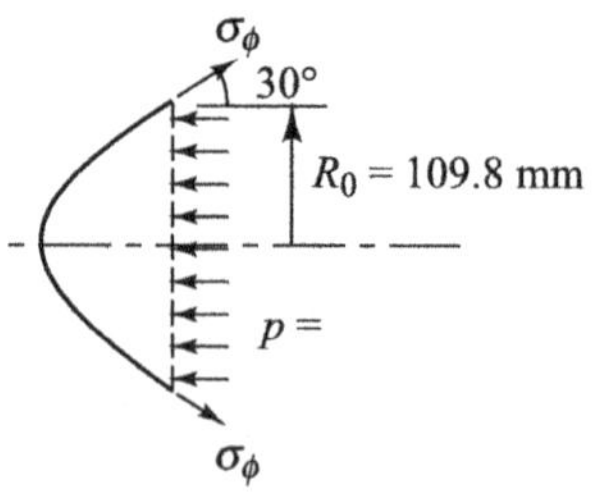

(b)

Fig. 13.8

Solution 1. At point *X*:

$R_\phi = 300$ mm

$$R_\theta = 300 - \frac{150}{\cos 30°} = 126.795 \text{ mm}$$

$$R_0 = 126.975 \cos 30° = 109.8 \text{ mm}$$

Consider the equilibrium of the free body diagram in Fig. 13.8(b) for meridional stress σ_ϕ as follows.

$$2\,\pi r_0 t \sigma_\phi \cos 30° = \pi r_0^2\, p$$

$$\text{or, } \sigma_\phi = \frac{r_0^2 p}{2t\cos 30°} = \frac{109.8 \times 5}{2 \times 5 \cos 30°} = 63.393 \text{ MPa}$$

Substituting in Eq. 13.1,

$$\frac{\sigma_\theta}{r_\theta} + \frac{\sigma_\phi}{r_\phi} = \frac{p}{t}$$

$$\text{or, } \frac{\sigma_\theta}{126.795} + \frac{63.393}{300} = \frac{5}{5}$$

or, $\sigma_\theta = 100$ MPa

2. At point Y:

$$\text{Hoop stress, } \sigma_\theta = \frac{pr_\theta}{t} = \frac{5 \times 150}{5}$$

$$= 150 \text{ MPa}$$

Longitudinal stress, $\sigma_\phi = \sigma_\theta/2 = 75$ MPa

Example 13.4 A simply supported concrete dome of radius 3.5 m is to bear its self weight 25 kN/m^3. Calculate meridional and hoop stress when $\phi = 30°$ and $\phi = 60°$.

Solution *Case I*: $\phi = 30°$

$$w = 25 \text{ kN/m}^3 = 25 \times 10^{-6} \text{ N/mm}^3$$

Substituting in Eq. (13.9) and Eq. (13.10)

$$\sigma_\phi = \frac{-wr}{1 + \cos\phi} = \frac{-25 \times 10^{-6} \times 3500}{1 + \cos 30°}$$

$$= -0.047 \text{ MPa (compressive)}$$

$$\sigma_\theta = wr\left(\frac{1}{1+\cos\phi} - \cos\phi\right)$$

$$= 3500 \times 25 \times 10^{-6}\left(\frac{1}{1+\cos 30°} - \cos 30°\right)$$

$$= -0.265 \text{ MPa (compressive)}$$

Case II :

$$\sigma_\phi = \frac{-25 \times 10^{-6} \times 3500}{1 + \cos 60°}$$

$$= -0.058 \text{ MPa (compressive)}$$

$$\sigma_\theta = 3500 \times 25 \times 10^{-6}\left(\frac{1}{1+\cos 60°} - \cos 60°\right)$$

$$= 0.525 \text{ MPa (tensile)}$$

It should be noted that meridional stress σ_ϕ is always +ve in case of domes and circumferential stress σ_θ changes sign from compression to tension when $\sigma_\theta = 0$ at $\phi = 51°50'$.

Example 13.5 A steel spherical shell of internal diameter 1 m, has its wall thickness 5 mm. An incompressible fluid is pumped in the shell till its pressure is 3 MPa. Determine the stress in the wall of the shell and the changes in internal volume. Take modulus of elasticity, $E = 210$ GPa and Poisson's ratio, $\mu = 0.3$ for steel.

Solution Using Eq. (13.2) for spherical shells,

$$\sigma = \frac{pr}{2t} = \frac{3 \times 500}{2 \times 5} = 150 \text{ MPa}$$

Change in volume using Eq. (13.7),

$$\Delta V = \varepsilon_v \times V = \frac{3pr}{2Et}(1-\mu) \times \frac{4\pi r^3}{3}$$

$$= \frac{2\pi \times 3 \times 500^4}{210 \times 10^3 \times 5}(1 - 0.3) = 785398 \text{ mm}^3$$

Example 13.6 A laminated cylinder consists of a thin steel shell shrunk fitted on an aluminium cylinder. The thickness of each shell is 2 mm and the mean diameter is 100 mm. The initial interference of diameters of the shells is 0.25 mm. Find the circumferential stress in each shell caused by this shrunk fit. Also calculate the circumferential stress in each shell due to internal fluid pressure of 2 MPa. Take $E = 210$ GPa for steel and $E = 70$ GPa for aluminium.

Solution **Stage 1:** Assuming no longitudinal stress arise ($\sigma_\phi = 0$) during shrunk fit, change in radius Δr due to interference pressure p is given by Section 13.2 as follow.

$$\Delta r = r\varepsilon_\theta = \frac{pr^2}{Et}$$

Total change in radius = Initial interference

i.e., $$\Delta r_1 + \Delta r_2 = \frac{0.25}{2} = 0.125 \text{ mm}$$

or, $$\frac{p \times 50^2}{210000 \times 2} + \frac{p \times 50^2}{70000 \times 2} = 0.125$$

or, $p = 5.25$ MPa

Inner cylinder will get compressed and outer cylinder will be in tension due to this interface pressure p as follows.

$$\sigma_s = -\sigma_a = \frac{pr}{t} = \frac{5.25 \times 50}{2} = 131.25 \text{ MPa}$$

Stage 2: Both inner and outer cylinders will get additional tensile stress due to internal fluid pressure of 2 MPa. Let the interface pressure be p_2 due to internal pressure $p_1 = 2$ MPa.

Increase in radius of steel cylinder due to internal pressure p_2,

$$\Delta r_s = \frac{p_2 r^2}{E_s t}$$

Increase in radius of aluminium cylinder due to internal pressure $p_1 = 2$ MPa and external pressure p_2,

$$\Delta r_a = \frac{p_1 r^2}{E_a t} - \frac{p_2 r^2}{E_a t}$$

Since change in radii of the two cylinders are equal,

$$\therefore \quad \Delta r_s = \Delta r_a$$

or, $$\frac{p_2 r^2}{E_s t} = \frac{p_1 r^2}{E_a t} - \frac{p_2 r^2}{E_a t}$$

or, $$p_2\left(\frac{1}{E_s} + \frac{1}{E_a}\right) = \frac{2}{E_a}$$

or, $p_2 = 1.5$ MPa

Hoop stress in aluminium cylinder due to internal pressure $p_1 = 2$ MPa and external pressure $p_2 = 1.5$ MPa

$$\sigma_\theta = \frac{p_1 r}{t} - \frac{p_2 r}{t} = (2 - 1.5) \times 50/2$$

$$= 12.5 \text{ MPa}$$

Hoop stress in steel cylinder due to internal pressure $p_2 = 1.5$ MPa,

$$\sigma_\theta = \frac{p_2 r}{t} = 1.5 \times 50/2 = 37.5 \text{ MPa}$$

Total stress in aluminium cylinder

= – 131.25 +12.5

= – 118.75 MPa

Total stress in steel cylinder = 131.25 + 37.5

= 168.75 MPa

Example 13.7 Determine the stresses in the compound cylinder of Example 13.6 if the temperature also rises by 10°C. Take the coefficient of expansion for aluminium and steel to be 25×10^{-6}/°C and 11×10^{-6}/°C.

Solution Increase in circumference of aluminium cylinder

$$= 2\pi r\alpha_a \Delta t = 2\pi \times 50 \times 25 \times 10^{-6} \times 10$$

$$= 0.07854 \text{ mm}$$

Increase in circumference of steel cylinder

$$= 2\pi r\alpha_s \Delta t = 2\pi \times 50 \times 11 \times 10^{-6} \times 10$$

$$= 0.03456 \text{ mm}$$

Increase in radius of aluminium cylinder

$$= \frac{0.07854}{2\pi} = 0.0125 \text{ mm}$$

Increase in radius of steel cylinder

$$= \frac{0.03456}{2\pi} = 0.0055 \text{ mm}$$

Total interference = 0.0125 – 0.0055 = 0.007 mm

There will be interference pressure due to differential radial expansion. The inner cylinder will be compressed and the outer cylinder will get additional tension. As in previous example, the interference pressure p is calculated as follows.

$$\frac{p \times 50^2}{210000 \times 2} + \frac{p \times 50^2}{70000 \times 2} = 0.007$$

or, $p = 0.294$ MPa

$$\sigma_s = -\sigma_a = \frac{pr}{t} = \frac{0.294 \times 50}{2} = 7.35 \text{ MPa}$$

Final stress in aluminium cylinder

$$= -118.75 - 7.35$$
$$= -126.1 \text{ MPa}$$

Final stress in steel cylinder

$$= 168.75 + 7.35$$
$$= 176.1 \text{ MPa}$$

Example 13.8 The outer diameter of wheels of a tractor is 1.5 m. Determine the stresses in the tyres due to internal pressure of 0.4 MPa if its radius is 150 mm and thickness is 20 mm.

Solution Longitudinal stress using Eq. (13.12),

$$\sigma_\theta = \frac{pR}{2t} = \frac{0.4 \times 150}{2 \times 20} = 1.5 \text{ MPa}$$

Circumferential stress using Eq. (13.13) at inner side of the tyre,

where, $r_0 = 750 - 300 = 450$ mm

and, $b = 750 - 150 = 600$ mm,

$$\sigma_\phi = \frac{pR}{tr_0}\left(\frac{r_0 + b}{2}\right)$$

$$= \frac{0.4 \times 150}{20 \times 450}\left(\frac{450 + 600}{2}\right) = 3.5 \text{ MPa}$$

On the outer edge, $r_0 = 750$ mm,

$$\sigma_\phi = \frac{pR}{tr_0}\left(\frac{r_0 + b}{2}\right)$$

$$= \frac{0.4 \times 150}{20 \times 750}\left(\frac{750 + 600}{2}\right) = 2.7 \text{ MPa}$$

On the sides, $r_0 = b = 600$ mm

$$\text{and,} \sigma_\phi = \frac{pR}{tr_0}\left(\frac{r_0 + b}{2}\right) = \frac{0.4 \times 150}{20 \times 600}\left(\frac{600 + 600}{2}\right)$$

$$= 3 \text{ MPa}$$

Example 13.9 A copper pipe of diameter 150 mm has 1.5 mm thick wall. It is closely wound by a steel wire of diameter 1.8 mm with initial tension of 10 N. Determine the stresses in the wire and the pipe when subjected to internal pressure of 2 MPa. Take Poisson's ratio $\mu = 0.33$, Young's Modulus $E = 100$ GPa for copper and $E = 200$ MPa for steel.

Solution (i) Initial pressure, $p = 0$

Initial tension in wire/mm

= Initial compression in pipe/mm

$$= \frac{10}{1.8} \text{ N/mm}$$

Initial tension in wire

$$= \frac{10}{(\pi/4)1.8^2} = 3.93 \text{ N/mm}^2$$

Initial compression in pipe

$$= \frac{10}{1.8 \times 1.5} = 3.7 \text{ N/mm}^2$$

(ii) Internal pressure, $p = 2$ MPa

Equating circumferential strain in copper pipe and steel wire,

$$\frac{1}{E_c}\left(\sigma_c - \mu\frac{pr}{2t}\right) = \frac{\sigma_s}{E_s}$$

where suffix c and s stand for copper and steel respectively.

$$\sigma_c - \frac{E_c}{E_s}\sigma_s = \mu\frac{pr}{2t}$$

$$\sigma_c - 0.5\sigma_s = \frac{0.33 \times 2 \times 75}{2 \times 1.5} = 16.5 \text{ N/mm}^2 \quad \text{(i)}$$

Consider 1.8 mm length of the pipe as shown in Fig. 13.9 and equate the resisting force with bursting force,

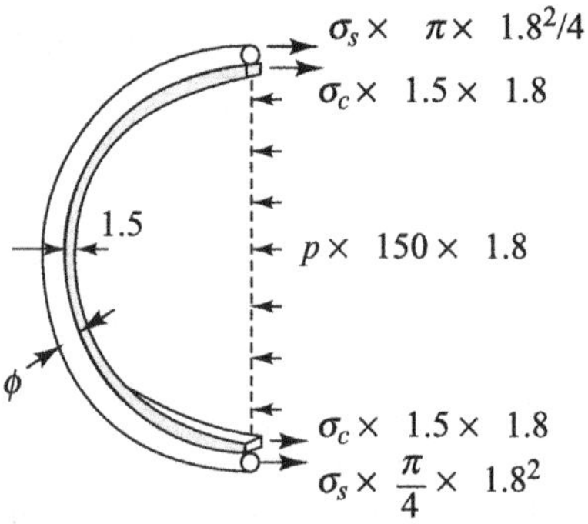

Fig. 13.9

$$2 \times \sigma_c \times 1.5 \times 1.8 + 2 \times \sigma_s \times \frac{\pi}{4} \times 1.8^2$$

$$= p \times 150 \times 1.8$$

or, $\sigma_c + 0.942\ \sigma_s = 100$ (ii)

Equations (i) and (ii) give,

$\sigma_c = 45.45$ MPa and $\sigma_s = 57.9$ MPa

Total stress in steel wire

= 57.9 + 3.93 = 61.83 MPa

Total stress in copper pipe

= 45.45 – 3.7 = 41.75 MPa

Example 13.10 A thin cylinder with closed ends has diameter 700 mm and wall thickness 5 mm. It is subjected to internal fluid pressure of 0.5 MPa along with external torsion 66×10^6 N.mm and moment 44×10^6 N.mm as shown in Fig. 13.10. Determine the principal stresses and their orientation for a top extreme point.

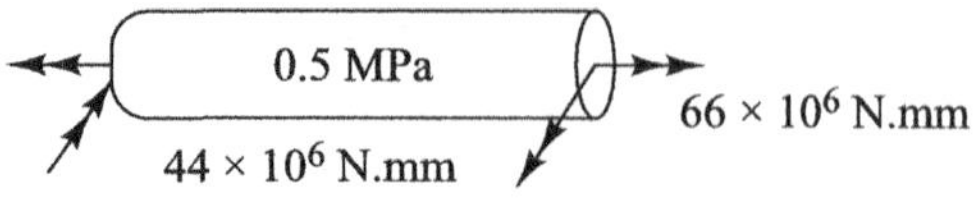

Fig. 13.10

Solution Hoop stress due to pressure

$$= \frac{pr}{t} = \frac{0.5 \times 350}{5} = 35 \text{ MPa}$$

Longitudinal stress due to pressure

$$= \frac{pr}{2t} = \frac{0.5 \times 350}{2 \times 5} = 17.5 \text{ MPa}$$

For cylinders, $I \approx \pi t r^3$

and, $J = 2I \approx 2\pi t r^3$

$$\text{Bending stress} = \frac{Mr}{I} = \frac{44 \times 10^6}{\pi \times 5 \times 350^2}$$

= 22.866 MPa (compression)

$$\text{Torsion shear} = \frac{Tr}{J} = \frac{66 \times 10^6}{2\pi \times 5 \times 350^2}$$

= 17.15 MPa

Longitudinal stress due to pressure + bending

= 17.5 – 22.866

= – 5.366 MPa

Principal stresses are given by Eq. 2.4,

$$\sigma_{x'\max/\min} = \frac{\sigma_x + \sigma_y}{2} \pm \sqrt{\left(\frac{\sigma_x - \sigma_y}{2}\right)^2 + \tau_{xy}^2}$$

$$= \frac{35 - 5.366}{2} \pm \sqrt{\left(\frac{35 + 5.366}{2}\right)^2 + 17.15^2}$$

= 14.817 ± 26.485

= 41.3 MPa, – 11.67 MPa

Principal plane angle θ_p is given by Eq. (2.3),

$$\tan 2\theta_p = \frac{-\tau_{xy}}{\left(\frac{\sigma_x - \sigma_y}{2}\right)} = \frac{-17.15}{\left(\frac{35 + 5.366}{2}\right)}$$

or, $2\theta_p = -40.355°$

Maximum shear stress is given by Eq. (2.6) as follows.

$$\tau_{x'y',\max} = \pm\sqrt{\left(\frac{\sigma_x - \sigma_y}{2}\right) + \tau_{xy}^2} = \pm\, 26.485 \text{ MPa}$$

Example 13.11 Derive the relations for stresses in a thick walled pressure vessel subjected to internal pressure p and compare with the relations for thin walled cylinders.

Solution A thick walled open ended cylinder of internal and external radius r_i and r_o respectively, subjected to internal pressure p is shown in Fig. 13.11(a). For thick walled pressure vessel the radial and circumferential stresses will vary along the radius. Consider an elementary thin walled concentric cylinder of thickness dr at a radius r for which internal pressure is σ_r and the external pressure is $\sigma_r + d\sigma_r$. The circumferential stresses $\sigma_{\theta 1}$ and $\sigma_{\theta 2}$ due to these external and internal pressures are given by Eq. (13.3) as follows.

$$\sigma_{\theta 1} = -\frac{\sigma_r r}{dr} \qquad \text{(compressive)}$$

$$\sigma_{\theta 2} = \frac{(\sigma_r + d\sigma_r)(r + dr)}{dr}$$

$$\approx \frac{\sigma_r r + r d\sigma_r + \sigma_r dr}{dr} \qquad \text{(tensile)}$$

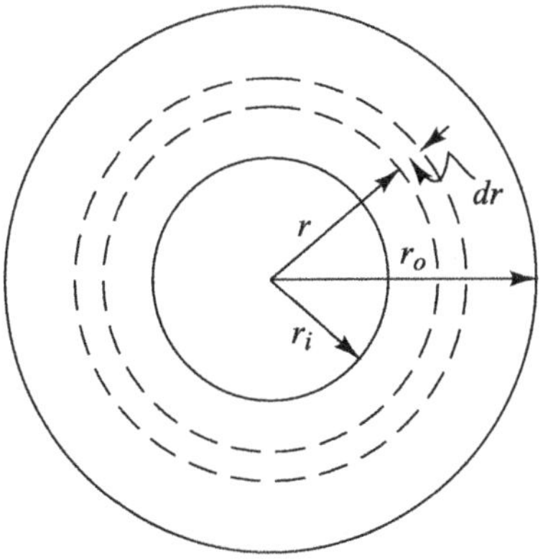

Fig. 13.11(a)

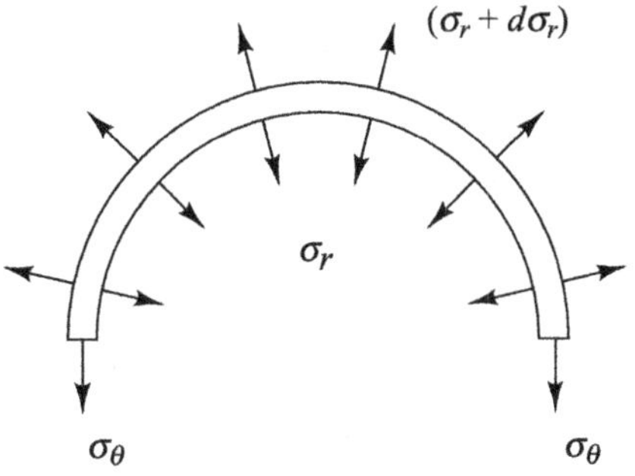

Fig. 13.11(b)

The total circumferential stress, $\sigma_\theta = \sigma_{\theta 1} + \sigma_{\theta 2}$

or, $$\sigma_\theta = \sigma_r + r\frac{d\sigma_r}{dr}$$

or, $$\frac{d\sigma_r}{dr} + \frac{\sigma_r - \sigma_\theta}{r} = 0 \qquad \text{(i)}$$

It is assumed that plane section will remain plane and longitudinal strain is constant. This holds well for open ended cylinders and at sections away from closed ends. Hook's law in polar coordinates gives the longitudinal strain as follows.

$$\varepsilon_l = \frac{1}{E}(\sigma_l - \mu(\sigma_r + \sigma_\theta)) = \text{ constant}$$

For open ended cylinders longitudinal stress $\sigma_l = 0$

$$\therefore \quad \sigma_r + \sigma_\theta = 2A \qquad \text{(ii)}$$

where, $2A$ = constant

Equations (i) and (ii) gives,

$$r\frac{d\sigma_r}{dr} + 2\sigma_r = 2A$$

or, $$\frac{d}{dr}\left(r^2\sigma_r\right) = 2rA$$

Integration of the above equation yields,

$$r^2\sigma_r = Ar^2 + B$$

or, $$\sigma_r = A + \frac{B}{r^2} \qquad \text{(iii)}$$

where, B is a constant of integration. Substituting σ_r in Eq. (ii), or, $$\sigma_\theta = A - \frac{B}{r^2} \quad \text{(iv)}$$

These are Lame's equations. The constants of integration are determined by boundary conditions as follows.

$$\sigma_r = -p \quad \text{at} \quad r = r_i$$

and $$\sigma_r = 0 \quad \text{at} \quad r = r_o$$

Substituting the boundary conditions in Eq. (iii and iv),

$$-p = A + \frac{B}{r_i^2} \text{ and, } 0 = A + \frac{B}{r_o^2}$$

which gives, $$A = \frac{p \times r_i^2}{r_o^2 - r_i^2} \text{ and, } B = \frac{-p \times r_i^2 \times r_o^2}{r_o^2 - r_i^2}$$

or $$\sigma_r = \frac{r_i^2 p}{r_o^2 - r_i^2}\left(1 - \frac{r_o^2}{r^2}\right) \quad \text{(v)}$$

$$\sigma_\theta = \frac{r_i^2 p}{r_o^2 - r_i^2}\left(1 + \frac{r_o^2}{r^2}\right) \quad \text{(vi)}$$

Equation (13.3) gives hoop stress in thin walled cylinders as follows.

$$\sigma_\theta = \frac{pr_\theta}{t} = \frac{pr_i}{r_o - r_i} \quad \text{(vii)}$$

The ratio S of Eq. (vi) and (vii) gives the extent to which Eq. (13.3) underestimates the maximum hoop stress at $r = r_i$ as follows.

$$S = \frac{r_o^2 + r_i^2}{r_i\left(r_o + r_i\right)} = \frac{K^2 + 1}{K + 1}$$

where, $K = r_o/r_i$. The comparison of the expressions for thick/thin cylinders is given in the following table.

$K =$	1.0	1.05	1.1	1.15	1.2
$S =$	1.0	1.026	1.052	1.08	1.109

This shows that when $K = 1.1$ i.e. wall thickness is about 1/10th of the radius, the error will be about 5% if the cylinder is treated as thin cylinder.

Example 13.12 A thin cylinder of radius R, thickness t and length L rotates at an angular velocity ω about its axis. Determine the circumferential stress due to rotation.

Solution Figure 13.12 shows an elementary portion of the cylinder which subtends a small angle $d\theta$ at the axis. If σ_θ is the hoop stress in the shell due to rotation about its axis, then equating its inward component with outward centrifugal force,

$$2\sigma_\theta Lt \sin\ d\theta/2 = \rho LtRd\theta\ \omega^2 R$$

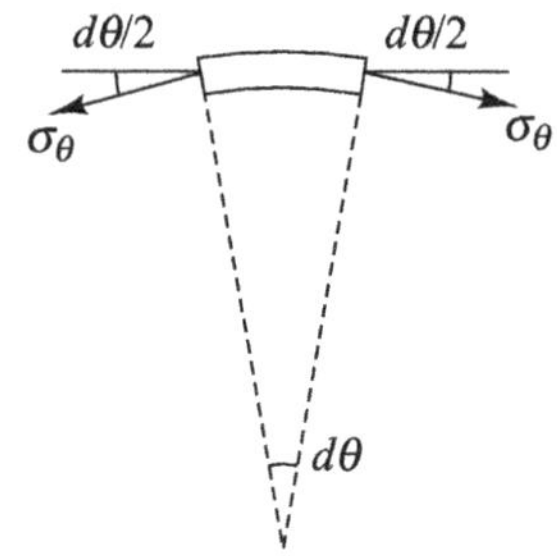

Fig. 13.12

where, ρ = density of the material of the ring

$LtRd\theta$ = volume of the elementary section

Since $\sin d\theta/2 \approx d\theta/2$ ($d\theta$ being a small angle),

$$\sigma_\theta = \rho R^2\ \omega^2 = \rho V^2$$

where, V = peripheral velocity.

(Note that the hoop stress σ_θ is independent of length and thicknes of the cylinder.)

EXERCISES

PROBLEMS

13.1 A boiler shell of inner diameter 1.5 m is made of 10 mm thick plate with double cover, double riveted butt joint. The rivets are staggered at a gauge distance of 100 mm. Determine the shearing stress in rivets and the tensile stress in boiler plate and cover plate due to internal pressure of 1.1 MPa if the diameter of rivets is 20 mm and thickness of cover plate is 6 mm.

13.2 A steel spherical shell of thickness 3 mm and diameter 1 m is made up of two hemispherical shells held together by 24-20 mm diameter high strength steel bolts at flanges of width 60 mm. Calculate the maximum pressure in the shell if the allowable tensile stress of steel is 150 MPa. Also calculate the compressive stress in the gasket of the flange under maximum pressure if the initial tension in the bolts is 95 kN before application of fluid pressure.

13.3 A spherical pressure vessel of steel has a diameter of 1 m and a wall thickness of 2 mm. If the maximum allowable tensile stress in steel is 150 MPa, calculate the maximum allowable pressure in the vessel. Also calculate the changes in diameter and internal volume of the sphere under maximum allowable pressue if the modulus of elasticity is 200 GPa and Poisson's ratio is 0.3 for steel.

13.4 A hollow sphere of radius 500 mm is placed inside a sphere of radius 550 mm. An incompressible liquid is filled in the gap between two spheres under atmospheric pressure. How much more liquid can be filled under gauge pressure 1.5 MPa if the thickness of both spheres is 2 mm. The modulus of elasticity and Poisson's ratio for both spheres are 210 GPa and 0.3 respectively.

13.5 A spherical steel shell of internal diameter 1 m, has its wall thickness 5 mm. An incompressible liquid is pumped in the shell till it is full under atmospheric pressure. How much more water can be pumped in so that the stress in the shell walls reaches its yield stress of 250 MPa. Take modulus of elasticity, $E = 210$ GN/m^2 and Poisson's ratio = 0.3 for steel.

13.6 Redo previous Problem 13.5 if the liquid is water having bulk modulus, $k = 2.3$ GPa.

13.7 A cylindrical pressure vessel with hemi-spherical ends of radius 500 mm is 3 m long overall. The thickness of cylindrical shell and the hemispherical ends are 5 mm and 2 mm respectively. It is made of copper having Young's modulus of elasticity, $E = 120$ GPa, and Poisson's ratio = 1/3. It is filled with oil upto atmospheric pressure. Neglecting the effect of gravity, determine the amount of oil which must be pumped in to increase the internal pressure to 0.5 MPa if the bulk modulus of oil is 2.5 GN/m^2.

13.8 A thin cylinder with closed ends has diameter 300 mm and wall thickness 3 mm. It is fixed rigidly at one end and at other free end a torque of 35×10^5 N.mm is applied along with a downward shear force 15 kN as shown in Fig. 13.13. If the internal fluid pressure of the cylinder is 1.5 MPa, determine the principal stresses and their orientation for a top extreme point A at a distance of 300 mm from free end.

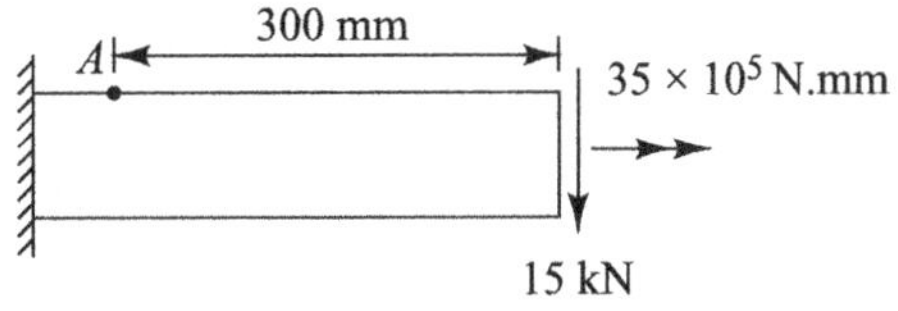

Fig. 13.13

13.9 A thin steel cylinder just fits over a copper cylinder. Determine the stresses in both the cylinders due to an internal pressure of 0.5 MPa if they are free to expand longitudinally. The average diameter of the cylinder is 1 m and the thickness of the cylinders is 2 mm. Take modulus of elasticity 200 GPa for steel and 90 GPa for copper.

13.10 A laminated steel cylinder of mean diameter 200 mm is fabricated by shrink fitting two cylinders of thickness 5 mm. The initial interference of diameters of two cylinders is 0.4 mm. Determine the stresses in the cylinders due to internal pressure of 5 MPa if modulus of elasticity for steel is 200 GPa.

13.11 A long copper pipe with closed ends has external diameter 146 mm and wall thickness 3 mm. It is reinforced by closely winding steel wire of diameter 1.6 mm. Calculate the initial tension required in the wire if an internal gauge pressure of 3 MPa in the pipe gives a tensile stress of 40 MPa in the wall of the pipe. Also calculate the final stress in the wire. Take Poisson's ratio $\mu = 0.3$, Young's Modulus $E = 120$ GPa for copper and $E = 200$ MPa for steel.

13.12 Determine the changes in length, diameter and thickness of the cylinder in Problem 13.17 if its length is 300 mm. Take E = 200 GPa and Poisson's ratio is 0.3.

OBJECTIVE QUESTIONS

13.13 An aluminium cylinder of diameter 300 mm is subjected to internal pressure of 4 MPa. If the permissible stress for the material of the cylinder is 150 MN/m^2, then the thickness of the cylinder should be

(a) 1 mm (b) 2 mm
(c) 4 mm (d) 8 mm

13.14 A mild steel pipe of diameter 200 mm has wall thickness 2 mm. If the yield stress of mild steel is 250 MPa and stress factor for maximum shear stress is 2.5, then the maximum allowable internal pressure is,

(a) 1 MPa (b) 2 MPa
(c) 2.5 MPa (d) 4 MPa

13.15 A thin cylinder has an internal diameter 200 mm and thickness 2 mm. Its ends are closed by pistons sliding in the cylinder. The circumferential and the longitudinal stresses in the cylinder due to internal pressure of 2 MPa will be,

(a) 100 MPa and 50 MPa
(b) 100 MPa and 0 MPa
(c) 200 MPa and 100 MPa
(d) 200 MPa and 0 MPa

13.16 Efficiency of thin cylinders to withstand high internal pressure is increased by using,

(a) higher thickness
(b) coaxial cylinders
(c) longer cylinders
(d) larger diameters

13.17 A cylindrical steel centrifuge with closed ends has external diameter 100 mm and 2 mm thick wall. It rotates at 6000 rpm about its axis and an external gauge pressure 0.4 MN/m^2 is applied on its external surface. If the density of steel is 7700 kg/m^3, the circumferential and longitudinal are,

(a) –2.7 MPa, – 5.1 MPa
(b) –5 MPa, 7.3 MPa
(c) 7.3 MPa, – 5 MPa
(d) 63 MPa, – 4.9 MPa

14

Thick Walled Cylinders

CONCEPT REVIEW

14.1 THICK WALLED CYLINDERS —LAME'S EQUATION

A pressure vessel can be treated as thick walled cylinder when its wall thickness is about $1/10^{th}$ of the radius. They are usually subjected to internal or external pressure and end loads if the ends are closed. The deformations are symmetric about z-axis, i.e. deformations are independent of θ. The tangetial deformations v and the shear stress $\tau_{r\theta}$ are therefore zero and the displacements due to internal or external pressure will be in the radial direction only.

In the absence of body forces, equilibrium equations in polar coordinates (Eq. (1.18) (a) and (b)) are reduced to,

$$\frac{d\sigma_r}{dr} + \frac{\sigma_r - \sigma_\theta}{r} = 0 \qquad (14.1)$$

It should be noted that total derivatives are used with respect to r since there are no variations in the direction of θ and z.

Strain-displacement equations (Eq. (1.21) (d) and (e)) are similarly reduced as follows.

$$\varepsilon_r = \frac{du}{dr}, \quad \varepsilon_\theta = \frac{u}{r}, \quad \gamma_{r\theta} = 0 \qquad (14.2,3,4)$$

Compatibility equation is obtained by substituting, $u = r\varepsilon_\theta$, in the first expression above.

$$\varepsilon_r = \frac{d(r\varepsilon_\theta)}{dr} = \varepsilon_\theta + r\frac{d\varepsilon_\theta}{dr}$$

or

$$\varepsilon_r - \varepsilon_\theta = r\frac{d\varepsilon_\theta}{dr} \qquad (14.5)$$

Hook's law in polar coordinates is written as,

$$\varepsilon_r = \frac{1}{E}[\sigma_r - \mu(\sigma_\theta + \sigma_z)] \qquad (14.6)$$

$$\varepsilon_\theta = \frac{1}{E}[\sigma_\theta - \mu(\sigma_r + \sigma_z)] \qquad (14.7)$$

$$\varepsilon_z = \frac{1}{E}[\sigma_z - \mu(\sigma_r + \sigma_\theta)] \qquad (14.8)$$

It is assumed that the cylinders are long enough to ensure that the plane sections will remain plane and the axial strain will be constant throughout, i.e.

$$\frac{d\varepsilon_z}{dr} = 0$$

or,

$$\frac{d\sigma_z}{dr} - \mu\left(\frac{d\sigma_r}{dr} + \frac{d\sigma_\theta}{dr}\right) = 0 \qquad (14.9)$$

Substituting values of ε_r and ε_θ in compatibility equation Eq. (14.5),

$$\sigma_r - \sigma_\theta - \mu(\sigma_\theta - \sigma_r) = r\left(\frac{d\sigma_\theta}{dr} - \mu\left(\frac{d\sigma_r}{dr} + \frac{d\sigma_z}{dr}\right)\right)$$

Substituting value of $d\sigma_z/dr$ from Eq. (14.9),

$$(\sigma_r - \sigma_\theta)(1+\mu) = r\left(\frac{d\sigma_\theta}{dr}(1-\mu^2) - \frac{\mu d\sigma_r}{dr}(1+\mu)\right)$$

or, $$\frac{\sigma_r - \sigma_\theta}{r} = \frac{d\sigma_\theta}{dr}(1-\mu) - \mu\frac{d\sigma_r}{dr}$$

By substituting Eq. (14.1), the above equation reduces to,

$$\frac{d\sigma_\theta}{dr} + \frac{d\sigma_r}{dr} = 0$$

Integrating the above equation yields,

$$\sigma_\theta + \sigma_r = 2A \qquad (14.10)$$

where 2A is a constant of integration. Substituting σ_θ in Eq. (14.1),

$$r\frac{d\sigma_r}{dr} + 2\sigma_r = 2A$$

or, $$\frac{d}{dr}(r^2\sigma_r) = 2rA$$

Integrating the above equation yields,

$$r^2\sigma_r = Ar^2 + B$$

or, $$\sigma_r = A + \frac{B}{r^2} \qquad (14.11)$$

where B is a constant of integration. Substituting σ_r in Eq. (14.10),

or, $$\sigma_\theta = A - \frac{B}{r^2} \qquad (14.12)$$

These are Lame's equations. The constants of integration are determined by boundary conditions as follows.

(1) *If only internal pressure p_i is acting,*

$$\sigma_r = -p_i \quad \text{at } r = r_i$$

and $$\sigma_r = -0 \quad \text{at } r = r_o$$

where r_i and r_o are the internal and external radii. Substituting the boundary conditions in Eq. (14.11) and Eq. (14.12).

$$-p_i = A + \frac{B}{r_i^2}$$

$$0 = A + \frac{B}{r_o^2}$$

which gives, $$A = \frac{p_i \times r_i^2}{r_o^2 - r_i^2}$$

$$B = -\frac{p_i \times r_i^2 \times r_o^2}{r_o^2 - r_i^2}$$

or, $$\sigma_r = \frac{r_i^2 p_i}{r_o^2 - r_i^2}\left(1 - \frac{r_o^2}{r^2}\right) \qquad (14.13)$$

$$\sigma_\theta = \frac{r_i^2 p_i}{r_o^2 - r_i^2}\left(1 + \frac{r_o^2}{r^2}\right) \qquad (14.14)$$

as $r_o > r$, the radial stress, σ_r is –ve i.e. compressive and the tangential or hoop stress, σ_θ is +ve i.e. tensile. The variation of σ_r and σ_θ on pressure walls is shown in Fig. 14.1.

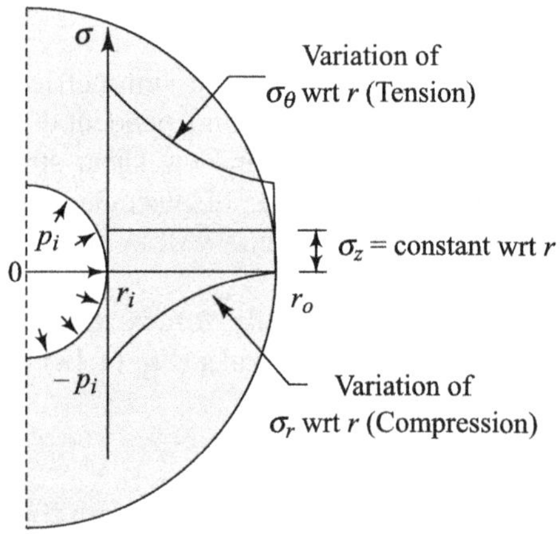

Fig. 14.1 *Stress distribution in thick wall cylinder with internal pressure p_i*

Axial stress σ_z in case of pressure vessel having open end will be zero as the end load $\pi r_i^2 p_i$ is borne by the plungers as shown in Fig. 14.2.

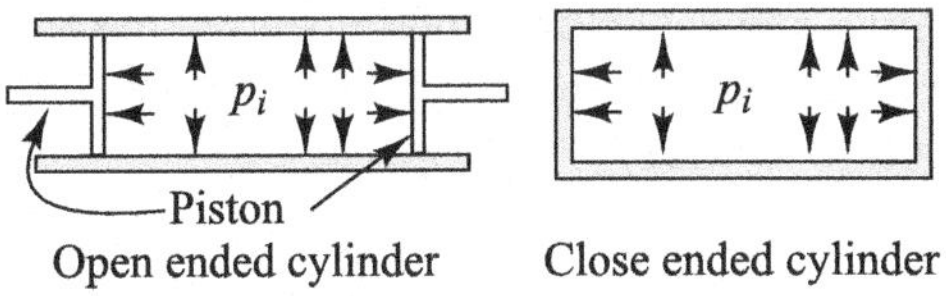

Fig. 14.2

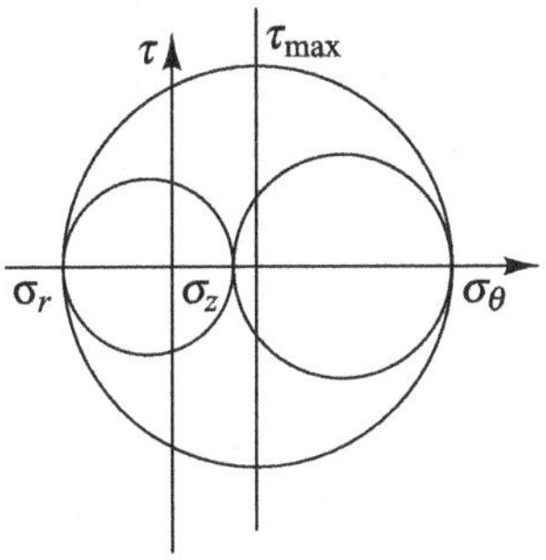

Fig. 14.3 *Mohr's circle for 3-dimensional stress in thick walled cylinder with internal pressure*

With closed ends, the axial stress σ_z will be tensile as follows.

$$\sigma_z = \frac{\pi r_i^2 p_i}{\pi(r_o^2 - r_i^2)} = \frac{r_i^2 p_i}{r_o^2 - r_i^2} \qquad (14.15)$$

(2) *If only external pressure p_o is acting,*

$$\sigma_r = 0 \quad \text{at } r = r_i$$

and

$$\sigma_r = p_o \quad \text{at } r = r_o$$

Substituting the boundary conditions in Eqs (14.11) and (14.12),

$$0 = A + \frac{B}{r_i^2}$$

$$-p_o = A + \frac{B}{r_o^2}$$

which gives,

$$A = \frac{-p_o \times r_o^2}{r_o^2 - r_i^2}$$

$$B = \frac{p_o \times r_i^2 \times r_o^2}{r_o^2 - r_i^2}$$

or,

$$\sigma_r = \frac{-r_o^2 p_o}{r_o^2 - r_i^2}\left(1 - \frac{r_i^2}{r^2}\right) \qquad (14.16)$$

$$\sigma_\theta = \frac{-r_o^2 p_o}{r_o^2 - r_i^2}\left(1 + \frac{r_i^2}{r^2}\right) \qquad (14.17)$$

since $r > r_i$, both radial stress, σ_r and tangential stress, σ_θ are –ve, i.e. compressive. The variation of σ_r and σ_θ on pressure walls is shown in Fig. 14.4.

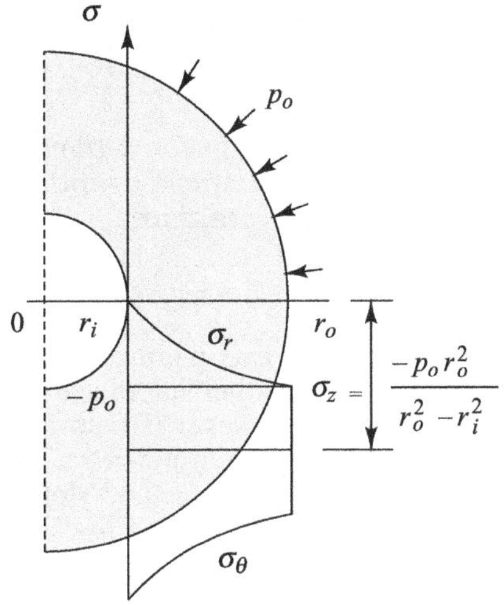

Fig. 14.4 *Stress distribution in thick wall cylinder with external pressure p_o*

If p_o is also applied on the closed ends of a cylinder, the axial stress σ_z will be compressive as follows.

$$\sigma_z = \frac{-\pi r_o^2 p_o}{\pi(r_o^2 - r_i^2)} = \frac{-r_o^2 p_o}{r_o^2 - r_i^2} \qquad (14.18)$$

(3) *If both internal and external pressures p_i and p_o are acting,*

For a general case of a thick walled cylinder subjected to both internal and external pressures p_i and p_o, respectively, particularly for compound pressure vessels made of several thick walled cylinders, the distribution of stress is given by simply summing the previous two cases as follows.

$$\sigma_r = \frac{r_i^2 p_i - r_o^2 p_o}{r_o^2 - r_i^2} - \frac{r_i^2 r_o^2 (p_i - p_o)}{r^2(r_o^2 - r_i^2)} \qquad (14.19)$$

$$\sigma_\theta = \frac{r_i^2 p_i - r_o^2 p_o}{r_o^2 - r_i^2} + \frac{r_i^2 r_o^2 (p_i - p_o)}{r^2(r_o^2 - r_i^2)} \qquad (14.20)$$

$$\sigma_z = \frac{r_i^2 p_i - r_o^2 p_o}{r_o^2 - r_i^2} \qquad (14.21)$$

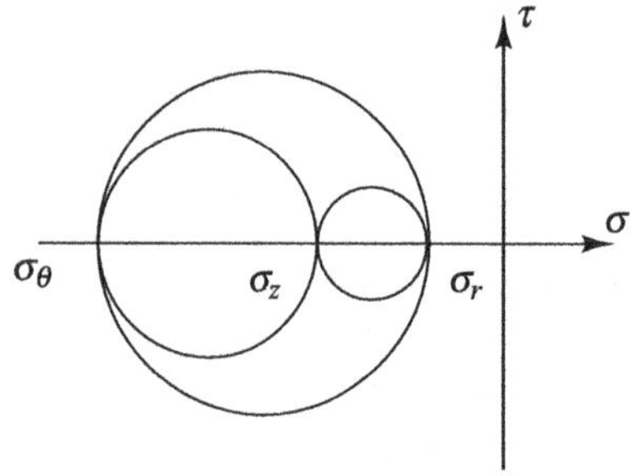

Fig. 14.5 *Mohr's circle for 3-dimensional stress in thick cylinders with external pressure*

The maximum numerical radial stress $\sigma_r = p_i$ at $r = r_i$ provided $p_i > p_o$. If $p_o > p_i$, the maximum value of $\sigma_r = p_o$ at $r = r_o$. The maximum tangential stress σ_θ occurs at $r = r_i$ if either internal or external pressure acts alone. However if both are acting then maximum σ_θ can occur at $r = r_o$. The shear stress in the walls can be obtained by Mohr's circle between σ_r, σ_θ and σ_z as shown in Figs. 14.3 and 5.

14.2 COMPOUND CYLINDERS

Figure 14.1 shows that the maximum stress occurs at the internal surface of the cylinder and the stresses are relatively low towards the outer surface. Thus utilization of material is poor in the outer layers of the cylinder. This effect is more pronounced as the ratio r_o/r_i is increased. The stresses can be distributed more uniformly by using a compound cylinder made by heating a larger cylinder and cooling a smaller cylinder. The smaller cylinder is then inserted into the larger cylinder. The outer cylinder will have tension and inner cylinder will have compression in hoop direction. When tensile hoop stress arises due to applied internal pressure p_i, the resulting stress distribution is relatively uniform and a better utilization of material is achieved. The axial stress σ_z induced during assembly is generally small and therefore neglected.

Consider a compound cylinder shown in Fig. 14.6. There is a small difference δ between the radii of the two cylinders at the interface before assembly as follows.

$$\delta = -u_i + u_o \tag{14.22}$$

where u_i and u_o are the displacements of the internal and external tubes at the interface. Note that u_i is negative.

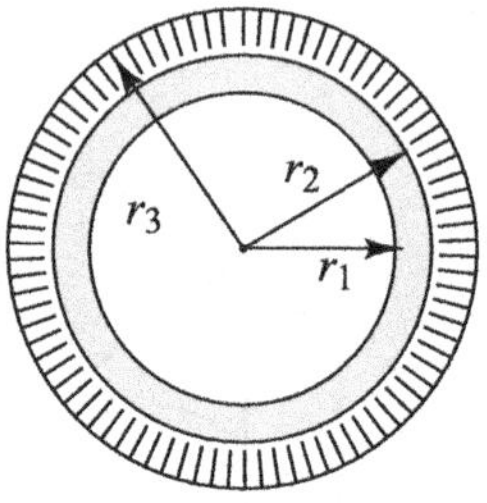

Fig. 14.6 *Compound cylinder*

Inner cylinder experiences external pressure $p_o = P$, where P is the pressure at the interface. The displacement at its external surface u_i is given by Eqs (14.3) and (14.7) as follows.

$$u_i = r_2\varepsilon_\theta = \frac{r_2}{E}(\sigma_\theta - \mu\sigma_r)$$

Substituting values of σ_θ and σ_r from Eqs (14.16) and (14.17) wherein $r_i = r_1$ and $r = r_o = r_2$,

$$u_i = \frac{r_2}{E}\left(-\frac{P}{r_2^2 - r_1^2}\left(r_2^2 + r_1^2\right) + \frac{\mu \times P}{r_2^2 - r_1^2}\left(r_2^2 - r_1^2\right)\right)$$

$$\text{or,}\quad u_i = \frac{r_2 \times P}{E}\left(\mu - \frac{r_2^2 + r_1^2}{r_2^2 - r_1^2}\right) \tag{14.23}$$

Outer cylinder will experience internal pressure $p_i = P$, and the displacement u_o of its internal surface is similarly given by substituting values of σ_θ and σ_r from Eqs (14.16) and (14.17) wherein $r = r_i = r_2$ and $r_o = r_3$, as follows.

$$u_o = r_2\varepsilon_\theta = \frac{r_2}{E}(\sigma_\theta - \mu\sigma_r)$$

or,

$$u_o = \frac{r_2}{E}\left(\frac{P}{r_3^2 - r_2^2}\left(r_2^2 + r_3^2\right) - \frac{\mu \times P}{r_3^2 - r_2^2}\left(r_2^2 - r_3^2\right)\right)$$

$$\text{or,}\quad u_o = \frac{r_2 \times P}{E}\left(\mu + \frac{r_3^2 + r_2^2}{r_3^2 - r_2^2}\right) \tag{14.24}$$

Substituting the values of u_i and u_o in Eq. (14.22),

$$\delta = \frac{2r_2 \times P\left(r_2^2 r_3^2 - r_1^2 r_2^2\right)}{E\left(r_3^2 - r_2^2\right)\left(r_2^2 - r_1^2\right)} \tag{14.25}$$

14.2.1 Autofrettage

Compound cylinders are used to achieve better utilization of the material in the outer layers of the cylinders. A similar effect is achieved by subjecting a thick cylinder to high internal pressure so that the inner layers are subjected to plastic deformation. As the internal pressure is released, the outer layers will be unable to regain their original shape due to permanent plastic deformation of the internal layers. This cylinder will have residual tensile stress in the outer layers and compressive stress in the inner layers. Subsequently higher inner pressure can be applied before yielding can occur again due to residual compressive stress in the inner layers. This strengthening of pressure vessels is known as autofrettage. For Tresca yield condition, σ_θ and σ_r are the principal stresses and σ_z is the intermediate stress. Thus for failure to occur,

$$\tau_{\max} = \frac{\sigma_y}{2} = \frac{\sigma_\theta - \sigma_r}{2}$$

or, $$\sigma_y = \sigma_\theta - \sigma_r \qquad (14.26)$$

If the internal pressure, $p_i = p_y$ at which yielding just starts at the inner surface of the cylinder, then from Eqs (14.13) and (14.14),

$$\sigma_y = \sigma_\theta - \sigma_r = \frac{r_i^2 p_y}{r_0^2 - r_i^2}\left(1 - 1 + \frac{r_o^2}{r_i^2} + \frac{r_o^2}{r_i^2}\right) = \frac{2r_o^2 p_y}{r_o^2 - r_i^2}$$

or, $$p_y = \frac{\sigma_y\left(r_o^2 - r_i^2\right)}{2r_o^2} \qquad (14.27)$$

Substituting Eq. (14.27) in the equation of equilibrium Eq. (14.1),

$$\frac{d\sigma_r}{dr} = \frac{\sigma_\theta - \sigma_r}{r} = \frac{\sigma_y}{r}$$

or, $$\sigma_r = \sigma_y \ln r + \text{constant} \qquad (14.28)$$

At elastic–plastic interface $r_i = r_c$, the radial stress to start yielding in the outer layers will be $\sigma_r = -p_y$, Eq. (14.27) gives,

or, $$\sigma_r = -p_y = \frac{-\sigma_y}{2}\left(1 - \frac{r_c^2}{r_o^2}\right)$$

Substituting in Eq. (14.28) with $r = r_c$,

$$\frac{-\sigma_y}{2}\left(1 - \frac{r_c^2}{r_o^2}\right) - \sigma_y \ln r_c = \text{constant} \qquad (14.29)$$

Substituting the value of the constant from Eq. (14.29) in Eq. (14.28) gives the stress in plastic zone, for $r_i < r < r_c$, as follows.

$$\sigma_r = \frac{\sigma_y}{2}\left(2\ln\frac{r}{r_c} - 1 + \frac{r_c^2}{r_o^2}\right) \qquad (14.30)$$

$$\sigma_\theta = \frac{\sigma_y}{2}\left(2\ln\frac{r}{r_c} + 1 + \frac{r_c^2}{r_o^2}\right) \qquad (14.31)$$

In elastic zone for $r_c < r < r_o$, the stresses are given by Eqs (14.13) and (14.14) by substituting $r_i = r_c$, and $p_y = \sigma_y(r_o^2 - r_i^2)/2r_o^2$ as follows.

$$\sigma_r = \frac{r_c^2\sigma_y}{2r_o^2}\left(1 - \frac{r_o^2}{r^2}\right) \qquad (14.32)$$

$$\sigma_\theta = \frac{r_c^2\sigma_y}{2r_o^2}\left(1 + \frac{r_o^2}{r^2}\right) \qquad (14.33)$$

14.3 ROTATING DISKS

Disks rotating at high speeds like rotors of turbines, flywheels etc. are examples of axis-symmetric problem where body force, F_r = centrifugal force unit volume, $\rho\omega^2 r$.

Thus the equilibrium equation Eq. (14.1) is written as,

$$\frac{d\sigma_r}{dr} + \frac{\sigma_r - \sigma_\theta}{r} + \rho\omega^2 r = 0 \qquad (14.34)$$

The axial stress σ_z may be assumed to be zero for a thin plate, then compatibility equation Eq. (14.5) and Hook's law Eqs (14.6) and (14.7) give,

$$\varepsilon_r - \varepsilon_\theta = r\frac{d\varepsilon_\theta}{dr}$$

or, $$\sigma_r - \sigma_\theta - \mu(\sigma_\theta - \sigma_r) = r\left(\frac{d\sigma_\theta}{dr} - \mu\frac{d\sigma_r}{dr}\right)$$

or, $$\frac{\sigma_r - \sigma_\theta}{r} = \frac{1}{1+\mu}\left(\frac{d\sigma_\theta}{dr} - \mu\frac{d\sigma_r}{dr}\right) \quad (14.35)$$

Substituting equilibrium equation Eq. (14.34) in Eq. (14.35),

$$\frac{d\sigma_\theta}{dr} + \frac{d\sigma_r}{dr} = -\rho\omega^2 r(1+\mu)$$

or, $$\frac{d}{dr}(\sigma_\theta + \sigma_r) = -\rho\omega^2 r(1+\mu)$$

Integrating,

$$\sigma_\theta + \sigma_r = 2A - \frac{(1+\mu)\rho\omega^2 r^2}{2} \quad (14.36)$$

where $2A$ = constant of integration. Adding Eq. (14.34) and (14.36),

$$2\sigma_r = 2A - r\frac{d\sigma_r}{dr} - \frac{(3+\mu)\rho\omega^2 r^2}{2}$$

or, $$2rA = 2r\sigma_r + r^2\frac{d\sigma_r}{dr} + \frac{(3+\mu)\rho\omega^2 r^3}{2}$$

or, $$2rA = \frac{d}{dr}\left(r^2\sigma_r\right) + \frac{(3+\mu)\rho\omega^2 r^3}{2}$$

Integrating,

$$Ar^2 = r^2\sigma_r + \frac{(3+\mu)\rho\omega^2 r^4}{8} + B \quad (14.37)$$

where B = constant of integration Eqs (14.36) and (14.37) give,

$$\sigma_r = A - \frac{B}{r^2} - \frac{(3+\mu)\rho\omega^2 r^2}{8} \quad (14.38)$$

and, $$\sigma_\theta = A + \frac{B}{r^2} - \frac{(1+3\mu)\rho\omega^2 r^2}{8} \quad (14.39)$$

Solid Disk

For a solid disk, $B = 0$ since otherwise the stresses σ_r and σ_θ are infinite at the centre. The radial stress $\sigma_r = 0$ at the outer radius r_2, Eq. (14.38) gives,

$$A = \frac{(3+\mu)\rho\omega^2 r_2^2}{8}$$

and, $$\sigma_r = \frac{(3+\mu)}{8}\rho\omega^2\left(r_2^2 - r^2\right) \quad (14.40)$$

$$\sigma_\theta = \frac{(3+\mu)}{8}\rho\omega^2\left(r_2^2 - \frac{1+3\mu}{3+\mu}r^2\right) \quad (14.41)$$

Annular Disk

For a disk of radius r_2 with a central hole of radius r_1, the constants of integration are given by the boundary conditions that $\sigma_r = 0$ at $r = r_1$ and $r = r_2$, which gives

$$A = \frac{(3+\mu)}{8}\rho\omega^2\,(r_1^2 + r_2^2) \text{ and}$$

$$B = \frac{(3+\mu)}{8}\rho\omega^2\, r_1^2 \times r_2^2$$

and,

$$\sigma_r = \frac{(3+\mu)}{8}\rho\omega^2\left(r_1^2 + r_2^2 - \frac{r_1^2 r_2^2}{r^2} - r^2\right) \quad (14.42)$$

$$\sigma_\theta = \frac{(3+\mu)}{8}\rho\omega^2\left(r_1^2 + r_2^2 + \frac{r_1^2 r_2^2}{r^2} - \frac{1+3\mu}{3+\mu}r^2\right) \quad (14.43)$$

It is seen that the stresses do not depend on the thickness of the plate. The radius at which σ_r is maximum, is found by differentiating Eq. (14.42) and equating the result to zero to get,

$$\frac{d\sigma_r}{dr} = \frac{(3+\mu)}{8}\rho\omega^2\left(\frac{2r_1^2 r_2^2}{r^3} - 2r\right) = 0$$

i.e. when, $r = \sqrt{r_1 r_2}$

At which,

$$\sigma_{r,\max} = \frac{(3+\mu)}{8}\rho\omega^2\left(r_2 - r_1\right)^2 \quad (14.44)$$

The radius for $\sigma_{\theta,\max}$ cannot be found by differentiating Eq. (14.43) as a real solution does no exist. $\sigma_{\theta,\max}$ occurs at the inside radius of the disc and its value is given by Eq. (14.42) as,

$$\sigma_{\theta,\max} = \frac{\rho\omega^2}{4}\left((1-\mu)r_1^2 + (3+\mu)r_2^2\right) \quad (14.45)$$

EXAMPLES

Example 14.1 A steel cylinder has external diameter 150 mm and internal diameter 100 mm. It is subjected to an internal pressure of 50 MPa. Plot the distribution of tangential and radial stresses across the wall of the pipe.

Solution Radial stress is given by Eq. (14.13) as,

$$\sigma_r = \frac{r_i^2 p_i \left(r^2 - r_o^2\right)}{r^2\left(r_o^2 - r_i^2\right)} = \frac{50^2 \times 50(r^2 - 75^2)}{r^2(75^2 - 50^2)}$$

The variation of radial stress across the wall of the cylinder is shown by lower curve (in compression) of Fig. 14.7. Maximum radial stress $\sigma_r = -50$ MPa (compressive) occurs at $r = r_i = 50$ mm,

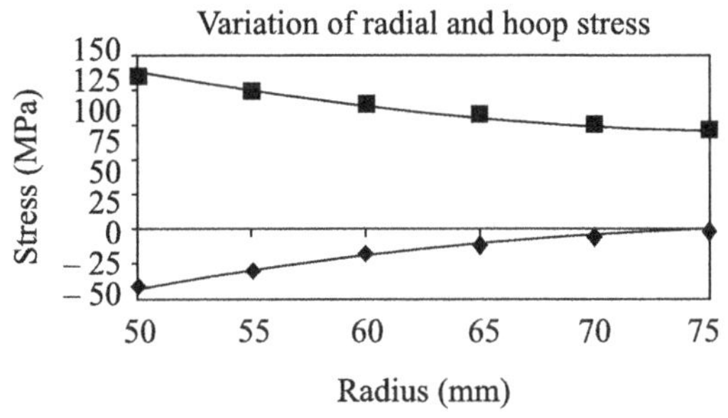

Fig. 14.7

Tangential stress is given by Eq. (14.14) as,

$$\sigma_\theta = \frac{r_i^2 p_i \left(r^2 + r_o^2\right)}{r^2\left(r_o^2 - r_i^2\right)} = \frac{50^2 \times 50(r^2 + 75^2)}{r^2(75^2 - 50^2)}$$

The variation of tangential stress across the wall of the cylinder is shown by the upper curve (in tension) of Fig. 14.7. Maximum tangential stress $\sigma_r = 130$ MPa (tensile) occurs at $r = r_i = 50$ mm,

Example 14.2 Redo Example 14.1 if the pressure applied is external instead of internal.

Solution Radial stress is given by Eq. (14.16) as,

$$\sigma_r = \frac{-r_o^2 p_o \left(r^2 - r_i^2\right)}{r^2\left(r_o^2 - r_i^2\right)} = \frac{-75^2 \times 50(r^2 - 50^2)}{r^2(75^2 - 50^2)}$$

The variation of radial stress across wall of the cylinder is shown by the upper curve (in compression) of Fig. 14.8. Maximum radial stress $\sigma_r = -50$ MPa (compressive) occurs at $r = r_o = 75$ mm,

Tangential stress is given by Eq. (14.17) as follows

$$\sigma_\theta = \frac{-r_o^2 p_o \left(r^2 + r_i^2\right)}{r^2\left(r_o^2 - r_i^2\right)} = \frac{-75^2 \times 50(r^2 + 50^2)}{r^2(75^2 - 50^2)}$$

The variation of tangential stress across the wall of the cylinder is shown by the lower curve (in compression) of Fig. 14.8. Maximum tangential stress $\sigma_r = -180$ MPa (compression) occurs at $r = r_i = 50$ mm,

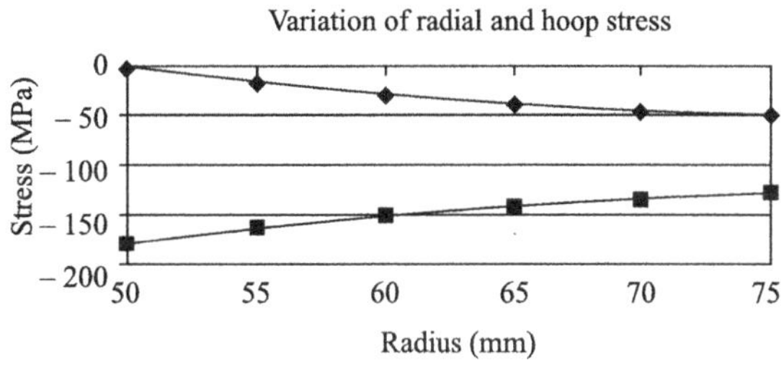

Fig. 14.8

Example 14.3 A thick steel cylinder is subjected to internal pressure of 50 MPa and external pressure of 10 MPa. The external and internal diameters of the cylinder are 200 mm and 100 mm, respectively. Find the maximum axial and shearing stresses in the cylinder. Also determine the change of external diameter. Take $E = 200$ GPa and $\mu = 0.3$.

Solution Maximum radial stress is given by Eq. (14.19) at $r = r_i = 50$ mm, if $p_i > p_o$.

$$\sigma_\theta = \frac{r_i^2 p_i - r_o^2 p_o}{r_o^2 - r_i^2} - \frac{r_i^2 r_o^2 (p_i - p_o)}{r^2 (r_o^2 - r_i^2)}$$

$$= \frac{50^2 \times 50 - 100^2 \times 10 - 100^2 \times 40}{100^2 - 50^2}$$

= –50 MPa (compressive)

$$\sigma_\theta = \frac{r_i^2 p_i - r_o^2 p_o}{r_o^2 - r_i^2} + \frac{r_i^2 r_o^2 (p_i - p_o)}{r^2 (r_o^2 - r_i^2)}$$

$$= \frac{50^2 \times 50 - 100^2 \times 10 + 100^2 \times 40}{100^2 - 50^2}$$

= 56.67 MPa (tensile)

Maximum shear stress, Eq. (2.6),

$$\tau_{\max} = \frac{\sigma_\theta - \sigma_r}{2} = \frac{56.67 + 50}{2} = 53.33 \text{ MPa}$$

The radial and tangential stresses at outer surface, $r = r_o = 100$ mm are as follow.

$$\sigma_r = \frac{50^2 \times 50 - 100^2 \times 10 - 50^2 \times 40}{100^2 - 50^2}$$

= – 10 MPa (Compressive)

$$\sigma_\theta = \frac{50^2 \times 50 - 100^2 \times 10 + 50^2 \times 40}{100^2 - 50^2}$$

= 16.66 MPa (Tensile)

The change in external radius by Eq. (14.3),

$$u = \varepsilon_\theta \times r_o = \frac{r_o}{E}[\sigma_\theta - \mu\sigma_r]$$

$$= \frac{100}{200000}(16.66 + 0.3 \times 10)$$

= 0.00983 mm

Therefore change in external diameter

= 2 × 0.00983 = 0.01966 mm

Example 14.4 A cast iron pipe 25 mm thick has external diameter 200 mm. It bursts under an internal fluid pressure of 50 MPa. Find the safe pressure using factor of safety of 4 if the wall thickness is increased to 35 mm without changing the internal diameter.

Solution Maximum raidal stress (at r = 75 mm), σ_r = –50 MPa

Maximum tangential stress (at r = 75 mm) stress,

$$\sigma_\theta = \frac{p_i(r_o^2 + r_i^2)}{r_o^2 - r_i^2} = \frac{50(100^2 + 75^2)}{100^2 - 75^2} = 178.57 \text{ MPa}$$

Maximum shear stress,

$$\tau_{\max} = \frac{\sigma_\theta - \sigma_r}{2} = \frac{178.57 + 50}{2} = 114.28 \text{ MPa}$$

When the wall thickness is increased to 35 mm,

$$\text{Allowable direct stress} = \frac{178.57}{4} = 44.64 \text{ MPa}$$

$$\text{Allowable shear stress} = \frac{114.28}{4} = 28.57 \text{ MPa}$$

Substituting these allowable values in their respective equations at r = 75 mm,

$$\sigma_\theta = 44.64 = \frac{p_i(110^2 + 75^2)}{110^2 - 75^2} \quad \text{or,} \quad p_i = 16.3 \text{ MPa}$$

$$\tau_{all} = 28.57 = \frac{\sigma_\theta - \sigma_r}{2} = \frac{1}{2}\left(p_i \frac{110^2 + 75^2}{110^2 - 75^2} + p_i\right)$$

or, p_i = 15.29 MPa

Thus internal pressure, $p_i \le 15.29$ MPa

Example 14.5 A tube has internal diameter 15 mm and external diameter 20 mm. It is 500 mm long and is closed at one end. Water is pumped into the open end till the pressure in the tube is 20 MPa. Calculate the excess valume of water pumped in the tube beyond what is required to fill it at gauge pressure zero. Take modulus of elasticity = 200 GPa, Poisson's ratio = 0.3 for pipe material and bulk modulus of elasticity for water = 2.3 GPa

Solution The change of internal radius by Eq. (14.3),

$$u = \varepsilon_\theta \times r_i = \frac{r_i}{E}[\sigma_\theta - \mu(\sigma_r + \sigma_z)]$$

where,

$$\sigma_r = \frac{r_i^2 p_i}{r_o^2 - r_i^2}\left(1 - \frac{r_o^2}{r_i^2}\right) = -p_i = -20 \text{ MPa}$$

$$\sigma_r = \frac{r_i^2 p_i}{r_o^2 - r_i^2}\left(1 + \frac{r_o^2}{r_i^2}\right) = \frac{20 \times (10^2 + 7.5^2)}{10^2 - 7.5^2}$$

$= 71.428$ MPa

$$\sigma_z = \frac{r_i^2 p_i}{r_o^2 - r_i^2} = \frac{7.5^2 \times 20}{10^2 - 7.5^2} = 25.71 \text{ MPa}$$

or,

$$u = \frac{7.5}{200000}(71.42 - 0.3(-20 + 25.71))$$

$= 0.00261$ mm

Longitudinal strain Eq. (14.8),

$$\varepsilon_z = \frac{1}{E}[\sigma_z - \mu(\sigma_r + \sigma_\theta)]$$

$$= \frac{1}{200000}(25.71 - 0.3(71.43 - 20))$$

$= 5.14 \times 10^{-5}$

Change in length of the tube,

$\Delta L = \varepsilon_z \times L = 5.14 \times 10^{-5} \times 500 = 0.0257$ mm

Initial volume of the tube $= \pi r^2 L = \pi \times 7.5^2 \times 500$

$= 88357.3$ mm^3

Final volume of the tube $= \pi(r + u)^2(L + \Delta L)$

$= \pi \times (7.5 + 0.00261)^2 \times (500 + 0.0257)$

$= 88423.46$ mm^3

Change in volume of the tube $= 88423.46 - 88357.3$

$= 66.16$ mm^3

Change in volume of water under pressure,

$$\Delta v = \frac{pv}{k} = \frac{20 \times 88357.3}{2300} = 768.32 \text{ mm}^3$$

Total water to be pumped $= 768.32 + 66.16$

$= 834.48$ mm^3

Example 14.6 A steel cylinder has external and internal diameters 100 mm and 60 mm, respectively. It has an internal bronze lining of internal diameter 50 mm. Calculate the greatest shearing stress produced in the cylinder and the lining when the temperature is increased by 100°C. Coefficient of expansion is 11.6×10^{-6}/°C for steel and 18.9×10^{-6}/°C for bronze. Take modulus of elasticity equal to 200 GPa for steel and 100 GPa for bronze and Poisson's ratio equal to 0.3 for both steel and bronze.

Solution Increase in external radius of bronze lining

$= 18.9 \times 10^{-6} \times 30 \times 100$ mm

Increase in internal radius of steel cylinder

$= 11.6 \times 10^{-6} \times 30 \times 100$ mm

Interference in radius of the two cylinders,

$\delta = (18.9 - 11.6) \times 10^{-6} \times 30 \times 100$

$= 0.0219$ mm

Displacement u_i of the external surface inner cylinder is given by Eq. (14.23),

$$u_i = \frac{r_2 \times P}{E}\left(\mu - \frac{r_2^2 + r_1^2}{r_2^2 - r_1^2}\right)$$

$$= \frac{30 \times P}{100 \times 10^3}\left(0.3 - \frac{30^2 + 25^2}{30^3 - 25^2}\right)$$

Displacement, u_o of the internal surface outer cylinder is given by Eq. (14.24),

$$u_o = \frac{r_2 \times P}{E}\left(\mu + \frac{r_3^2 + r_2^2}{r_3^2 - r_2^2}\right)$$

$$= \frac{30 \times P}{200 \times 10^3}\left(0.3 + \frac{50^2 + 30^2}{50^2 - 30^2}\right)$$

where P = interference pressure. Substituting the above values in Eq. (14.22),

$$\delta = -u_i + u_o$$

$$\text{or,}\quad 0.0219 = \frac{30 \times P}{100 \times 10^3}\left(0.3 - \frac{30^2 + 25^2}{30^3 - 25^2}\right)$$

$$+ \frac{30 \times P}{200 \times 10^3}\left(0.3 + \frac{50^2 + 30^2}{50^3 - 30^2}\right)$$

or, $P = 11.3$ MPa

For bronze lining, the maximum stress occurs at $r = r_i = 25$ mm by Eqs (14.16) and (14.17), where, $\sigma_r = \dfrac{-r_o^2 P}{r_o^2 - r_i^2}\left(1 - \dfrac{r_i^2}{r_i^2}\right) = 0$

$$\sigma_\theta = \frac{-r_o^2 P}{r_o^2 - r_i^2}\left(1 + \frac{r_i^2}{r_i^2}\right)$$

$$= \frac{-11.3 \times 30^2 \times 2}{30^2 - 25^2} = -73.96\text{ MPa}$$

Maximum shear stress in bronze,

$$\tau_{\max} = \frac{\sigma_\theta - \sigma_r}{2} = \frac{-73.96}{2} = 36.98\text{ MPa}$$

For steel cylinder, the maximum stress occurs at $r = r_i = 30$ mm by Eqs (14.13) and (14.14),

where, $\sigma_r = \dfrac{r_i^2 P}{r_o^2 - r_i^2}\left(1 - \dfrac{r_o^2}{r_i^2}\right)$

$$= -P = -11.3\text{ MPa}$$

$$\sigma_\theta = \frac{r_i^2 P}{r_o^2 - r_i^2}\left(1 + \frac{r_o^2}{r_i^2}\right) = \frac{11.3 \times (30^2 + 25^2)}{30^2 - 25^2}$$

$$= 24.02\text{ MPa}$$

Maximum shear stress in steel,

$$\tau_{\max} = \frac{\sigma_\theta - \sigma_r}{2} = \frac{24.02 + 11.3}{2}$$

$$= 17.66\text{ MPa}$$

Example 14.7 A steel cylinder has external diameters 250 mm and it is 25 mm thick. Another cylinder of internal diameter 249.8 mm and external diameter 300 mm is heated and shrunk fitted on first cylinder. Determine the stresses in the coaxial cylinders due to internal fluid pressure of 100 MPa. Take modulus of elasticity equal to 200 GPa for steel.

Solution Interference in radius of the two cylinders,

$$\delta = \frac{250 - 249.8}{2} = 0.1\text{ mm}$$

Interference pressure using Eq. (14.25),

$$P = \frac{\delta E\left(r_3^2 - r_2^2\right)\left(r_2^2 - t_1^2\right)}{2r_2\left(r_2^2 r_3^2 - r_1^2 r_2^2\right)}$$

$$= \frac{0.1 \times 200 \times 10^3\left(150^2 - 125^2\right)\left(125^2 - 100^2\right)}{2 \times 125\left(150^2 \times 125^2 - 125^2 \times 100^2\right)}$$

$$= 15.84\text{ MPa}$$

This interference pressure will act as external pressure on the inner cylinder and will produce tangential stress given by Eq. (14.17),

at $r = 100$ mm, $\sigma_\theta = \dfrac{-r_o^2 p}{r_o^2 - r_i^2}\left(1 + \dfrac{r_i^2}{r_i^2}\right)$

$$= \frac{-15.84 \times 125^2 \times 2}{125^2 - 100^2}$$

$$= -88\text{ MPa}$$

at $r = 125$ mm, $\sigma_\theta = \dfrac{-r_o^2 p}{r_o^2 - r_i^2}\left(1 + \dfrac{r_i^2}{r_o^2}\right)$

$$= \frac{-15.84 \times (125^2 + 100^2)}{125^2 - 100^2}$$

$$= -72.16\text{ MPa}$$

The interference pressure will act as internal pressure on the outer cylinder and will produce tangential stress given by Eq. (14.14),

at $r = 125$ mm, $\sigma_\theta = \dfrac{r_i^2 p}{r_o^2 - r_i^2}\left(1 + \dfrac{r_o^2}{r_i^2}\right)$

$$= \frac{15.84 \times (125^2 + 150^2)}{150^2 - 125^2}$$

$$= 87.84\text{ MPa}$$

at $r = 150$ mm, $\sigma_\theta = \dfrac{r_i^2 p}{r_o^2 - r_i^2}\left(1 + \dfrac{r_o^2}{r_o^2}\right)$

$$= \frac{15.84 \times 125^2 \times 2}{150^2 - 125^2}$$

$$= 72\text{ MPa}$$

The two cylinders are considered as one thick walled cylinder when internal pressure of 100 MPa is applied, the tangential stresses are given by Eq. (14.14),

$$\text{at } r = 100 \text{ mm}, \quad \sigma_\theta = \frac{r_i^2 p}{r_o^2 - r_i^2}\left(1 + \frac{r_o^2}{r^2}\right)$$

$$= \frac{100 \times (150^2 + 100^2)}{150^2 - 100^2}$$

$$= 260 \text{ MPa}$$

$$\text{at } r = 125 \text{ mm}, \quad \sigma_\theta = \frac{100^2 \times 100 \times (125^2 + 150^2)}{125^2 \times (150^2 - 100^2)}$$

$$= 195.2 \text{ MPa}$$

$$\text{at } r = 150 \text{ mm}, \quad \sigma_\theta = \frac{100 \times 100^2 \times 2}{150^2 - 100^2}$$

$$= 160 \text{ MPa}$$

Total tangential stresses in the inner cylinder,

at $r = 100$ mm, $\sigma_\theta = -88 + 260 = 172$ MPa

at $r = 125$ mm, $\sigma_\theta = -72.16 + 195.2$

$= 123.04$ MPa

Total tangential stresses in the outer cylinder,

at $r = 125$ mm, $\sigma_\theta = 87.84 + 195.2$

$= 283.04$ MPa

at $r = 150$ mm, $\sigma_\theta = 72 + 160 = 232$ MPa

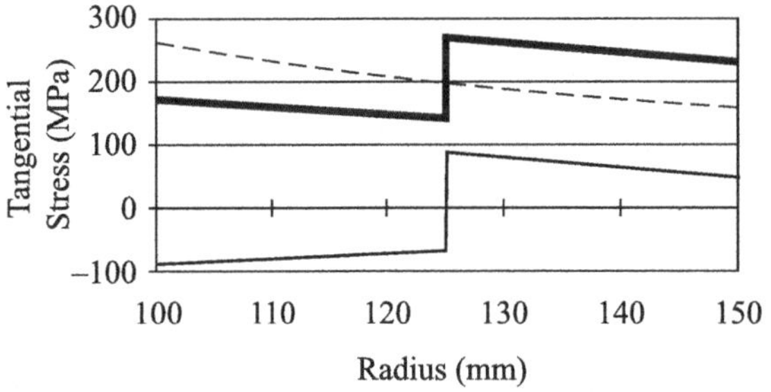

Fig. 14.9

The variation of tangential stress in 1st stage is shown by thin lower curve and 2nd stage is shown by upper dotted curve while their sum is shown by thick middle curve in Fig. 14.9. It should be noted that inner cylinder is under compression in first stage so that its maximum stress is reduced from 260 MPa to 172 MPa in final stage.

Example 14.8 A high yield steel pipe has external diameters 300 mm and it is 40 mm thick. Find the internal pressure to cause yielding of the pipe material at the bore. Determine the internal pressure which causes yielding of the pipe material up to half its thickness and plot the stresses. Also plot the residual stresses when the internal pressure is removed. Take the yield stress of the pipe material to be 500 MPa.

Solution Equation (14.27) gives the yield pressure as follows.

$$p_y = \frac{\sigma_y (r_o^2 - r_i^2)}{2 r_o^2} = \frac{500 \times (300^2 - 260^2)}{2 \times 300^2}$$

$$= 62.22 \text{ MPa}$$

Eqs (14.30) and (14.31) give the stresses in plastic zone, for $260 < r < 280$, as follows.

$$\sigma_r = \frac{\sigma_y}{2}\left(2 \ln \frac{r}{r_c} - 1 + \frac{r_c^2}{r_o^2}\right)$$

$$= 250 \times (2 \ln (r/280) - 1 + (280/300)^2)$$

$$\sigma_\theta = \frac{\sigma_y}{2}\left(2 \ln \frac{r}{r_c} + 1 + \frac{r_c^2}{r_o^2}\right)$$

$$= 250 \times (2 \ln (r/280) + 1 + (280/300)^2)$$

The stresses in elastic zone, for $280 < r < 300$, are given by Eqs (14.32) and (14.33) as follows.

$$\sigma_r = \frac{r_c^2 \sigma_y}{2 r_o^2}\left(1 - \frac{r_o^2}{r^2}\right) = \frac{280^2 \times 500}{2 \times 300^2}\left(1 - \frac{300^2}{r^2}\right)$$

$$\sigma_\theta = \frac{r_c^2 \sigma_y}{2 r_o^2}\left(1 + \frac{r_o^2}{r^2}\right) = \frac{280^2 \times 500}{2 \times 300^2}\left(1 + \frac{300^2}{r^2}\right)$$

The values of the radial and tangential stresses given by these equations are given in Tables 14.1 and 14.2.

Table 14.1

Radial stress in elastic-plastic range			
Radius (mm)	*Elastic-plastic range*	*Fictitious elastic range*	*Difference*
260	–69.28	–69.28	0.00
265	–59.75	–58.87	–0.88
270	–50.41	–49.04	–1.37
275	–41.23	–39.74	–1.49
280	–32.22	–30.93	–1.29
285	–23.53	–22.59	–0.94
290	–15.28	–14.67	–0.61
295	–7.44	–7.15	–0.30
300	0.00	0.00	0.00

Table 14.2

Tangential stress in elastic-plastic range			
Radius (mm)	*Elastic-plastic range*	*Fictitious elastic range*	*Difference*
260	430.72	487.40	–56.68
265	440.24	477.00	–36.75
270	449.59	467.17	–17.57
275	458.76	457.86	0.89
280	467.77	449.06	18.71
285	459.08	440.71	18.36
290	450.83	432.79	18.03
295	443.00	425.27	17.72
300	435.55	418.13	17.42

The internal pressure p_y to cause yielding up to half the thickness is equal to radial stress $\sigma_r = 69.28$ MPa at the bore, $r = 260$ mm. When the internal pressure is removed, then the residual stresses arise due to the plastic deformation of the inner half shell. The residual stresses are given as the difference, between the stresses in elastic-plastic range and the stress due to a fictitious internal pressure of 69.28 MPa assuming no plastic deformation using usual Eqs. (14.13) and (14.14) as follows,

$$\sigma_r = \frac{r_i^2 p_i \left(r^2 - r_o^2\right)}{r^2\left(r_o^2 - r_i^2\right)} = \frac{260^2 \times 69.28\left(r^2 - 300^2\right)}{r^2\left(300^2 - 260^2\right)}$$

$$\sigma_\theta = \frac{r_i^2 p_i \left(r^2 + r_o^2\right)}{r^2\left(r_o^2 - r_i^2\right)} = \frac{260^2 \times 69.28\left(r^2 + 300^2\right)}{r^2\left(300^2 - 260^2\right)}$$

The variation of the radial and tangential stresses as per these equations is shown in Figs. 14.10 (a) and (c). Figures 14.10 (b) and (d) show the residual stresses when the pressure is removed as per last column of the tables as above. The difference of *tangential stress* and *radial stress* $\sigma_r - \sigma_\theta = \sigma_y = 500$ MPa in plastic zone.

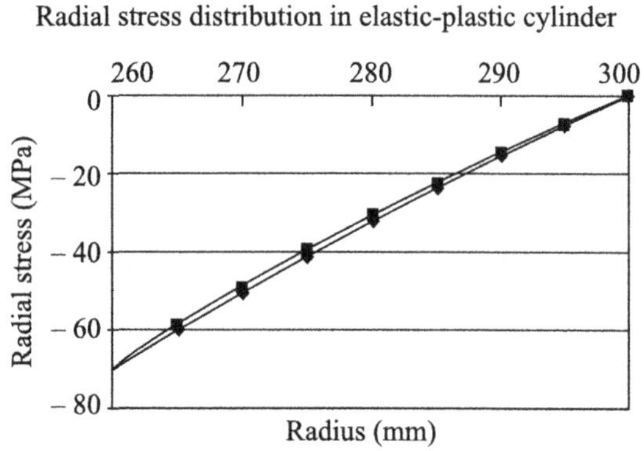

Fig. 14.10(a)

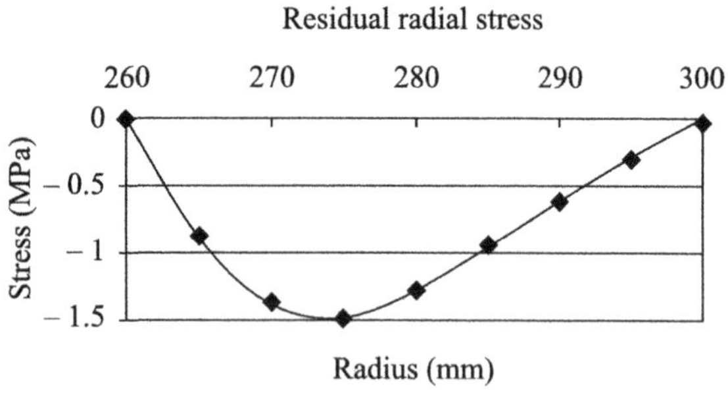

Fig. 14.10(b)

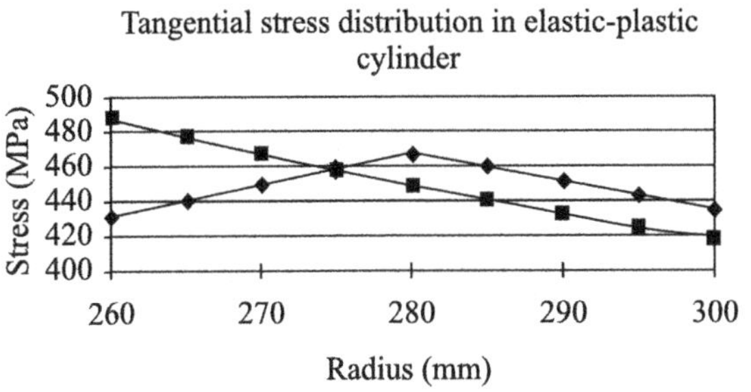

Fig. 14.10(c)

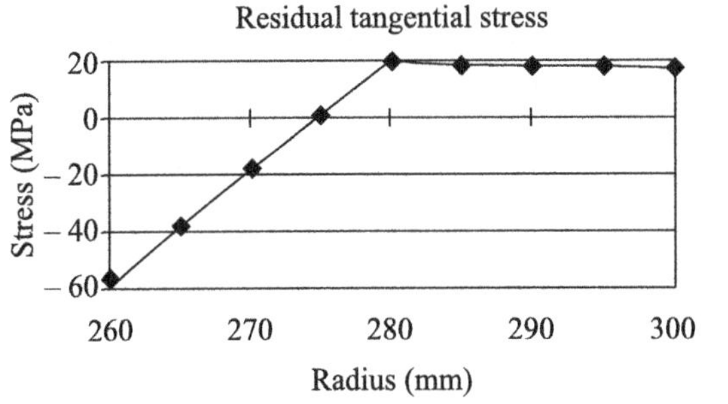

Fig. 14.10(d)

Example 14.9 A steel pipe of internal diameter 200 mm and external diameter 300 mm, is subjected to an internal pressure twice than the external pressure. Calculate the internal pressure at which yielding will just start as per various theories of failure if the yield stress of steel is 250 MPa and Poisson's ratio is 0.3.

Solution Equations 14.19 and 20 give the maximum stresses at the bore, $r = r_i = 100$ mm as follow.

$$\sigma_r = \frac{100^2\ p_i - 150^2\ p_i/2 - 150^2\ p_i/2}{150^2 - 100^2} = -p_i$$

$$\sigma_\theta = \frac{100^2\ p_i - 150^2\ p_i/2 + 150^2\ p_i/2}{150^2 - 100^2} = 0.8\ p_i$$

Internal pressure at which yielding will start as per various theories of failure is as follow.

(i) *Maximum principal stress theory*:

$$p_i = -\sigma_\theta = -\sigma_y = 250 \text{ MPa}$$

(Assuming Yield stress to be equal in tension and compression)

(ii) *Maximum shear stress theory*:

$$\frac{\sigma_y}{2} = \tau_y = \frac{\sigma_\theta - \sigma_r}{2} = \frac{(0.8\ p_i + p_i)}{2}$$

$$= 0.9\ p_i$$

$$\text{or,}\quad p_i = \frac{250}{2 \times 0.9} = 138.9 \text{ MPa}$$

(iii) *Maximum Principal strain theory*:

$$\sigma_\theta - \mu\sigma_r = \sigma_y$$

$$\text{or,}\quad 0.8\ p_i + 0.3\ p_i = 250 \text{ MPa}$$

$$\text{or,}\quad p_i = 227.27 \text{ MPa}$$

(iv) *Energy of Distortion theory* and *Octahedral shear stress theory*:

$$(\sigma_\theta^2 + \sigma_r^2 - \sigma_\theta\sigma_r)^{1/2} = \sigma_y$$

$$\text{or,}\ (0.64 + 1 - 0.8)^{1/2}\ p_i = 250 \text{ MPa}$$

$$p_i = \frac{250}{0.9165} = 272.72 \text{ MPa}$$

The internal pressure should be restricted to 138.9 MPa divided by a factor of safety as per *Maximum shear stress theory*.

Example 14.10 A circular disk of uniform thickness has diameter 1 m rotates at an angular speed of 3000 rpm. Plot the radial and tangential stresses if the density and Poisson's ratio are 7850 kg/m^3 and 0.3, respectively.

Solution Angular speed = 3000 × 2π/60
= 100 π rad/s

Radial and tangential stresses are given by Eqs (14.40) and (14.41) and their plots are shown in Figs. 14.11 (a) and (b) as follows.

$$\sigma_r = \frac{(3+\mu)}{8}\rho\omega^2\left(r_2^2 - r^2\right)$$

$$= \frac{(3+0.3)}{8\times10^6}\times 7850 \times (100\pi)^2 (0.5^2 - r^2)$$

$$= 319.6 \times (0.25 - r^2) \text{ MPa}$$

$$\sigma_\theta = \frac{(3+\mu)}{8}\rho\omega^2\left(r_2^2 - \frac{1+3\mu}{3+\mu}r^2\right)$$

$$= \frac{3.3\times 7850\times(100\pi)^2}{8\times10^6}\left(0.5^2 - \frac{1.9}{3.3}\times r^2\right)$$

$$= 319.6 \times (0.25 - 0.57r^2) \text{ MPa}$$

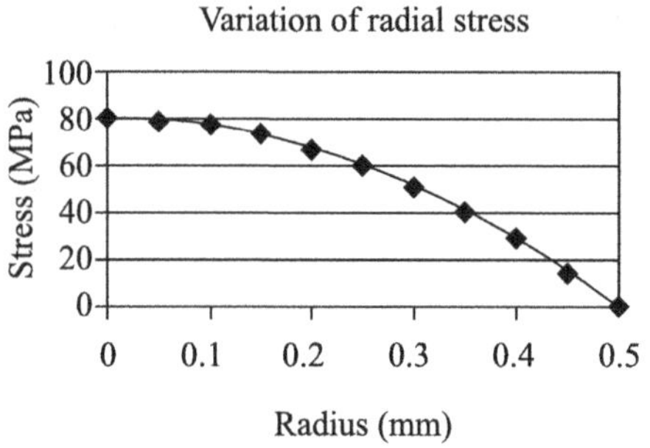

Fig. 14.11(a)

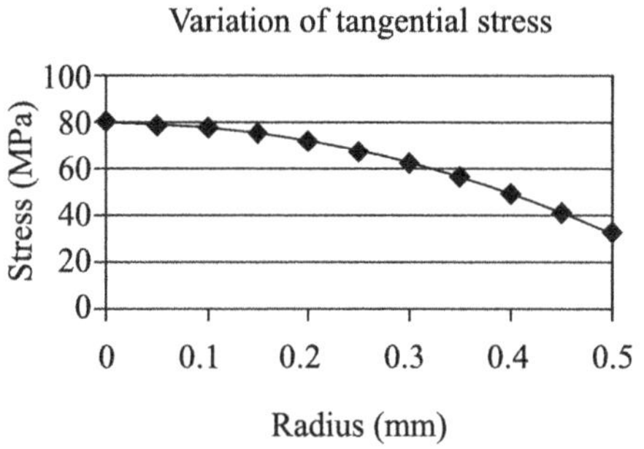

Fig. 14.11(b)

Example 14.11 Re-do Example 14.10 if the disk has an inner hole of diameter 0.25 m. Also calculate maximum stresses.

Solution Radial and tangential stresses are given by Eqs (14.42) and (14.43) and their plot are shown in Figs. 14.12(a) and (b) as follows.

$$\sigma_r = \frac{(3+\mu)}{8}\rho\omega^2\left(r_1^2 + r_2^2 - \frac{r_1^2 r_2^2}{r^2} - r^2\right)$$

$$= \frac{3.3 \times 7850 \times (100\pi)^2}{8 \times 10^6}$$

$$\left(0.125^2 + 0.5^2 - r^2 - \frac{(0.125 \times 0.5)^2}{r^2}\right)$$

$$= 319.6 \times \left(0.266 - r^2 - \frac{0.0039}{r^2}\right)$$

$$\sigma_\theta = \frac{(3+\mu)}{8}\rho\omega^2\left(r_1^2 + r_2^2 + \frac{r_1^2 r_2^2}{r^2} - \frac{1+3\mu}{3+\mu}r^2\right)$$

$$= 319.6 \times \left(0.125^2 + 0.5^2 + \frac{(0.125 \times 0.5)^2}{r^2} - \frac{1.9r^2}{3.3}\right)$$

$$= 319.6\left(0.266 + \frac{0.0039}{r^2} - 0.576r^2\right)$$

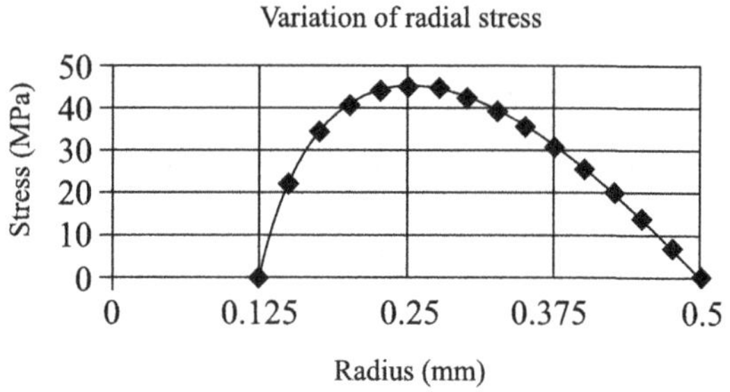

Fig. 14.12(a)

Maximum radial stress is given by Eq. (14.44), as follows.

$$\sigma_{r,\,\max} = \frac{(3+\mu)}{8}\rho\omega^2\left(r_2 - r_1\right)^2$$

$$= 319.6 \times (0.5 - 0.125)^2 = 44.9 \text{ MPa}$$

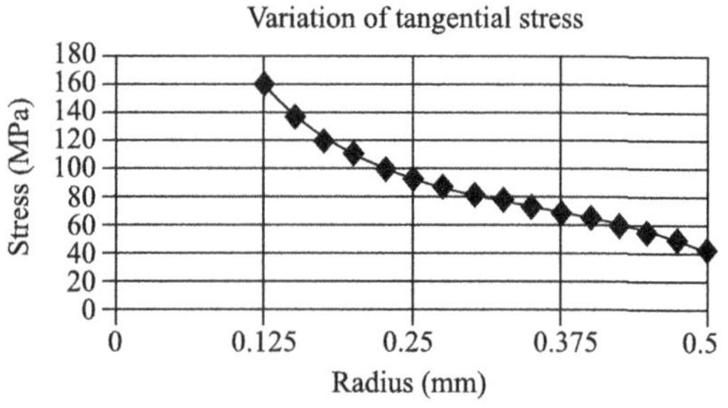

Fig. 14.12(b)

Maximum tangential stress is given by Eq. (14.45) as follows.

$$\sigma_{\theta,\,\max} = \frac{\rho\omega^2}{4}\left((1-\mu)r_1^2 + (3+\mu)r_2^2\right)$$

$$= \frac{7850 \times (100\pi)^2}{4 \times 10^6}$$

$$(0.7 \times 0.125^2 + 3.3 \times 0.5^2) = 161.9 \text{ MPa}$$

EXERCISES

PROBLEMS

14.1 A steel pipe 25 mm thick has external diameter 100 mm. It is subjected to an internal pressure of 15 MPa and external pressure of 5 MPa. Plot a curve showing distribution of tangential and radial stresses across the wall of the pipe.

14.2 A steel cylinder has external diameter 400 mm and internal diameter 300 mm. If the permissible working stress for steel is 150 MPa, determine (a) the maximum permissible internal pressure and (b) the maximum permissible external pressure. Also determine the change in internal diameter in both cases if the modulus of elasticity is 200 GPa and Poisson's ratio is 0.3.

14.3 A cylinder has external diameter 250 mm and internal diameter 150 mm. It is subjected to an internal pressure of 100 MN/m^2 and an external pressure of 10 MN/m^2. Determine the greatest hoop stress and plot the variation of hoop stress across the wall of the cylinder.

14.4 A steel cylinder of external diameter 400 mm just fits over a copper cylinder of external diameter 300 mm and internal diameter 200 mm. Determine the stresses due to internal pressure of 200 MPa if the modulus of elasticity and Poisson's ratio for steel are 200 GPa and 0.3, respectively and for copper are 120 GPa and 0.33, respectively.

OBJECTIVE QUESTIONS

14.5 A thick walled cylinder has inner radius r_i and outer radius r_o. It is subjected to internal pressure p_i and external pressure p_o, then the maximum shear stress at a radius r is given by,

(a) $\dfrac{r^2(p_o - p_i)}{r_o^2 - r_i^2}$ (b) $\dfrac{r_o^2 r_i^2(p_o - p_i)}{r^2(r_o^2 - r_i^2)}$

(c) $\dfrac{p_o r_o^2 - p_i r_i^2)}{r_o^2 - r_i^2}$ (d) $\dfrac{r^2(p_o r_o^2 - p_i r_i^2)}{r_o^2 - r_i^2}$

14.6 A compound cylinder is made of two cylinders by shrunk fitting. The ratio of shrinkage allowance and radius at the junction is equal to the ratio of,

(a) sum of radial stress in the two cylinder at the junction and modulus of elasticity,
(b) sum of tangential stress in the two cylinder at the junction and modulus of elasticity,
(c) difference of tangential stress in the two cylinder at the junction and modulus of elasticity,
(d) sum of tangential stress in the two cylinder at the junction and Poisson's ratio.

15

Plastic Theory of Structures

CONCEPT REVIEW

15.1 STRESS–STRAIN RELATION OF MILD STEEL

Stress–strain relation of mild steel from a typical tensile test was shown in Fig. 1.2.

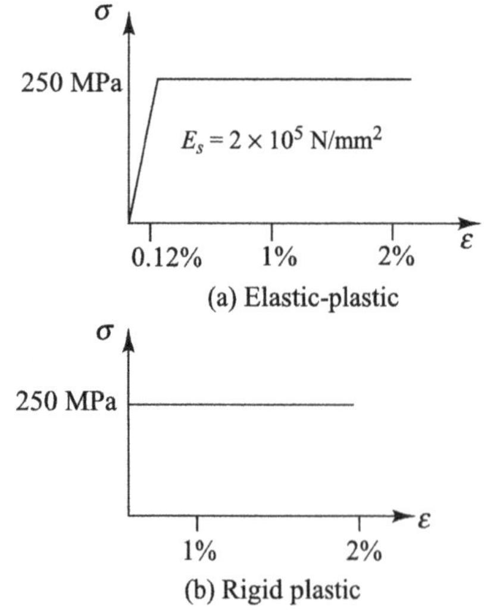

Fig. 15.1 *Idealized Stress-strain Relation for Mild-steel*

A simplified stress–strain curve is used in plastic design theory of structures which neglects the strain hardening of members as shown in Fig. 15.1. It is on the safer side to assume the material to be perfectly elastic-plastic or rigid-plastic ignoring strain hardening.

15.2 ELASTIC DESIGN VERSUS ULTIMATE LOAD DESIGN

The basis of elastic theory of design is that the stresses in the structures under design load (working or service load) are less than the allowable working stresses. The theory of elasticity can be used for rational analysis of structures under any system of load within the elastic zone. However, at ultimate load the stresses at several points will exceed the elastic limit. Therefore the theory of elasticity will not be applicable for structural analysis at the time of failure. The main disadvantage of the elastic method is that it does not provide an uniform overload capacity for all parts of the structures.

The ultimate load design or plastic design is nowadays preferred over the elastic theory since it offers more simplicity and economy in the design of structures.

15.3 PLASTIC BENDING OF BEAMS

Let an increasing bending moment be applied on a beam section as shown in Fig. 15.2. The stress distribution will change as the bending moment increases as follows:

(i) **Elastic stage** ($M < M_y$): For low values of bending moment, the maximum extreme fibre stress will be within the elastic range.

(ii) **Elastic-plastic stage** ($M_y < M < M_p$): With further increase in bending moment, yielding of extreme fibres will start. The neutral axis no longer passes through the centroid of the section. Location of neutral axis in any case is such that the bending tensile and compressive forces on the section are equal.

(iii) **Plastic stage** ($M = M_p$): With further increase in bending moment, a stage will come when the cross-section will yield and act as a plastic hinge. The corresponding bending moment is called the **plastic moment of resistance**, or simply **plastic moment** M_p. Neglecting strain-hardening, the beam cannot resist bending moment more than M_p. The neutral axis of fully plastic section passes through the equal area axis.

15.3.1 Assumptions

In plastic analysis of beams, assumptions are made as follows:

(1) Plane section normal to the axis of bending remains plane after bending.

(2) The material obeys ideal stress–strain relationship as shown in Fig. 15.2 and the increase in strength due to strain hardening is neglected.

(3) There is no axial load on the beam.

(4) Shear strains are neglected.

(5) The deformations at the limit state of collapse are small and the geometry of the undeformed structures is used in formulating the equations of equilibrium.

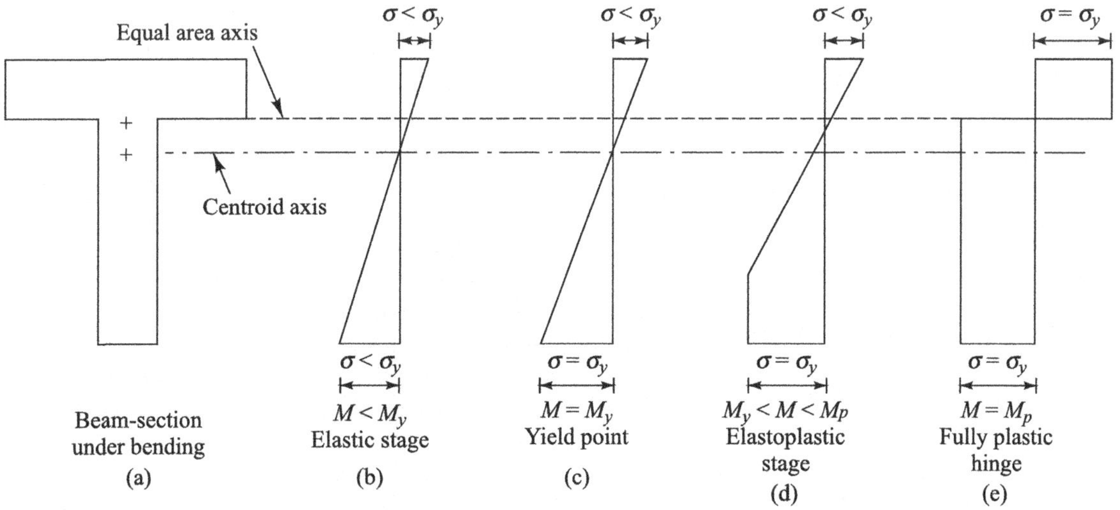

Fig. 15.2 *Plastic Bending of Beam*

(6) Stress–strain relation for tension is same as that for compression.

15.4 SHAPE FACTOR

Consider a beam section subjected to fully plastic moment M_p as shown in Fig. 15.3. Let A_1 and A_2 be the areas of cross-section above and below the neutral axis while y_1 and y_2 are the distances of their respective centres of gravity from the neutral axis.

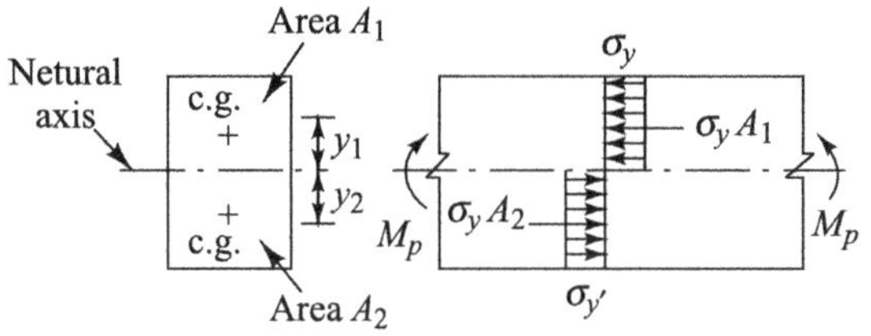

Fig. 15.3 *Fully Plastic Moment of Section*

Compressive force on the cross-section = $A_1\sigma_y$

Tensile force on the cross-section = $A_2\sigma_y$

For equilibrium, Compressive force = Tensile force

Therefore, $A_1\sigma_y = A_2\sigma_y$

or, $$A_1 = A_2 \qquad (15.1)$$

It follows from Eq. (15.1) that neutral axis passes through the middle of the section.

If A = total cross-sectional area then,

$$A_1 = A_2 = A/2 \qquad (15.2)$$

The couple formed tensile and compressive forces to external moment M_p,

$$M_p = A_1\sigma_y y_1 + A_2\sigma_y y_2 = \sigma_y(y_1 + y_2)A/2 = \sigma_y Z_p$$

where, Z_p = Plastic sectional modules $= (y_1 + y_2)A/2$

The ratio of plastic moment and yield moment is called the shape factor denoted by S.

$$\text{Shape factor, } S = \frac{M_p}{M_y} = \frac{\sigma_y Z_p}{\sigma_y Z} = \frac{Z_p}{Z} \qquad (15.3)$$

The shape factor relates the plastic and elastic moment capacity of a section and it depends only on the shape of the section.

15.5 LOAD FACTOR

Load factor it is the ratio of ultimate collapse load P_u and working load P_w.

Load factor,

$$\lambda = \frac{P_u}{P_w} = \frac{M_p}{M_w} = \left(\frac{Z_p}{Z}\right) \times \left(\frac{\sigma_y}{\sigma_{bc}}\right) \qquad (15.4)$$

= Shape factor × Factor of safety

Load factor is usually constant. As the shape factor is dependent on shape of section only, therefore factor of safety is not constant for all sections.

15.6 PLASTIC ANALYSIS

In this chapter M_p and P_u are referred as the ultimate capacity of structure to resist moment and load, respectively, whereas M and P in general denotes externally applied loads and moments. The aim of plastic analysis is to find collapse load P_u (or to find load factor λ).

In plastic analysis three basic conditions are satisfied as follows.

Yield condition: The plastic moment M_p must not exceed anywhere in the structure.

Equilibrium: Sum of the forces and moments (external and internal) acting on the structure must be equal to zero.

Mechanism: At ultimate collapse load, the number of plastic hinges must be just enough to form a mechanism.

There are three fundamental theorems which satisfy the above three conditions as displayed below:

Uniqueness theorem

→ [Mechanism condition, Equilibrium condition, Yield condition]

- Mechanism condition + Equilibrium condition — Upper bound theorem
- Equilibrium condition + Yield condition — Lower bound theorem

15.6.1 Lower Bound Theorem

It is also called **Static theorem** or **Safe theorem.** It states that if a structure is in equilibrium condition under the external load and the applied moment does not exceed the plastic moment of resistance anywhere in the structure ($M \leq M_p$) then the structure will not collapse. In other words if at any load factor λ', a distribution of bending moments can be found which satisfies both the equilibrium condition and yield condition, then λ' is less than or equal to the true collapse load factor λ. Hence in safe theorem applied load will never exceed the collapse load ($\lambda' \leq \lambda$).

15.6.2 Upper Bound Theorem

It is also called **Kinetic** or **Unsafe theorem**. It states that for a structure subjected to a loading corresponding to any assumed collapse mechanism must be either greater than or equal to, but cannot be less than the true collapse load. In terms of load factor, the Unsafe theorem states that the load factor λ' obtained from the work equation written for any arbitrary assumed mechanism is greater than, or at least equal to, the true collapse load factor λ. It is called Unsafe theorem because the value of the plastic moment M_p obtained on the basis of an arbitrarily assumed collapse mechanism is smaller than, or at best equal to the actual collapse load ($\lambda' \geq \lambda$).

15.6.3 Uniqueness Theorem

It states that if at any load factor λ', a distribution of bending moment can be found which simultaneously satisfies (1) the mechanism condition, (2) the equilibrium condition and (3) the yield condition, then the load factor λ' is the true load factor λ. Thus for complete safety and economy in design, all three conditions are checked, i.e. Uniqueness theorem is applied.

15.7 METHOD OF PLASTIC ANALYSIS

Both statical and kinematic methods are suitable for simple beams and frames but for complicated indeterminate structures, the kinematic method is simpler.

15.7.1 Statical or Lower Bound Method

Lower bound collapse load is determined using the following steps:

1. Convert the structure into statically determinate structure by removing redundant forces.
2. Draw the bending moment diagram for the determinate structure.
3. Draw the bending moment diagram for the determinate structure which is loaded by the redundant forces.
4. Draw the combined bending moment diagram in such a way that a mechanism is formed.
5. Calculate the value of ultimate collapse load by solving equilibrium equations.
6. Check that the bending moment at every point in the structure is less than or equal to the plastic moment capacity M_p.

15.7.2 Kinematic or Mechanism or Upper Bound Method

Kinematic method is carried out as follows:

1. Select the possible locations of plastic hinges such as load points, frame joints, maximum bending moment points etc.
2. Select the independent and combined mechanisms. Joint, beam, portal and gable frame mechanisms are independent mechanisms as illustrated later in the examples. Number of independent mechanisms

$= H - R$

where H is the number plastic hinges and R is the number of redundant.

3. Form and solve the equilibrium equations for virtual work. Find the maximum collapse or ultimate load for forming a mechanism. The lowest load forming a mechanism is the upper bound of collapse load.
4. Check that the bending moment at every point in the structure is less than or equal to the plastic moment M_p.

15.8 EFFECT OF AXIAL FORCE ON PLASTIC MOMENT

When a section is subjected to axial force P, the plastic moment capacity is reduced. The reduction depends on the shape of the section, length of beam column and intensity of compressive force. Figure 15.4 shows the stress distribution in a beam at various stages of deformation caused by an axial force and moment. Due to the axial force, yielding occurs on the compression side prior to the tension side. Eventually plastification occurs, but since a part of the area resist axial force, the stress block no longer contains equal compression and tension areas as it was observed with pure bending. Thus, as shown in Fig. 15.4 the total stress distribution may be divided into two parts: (i) the part associated with axial force and (ii) part corresponding to bending moment.

The axial force P is given by

$$P = \sigma_y\, bkd \tag{15.5}$$

or,
$$P = kP_u \tag{15.6}$$

in which $P_u = \sigma_y bd$, is fully yield force capacity of section.

The bending moment,

$$M = \sigma_y b\,(d - kd)\,(d + kd)/4$$

or,
$$M = \sigma_y bd^2\,(1 - k^2)/4 \tag{15.7}$$

or,
$$M = M_p\,(1 - k^2) \tag{15.8}$$

Eliminating, k from Eqs (15.6) and (15.8) gives

$$\frac{M}{M_P} + \left[\frac{P}{P_u}\right]^2 = 1 \tag{15.9}$$

This is an interaction equation between moment and axial force capacity of a rectangular section as plotted in Fig. 15.5. When the axial force is zero, $M = M_p$. When the axial force reaches the value $P = P_u$, then the moment capacity is zero. However, for low values of P/P_u (< 0.15) the effect of axial force is neglected. For $P/P_u = 0.15$, the reduction in plastic moment capacity of a rectangular section is 2.25%.

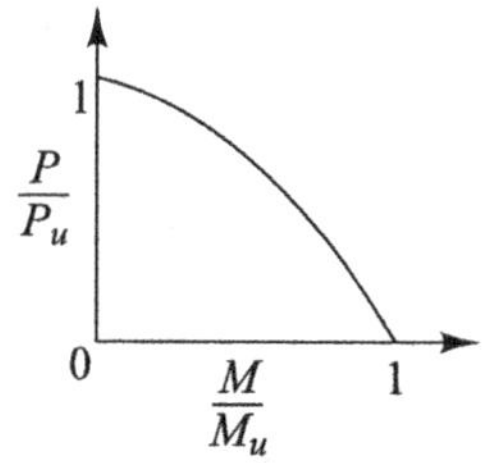

Fig. 15.5 *Interaction Curve*

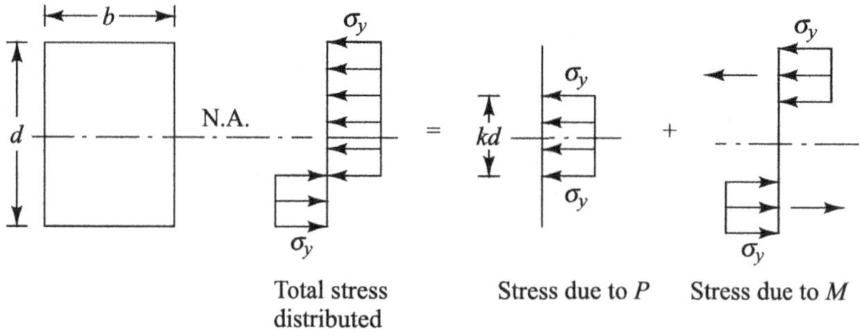

Fig. 15.4 *Stress Distribution due to Moment M and Thrust P*

EXAMPLES

Example 15.1 Determine the shape factor for the sections shown in Fig. 15.6(a–c).

Solution *Rectangular section:* Fig. 15.6(a)

Yield moment,

$$M_y = \sigma_y \times \frac{I}{y} = \sigma_y \times \frac{bd^3}{12} \times \frac{2}{d} = \frac{\sigma_y bd^2}{6}$$

Plastic moment,

$$M_p = \frac{\sigma_y (y_1 + y_2) A}{2} = \sigma_y \times \left(\frac{d}{4} + \frac{d}{4}\right) \times \frac{bd}{2}$$

$$= \frac{\sigma_y bd^2}{4}$$

Shape factor,

$$S = \frac{M_p}{M_y} = 1.5$$

Diamond section: Fig. 15.6(b)

Yield moment,

$$M_y = \sigma_y \times 2 \times \frac{bh^3}{12} \times \frac{1}{h} = \sigma_y \times \frac{bh^2}{6}$$

Plastic moment,

$$M_p = \frac{\sigma_y A(y_1 + y_2)}{2}$$

$$= \sigma_y \frac{bh}{2}\left(\frac{h}{3} + \frac{h}{3}\right) = \sigma_y \times \frac{bh^2}{3}$$

Shape factor,

$$S = \frac{\sigma_y bh^2/3}{\sigma_y bh^2/6} = 2$$

Tee section: To locate the centroidal axis, take moment of the area about bottom fibre, Fig. 15.6(c),

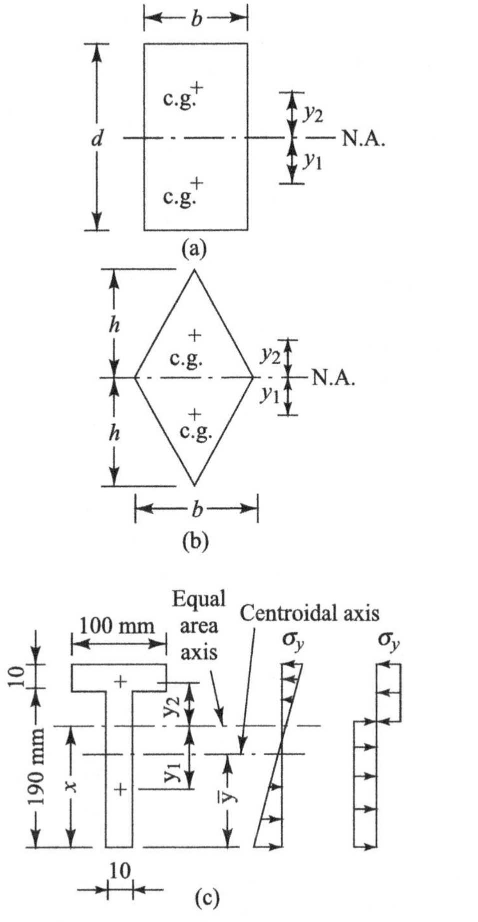

Fig. 15.6 *Plastic Moment Capacity of Various Sections*

$$\bar{y}(190 \times 10 + 100 \times 10)$$

$$= 10 \times 190 \times 190/2 + 100 \times 10(190 + 10/2)$$

or, $$\bar{y} = 129.48 \text{ mm}$$

Yield moment of the section, $M_y = \sigma_y \times \frac{I}{y}$

where $I = \frac{100 \times 10^3}{12} + 100 \times 10\left(\frac{190}{2} + \frac{10}{2} - 129.48\right)^2 + \frac{10 \times 190^3}{12} + 10 \times 190\left(\frac{190}{2} - 129.48\right)^2$

$= 12275890 \text{ mm}^4$

or, $M_y = \sigma_y \times \frac{12275890}{129.48} = 94809\ \sigma_y$

To locate the equal area axis, equating area on both sides of the axis,

$100 \times 10 + 10\,(190 - x) = 10x$

or, $x = 145$ mm

Plastic moment of the section,

$$M_p = \sigma_y A(y_1 + y_2)/2$$

$$A = 100 \times 10 + 190 \times 10 = 2900 \text{ mm}^2$$

$$y_1 = 145/2 = 72.5 \text{ mm}$$

$$y_2 = \frac{100 \times 10(55 - 5) + 10 \times 45 \times 45/2}{100 \times 10 + 10 \times 45}$$

$$= 41.46 \text{ mm}$$

Hence, $M_p = \sigma_y \times 2900 \times (72.5 + 41.46)/2$

$= 165242\ \sigma_y$

Therefore, shape factor,

$$S = \frac{165242\ \sigma_y}{94809\ \sigma_y} = 1.74$$

Example 15.2 Derive an expression for the bending moment in a rectangular section where the stress distribution is partially elastic and partially plastic. Also sketch the moment curvature curve.

Solution From stress diagram of partly elastic-plastic rectangular section shown in Fig. 15.7, the moment of resistance,

$$M = (\text{Force} \times \text{Lever arm})_{\square + \Delta \text{ portion}}$$

$$M = \sigma_y \times \frac{b(d - kd)}{2} \times \frac{(d + kd)}{2} + \frac{1}{6} \times \sigma_y \times b(kd)^2$$

$$= \frac{1}{4}\sigma_y bd^2\left(1 - \frac{1}{3}k^2\right) = M_p\left(1 - \frac{1}{3}k^2\right) \quad \text{(i)}$$

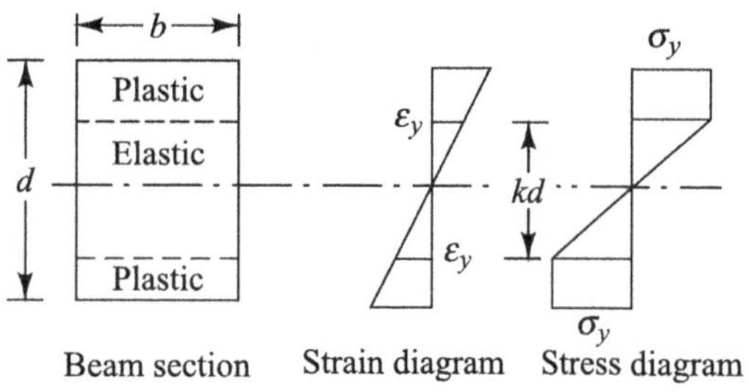

Fig. 15.7

For fully plastic section, $k = 0$ and

$$M = M_p = \sigma_y bd^2/4 \quad \text{(ii)}$$

For fully elastic section, $k = 1$ and

$$M = M_y = 2M_p/3 \quad \text{(iii)}$$

From usual moment curvature equation,

$$\frac{E}{R} = \frac{\sigma}{y}$$

$$\frac{E}{R} = \frac{\sigma_y}{kd/2} \quad \text{(iv)}$$

Substituting for σ_y from (i) in (iv),

$$\frac{1}{R} = \frac{4M}{Ebd^2}\frac{2}{kd\left(1 - k^2/3\right)}$$

or, $$\frac{1}{R} = \frac{2M}{EIk\left(3 - k^2\right)} \quad \text{(v)}$$

The moment curvature diagram is shown in Fig. 15.8.

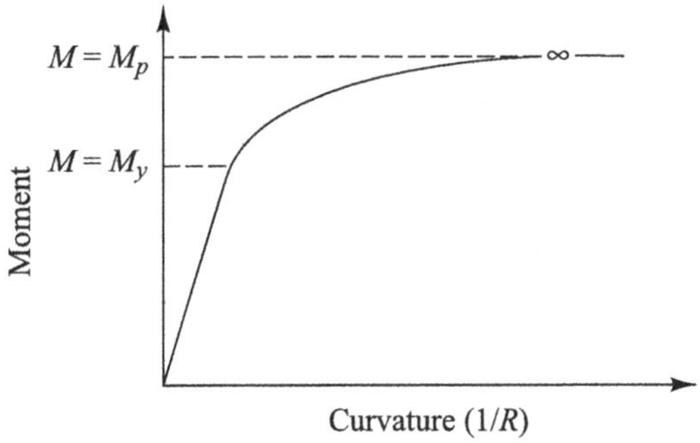

Fig. 15.8

Note: It should be noted that full plastic moment can be reached when extreme fibre strain is infinity. However M approaches M_p at $k = 1/15$ as shown in Table 15.1. For mild steel $\varepsilon_y \approx 0.12\%$ and it remains in plastic state for $\varepsilon \approx 15 \times \varepsilon_y = 1.8\%$. Thus for practical purpose a fully plastic hinge can develop.

Table 15.1 *Relation between k and M*

k	M
1	$0.667\ M_p$
1/10	$0.997\ M_p$
1/15	$0.999\ M_p$

Example 15.3 A simply supported rectangular beam 3 cm × 4 cm carries a concentrated load W at mid point of span 3 m.

(1) Determine W and draw the shape of plastic zone when strain at the extreme fibre is $4\varepsilon_y$ (4 times yield strain).

(2) Determine the distribution of curvature along the beam.

Solution

(1) Referring to Example 15.2, the moment of resistance for elastic-plastic zone is given by Eq. (i),

$$M = M_p\left(1 - \frac{1}{3}k^2\right)$$

or, $$k = [3\,(1 - M/M_p)]^{1/2}$$

where, $M = \left.\dfrac{Wx}{2}\right|_{x \le L/2}$ and $M_p = \sigma_y \times \dfrac{bd^2}{4}$

$$k = \left[3\left(1 - \frac{2\,Wx}{\sigma_y\, bd^2}\right)\right]^{1/2} \quad \text{at } x \le L/2$$

From stress diagram at mid span, Fig. 15.9(b), $k = 1/4$

or, $$k = \left[3\left(1 - \frac{2\,Wx}{\sigma_y\, bd^2}\right)\right]^{1/2} = 1/4$$

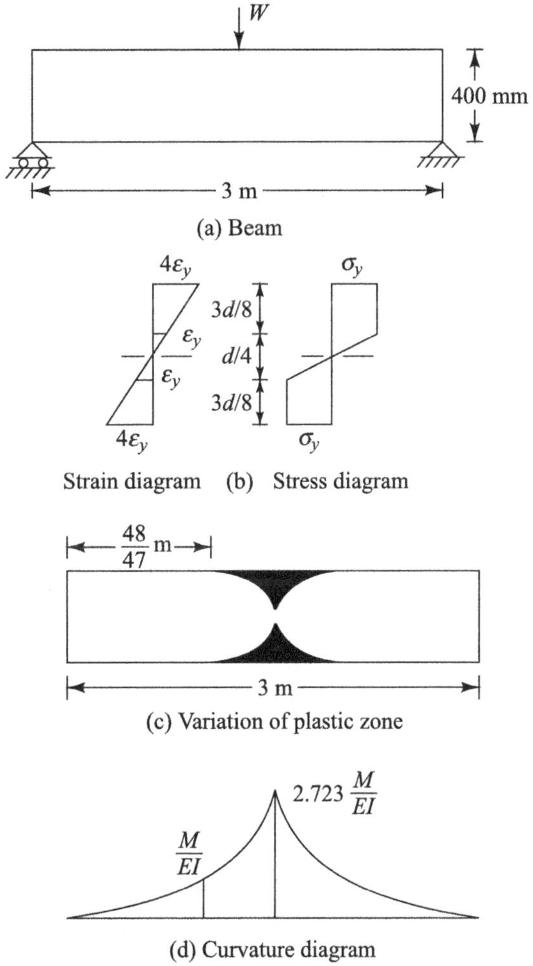

Fig. 15.9

Substituting, $L = 3000$ mm, $b = 30$ mm,
$d = 40$ mm, $\sigma_y = 250$ MPa

$$W = \left(1 - \frac{1}{16\times 3}\right)\times\frac{30\times 40^2\times 250}{3000} = 3917 \text{ N}$$

For fully elastic section, $k = 1$,

$$\text{or,} \quad k = \left[3\left(1 - \frac{2Wx}{\sigma_y bd^2}\right)\right]^{1/2} = 1$$

$$\text{or,} \quad x = \left(1 - \frac{1}{3}\right)\times\frac{30\times 40^2\times 250}{2\times 3917} = 1021.28 \text{ mm}$$

Figure 15.9(c) shows the variation of plastic zone along the span of the beam.

(2) The variation of curvature is given by Eq. (v) of Example 15.2,

$$\frac{1}{R} = \frac{2M}{EIk\left(3 - k^2\right)}$$

Within elastic zone $0 < x < 1021.28$ mm, $k = 1$

$$\text{or,} \quad \frac{1}{R} = \frac{M}{EI}$$

At mid span, $k = 1/4$

$$\text{or,} \quad \frac{1}{R} = 2.723\frac{M}{EI}$$

Figure 15.9(d) shows the curvature diagram.

Example 15.4 A rigid horizontal beam is supported by three equally spaced tension bars of same cross-sectional area A as shown in Fig. 15.10. The length of middle bar is half of the outer bars. Find the ultimate load P_u on the horizontal rigid beam if the load is applied at the centre. Draw the load deflection curve.

Solution Let T_1 and T_2 = tension in outer and inner bars, respectively.

Total load, $$P = 2\,T_1 + T_2 \qquad \text{(i)}$$

Elastic stage: Since elongation of outer and inner bars are same

$$\delta_1 = \delta_2$$

$$\text{or,} \quad \frac{T_1 L_1}{A\,E} = \frac{T_2 L_2}{A\,E} \qquad \text{(ii)}$$

From Eqs (i) and (ii),

$$T_2 = 2T_1 = P/2 \qquad \text{(iii)}$$

Since, $T_2 > T_1$, the middle bar will reach the yield stress earlier than the outer and the corresponding load P is called the yield load P_y. From Eq. (iii) at yield stage,

$$P_y = 2T_2 = 2\sigma_y A$$

Load deflection at yield stage,

$$\delta_y = \frac{T_2 L_2}{A\,E} = \frac{\sigma_y AL/2}{AE} = \frac{\sigma_y L}{2E}$$

Plastic stage: At plastic stage all three bars have yield stress and the corresponding load is called the ultimate load P_u. From Eq. (i) at plastic stage,

$$P_u = 2T_1 + T_2 = 2\sigma_y A + \sigma_y A$$

$$\text{or,} \quad P_u = 3\,\sigma_y A$$

Load deflection at fully plastic stage,

$$\delta_u = \frac{\sigma_y L}{E}$$

Figure 15.10(b) shows the load–deflection curve.

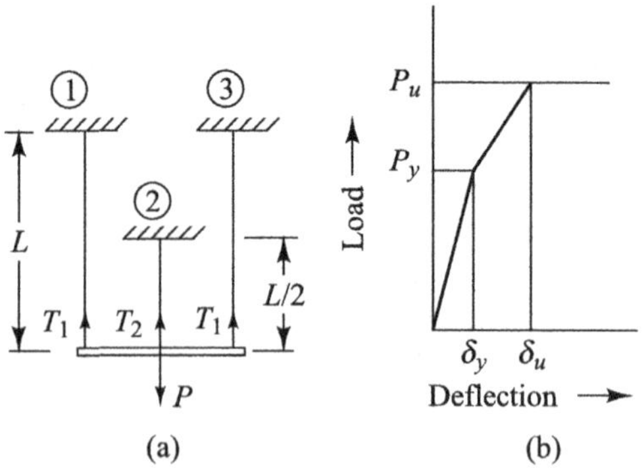

Fig. 15.10

Example 15.5 Determine the collapse load W_u by static and kinematic method for a beam shown in Fig. 15.11(a) carrying an uniformly distributed load W on the span of length L if the plastic moment capacity of the beam is equal to M_p.

Solution

Static method: Maximum bending moment at mid span = $WL/8$

Equilibrium equation when plastic hinge forms at mid span,

$$\frac{W_u L}{8} = M_p$$

or,
$$W_u = \frac{8M_p}{L}$$

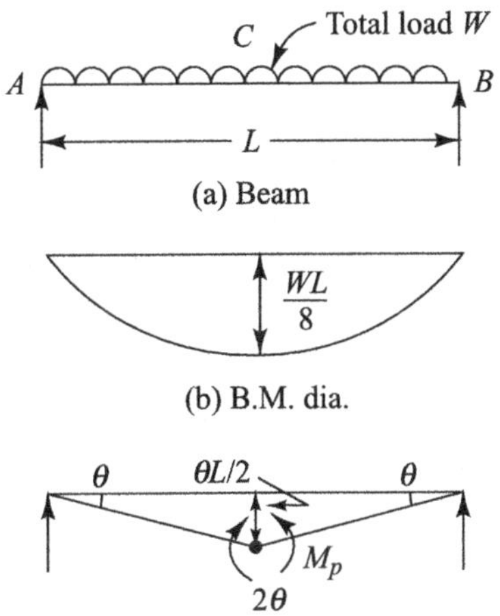

(a) Beam

(b) B.M. dia.

(c) Beam mechanism

Fig. 15.11

Kinematic method: Assuming the collapse mechanism show in Fig. 15.11(c), average work done by uniformly distributed collapse load,

$$= W_u \times \frac{1}{2} \times \frac{L\theta}{2}$$

Internal work done by plastic moment = $M_p \times 2\theta$

Equating work done by external load and work done by plastic moment

$$W_u \times \frac{L\theta}{4} = 2\,M_p\,\theta$$

or,
$$W_u = \frac{8M_p}{L}$$

Example 15.6 Determine the collapse load for a propped cantilever shown in Fig. 15.12(a) by static and kinematic methods.

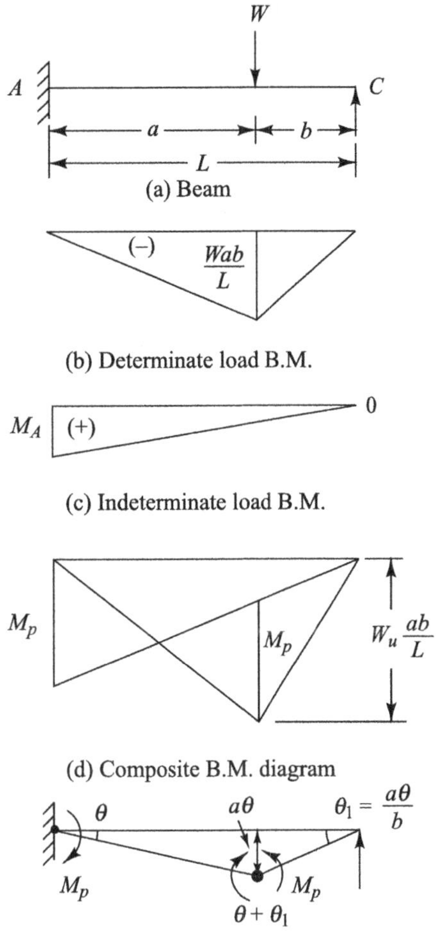

(a) Beam

(b) Determinate load B.M.

(c) Indeterminate load B.M.

(d) Composite B.M. diagram

(e) Beam mechanism

Fig. 15.12

Solution

Static method: Figure 15.12(b) shows bending moment diagram for the determinate structure Figure 15.12(c) shows bending moment diagram of the redundant (fixed end moment). Figure 15.12(d) is the combined bending moment diagram.

From Fig. 15.12(d), the equilibrium equation at ultimate load W_u,

$$W_u \times \frac{ab}{L} = M_p + M_p \frac{b}{L}$$

or, $$W_u = M_p \times \frac{(L+b)}{ab}$$

Kinematic method: Figure 15.12(e) shows the collapse beam mechanism. Equating the work done by external forces by work done by internal forces,

or, $$W_u\,(a\theta) = M_p\theta + M_p\,(\theta + a\theta/b)$$

or, $$W_u = M_p\left(\frac{2}{a} + \frac{1}{b}\right)$$

or, $$W_u = M_p \times \frac{(L+b)}{ab}$$

which is same as calculated earlier by static method.

Example 15.7 A fixed beam of span L carries a uniformly distributed load W (total) on the left half portion. Using plastic theory, determine the value of W at collapse. The plastic moment of resistance of beam is M_p.

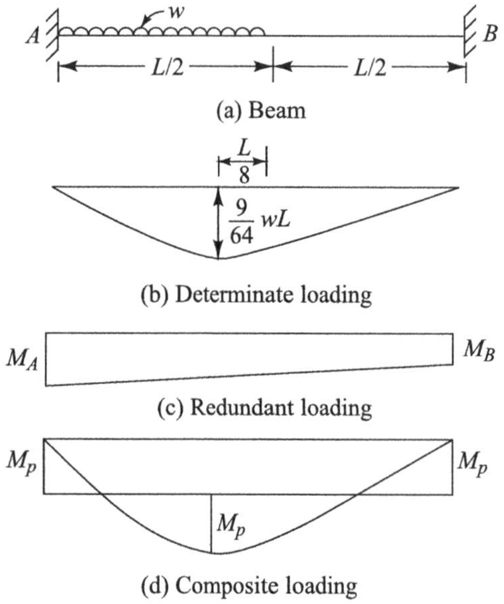

(a) Beam
(b) Determinate loading
(c) Redundant loading
(d) Composite loading

Fig. 15.13

Solution Figure 15.13(b) and (c) show bending moment diagram for the determinate structure and the redundant, respectively. From combined bending moment diagram in Fig. 15.13(d),

$$\frac{9}{64}\,W_u L = 2\,Mp$$

or, $$W_u = \frac{128}{9} \times \frac{M_p}{L}$$

Note: It is simpler to use static method in the problems of uniformly distributed loads or when location of plastic hinge is unknown without bending moment diagram. Hit and trial or differentiation could be used for kinematic method.

Example 15.8 A propped cantilever carries an uniformly distributed load W over span L. Determine the ultimate collapse load W_u, if the plastic moment capacity of the beam is M_p.

Solution *Static Method:* Figure 15.14(b) shows the combined bending moment diagram of the determinate structure and the redundant. At collapse the plastic hinges will form at the fixed end and a distance of x from the free end where the difference of parabolic and triangular bending moment diagram is maximum.

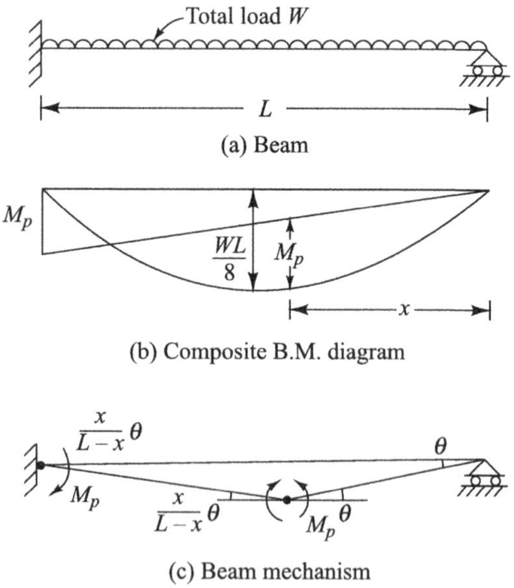

(a) Beam
(b) Composite B.M. diagram
(c) Beam mechanism

Fig. 15.14

From Fig. 15.14(b) at a distance x from hinged end,

$$M_p + \frac{x}{L} M_p = \frac{W_u}{2} x - \frac{W_u}{L} \times \frac{x^2}{2}$$

or, $$W_u = 2M_p \frac{L + x}{x(L - x)}$$

For maximum value of W_u, $\frac{dW_u}{dx} = 0$

or, $$2M_p \times \frac{[x(L-x)1-(L+x)(L-2x)]}{x^2(L-x)^2} = 0$$

or, $x^2 + 2Lx - L^2 = 0$

or $x = 0.414\ L$

Substituting x in the equilibrium Eq. (i),

or, $W_u = 11.66\ M_p/L$

Kinematic method: For beam mechanism shown in Fig. 15.14(c),

$$W_u \frac{x\theta}{2} = M_p \left(\frac{x}{L-x}\theta + \theta + \frac{x}{L-x}\theta \right)$$

$$W_u = \frac{2M_p}{L} \frac{L+x}{x(L-x)}$$

For maximum W_u, $x = 0.414L$ (as calculated earlier). Thus,

$$W_u = 11.66\ M_p/L$$

Example 15.9 Determine the minimum load W that will cause collapse of the propped beam shown in Fig. 15.15(a) if the yield stress is 250 MPa both in tension and compression.

Solution Plastic moment,

$$M_p = \sigma_y A\ (y_1 + y_2)/2$$
$$= 250 \times 1000 \times (25 + 5)/2$$
$$= 3.75 \text{ kN m}$$

The virtual work equation of the collapse mechanism shown in Fig. 15.15(c), is

$$W(x\theta) = M_p \left(\frac{x}{1-x} + 1 + \frac{x}{1-x} \right) \theta$$

$$W = M_p \frac{1+x}{x(1-x)} \qquad \text{(i)}$$

For minimum value of W differentiating w.r.t. x,

$$\frac{dW}{dx} = M_p \frac{[x(1-x)1-(1+x)(1-2x)]}{x^2(1-x)^2} = 0$$

or, $x^2 + 2x - 1 = 0$

or, $x = 0.414$ m

Substituting x and M_p in Eq. (i), $W_{min} = 21.86$ kN

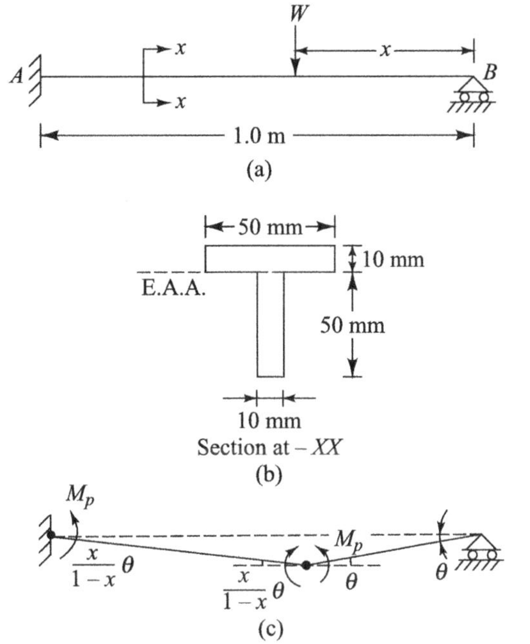

Fig. 15.15

Example 15.10 Find the collapse load for the portal frame shown in Fig. 15.16(a).

Solution Number of independent mechanisms,

$$N = H - R$$

where, H = plastic hinges possible = 4

and, R = number of redundants = 2

Therefore, number of independent mechanisms,

$$N = 4 - 2 = 2$$

Figures 15.16(b) through (d) show two basic mechanisms and one combined mechanism. Forming the virtual work equation for,

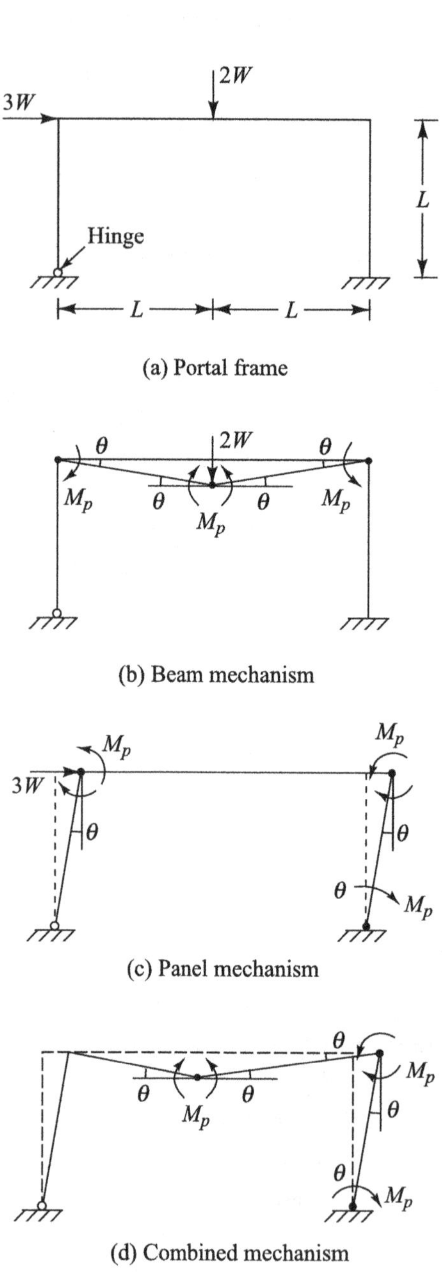

(a) Portal frame

(b) Beam mechanism

(c) Panel mechanism

(d) Combined mechanism

Fig. 15.16(a-d)

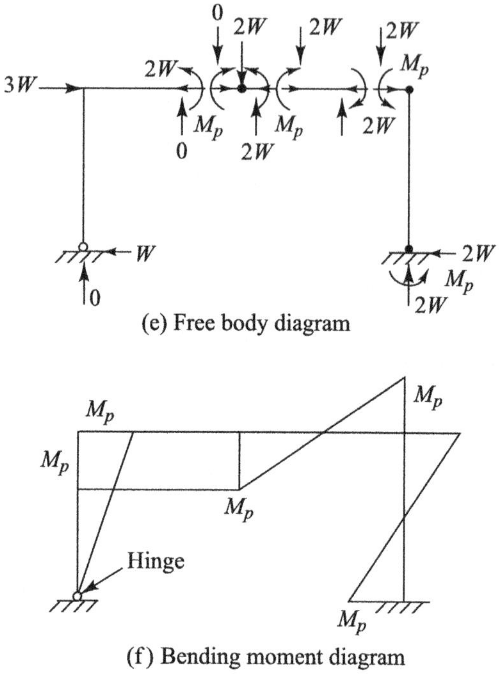

(e) Free body diagram

(f) Bending moment diagram

Fig. 15.16(e-f)

(i) Beam Mechanisms,

$$2W\,(L\theta) = M_p\,(\theta + 2\theta + \theta)$$

or, $W = 2\,M_p/L$ (i)

(ii) Panel Mechanism,

$$3W\,(L\theta) = M_p\,(\theta + \theta + \theta)$$

or, $W = M_p/L$ (ii)

(iii) Combined Mechanism: One hinge will cancel from each mechanism (i) and (ii), when they combine as in Fig. 15.16(d),

	Mechanism	*Work Equation*	*Hinges Cancelled*
(i)	Beam	$2WL\theta = 4M_p\theta$	$-M_p\theta$
(ii)	Panel	$3WL\theta = 3M_p\theta$	$-M_p\theta$
(iii)	Combined	$5\,WL\theta = 7\,M_p\theta$	$-2\,M_p\theta$

or, $5WL\theta = 5\,M_p\theta$

or, $W = M_p/L$ (iii)

Therefore, true collapse load, $W_u = M_p/L$ and corresponding bending moment diagram at collapse is shown in Fig. 15.16(f).

Example 15.11 Plot a load/mechanism interaction curve diagram for the portal frame shown in Fig. 15.17(a).

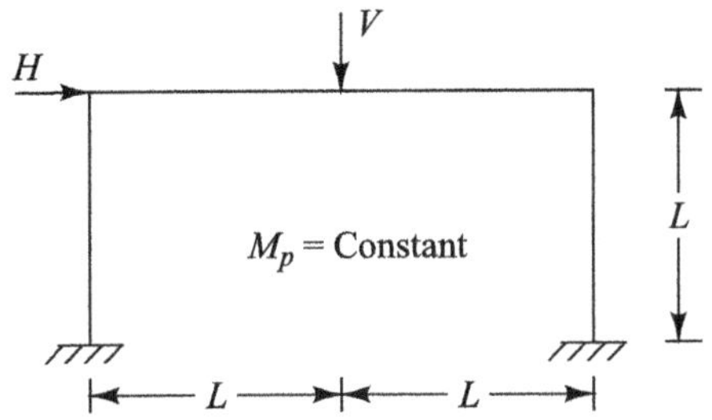

(a)

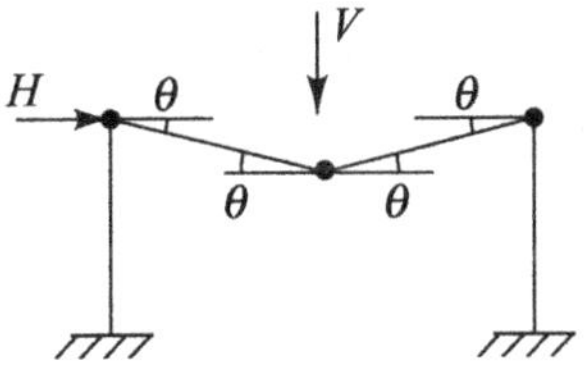

(b) Beam mechanism

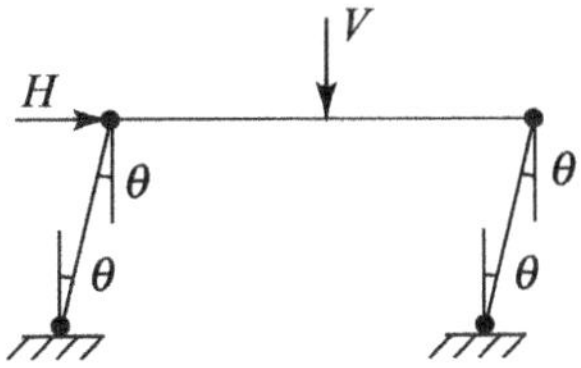

(c) Sway mechanism

Fig. 15.17(a-c)

Solution Writing virtual work equation for,

(i) Beam mechanism: Fig. 15.17(b),

$$V(L\theta) = M_p\,(\theta + 2\theta + \theta)$$

or, $V = 4\,M_p/L$ (i)

(ii) Sway mechanism: Fig. 15.17(c),

$$H\,(L\theta) = M_p\,(\theta + \theta + \theta + \theta)$$

or, $H = 4\,M_p/L$ (iii)

(iii) Combined mechanism: One hinge will cancel out from beam mechanism and sway mechanism as shown in Fig. 15.17(d).

	Mechanism	*Work Equation*	*Hinges Cancelled*
(i)	Beam	$VL\theta = 4M_p\theta$	$-M_p\theta$
(ii)	Panel	$HL\theta = 4M_p\theta$	$-M_p\theta$
(iii)	Combined	$(V + H)\,L\theta = 8\,M_p\theta$	$-2\,M_p\theta$

or, $(V + H) = 6\,M_pL$

Figure 15.17(e) shows load/mechanism interaction diagram.

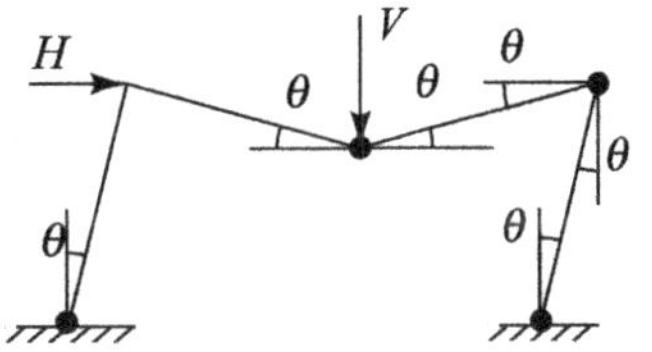

(d) Combined mechanism

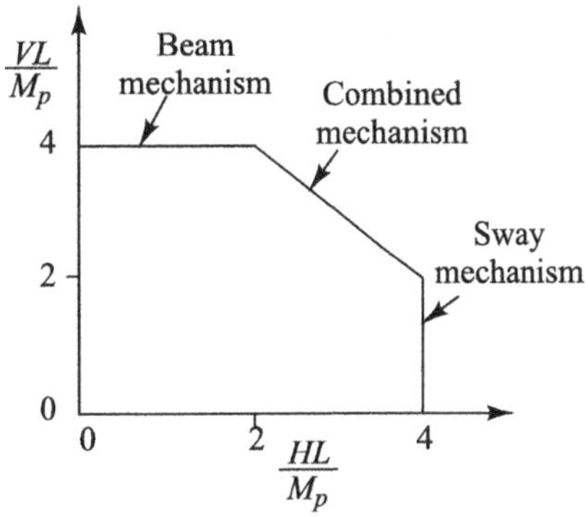

(e) Intraction diagram

Fig. 15.17(d-e)

EXERCISES

PROBLEMS

15.1 What do you understand by plastic hinge? How it differs from real hinge?

15.2 Define shape factor and load factor.

15.3 Why no internal strain energy of the beams is involved in virtual work equation?

15.4 What is the difference between elastic theory and plastic theory of design?

15.5 Why the effect of axial force on moment capacity is neglected when P/P_u is less than 0.15?

15.6 Calculate the shape factor of the section shown in Fig. 15.18.

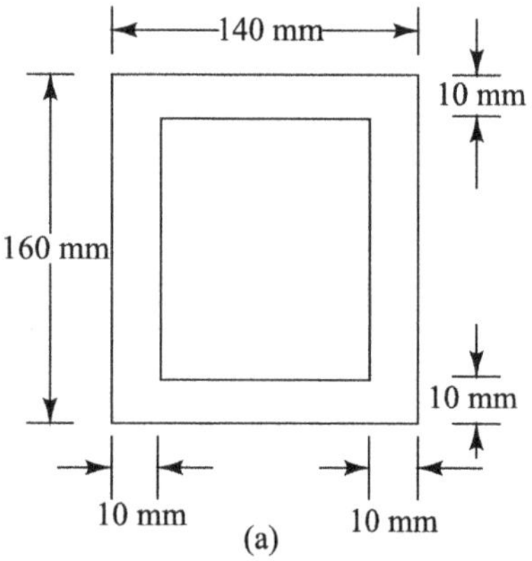

(a)

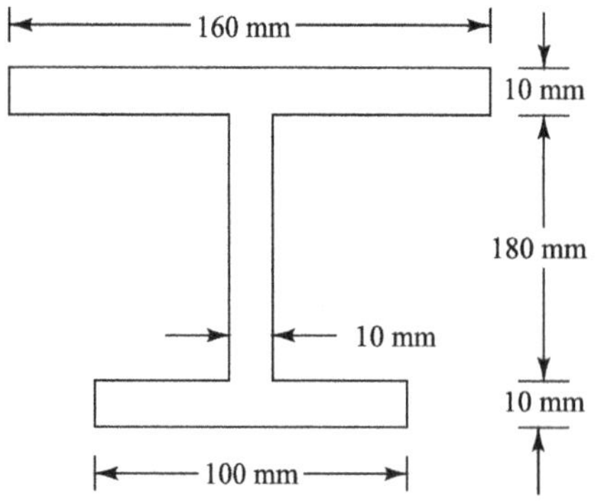

(b)

Fig. 15.18(a-b)

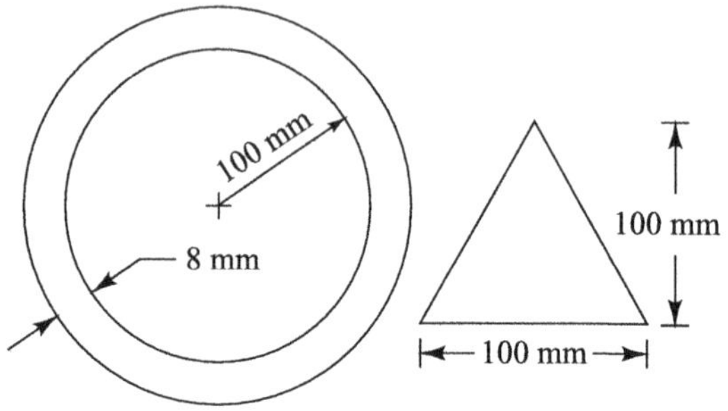

(c) (d)

Fig. 15.18(c-d)

15.7 Determine the shape factor of the rectangular section if strain in extreme fibre is 12 times ε_y. The idealized stress–strain curve is shown in Fig. 15.19.

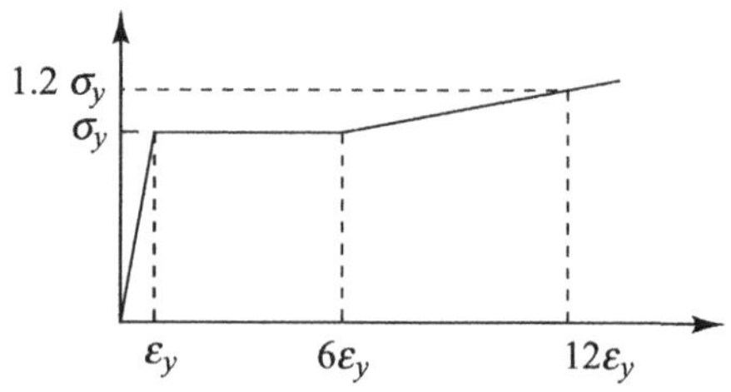

Fig. 15.19

15.8 Compute the ultimate load P_u for the structure shown in Fig. 15.20.

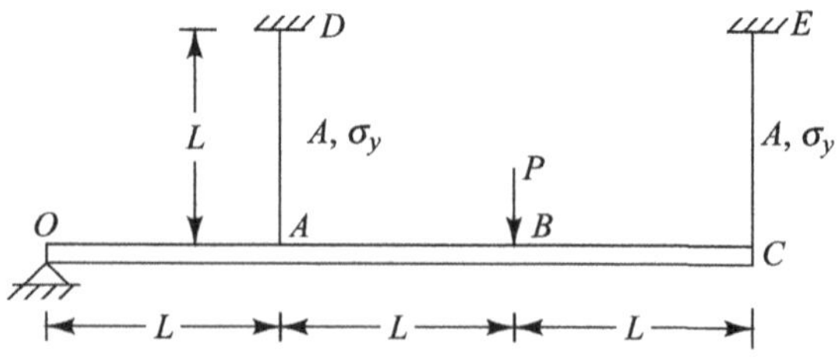

(a)

Fig. 15.20(a)

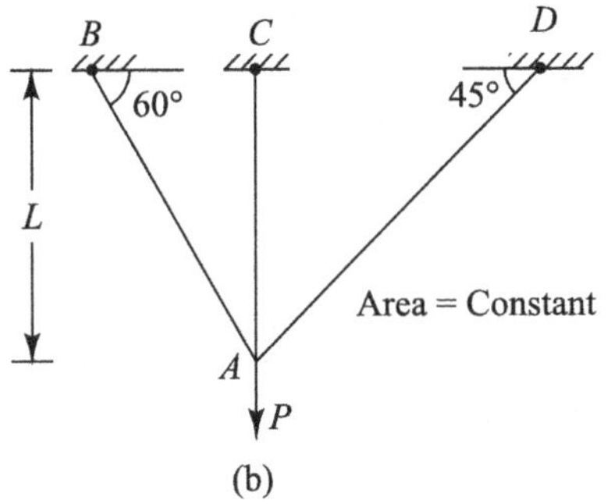

Fig. 15.20(b)

15.9 Derive the axial force vs. moment capacity interaction equation for circular section.

15.10 Calculate W_u for the ring shown in Fig. 15.21. The plastic moment capacity of the ring is M_p.

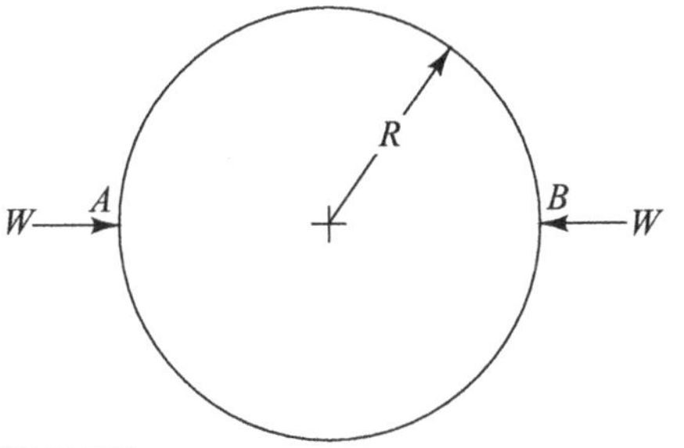

Fig. 15.21

15.11 A fixed beam has a plastic moment of resistance $3M_p$ at one end and $2M_p$ at the other (vary linearly). Determine the ultimate uniformly distributed load which the beam can carry.

15.12 Calculate the ultimate load, W_u of rectangular portal frames shown in Fig. 15.22. Also, plot the bending moment diagram at the ultimate load.

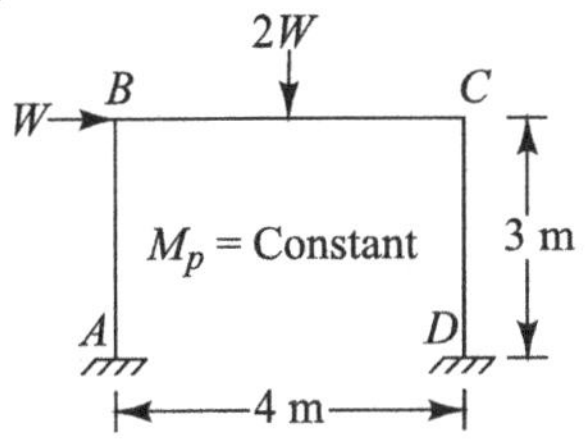

Fig. 15.22

OBJECTIVE QUESTIONS

15.13 In a statically indeterminate structure, the formation of first plastic hinge will reduce the number of redundancy by
(a) 0 (b) 1 (c) 2 (d) 3

15.14 It can be stated in general that if a statically indeterminate structure has R redundancy then it will become statically determinate structure if the number of plastic hinge formed is equal to
(a) R (b) $R + 1$
(c) $R - 1$ (d) None of above

15.15 Which of the following sections will have maximum shape factor ?
(a) Square (b) Circular
(c) Diamond (d) Triangle

15.16 The assumption that plane section remains plane under pure bending which is valid for linear elastic material it is not exactly true for plastic bending
(a) True
(b) False
(c) True if material is non-linear elastic
(d) True if deformation are continuous

15.17 The value of shape factor for any section is always,
(a) greater than 1
(b) less than 1
(c) equal to 1
(d) None of above

15.18 The relation between moment and curvature at the plastic hinge is
(a) linear
(b) non-linear
(c) curvature is constant for all moments
(d) none of above

15.19 The moment capacity of a section at plastic hinge is
(a) zero
(b) yield moment
(c) twice of yield moment
(d) fully plastic moment

15.20 The shape factor of a rectangular section is
(a) 0.5 (b) 1 (c) 1.5 (d) 2

15.21 The shape factor for a typical I-section is
(a) 1 (b) 1.15 (c) 1.5 (d) 2

15.22 The ultimate moment capacity of a mild-steel section is,
(a) greater than M_p
(b) less than M_p
(c) equal to M_p
(d) greater than M_y

15.23 The plastic analysis of structure is applicable to structure made of
(a) brittle materials
(b) ductile materials
(c) any type of structural material
(d) brittle and ductile materials

15.24 The independent mechanism possible in the plastic analysis of plane frames are,
(a) beam mechanism
(b) beam mechanism, sway mechanism
(c) beam mechanism, joint mechanism
(d) all the above mechanisms

15.25 A structure becomes a mechanism if the virtual work done by the external load is,
(a) equal,
(b) less or equal,
(c) more or equal,
(d) less than,
the internal dissipation of energy at the hinge.

16

Springs

CONCEPT REVIEW

16.1 CLOSE COILED SPRING SUBJECTED TO AXIAL LOAD

Figure 16.1 shows a close coiled helical spring of mean radius R and carrying an axial load P. Let d be the diameter of the wire of the spring.

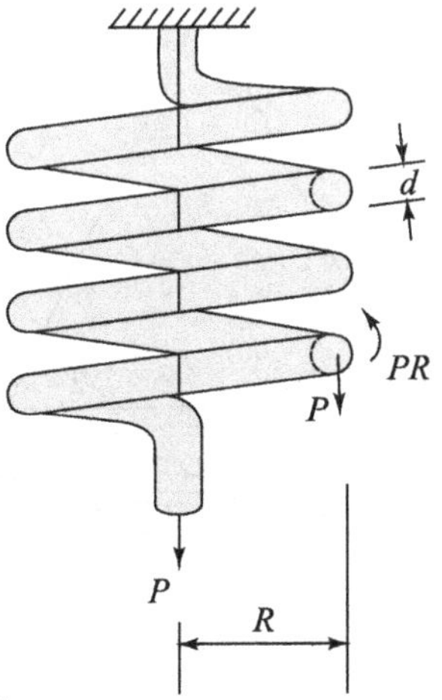

Fig. 16.1

$$\text{Direct shear stress} = \frac{P}{A} = \frac{P}{\pi d^2/4}$$

$$\text{Torsional shear stress} = \frac{PR}{\pi d^3/16}$$

These stresses add at the innermost point and the maximum shear stress is as follows:

$$\tau = \frac{P}{A} + \frac{16PR}{\pi d^3}$$

$$= \frac{16PR}{\pi d^3}\left(1 + \frac{d}{4R}\right) \qquad (16.1)$$

P/A will be small if the ratio $d/4R$ is small and can be neglected.

The deflection of spring is mainly due to deflection of the spring wire under torsion PR as follows:

$$\frac{T}{I_p} = \frac{G\theta}{L}$$

or, $$\theta = \frac{TL}{I_p G} = \frac{PR \times 2\pi Rn}{G \times \pi d^4/32}$$

where, n = number of turns

Deflection of spring, $\delta = R\theta = \dfrac{64PR^3n}{Gd^4}$ (16.2)

Spring constant, $k = \dfrac{P}{\delta} = \dfrac{Gd^4}{64R^3n}$ (16.3)

Energy stored in spring due to gradually applied load $= \dfrac{1}{2}P\delta = \dfrac{32P^2R^3n}{Gd^4}$ (16.4)

Resilience in terms of maximum permisible stress τ_{max} by using Eq. (16.1)

$$= \frac{32R^3n}{Gd^4} \times \frac{\pi^2 d^6 \tau_{max}^2}{16^2 R^2}$$

$$= \frac{\pi^2 nRd^2}{8G}\tau_{max}^2 \quad (16.5)$$

$$= \frac{\tau_{max}^2}{4G} \times \text{volume of coil} \quad (16.6)$$

Thus all springs will have same volume for equal resilience, τ_{max} and G.

16.2 SPRINGS IN SERIES

Two springs of stiffness k_1 N/mm and k_2 N/mm are attached in series as shown in Fig. 16.2. Both springs will deflect under same load P.

Total deflection, $\delta = \delta_1 + \delta_2$

or, $$\frac{P}{k} = \frac{P}{k_1} + \frac{P}{k_2}$$

or, $$\frac{1}{k} = \frac{1}{k_1} + \frac{1}{k_2} \quad (16.7)$$

where, $\delta_1 = \dfrac{P}{k_1}$, $\delta_2 = \dfrac{P}{k_2}$

and, k = stiffness of the combined spring

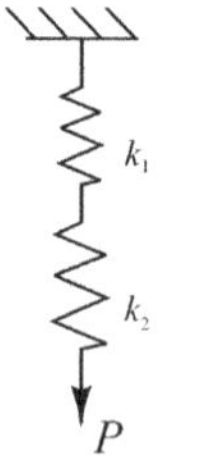

Fig. 16.2

16.3 SPRINGS IN PARALLEL

If two springs of stiffness k_1 N/mm and k_2 N/mm are attached in parallel and subjectd to a load P such that the deflection of all springs is same, then, the load will be shared by two springs as follows:

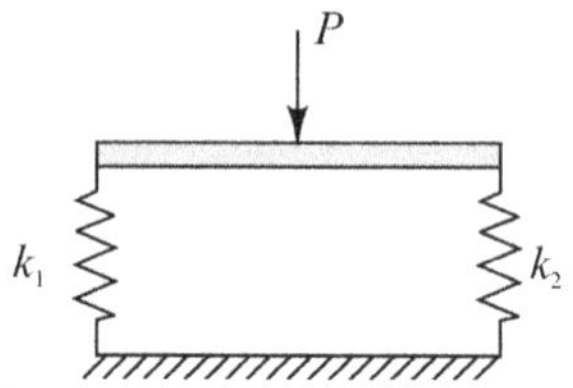

Fig. 16.3

$$P = P_1 + P_2$$

or, $$\delta k = \delta k_1 + \delta k_2$$

or, $$k = k_1 + k_2 \quad (16.8)$$

where, k is the stiffness of the combined system.

16.4 APPROXIMATE THEORY OF LEAF SPRING

Consider the leaf spring shown in Fig. 16.4 consisting of n number of metal strips of width b and thickness t. It is assumed that all the strips are initially bent to same radius R so that they will touch each other at the ends only. The metal strips overhang by an amount $a = \dfrac{L}{2n}$.

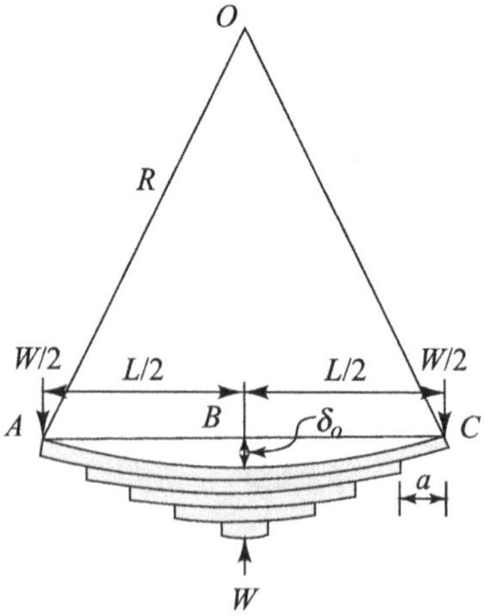

Fig. 16.4

Thus the overlap portion will have constant bending moment (see Fig. 16.5) as follows:

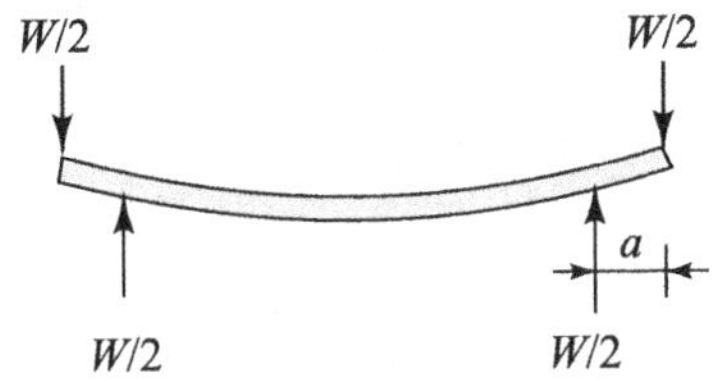

Fig. 16.5

$$M = \frac{W}{2} \times a$$

The new radius of curvature r for overlap portion is given by,

$$\frac{1}{R} - \frac{1}{r} = \frac{M}{EI} = \frac{6Wa}{Ebt^3} \quad (16.9)$$

The load W_0 which will straighten out all the plates is called proof load and is obtained by equating $r = \infty$ as follows:

$$W_0 = \frac{Ebt^3}{6aR} = \frac{nEbt^3}{3LR} \quad (16.10)$$

From triangle ABO,

$$AO^2 = AB^2 + BO^2$$

or,
$$R^2 = \left(\frac{L}{2}\right)^2 + (R-\delta_0)^2$$

Neglecting δ_0^2,

or,
$$\delta_0 = \frac{L^2}{8R} \quad (16.11)$$

Substituting in Eq. (16.10), proof load is given by,

$$W = \frac{8nEbt^3\delta_0}{3L^3} \quad (16.12)$$

Stress in each plate,

$$\sigma = \frac{M}{Z} = \frac{Wa}{2bt^2/6} = \frac{3WL}{2nbt^2} \quad (16.13)$$

When W = proof load W_0, the corresponding stress,

$$\sigma_0 = \frac{3W_0L}{2nbt^2} = \frac{4Et\delta_0}{L^2} \quad (16.14)$$

Since σ_0 and E are fixed for a material, Eq. (16.14) gives relation between initial dip δ_0 and thickness t of the plates.

Dip δ for a load W is given by Eq. (16.9),

where $\delta = L^2/8\,r$, as follows,

$$\delta_0 - \delta = \frac{6Wa}{Ebt^3} \times \frac{L^2}{8} = \frac{3WL^3}{8nEbt^3} \quad (16.15)$$

EXAMPLES

Example 16.1 A close coiled helical spring is made of steel wire 3 mm and the centre line of the wire form a cylinder of diameter 50 mm. The two ends are pulled apart by an axial pull of 10 N. What is the shearing stress and elongation of the spring if the spring has 20 number of turns. Take G = 80 GPa.

Solution Torsional shear stress (Eq. 16.1)

$$\tau_{max} = \frac{16PR}{\pi d^3}\left(1 + \frac{d}{4R}\right)$$

$$= \frac{16 \times 10 \times 25}{\pi \times 3^3}\left(1 + \frac{3}{4 \times 25}\right)$$

$$= 48.57 \text{ N/mm}^2$$

Deflection of spring by Eq. (16.2),

$$\delta = \frac{64PR^3n}{Gd^4}$$

$$= \frac{64 \times 10 \times 25^3 \times 20}{80000 \times 3^4}$$

$$= 30.86 \text{ mm}$$

Example 16.2 A close coiled helical spring is made by 5 mm diameter steel wire having shear strength 500 MPa. The spring has mean diameter 50 mm. A prestress of 60 MPa causes the coils to be closed under no-load condition. Calculate the no load force in the spring and the maximum load that can be applied to the spring. Also calculate the maximum extension of spring having 30 number of coils of the modulus of rigidity is 80 GPa.

Solution Using Eq. (16.1)

$$\tau = \frac{16PR}{\pi d^3}\left(1 + \frac{d}{4R}\right)$$

For no load condition force P is given by,

$$60 = \frac{16P \times 25}{\pi \times 5^3}\left(1 + \frac{5}{4 \times 25}\right)$$

or, $$P = \frac{60 \times \pi \times 125}{16 \times 25 \times 1.05} = 56.1 \text{ N}$$

The maximum load is obtained when $\tau = 500$ MPa as follows:

$$P = \frac{500\pi \times 125}{16 \times 25 \times 1.05} = 467.5 \text{ N}$$

Maximum deflection,

$$\delta = \frac{64nPR^3}{Gd^4}$$

$$= \frac{64 \times 30 \times 467.5 \times 25^3}{80000 \times 5^4}$$

$$= 109.57 \text{ mm}$$

Example 16.3 A safety valve of diameter 50 mm is to blow off at a gange pressure of 0.5 MPa. It is held by a close coiled compression spring of average diameter 70 mm and the initial compression of the spring is 20 mm. Determine the diameter of the spring wire and the number of turns if the allowable shear stress is 190 MPa and the shear modulus is 80 GPa.

Solution Force required to lift the valve

$$= p \times \frac{\pi}{4} d_v^2 = 0.5 \times \frac{\pi}{4} \times 50^2$$

$$= 981.75 \text{ N}$$

Using Eq. (16.1) (neglecting $d/4R$ if $4R >> d$)

$$\tau = \frac{16PR}{\pi d^3}$$

or, $$190 = \frac{16 \times 981.75 \times 35}{\pi d^3}$$

or, $$d = 9.73 \text{ say } 10 \text{ mm}$$

Number of coils are given by Eq. (16.2),

$$n = \frac{\delta G d^4}{64PR^3}$$

$$= \frac{20 \times 80000 \times 10^4}{64 \times 981.75 \times 35^3}$$

$$= 5.94 \text{ say } 6 \text{ turns}$$

Example 16.4 A railway wagon weighing 200 kN is stopped by a buffer of 4 springs while moving at a speed of 10 km/hr. If the maximum compression of springs is 150 mm, find the diameter of the spring rod if the mean diameter of the spring coil is 200 mm and 8 number of turns are in a coil. Take modulus of rigidity = 80 GPa.

Solution Kinetic energy $= \frac{1}{2} mV^2$

$$= \frac{1}{2} \times \frac{200000}{9.81} \times \left(\frac{10000}{3600}\right)^2 = 78655 \text{ N m}$$

Energy absorbed per spring = 78655/4

= 19663.73 N m

Work done = Energy absorbed

$$P \times \frac{\delta}{2} = 19.663.73$$

or $$P = \frac{2 \times 19663.73}{0.15} = 262183 \text{ N}$$

Substituting in Eq. (16.2)

$$d = \sqrt[4]{\frac{64nPR^3}{G\delta}}$$

$$= \sqrt[4]{\frac{64 \times 8 \times 262183 \times 100^3}{80000 \times 150}}$$

$$= 53.8 \text{ mm say } 54 \text{ mm}$$

Example 16.5 A close coiled spring has a stiffness of 5 N/mm. The solid length of spring is 60 mm when the coils are touching each other. Determine (i) the mean diameter of the coil, and (ii) the diameter of the wire if the maximum load is 100 N corresponding to allowable shearing stress is 150 MPa. Take G = 80 GPa.

Solution Stiffness of the spring, $\frac{P}{\delta} = 5$ N/mm

Using Eq. (16.1) (neglecting $d/4R$),

$$\tau = \frac{16PR}{\pi d^3}$$

or, $$R = \frac{150 \times \pi \times d^3}{16 \times 100} = 0.294\, d^3$$

Substituting in Eq. (16.3),

$$\frac{P}{\delta} = \frac{Gd^4}{64R^3 n}$$

where, $$n = \frac{60}{d}$$

or, $$5 = \frac{80000\, d^4}{64 \times (0.294d^3)^3 \times 60/d}$$

or, $$d = 3.57 \text{ mm}$$

and, $$R = 13.44 \text{ mm}$$

Example 16.6 A close coiled helical spring has a stiffness of 4 N/mm. Its solid length is 160 mm when the adjecent coils touch each other. Determine the mean diameter of the spring and the diameter of the wire if their ratio is 10:1. Take modulus of rigidity equal to 80 GPa. If the gap between the coils is 5 mm, what is the load required to close it and what is the corresponding shear stress.

Solution Stiffness, $\frac{P}{\delta} = 4$ N/mm

Mean coil radius, $R = 5d$

Number of coils, $n = \frac{160}{d}$

Using Eq. (16.3), $$\frac{P}{\delta} = \frac{Gd^4}{64R^3 n}$$

$$4 = \frac{80000 \times d^4}{64 \times 125d^3 \times 160/d}$$

or, Diameter of wire, $d = 8$ mm

Mean diameter of spring = 80 mm

Number of turns, $n = \frac{160}{8} = 20$

Deflection required to close 5 mm gap between the coils

$$\delta = 20 \times 5 = 100 \text{ mm}$$

Force required for 100 mm deflection,

$$P = 100 \times 4 = 400 \text{ N}$$

The corresponding shear stress,

$$\tau = \frac{16PR}{\pi d^3}\left(1 + \frac{d}{4R}\right)$$

$$= \frac{16 \times 400 \times 40}{\pi \times 8^3}\left(1 + \frac{8}{4 \times 40}\right)$$

$$= 158.75 \text{ MPa}$$

Example 16.7 A weight of 200 N drops axially on a close coiled helical spring. The spring has 15 coils of mean diameter 200 mm. The spring wire has diameter 25 mm. Determine the height of drop if the instantaneous compression is 120 mm and the modulus of rigidity is 80 GPa.

Solution Equivalent static load P_e for 120 mm deflection is given by equation (16.3) as follows:

$$P_e = \frac{Gd^4\delta}{64R^3 n} = \frac{80000 \times 25^4 \times 120}{64 \times 100^3 \times 15}$$

$$= 3906.25 \text{ N}$$

Equating work done by falling load and the strain energy stored,

$$P(h + \delta) = \frac{1}{2} P_e\, \delta$$

or, $$200(h + 120) = \frac{1}{2} \times 3906.25 \times 120$$

or, $$h = 1052 \text{ mm}$$

Example 16.8 A 50 N weight is dropped on a close coiled helical spring from a height of 150 mm. The spring has 10 coils of mean diameter 30 mm. If the diameter of the spring wire is 5 mm, determine the deflection and stress in the spring. Take modulus of rigidity equal to 80 GPa.

Solution Work done by impact load = 50 (150+δ) where, δ = deflection due to impact load 50 N.

If P_e is the equivalent static load which will produce δ, then energy stored in the spring

$$= \frac{P_e\delta}{2}$$

Equating,

Work done = Energy stored

$$50\,(150 + \delta) = \frac{1}{2}\,P_e\delta$$

Substitute P_e from Eq. (16.2),

$$\text{or} \quad 50(150 + \delta) = \frac{1}{2}\frac{\delta G d^4}{64R^3 n}\delta$$

$$= \frac{1}{2}\times\frac{80000\times 5^4\times\delta^2}{64\times 15^3\times 10}$$

or, $23.148\,\delta^2 - 100\delta - 15000 = 0$

or, Deflection, δ = 27.7 mm

Equivalent static load

$$P_e = \frac{\delta G d^4}{64R^3 n} = \frac{27.7\times 80000\times 5^4\delta^2}{64\times 15^3\times 10}$$

$$= 641.2 \text{ N}$$

$$\text{Shear stress,} \quad \tau = \frac{16PR}{\pi d^3}\left(1+\frac{d}{4R}\right)$$

$$= \frac{16\times 641.2\times 15}{\pi\times 5^3}\left(1+\frac{5}{60}\right)$$

$$= 424.5 \text{ MPa}$$

Example 16.9 Two springs of mean diameters 15 mm and 25 mm are connected in series. Each spring has 10 coils and the diameter of spring wire is 3 mm. Determine the deflection of the spring under maximum load if the permissible shear stress is 160 MPa and the shear modulus is 80 GPa.

Solution Using Eq. (16.1), maximum load is given by

$$P = \frac{\tau\pi d^3}{16R\left(1+\frac{d}{4R}\right)}$$

where R = 7.5 mm,

$$P = \frac{160\times\pi\times 3^3}{16\times 7.5\times\left(1+\frac{3}{30}\right)} = 102.8 \text{ N}$$

when R = 12.5 mm

$$P = \frac{160\times\pi\times 3^3}{16\times 12.5\times\left(1+\frac{3}{50}\right)} = 64 \text{ N}$$

Since both springs carry equal load, it should not exceed 64 N. The deflection corresponding to 64 N is given by Eq. (16.2) as follows

$$\delta = \frac{64nP\left(R_1^3 + R_2^3\right)}{Gd^4}$$

$$= \frac{64\times 10\times 64\times\left(7.5^3 + 12.5^3\right)}{80000\times 34}$$

$$= 15 \text{ mm}$$

Example 16.10 Repeat Example 16.9 if the smaller coil is placed inside the larger coil.

Solution Stiffness of smaller spring,

$$\frac{P_1}{\delta} = \frac{80000\times 3^4}{640\times 7.5^3} = 24 \text{ N/mm}$$

Stiffness of larger spring,

$$\frac{P_2}{\delta} = \frac{80000\times 3^4}{640\times 12.5^3} = 5.184 \text{ N/mm}$$

Since, stiffness is more for smaller coil, it will take more load for same deflection and it will reach permissible shear earlier.

$$P_1 = \frac{\tau\pi d^3}{16R\left(1+\frac{d}{4R}\right)}$$

$$= \frac{160 \times \pi \times 3^3}{16 \times 7.5 \times \left(1 + \frac{3}{30}\right)} = 102.8 \text{ N}$$

$$\delta = \frac{102.8}{24} = 4.283 \text{ mm}$$

$$P_2 = \delta \times k_2 = 4.283 \times 5.184 = 22.2 \text{ N}$$

Example 16.11 A leaf spring carries a maximum central load of 10 kN. It is built-up with 1.5 m long, 100 mm wide and 12 mm thick plates. Determine the number of plates and the initial dip if the permissible bending stress is 275 MPa and the modulus of elasticity is 200 GPa.

Solution Number of plates and initial dip is given by Eq. (16.14),

$$n = \frac{3W_o L}{2\sigma_o bt^2} = \frac{3 \times 10000 \times 1500}{2 \times 275 \times 100 \times 12^2}$$

$$= 5.68 \text{ say 6 plates}$$

$$\delta_0 = \frac{\sigma_0 L^2}{4Et} = \frac{275 \times 1500^2}{4 \times 200000 \times 12} = 64 \text{ mm}$$

Example 16.12 A laminated spring is to carry a central load of 5 kN. It is made up of 16 plates 75 mm wide, 6 mm thick. If the central dip is 40 mm and the permissible bending stress is 190 MPa, determine the length and curvature for the leaf spring.

Solution Using Eq. (16.12)

$$\sigma_0 = \frac{3W_0 L}{2nbt^2}$$

$$L = \frac{190 \times 2 \times 16 \times 75 \times 6^2}{3 \times 5000}$$

$$= 1094.9 \text{ mm say 1100 mm}$$

Radius of curvature is given by Eq. (16.9)

$$R = \frac{L^2}{8\delta_0} = \frac{1100}{8 \times 40} = 3.78 \text{ m}$$

EXERCISES

PROBLEMS

16.1 A close coilded helical spring is to carry a load of 500 N. Determine the diameter of the spring wire if the maximum permissible shear stress is 170 MPa and the mean diameter of the spring coil is 8 times the diameter of the spring wire.

16.2 A close helical spring has striffness 10 N/mm. It is to carry a maximum load of 500 N and the permissible shear stress is 200 N/mm^2. Find the diameter and length of the spring wire required for the spring if the mean diameter of the spring in 35 mm. Take the modulus of rigidity equal to 80 GPa.

16.3 A close coiled helical spring of striffness 1 N/mm has solid length 100 mm when the adjecent coils touch each other. The coil carries a maximum load of 100 N when the shear stress is 150 MPa. Determine the diameter of the wire and mean diameter of the coil. Also determine the number of turns if the modulus of rigidity is equal to 80 GPa.

16.4 A helical spring is made of 3 mm diameter wire and the mean diameter of each coil is 20 mm. Determine the maximum axial load and the corresponding deflection if the maximum permissible shear stress is 216 MPa. The solid length of the spring when the adjecent coils touch each other is 90 mm. Also determine the stiffness of the spring. Take the modulus of rigidity as 80 GPa.

16.5 A close coiled helical spring has a stiffness of 2 N/mm. Its solid length when adjecent coils are touching each other is 90 mm. The modulus of rigidity is 80 GPa. Determine the diameter of the wire and mean diameter of the coil if their ratio is 1/8. If the gap between the adjecent coils is 5 mm, what is the compressive load which will make the spring solid, i.e, when the adjecent coils touch each other.

16.6 Two coiled concentric helical springs of same length support a compressive load of 500 N. The inner spring consists of 16 turns of mean diameter 25 mm and the outer spring consists of 14 turns of mean diameter 50 mm. Calculate the maximum stress produced in each spring if the diameter of wire is 6 mm.

16.7 A railway wagon weighing 150 kN is stopped by a buffer of four springs while moving with a speed of 8 km/h. The diameter of spring wire is 25 mm and the mean diameter of the spring is 150 mm. Determine the deflection if the modulus of rigidity is 80 GPa and each spring has 10 coils.

16.8 A close coiled helical spring is made of 5 mm diameter steel wire. The mean diameter of the coils is 20 mm. If the solid length of the spring is 60 mm when the coils touch each other, find the deflection and shear stress under the load of 500 N. Also calculate the strain energy stored in the spring. Take $G = 80$ GPa.

16.9 A weight of 500 N drops on a close coiled compression spring made of 20 mm diameter steel wire. The mean diameter of the spring coil is 80 mm and the number of coils is 20. If the instant compression is 100 mm, calculate the height of drop. Take $G = 80$ GPa.

16.10 A safety value of diameter 75 mm is to blow off at a gauge pressure of 1.2 MPa. It is held is position by a close coiled spring of mean diameter 100 mm and the initial compression is 25 mm. Determine the diameter of the spring wire and the number of turns required if the maximum shear stress is not to exceed 100 MPa. Take $G = 80$ GPa.

16.11 A horizontal rigid bar of negligible weight transmits load to three springs of same material spaced at 1 m centre to centre as shown in Fig. 16.6. Spring A is made of 5 mm diameter bar having 10 coils of mean diameter 20 mm. Spring B is made of 6 mm diameter bar having 8 coils of mean diameter 24 mm and spring C is made of 8 mm diameter bar having 6 coils of mean diameter 32 mm. Find the location of 500 N weight so that the bar remain horizontal.

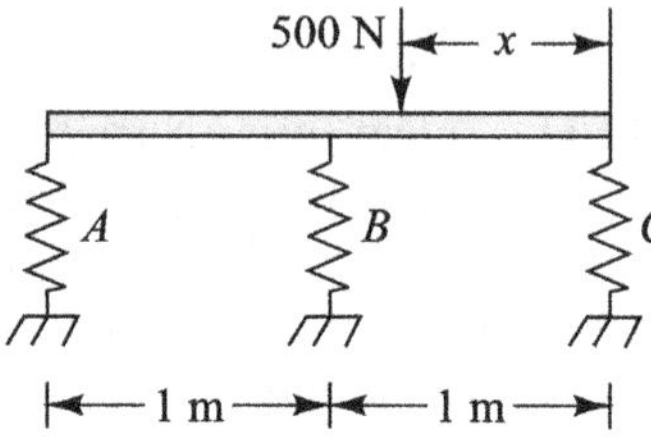

Fig. 16.6

16.12 A carriage spring of length 800 mm is built by steel plates 75 mm wide, 10 mm thick. How many plates are required to support a central load of 10 kN if the stress is not to exceed 200 MPa. Calculate the central deflection if the modulus of elesticity is equal to 200 GPa.

16.13 A steel carriage spring 1 m long is made by 16 plates 75 mm wide and 6 mm thick. Its initial dip is 37.5 mm. What will be the deflection under a central load of 2000 N if $E = 200$ GPa.

OBJECTIVE QUESTIONS

16.14 Calculate the number of turns required for a deflection of 10 mm in a steel spring of mean diameter 50 mm for a load of 400 N. Take modulus of rigidity $G = 80$ GPa and diameter of spring wire = 10 mm.
(a) 10 (b) 15 (c) 20 (d) 25

16.15 The stiffness of a spring will be higher if
(1) mean diameter of the spring coil is more,
(2) diameter of wire is more,
(3) number of turns is less,
Of these statements,
(a) only 1 is correct
(b) 1 and 2 are correct
(c) 1 and 3 are correct
(d) 2 and 3 are correct

16.16 Deflection of a helical spring will be more if,
(1) mean diameter of the spring coil is more,
(2) modulus of rigidity is more,
(3) number of turns are more,
Of these statements,
(a) only 1 is correct
(b) 1 and 2 are correct
(c) 1 and 3 are correct
(d) 2 and 3 are correct

16.17 A close coiled helical spring of mean diameter 50 mm is made of 5 mm diameter bar has 10 number of turns. Find the height through which a weight of 50 N is dropped on this spring so that the deflection of the spring is 50 mm. Take modulus of rigidity equal to 80 GPa.
(a) 25 mm (b) 50 mm
(c) 75 mm (d) 100 mm

16.18 A close coiled helical spring of 5 mm diameter steel wire has 10 turns. The mean diameter of the coil is 50 mm. If the modulus of rigidity is equal to 80 GPa, the deflection of the spring under 100 N load will be,
(a) 10 mm (b) 20 mm
(c) 160 mm (d) 200 mm

16.19 Stiffness of a close coiled helical spring having n number of turns of mean diameter D made of circular wire of diameter d is given by,

(a) $\dfrac{64\,D^4 n}{Gd^3}$ (b) $\dfrac{G\,d^4}{8\,D^3 n}$

(c) $\dfrac{G\,d^4}{32\,D^3 n}$ (d) $\dfrac{32\,D^3 n}{Gd^4}$

where, G = modulus of rigidity

16.20 A leaf spring 1 m long is to carry a central load of 8 kN. It is built with 100 mm by 10 mm plates. Determine the number of plates required for the spring if the permissible stress is 150 MPa.
(a) 6 (b) 8 (c) 10 (d) 11

16.21 A leaf spring 1 m long has initial deflection 60 mm. If the modulus of elesticity is 200 GPa and the thickness of each plate is 10 mm, then the proof stress will be,
(a) 160 MPa (b) 240 MPa
(c) 480 MPa (d) 620 MPa

16.22 All the strips in the leaf spring are bent to the same radius so that,
(a) they make a complete circle,
(b) they touch each other all along the length,
(c) they share equal bending moment,
(d) they touch each other at the ends only.

16.23 Two close coiled concentric helical spring of same length and wire diameter, support a compressive load of 300 N. The inner spring consists of 16 turns of mean diameter 30 mm and the outer spring consists of 10 turns of mean diameter 60 mm. The load shared by inner spring will be,
(a) 50 kN (b) 100 kN
(c) 150 kN (d) 250 kN

16.24 A load of 200 N is applied on a close coiled helical spring as shown in Fig. 16.7. If the number of turns, $n_1 = 4$ above the load and $n_2 = 6$ below the load. The load shared by the two parts of the spring above and below the load are,

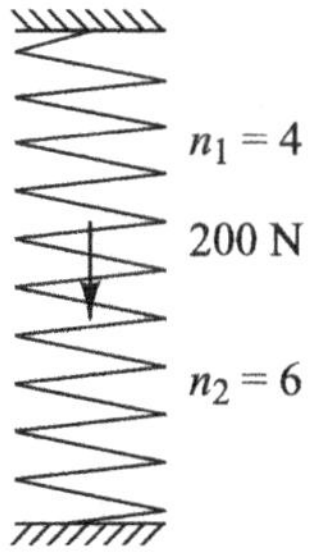

Fig. 16.7

(a) 40 N, 160 N (b) 160 N, 40 N
(c) 80 N, 120 N (d) 120 N, 80 N

16.25 A load of 153 N is applied at the junction of two springs of diameter 50 mm and 100 mm as shown in Fig 16.8. Both the springs have 6 coil of steel wire of diameter 5 mm. The load shared by two springs are,

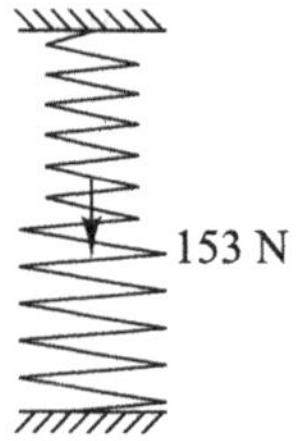

Fig. 16.8

(a) 51 N, 102 N (b) 30.6 N, 122.4 N
(c) 17 N, 136 N (d) 9 N, 144 N

16.26 A close coiled helical spring is cut into two equal pieces and combined as a parallel spring as shown in Fig. 16.9. The ratio of the maximum angular twist in two situations due same load W will be,

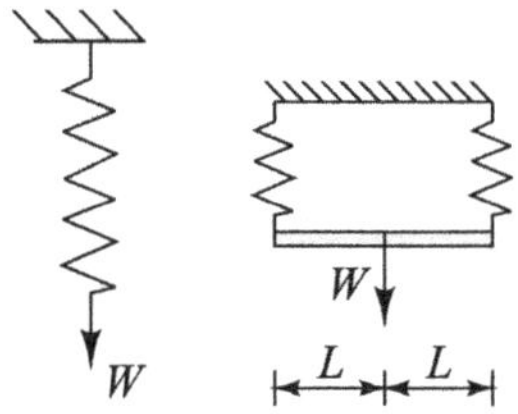

Fig. 16.9

(a) 1/2 (b) 1 (c) 1/4 (d) 1/8

16.27 For the system of springs shown in Fig. 16.10, the equivalent spring stiffnes is,

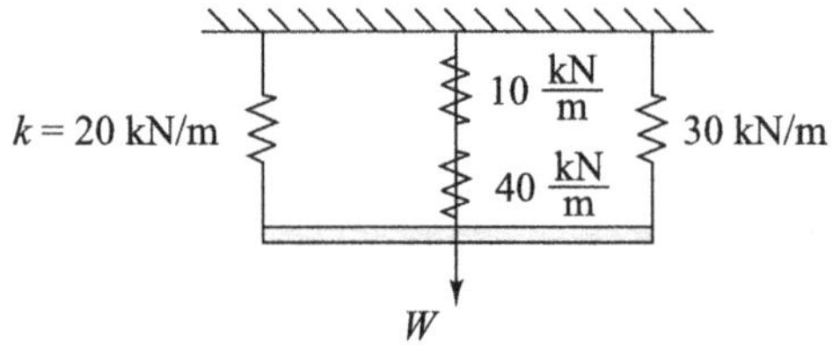

Fig. 16.10

(a) 58 kN/m (b) 62 kN/m
(c) 78 kN/m (d) 90 kN/m

16.28 A rigid bar AB is supported by a spring as shown in Fig. 16.11. What is the deflection of point B.

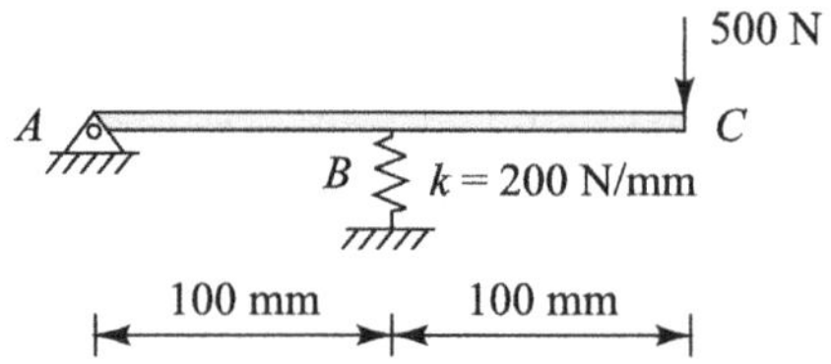

Fig. 16.11

(a) 25 mm (b) 20 mm
(c) 15 mm (d) 10 mm

16.29 A rigid beam shown in Fig. 16.12 is hinged at A and supported by two springs at C of stiffness k_2 and at D of stiffness k_1. A vertical load P is applied at B. If the ratio of stiffness $k_2/k_1 = 2$, then the ratio of forces in spring at D and C is,

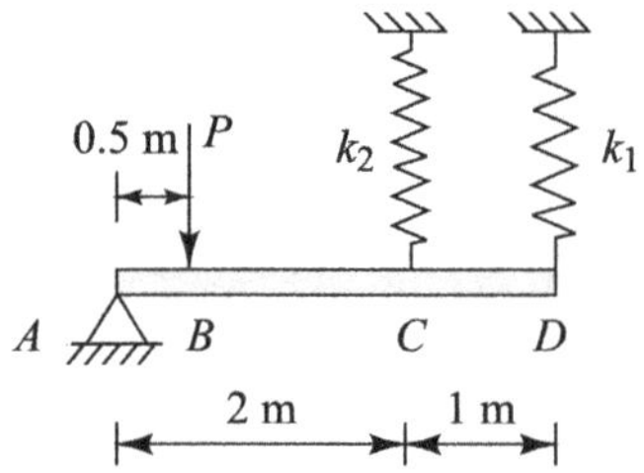

Fig. 16.12

(a) 3/4 (b) 1 (c) 4/3 (d) 2

Answers

CHAPTER 1

1.1 0.225 mm/mm^3, 123.4 N.mm/mm^3, 28%, 53%

1.2 100 GPa, 185 MPa, 245 MPa, 370 MPa, 174 N.mm/mm^3

1.3 7.5×10^{-3} mm

1.4 625250 mm^3

1.5 0.038 mm

1.6 0.639 mm

1.7 8037 mm

1.8 86 kN

1.9 404 mm

1.10 2.31 kN

1.11 0.4034 wL^2/E

1.12 0.407 mm, 63.66 MPa (tension), 25.46 MPa (tension)

1.13 50.2°C

1.14 112.5 MPa, 187.5 MPa

1.15 100.8 MPa

1.16 37.87 kN (tension), 7.87 kN (compression)

1.17 75 MPa (compression), 25 MPa (compression), 41.77°C (decrease)

1.18 12 kN

1.19 39.27 kN (tension)

1.20 10.24 MPa, 14.75 MPa

1.21 0.8 MPa, 56.42 MPa

1.22 30 MPa, – 60 MPa, 120 MPa

1.23 31.29 kN, 41.72 kN

1.24 0.0927 mm, 0.1545 mm

1.25 17.114

1.26 $(1 - 2\mu)\,(W + wAL/2)\,L/E$

1.27 $(1 + \mu)\,(1 - 2\mu)\,FL/(E(1- \mu))$

1.28 112.6 MPa, 108.8 MPa

1.29
$$\frac{\partial^2 \varepsilon_x}{\partial y^2} + \frac{\partial^2 \varepsilon_y}{\partial x^2} = \frac{\partial^2 \gamma_{xy}}{\partial x \partial y},$$

$$2\frac{\partial^2 \varepsilon_x}{\partial y \partial z} = \frac{\partial}{\partial x}\left(-\frac{\partial \gamma_{yz}}{\partial x} + \frac{\partial \gamma_{xz}}{\partial y} + \frac{\partial \gamma_{xy}}{\partial z}\right)$$

$$\frac{\partial^2 \varepsilon_y}{\partial z^2} + \frac{\partial^2 \varepsilon_z}{\partial y^2} = \frac{\partial^2 \gamma_{yz}}{\partial y \partial z},$$

$$2\frac{\partial^2 \varepsilon_y}{\partial z \partial x} = \frac{\partial}{\partial y}\left(\frac{\partial \gamma_{yz}}{\partial x} - \frac{\partial \gamma_{xz}}{\partial y} + \frac{\partial \gamma_{xy}}{\partial z}\right)$$

$$\frac{\partial^2 \varepsilon_z}{\partial x^2} + \frac{\partial^2 \varepsilon_x}{\partial z^2} = \frac{\partial^2 \gamma_{zx}}{\partial z \partial x},$$

$$2\frac{\partial^2 \varepsilon_z}{\partial x \partial y} = \frac{\partial}{\partial z}\left(\frac{\partial \gamma_{yz}}{\partial x} + \frac{\partial \gamma_{xz}}{\partial y} - \frac{\partial \gamma_{xy}}{\partial z}\right)$$

1.30 275.25 MPa, 4.02 mm

1.31 2.13 mm

1.32 3.42 MPa

1.33 4.3 kN, 41.41 MPa

1.34 11.05

1.35 27.5 N/mm

1.36 127.66 MPa, 100.9 MPa

1.37 20.11 MPa, 60.33 MPa

1.38 (d) **1.39** (b) **1.40** (c) **1.41** (c)

1.42 (d) **1.43** (c) **1.44** (a) **1.45** (d)

1.46 (d) **1.47** (a)

CHAPTER 2

2.1 (i) 87.32 MPa, 7.32 MPa
(ii) 38.43 MPa, 14.2 MPa

2.2 95.39 MPa

2.3 29.49°, 75.14 MPa, 23.67 MPa

2.4 11588 N

2.5 $\sigma_{\max/\min}$ = 51.83 MPa, –1.85 MPa, $\tau_{\max}$ = 25.92 MPa

2.6 $\sigma_{\max/\min}$ = 100.57 MPa, –34.55 MPa, $\tau_{\max}$ = 67.56 MPa

2.7 33.1 MPa, 17.38 MPa

2.8 σ_I = 46.1 MPa, c_{nx} = + 0.392,
c_{ny} = + 0.7076, c_{nz} = + 0.588
σ_{II} = 10 MPa, σ_{III} = – 26.06 MPa
τ_I = 18.05 MPa, τ_{II} = 36.1 MPa,
τ_{III} = 18.05 MPa
$\tau_{nn,\text{oct}}$ = 10 MPa, $\tau_{ns,\text{oct}}$ = 29.4 MPa
τ_{nn} = 43.3 MPa

2.9 $\tau_{ns,\text{oct}} = \dfrac{\sqrt{2}}{3}(I_1^2 - 3I_2)^{1/2}$

2.10 $\begin{vmatrix} 5 & 0 & 0 \\ 0 & 2 & 0 \\ 0 & 0 & -1 \end{vmatrix}$

2.11 $\tau_{nn,\text{cot}}$ = 6 MPa, $\tau_{ns,\text{oct}}$ = 9.59 MPa

2.12 (c) **2.13** (c) **2.14** (d) **2.15** (a)
2.16 (a) **2.17** (b) **2.18** (a) **2.19** (b)
2.20 (c) **2.21** (a) **2.22** (a)

CHAPTER 3

3.1 435×10^{-6}, -735×10^{-6},
Shear = 1170×10^{-6}, 35°

3.2 47.18 MPa, –132.89 MPa
Shear = 90 MPa

3.3 97.15 MPa, –78.1 MPa,
Max. Shear = 87.63 MPa, –27.1°

3.4 – 0.0002

3.5 82 N/mm

3.6 2.416 MPa

3.8 191.1×10^{-6}, – 0.8894, 0.4051, 0.212

3.9 $u = E/2\,(1+\mu)$, $n = \mu E/(1+\mu)\,(1-2\mu)$

3.10 $\begin{vmatrix} 0 & 32 & 0 \\ 32 & -64 & 16 \\ 0 & 16 & -16 \end{vmatrix}$ MPa

3.11 $J_1 = 200 \times 10^{-6}$, $J_2 = -\,3.156 \times 10^{-7}$,
$J_3 = -\,3.44 \times 10^{-11}$.

3.13 (c) **3.14** (d) **3.15** (d) **3.16** (c)
3.17 (a) **3.18** (c) **3.19** (c) **3.20** (?)
3.21 (c)

CHAPTER 4

4.1 15.7 kN m

4.2 80.95 MPa

4.3 40.74 MPa, 0.35°

4.4 12.09 MPa, $\theta_{A\backslash B}$ = 0.29° clockwise, $\theta_{B/C}$ = 0.59° clockwise, $\theta_{C/D}$ = 0.346° anticlockwse

4.5 AB = 46.53 mm, BC = 35.66 mm,
DE = 51.74 mm, θ_A = 0.1128 rad

4.6 $T_A = t_0L/6$, $T_B = -t_0L/3$, $\theta_{\max} = t_0L^2/9\sqrt{3}\,GJ$

4.7 0.5 kN m, 0.0124 rad, 0.00233 rad

4.8 τ_{AB} = 30.18 MPa, τ_{CD} = 32.33 MPa, 0.0554 rad

4.9 76.92 MPa, $\theta = 2.85 \times 10^{-5}x$ rad for AB, $\theta = 1.418 \times 10^{-8}\,(3000x - x^2) - 0.007$ rad for BC

4.10 1200, 6

4.11 $T/\theta = GL\,(L^2t_2 + t_1^{\,3})$

4.12 $T/\theta = GB^{\,3}H/12$

4.13 17.32 N/mm^2, 7.4×10^{-6} rad/mm

4.14 116.5 kN m

4.15 122.55 kN m, 78.35 MPa, 44.94 MPa, 0.119 rad/m

4.16 44.33 MPa, 0.0191 rad/mm

4.17 (b) **4.18** (a) **4.19** (a) **4.20** (c)
4.21 (a) **4.22** (b) **4.23** (d) **4.24** (d)
4.25 (a) **4.26** (c) **4.27** (a)

CHAPTER 5

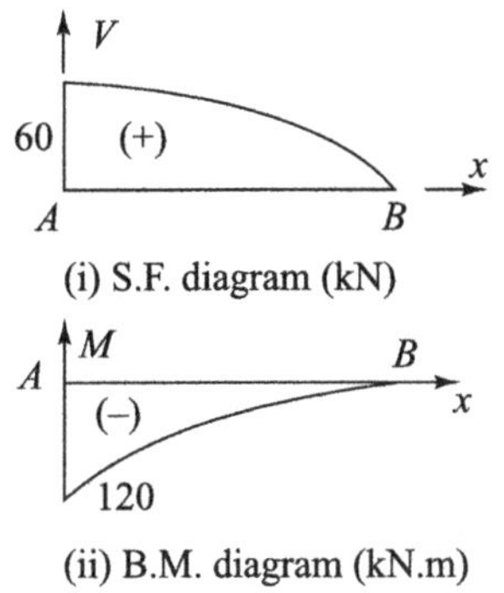

Fig. 5.37 *Ans 5.1*

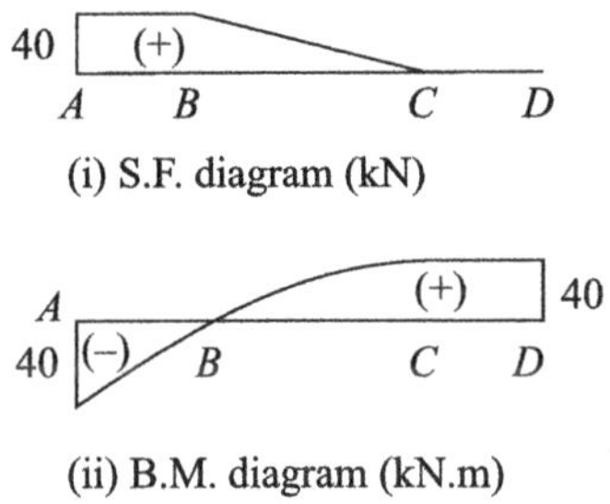

Fig. 5.38 *Ans 5.2*

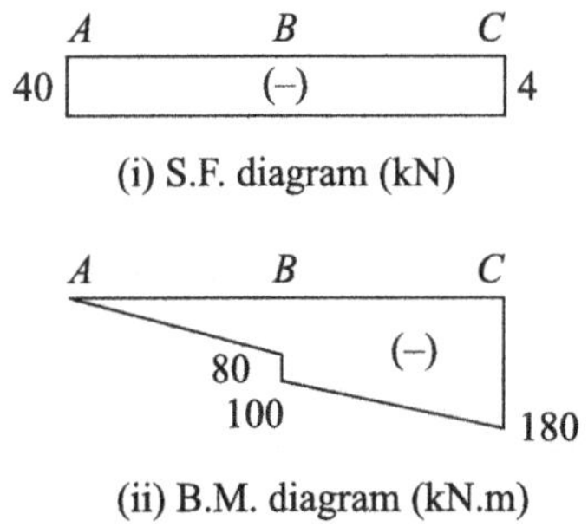

Fig. 5.39 *Ans 5.3*

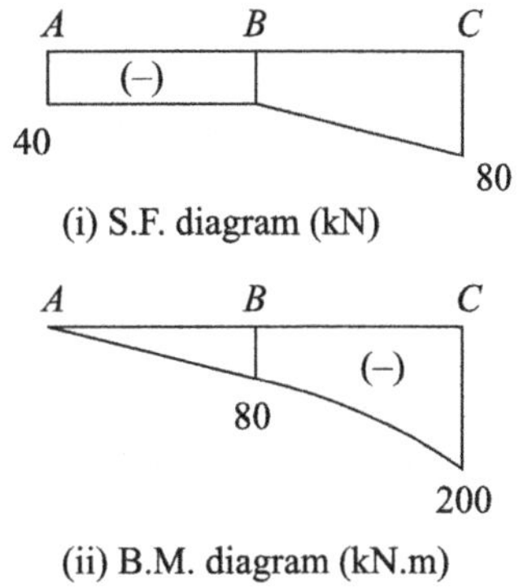

Fig. 5.40 *Ans 5.4*

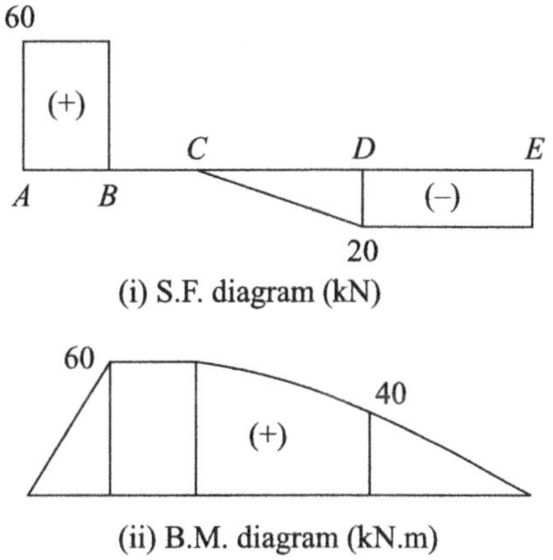

Fig. 5.41 *Ans 5.5*

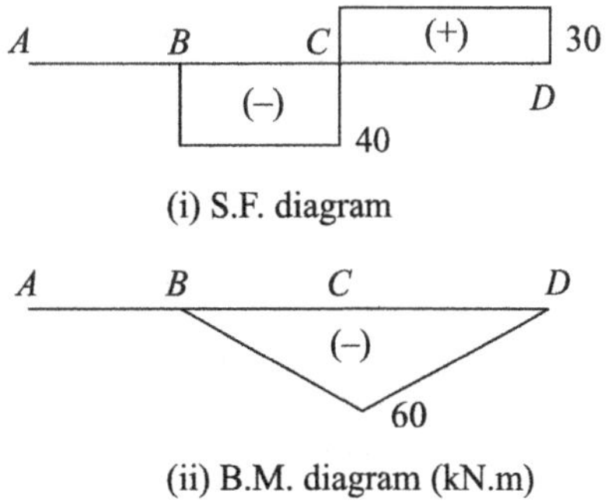

Fig. 5.42 *Ans 5.6*

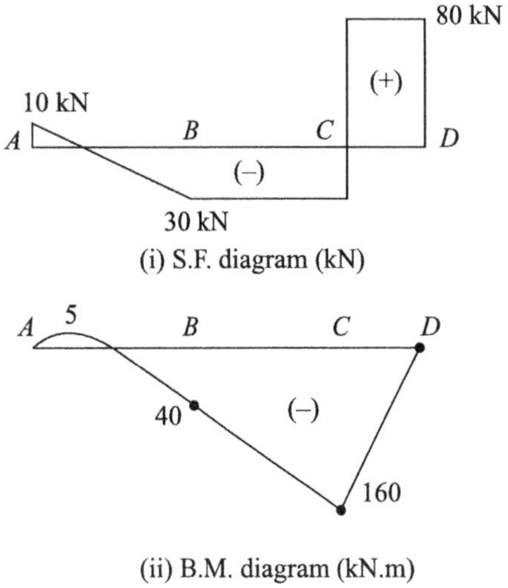

Fig. 5.43 *Ans 5.7*

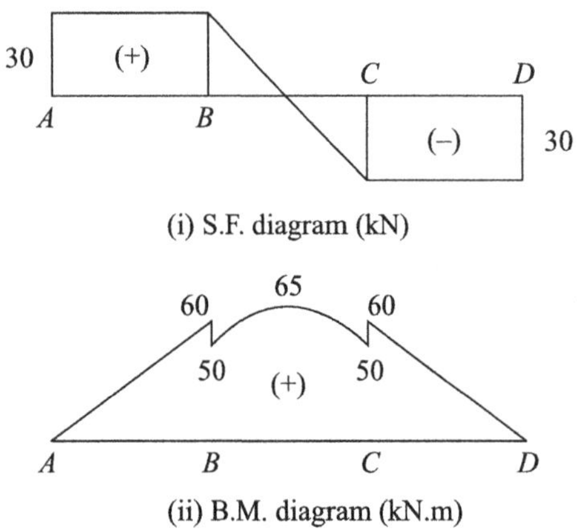

Fig. 5.44 *Ans 5.8*

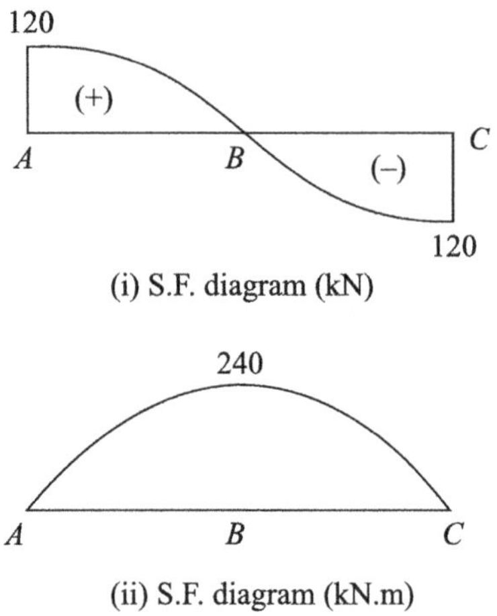

Fig. 5.45 *Ans 5.9*

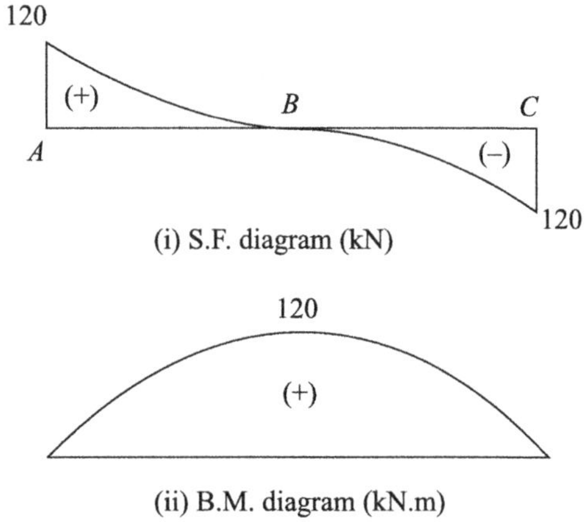

Fig. 5.46 *Ans 5.10*

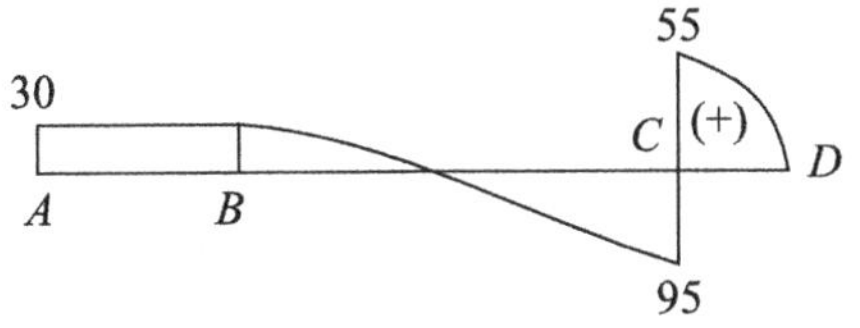

(i) S.F. diagram (kN)

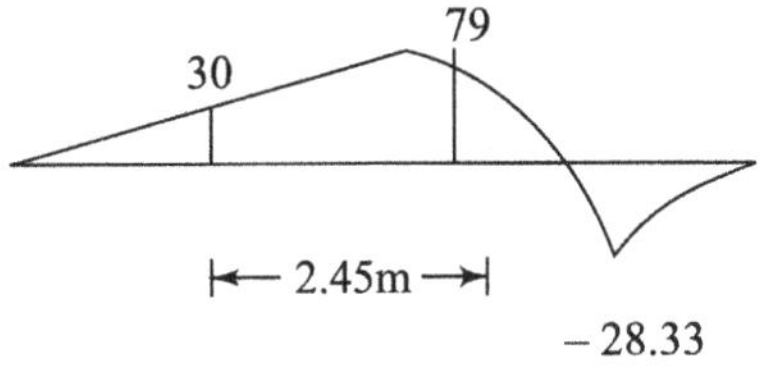

(ii) B.M. diagram (kN.m)

Fig. 5.47 *Ans 5.11*

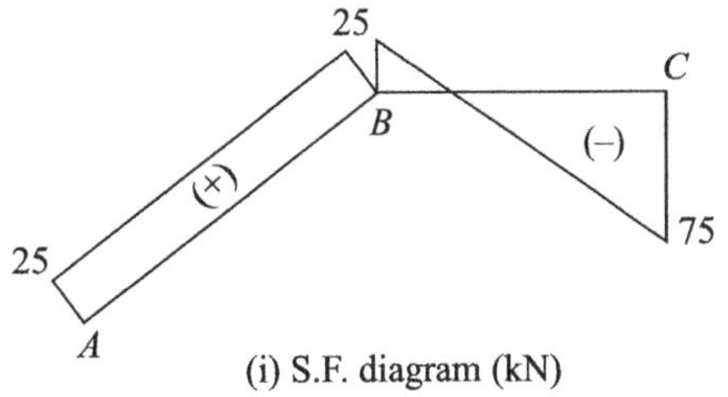

(i) S.F. diagram (kN)

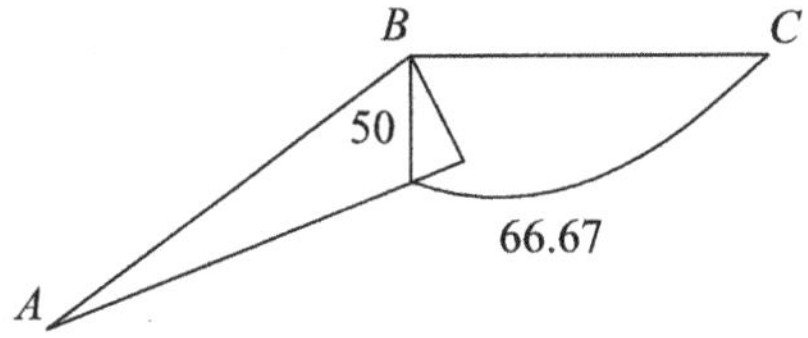

(ii) B.M. diagram (kN.M)

Fig. 5.49 *Ans 5.13*

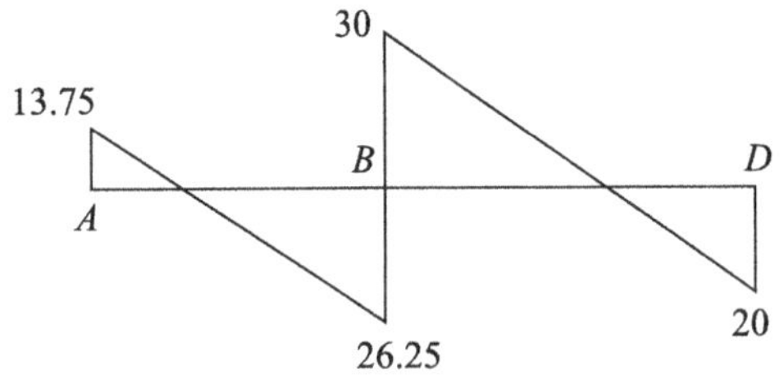

(i) S.F. diagram (kN)

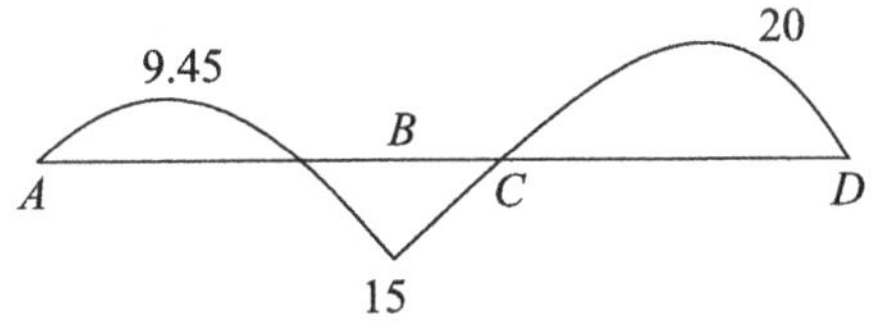

(ii) B.M. diagram (kN.m)

Fig. 5.48 *Ans 5.12*

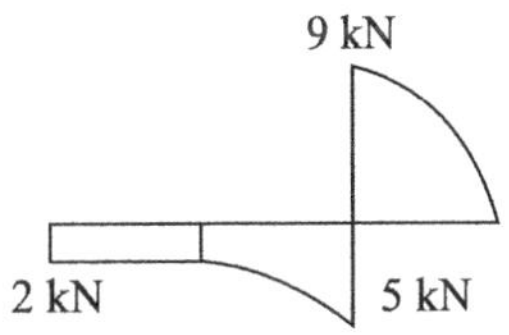

(i) S.F. diagram

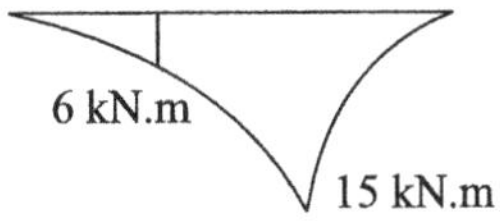

(ii) B.M. diagram

Fig. 5.50 *Ans 5.14*

5.16 (a) **5.17** (a) **5.18** (c) **5.19** (b)
5.20 (c) **5.21** (b) **5.22** (d) **5.23** (b)
5.24 (a) **5.25** (d) **5.26** (b)

CHAPTER 6

6.1 80 mm, 60 mm
6.2 81.71 mm, 48.76 mm
6.3 0 mm, 50.6 mm
6.4 55.26 mm, 36.31 mm
6.5 4.64×10^6 mm^4, 4.87×10^6 mm^4
6.6 10.15×10^6 mm^4
6.7 $\dfrac{ab^3}{28}, \dfrac{a^3b}{20}$
6.8 4.06×10^6 mm^4, 137.93×10^6 mm^4
6.9 418.277×10^6 mm^4, 195.83×10^6 mm^4
6.10 13.81×10^6 mm^4, 6.42×10^6 mm^4
6.11 392.89×10^6 mm^4, 25.63×10^6 mm^4
6.12 54.25 mm, 18.16 mm
6.13 26.53 mm, 10.75 mm
6.14 The C++ Program is as follows

```
#include<iostream.h>
#include<math.h>
#include<conio.h> //to clear output screen
#include<fstream.h> // to open output file
void main ()
{  clrscr () ;
  ofstream outfile ("iner3.cpp",ios: : out);
cout<< "Calculate centroid distance"
    <<"x, y and Ix wrt bottom, Iy"
    <<"wrt extreme left or right for"
    <<"rectangular sections" <<endl;
float  x1[10],y1[10],x3[10],y3[10],x,y,
    b[10],t[10],A=0, Ix=0, Iy=0, Ixcg,
    Iycg, Qx=0, Qy=0, Ixy=0, Imax, Imin,
    rmax, rmin, theeta;
float pie = 355./113.;
int N;
  cout<<"\n input Number of rectangles=";
            cin >> N;
for (int i=0; i< N; i++)
{cout <<"\n input for"
    <<(i+1)<<"rectangle"<<endl;
cout<<"\n input x and y coordinates"
    <<"of right top corner";
      cin>>x3[i]>>y3[i];
cout<<"\n input x and y coordinates"
  <<"of left bottom corner";
    cin>>x1[i]>>y1[i];
    b[i]=x3[i]-x1[i];
    t[i]=y3[i]-y1[i];
Ix=Ix + b[i]*pow (t[i],3)/12. +
    b[i]*t [i]*(y3[i]-t[i]/2)
    *(y3[i]- t[i]/2);
Iy=Iy + t[i]*pow(b[i],3)/12. +
    b[i]*t[i]*(x3[i]-b[i]/2)
    *(x3[i]-b[i]/2);
Qx=Qx + b[i]*t[i]*(y3[i]-t[i]/2);
Qy=Qy + b[i]*t[i]*(x3[i]-b[i]/2);
      A=A + b[i]*t[i];
}
      y=Qx/A;
      x=Qy/A;
      Ixcg=Ix - A*y*y;
      Iycg=Iy - A*x*x;
for(i=0; i < N; i++)
{Ixy=Ixy + b[i]*t[i]*(y-(y1[i]+t[i]/2))*
    (x-(x1[i]+b[i]/2));
}
Imax = (Ixcg+Iycg)/2+sqrt (pow((Ixcg-Iycg)/
    2,2) + Ixy*Ixy);
Imin = (Ixcg+Iycg)/2-sqrt (pow((Ixcg-Iycg)/
    2,2) + Ixy*Ixy);
rmax = sqrt (Imax/A);
rmin = sqrt (Imin/A);
theeta = (180./pie)*0.5*atan(-Ixy/((Ixcg-
    Iycg)/2.));
outfile << " area = "<<A<<endl;
outfile << " y = " <<y<<endl;
outfile << " x = " <<x<<endl;
outfile << " Qx = " <<Qx<<endl;
outfile << " Qy = " <<Qy<<endl;
outfile << " Ix = " <<Ix<<endl;
outfile << " Iy = " << Iy<<endl;
outfile << " Ixcg = " << Ixcg << endl;
outfile << " Iycg = " << Iycg<< endl;
outfile << " Ixy = " << Ixy<< endl;
outfile << " Imax = " << Imax<< endl;
outfile << " Imin = " <<Imin<< endl;
```

```
outfile << " rmax = " << rmax<<endl;
outfile << " rmin = " << rmin<<endl;
outfile << " theeta (in degrees) = "
          << theeta<<endl;
getch();
outfile. close();
}
```

6.15 c **6.16** c **6.17** d **6.18** b

CHAPTER 7

7.1 236.43 kN.m, 17.34 kN.m
7.2 6.66 kN/m,
7.3 97.5 MPa, 2.25 MPa
7.4 3.376 kN.m, 571 m
7.5 1.3 MPa, 0.08 MPa
7.6 198.2 mm
7.7 2.7 MPa, 50.79 MPa, 58 mm
7.8 3.2 m
7.9 50.4 MPa, 37.8 MPa
7.10 102.71 MPa, – 0.056 MPa
7.11 50 mm Upper face, 75 mm Lower face

7.12 (d)	**7.13** (a)	**7.14** (c)	**7.15** (d)
7.16 (d)	**7.17** (d)	**7.18** (d)	**7.19** (c)
7.20 (b)	**7.21** (b)	**7.22** (c)	**7.23** (d)
7.24 (d)	**7.25** (c)	**7.26** (d)	**7.27** (b)

CHAPTER 8

8.1 $F_{12} = -8$ kN, $F_{23} = -F_{34} = 6$ kN, $F_{24} = -5.2$ kN
8.2 $F_{12} = F_{24} = -F_{23} = -6.55$ kN, $F_{34} = -F_{45} = 16.55$ kN
8.3 $F_{12} = -F_{13} = -10.61$ kN, $F_{23} = F_{24} = -7.5$ kN, $F_{34} = 0$
8.4 $F_{12} = -30$ kN, $F_{13} = F_{24} = F_{34} = 42.43$ kN, $F_{23} = F_{25} = -60$ kN
8.5 $F_{12} = F_{34} = F_{45} = 15$ kN, $F_{24} = F_{35} = -21.21$ kN, $F_{14} = 30$ kN
8.6 $F_{12} = F_{23} = F_{25} = -P$, $F_{34} = -1.5P$, $F_{35} = 1.118P$, $F_{15} = 0.37P$
8.7 $F_{12} = F_{23} = -2.5$ kN, $F_{14} = F_{46} = 6.83$ kN, $F_{15} = 3.53$ kN
8.8 $F_{13} = 12$ kN, $F_{14} = 7.21$ kN, $F_{12} = 0$, $F_{35} = -F_{57} = 6.33$ kN
8.9 $F_{12} = F_{15} = 12$ kN, $F_{23} = F_{24} = F_{36} = 6$ kN, $F_{25} = F_{34} = -F_{46} = -F_{57} = -8.49$ kN
8.10 $F_{12} = F_{24} = -3.68$ kN, $F_{23} = F_{34} = F_{45} = 0$, $F_{13} = F_{35} = -1.7$ kN, $F_{67} = -2.5$ kN
8.11 $F_{AC} = F_{CD} = F_{CJ} = F_{JK} = F_{CK} = 0$, $-F_{AD} = F_{EK} = 20$ kN, $F_{ED} = F_{KL} = -16$ kN
8.12 $F_{AB} = F_{CD} = F_{BD} = F_{BC} = F_{EC} = 0$, $F_{DE} = F_{EF} = -10$ kN, $F_{AC} = F_{CF} = -F_{1F} = -F_{1K} = 6.25$ kN
8.13 $F_{12} = -54.12$ kN, $F_{16} = 20.71$ kN, $F_{26} = 41.42$ kN
8.14 $F_{AB} = -F_{CD} = -4$ kN, $F_{BC} = -9$ kN, $F_{BD} = -7.5$ kN
8.15 $F_{AB} = -6.4$ kN, $F_{AD} = 4$ kN, $F_{BD} = -5.83$ kN, $F_4 = -13.72$ kN
8.16 $F_{AB} = 4$ kN, $F_{AE} = F_{BE} = -F_{CE} = -F_{BF} = -5$ kN
8.17 $F_{12} = F_{24} = F_{23} = 0$, $F_{35} = F_{14} = 7$ kN, $F_{15} = -8.06$ kN
8.18 $F_{AB} = F_{BC} = F_{CD} = F_{DE} = -18$ kN, $F_{CK} = F_{CM} = 4.47$ kN, $F_{CL} = -8$ kN, $F_{EM} = 13.42$ kN, $E_{LM} = 8.94$ kN
8.19 $F_{AB} = -22.92$ kN, $F_{AD} = 8$ kN, $F_{BC} = -102.49$ kN, $F_{BD} = 94.49$ kN
8.20 $F_{AB} = 45$ kN, $F_{AF} = 61.85$ kN, $F_{BC} = -F_{AD} = 33.54$ kN

8.22 (b)	**8.23** (c)	**8.24** (d)	**8.25** (a)
8.26 (c)	**8.27** (d)	**8.28** (c)	**8.29** (d)
8.30 (d)	**8.31** (c)	**8.32** (c)	**8.33** (c)
8.34 (b)	**8.35** (c)	**8.36** (d)	**8.37** (a)
8.38 (a)	**8.39** (b)	**8.40** (c)	**8.41** (d)

CHAPTER 9

9.1 $P_{cr} = 10\ k/2L$
9.2 $kD^2/\ 2L$
9.3 3862.8 kN, 1931.38 kN
9.4 (i) 496.25 kN, 439.76 kN
(ii) 443.23 kN, 443.23 kN
9.5 1.2 mm
9.6 5 : 3
9.7 91

9.8 21.7 cm, 1961.28 kN
9.9 31.7 cm, 845.65 kN
9.10 6.58 kN, 1.82
9.11 609.2 N, 1218.4 N, 152.3 N, 609.3 N
9.12 205.37
9.13 $\delta = (4 - \pi) L/4\pi$
9.17 3.96 kN
9.18 180.27 N/mm^2, –18.59 N/mm^2, 16.26 mm
9.19 60 mm
9.20 1983 kN
9.21 14.7 EI/L^2
9.22 279.75 kN
9.23 226.25 kN
9.24 88.43 MPa
9.25 $P_{cr} = \dfrac{\pi^3 EI}{(3\pi - 2)L^2}$
9.26 $P_{cr} = 3.47\ EI/L^2$
9.27 $P_{cr} = 18.05\ EI_0/l^2$
9.28 (a) **9.29** (c) **9.30** (d) **9.31** (c)
9.32 (c) **9.33** (d) **9.34** (b) **9.35** (c)
9.36 (a) **9.37** (a)

CHAPTER 10

10.1 129.14 MPa
10.2 278.23 kN
10.3 0, 0.107 MPa, 0.16 MPa, 0.267 MPa
10.4 19.5 MPa, –23.77 MPa (Comp.)
10.5 0.744 MPa, –0.144 MPa (tension)
10.6 47.39 kN/m^2, 181.4 kN/m^2
10.7 (d) **10.8** (b) **10.9** (d) **10.10** (c)
10.11 (d) **10.12** (c) **10.13** (a) **10.14** (c)
10.15 (b) **10.16** (a)

CHAPTER 11

11.1 1.52
11.2 1.5×10^7 N compression
11.3 264600 $\nless 2(\sigma_y/\text{F.S.})^2$ unsafe
11.4 231280 $< 2(\sigma_y/\text{F.S.})^2$ safe
11.5 (a) 315 MPa (b) 364.5 MPa
(c) 470 MPa (d) 422.4 MPa
11.6 144.7 MPa
11.7 (c) **11.8** (a) **11.9** (d) **11.10** (a)
11.11 (a) **11.12** (a) **11.13** (c) **11.14** (b)

CHAPTER 12

12.2 1/4, 5/16
12.5 $\theta_C = -WL^2/16EI$, $\delta_C = WL^2\ h/16EI$
12.6 $\theta_A = -wL^3/3EI$, $y = -5\ wL^4/24EI$
12.7 $M_A = 5WL/54$, $M_B = 4\ WL/27$,
$5WL^3/1458\ EI$
12.8 6.4 mm
12.10 $6\ PL^3/Ebd^3$
12.11 12.94 mm
12.12 11.08 kN
12.13 19.6 kN
12.14 $WL^3/4EI$, $5WL^3/2EI$
12.15 2.667 kN
12.16 $EIy = w(L/\pi)^4 \sin \pi x/L$
12.17 $EIy = (3x^5 - 15Lx^4 + 20L^2x^3 - 8L^4x)w/360L$
12.18 $EIy = \dfrac{wL^4}{6\pi^4}\left[6\cos\dfrac{\pi x}{L} - 2\pi^2\left(\dfrac{x}{L}\right)^3 + 3\pi^2\left(\dfrac{x}{L}\right)^2 - (\pi^2 - 12)\dfrac{x}{L} - 6\right]$
12.19 $EIy = wx^5/120L - wL^2\ x^2/48 + wL^4/80$
12.20 $WL^3/60EI$
12.21 $(8L^3 + 7aL^2 - 4a^2L - 4a^3)\ W/480\ EI$
12.22 $3\ WL^3/320\ EI$
12.23 $11\ wL/40$
12.24 $wL/24$
12.25 $13\ wL/64$
12.26 $(\pi^2 - 3)\ wL/\pi^3$
12.27 $\Delta x_B = 5.33 \times 10^{-3}$ m, $\Delta y_B = -3.21 \times 10^{-2}$ m
12.28 $\Delta x_B = -2.67 \times 10^{-2}$ m, $\Delta y_B = -7.03 \times 10^{-2}$ m
12.29 (c) **12.30** (b) **12.31** (c) **12.32** (a)
12.33 (c) **12.34** (c)

CHAPTER 13

13.1 131.3 MPa, 103.125 MPa, 85.94 MPa
13.2 1.8 MPA, 153.9 kN
13.3 1.2 MPa, 824668.07 mm^3, 0.5247 mm
13.4 2.419×10^6 mm^3

13.5 5 MPa, 1.309×10^6 mm^3
13.6 1.6×10^6 mm^3
13.7 2164208.27 mm^3
13.8 78.25 MPa, 55.27 MPa, 40.355°
13.9 38.793 MPa, 86.207 MPa
13.10 60 MPa, 40 MPa
13.11 13.22 MPa, 71.62 MPa
13.12 – 0.0064 mm, – 0.000585 mm, 2.34×10^{-5} mm
13.13 (c) **13.14** (d) **13.15** (b) **13.16** (b)
13.17 (a)

CHAPTER 14

14.2 42 MPa, 0.2439 mm, 32.8 MPa, 0.225 mm
14.3 181.25 MPa, tensile
14.4 269.17 MPa, 138.85 MPa, 248.83 MPa, 179.16 MPa
14.5 (b) **14.6** (c)

CHAPTER 15

15.6 (i) 1.211 (ii) 1.3031
(iii) 1.3209 (iv) 2.3431
15.7 1.625
15.8 (i) $2\sigma_y A$ (ii) $2.573\ \sigma_y A$
15.10 $4\ M_P / R$
15.11 $43.16\ M_p / L^2$
15.12 $6\ M_p / 7$
15.13 (b) **15.14** (a) **15.15** (d) **15.16** (b)
15.17 (a) **15.18** (d) **15.19** (c) **15.20** (b)
15.21 (c) **15.22** (b) **15.23** (d) **15.24** (d)
15.25 (c)

CHAPTER 16

16.1 8 mm **16.2** 6 mm, 3.32 m
16.3 4.7 mm, 30.6 mm, 21.2
16.4 106.5 N, 31.56 mm, 3.375 N/mm
16.5 3 mm, 24 mm, 120 N
16.6 39.05 MPa, 144.42 MPa
16.7 12.77 mm
16.8 229.18 MPa, 1920 N mm
16.9 155.25 mm
16.10 24 mm, 15.6 **16.11** 0.677 m
16.12 8, 16 mm **16.13** 23 mm
16.14 (c) **16.15** (d) **16.16** (c) **16.17** (c)
16.18 (b) **16.19** (b) **16.20** (b) **16.21** (c)
16.22 (d) **16.23** (d) **16.24** (d) **16.25** (c)
16.26 (c) **16.27** (a) **16.28** (d) **16.29** (a)

Appendix A

Typical properties of some common engineering materials

Material	*Unit Weigh, w* (kN/m^3)	*Modulus of Elasticity, E* (GPa)	*Shear Modulus G* (GPa)	*Yield Stress* σ_y (MPa)	*Ultimate Stress* σ_{ult} (MPa)	*Poisson's ratio* μ
Steel (Mild)	77	200	79	250	400	0.27
Steel (High carbon)	77	200	79	345	450	0.27
Cast iron	72.3	103	41	–	700(c) 200(t)	0.25
Copper	80.6	117	41	245	345	0.33
Aluminium alloy	27	70	40	300	450	0.33
Brass	82.5	103	40	100	270	0.35
Concrete	22	14	–	0	30(c)	0.1
Timber	6	11	–	–	50(c)	–

Appendix B

Subroutines for matrix multiplication and invertion in Fortran.

```
      SUBROUTINE MUL (A,B,C,N,L,M)
      DIMENSION A (N,L), B (L,M), C(N,M)_
      Do 1 I=1,N
      Do 1 J=1,M
    1 C (I,J) = 0.0
      Do 11 I=1,N
      Do 11 J=1,M
      C (I,J)=0.0
      Do 11 K=1, L
   11 C  (I,J)=C(I,J)+A(I,K)*B(K,J)
      RETURN
      END
      SUBROUTINE  INV(A,B,N)
      DIMENSION A (N,N), B(N,N)
      DO 10 I=1,N
      DO 10 J=1,N
   10 B(I, J)=0.0
      DO 15 I=1,N
   15 B(I,I)=1.0
      DO 60 I=1,N
      TEMP=A(I,I)
      DO 20 J=1,N
      A(I,J)=A  (I,J)/TEMP
      B(I,J)=B  (I,J)/TEMP
   20 CONTINUE
      DO 50 J=1,N
      IF (I-J) 30,50,30
   30 RT=A(J,I)
      DO 40 K=1,N
      A(J,K)=A (J,K)-A (I,K)*RT
   40 B(J,K)=B (J,K)-B (I,K)*RT
   50 CONTINUE
   60 CONTINUE
      RETURN
      END
```

Appendix C

EXAMPLE 5.11 IN C++

```
#include<iostream.h>
#include<math.h>
#include<conio.h> //to clear output screan
#include<fstream.h> //to open output file
#include<stdlib.h>  // to define exit ()
                    // function
void main ()
{  clrscr ();
{ofstream outfile ("BM.CPP",ios::out);
if (!outfile) {//if outfile is not open
cerr <<"\tError: output file could not"
     <<"be opened.\n";
  exit (1);
}
cout << " Calculate Bending moment,"
     << "shear force and maximum B.M."
     << "for a beam" <<endl;
int  N,I,NODE,NM,N1,N2,NP,NDS,NS;
float  L[10], M[10], X[20], DL[20][2],
       DY[20], DF[20], P[10], BM[10][2],
       SF[10][2],BMAX[10], SUM=0,R1,R2,
       DM,DR2,X0,X1,X2,A,D,DD,DR;
cout << "NUMBER OF NODES = ?" <<endl;
cin>> N;
cout << "input: X-COORDINATES OF EACH"
<<"NODE WITH ZERO AT LEFT END"<<endl;
for (I=1; I<= N; I++)
       {cin>> X[I];
P[I]=0;M[I]=0;DF[I]=0;DY[I]=0;
BMAX[I]=0;//Initialise null matrix
       }
for (I = 1; I<=N - 1;I++)
      {
L[I]=X[I + 1] - X[I];
DL[I][1]=0; //Initialise null matrix
DL[I][2]=0; //Initialise null matrix
      }
cout <<  "\nNUMBER OF POINT LOAD = ? ";
cin >> NP;
for ( I = 1; I<=NP; I++)
       {
cout <<"\nInput: NODE NUMBER, POINT LOAD"
    <<"(DOWN +ve) = ? ";
cin>>  NODE;
cin>>  P[NODE] ;
      }
cout << "\nNUMBER OF EXTERNAL MOMENTS = ?";
cin>>  NM ;
for ( I = 1; I<=NM; I++)
      {
cout <<"\nInput: NODE NUMBER, EXTERNAL"
<<"MOMENT (ANTICLOCKWISE +ve) = ? ";
cin>> NODE;
cin>>M[NODE] ;
      }
cout <<  "\nNUMBER OF DISTRIBUTED LOAD"
     <<  "(DOWN +ve) = ? ";
cin>> NDS ;
for ( I = 1; I<=NDS; I++)
       {
cout << "\nInput: SEGMENT NUMBER,"
     << "LOADLEFT, LOADRIGHT= ? ";
cin>> NS;
cin>> DL[NS][1]>>DL[NS][2]  ;
DF[NS] = (DL[NS][1] + DL[NS][2]);
if (DL [NS] [1]>DL[NS][2])
DY [NS] = (DL[NS][ 1] + DF[NS]) *
        L[NS]/(3 * DF[NS]);
else
DY[NS] = L[NS]-(DL[NS] [ 2] + DF[NS])
        * L[NS]/(3 * DF[NS]);
        }
```

```
//**** CALCULATE SIMPLE SUPPORT REACTIONS
       // R1 AND R2 ****
cout <<"\nInput: NODE NUMBER OF SIMPLE"
      <<"SUPPORTS, N1, N2 = ?";
cin>>  N1>> N2;
       SUM = P[1];
       DR2 = M[1] + P[1] * (X[N2] -X[1]) ;
for (I = 1; I<=N-1 ; I++)
       {
       if ( I> N2)
       {
SUM =SUM + P[I+1] + DF[I] * L[I]/2;
DM = (DF[I]/2) * L[I] * (X[I+1]
      - X[N2] - DY[I]);
DR2 =DR2 + M[I + 1] - P[I + 1] *
      (X[I + 1] - X[N2]) - DM ;
      }
      else
      {
SUM =SUM + P[I + 1] + DF[I] * L[I]/2;
DM = (DF[I] / 2) * L[I] * (DY[I] +
      X[N2] - X[I + 1]);
DR2 =DR2 + M[I + 1] - P[I + 1] *
      (X[I + 1] - X[N2]) + DM;
      }}
R1 = DR2 / (X[N2] - X[N1]);
R2 = SUM - R1;
outfile << "SUPPORT REACTION R1 = "
        << R1<<endl;
outfile << "SUPPORT REACTION R2 = "
        << R2<<endl;
        P[N1] = P[N1] - R1;
        P[N2] = P[N2] - R2;
//**** CALCULATE SHEAR FORCE AND BENDING
   // MOMENTS ****//
       SF[1] [ 2] = P[1] ;
       BM[1] [ 2] = M[1];
for (I = 1; I<=N-1 ; I++)
      {
SF[I+1][1] = SF[1][2] + DF[I] * L[I]/2;
SF[I+1][2] = SF[I+1][1] + P[I+1];
DR = DF[I] * L[I] * DY[I]/2.
BM[I+1][1] = BM[I][2]+SF[I][2]*L[I]+DR;
BM[I+1][2] = BM[I+1][1] + M[I+1];
        }
outfile <<"SEG, SHEAR LEFT, SHEAR RIGHT,"
       <<"MOMENT LEFT, MOMENT RIGHT"<<endl;
for (I=1; I<=N-1; I++)
       {
outfile.width (4);
outfile << I ;
outfile.width (10);
outfile << -SF[I][2];
outfile.width (12);
outfile << -SF[I+1][ 1];
outfile.width (12);
outfile <<-BM[I][2];
outfile.width (12);
outfile <<-BM[I+1] [1]<<endl;
       }
//***SEARCH FOR MAX BM. WITHIN A SEGMENT
  // WHERE SF. IS ZERO***//
for (I = 1; I<= N-1; I++){
if (SF[I][2] * SF[I+1][1] >=0) goto escl;
       else
if (DL[I][1]== DL[I][2])
       {
       X0 = -SF[I][2]/DL[I][1];
       goto esc;
       }
       else
       {
A = (DL[I][2]-DL[I][1])/L[I];
D = 4 * DL[I][1]*DL[I][1]-8*SF[I][2]*A;
X1 = (-2 * DL[I][1]+sqrt(D))/(2 * A);
X2 = (-2 * DL[I][1]-sqrt(D))/(2 * A);
       if (X1> 0 && X1< L[I])
       X0 = X1;
       else
       X0 = X2;
       }
esc:   DD = DL[I][2]-DL[I][1];
BMAX[I] = BM[I][2]+SF[I][2]*X0 +
         (3*DL[I][1]+DD*X0/L[I])*X0*X0/6;
outfile<<"SEGMENT   MAX BM"
       <<"DIST FROM LEFT"<<endl;
outfile<< I<<"          "
       <<-BMAX[I] <<"     "<< X0;
```

```
escl: continue;
}
getch ();
outfile.close();
}}
```

The output file for Example 5.12 will be as follow.

```
  SUPPORT REACTION R1 = 11
  SUPPORT REACTION R2 = 29

SEG.    SHEAR LEFT.    SHEAR RIGHT.  MOMENT LEFT.  MOMENT RIGHT
  1            11             -13          0              -4
  2           -13             -13         -4             -30
  3           -13             -13         10             -16
  4            16              16        -16               0
SEGMENT        MAX BM         DIST FROM LEFT
1              10.083333         1.833333
```

Appendix D

The computer program for determinate truss analysis writes equilibrium equations for each joint e.g. 6 equilibrium equations for the truss shown in Fig. D-1 can be written as follow.

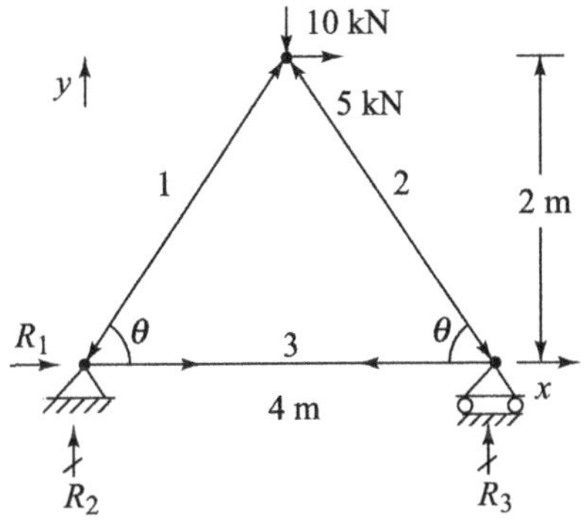

Fig. D-1

Joint 1

$$\Sigma F_x = 0, \qquad -F_1 \cos\theta + F_3 + R_1 = 0$$

$$\Sigma F_y = 0, \qquad -F_1 \sin\theta + R_2 = 0$$

Joint 2

$$\Sigma F_x = 0, \qquad F_1 \cos\theta - F_2 \cos\theta + 5 = 0$$

$$\Sigma F_y = 0, \qquad F_1 \sin\theta + F_2 \sin\theta - 10 = 0$$

Joint 3

$$\Sigma F_x = 0, \qquad F_2 \cos\theta - F_3 = 0$$

$$\Sigma F_y = 0, \qquad -F_2 \sin\theta + R_3 = 0$$

These are written in matrix form $[C]\{A\} = \{F\}$ as follow.

$$\begin{bmatrix} -\cos\theta & 0 & 1 & 1 & 0 & 0 \\ -\sin\theta & 0 & 0 & 0 & 1 & 0 \\ \cos\theta & -\cos\theta & 0 & 0 & 0 & 0 \\ \sin\theta & \sin\theta & 0 & 0 & 0 & 0 \\ 0 & \cos\theta & -1 & 0 & 0 & 0 \\ 0 & -\sin\theta & 0 & 0 & 0 & 1 \end{bmatrix} \begin{bmatrix} F_1 \\ F_2 \\ F_3 \\ R_1 \\ R_2 \\ R_3 \end{bmatrix} = \begin{bmatrix} 0 \\ 0 \\ -5 \\ 10 \\ 0 \\ 0 \end{bmatrix}$$

These equations are then solved by gausselemination method.

```
#include<iostream.h>
#include<math.h>
#include<conio.h>

void main()
{
clrscr ()
cout.precision(2);
float x[30], y[30], ang[6], L, fx[20],
fy[20], d, angl, pi = 355./113.
cosine, sine, c[20][21], a[20];
int i,j,k, n, m, np[3], nr[3], ps, rs,
jf, nf[20], nodl[30], nod2[30], p,g,h;

cout<<"Input no. of joints"
<<"(less than 15) =": cin>>n;
//initialize force coefficients
for (g = 1; g<=2*n; g++)
{
for (h = 1; h<=2*n+1; h++)
c[g][h] = 0;}
for (g = 1; g<=2*n; g++)
{       fx[g] = 0;
        fy[g] = 0;
        a[g] = 0;
```

```
}
for (i=1; i <=n; i++)
{
cout<<"\n Input x and y coordinate"
<<" of joint no." <<i<<endl;
cin >> x[i] >> y[i];
}
count<<"\n Input no. of members =";
cin>>m;
for (i = 1; i <=m; i++)
{
cout<<"\n Input initial and end"
<<"node no. of member no. "<<i<<endl;
cin>>nod1[i]>>nod2[i];
}
cout<<"\n How many joints have pin"
<<" support reaction = ?"; cin>>ps;
if (ps>0)
{
for (i = 1; i<=ps; i++)
{
{cout<<"\n Input pin support joint no.";
cin>>np[i];
}}
cout<<"\n How many roller supports?";
cin>>rs;
if (rs>0){
for (i = 1; i<=rs; i++)
{
cout<<"\n Input"
<<" 1: joint no. of roller support"
<<"\n 2: angle (in degrees) at which"
<<"the roller reaction is"
<<"\n inclined counterclockwise"
<<"\n wrt +ve y-axis =" <<endl;
cin>>nr[i]>> ang[i];
}}
cout<<"\n How many joints have"
    <<" external forces?";
cin>>jf;
for (i = 1; i<=jf; i++)
{
cout<<"\n Input the joint no. =";
cin>>nf[i];
cout<<"\n Input x direction force =";
cin>>fx[nf[i]];
cout<<"\n Input y direction force =";
cin>>fy[nf [i]];
}
//Direction cosine for each member
for (k = 1; k <= m; k++)
{
i=nod1 [k]; j=nod2[k];
L=sqrt (pow(x[j]-x[i],2)
    +pow (y[j]-y[i],2));
cosine=(x[j]-x[i])/L;
sine=(y[j]-y[i])/L;
c[2*i-1][k] = cosine;
c[2*i][k] = sine;
c[2*j-1][k] = -cosine;
c[2*j][k] = -sine;
}
//Direction cosine of each pin support
if (ps>0){
for (k = 1; k<= ps; k++)
{
i = np[k];
c[2*i-1][m+2*k-1] = 1;
c[2*i][m+2*k] = 1;
}}
//Direction cosine of roller support
if (rs>0)
{
for (k = 1; k<= rs; k++)
{
i = nr[k];
angl = ang[k]*pi/180.;
c[2*i-1] [2*n] = -sin (angl);
c[2*i][2*n] = cos (angl);
}}
//External force at nodes
for (k = 1; k<= jf; k++)
{
i = nf [k];
c[2*i - 1][2*n+1] = -fx[i];
c[2*i][2*n+1] = -fy[i];
}
//Gauss Elimination
n*=2;
for (i = 1; i <n; i++)
{
j = i;
```

```
for (int nn=i+1; nn<=n; nn++)
{
if (fabs (c[j][i])<fabs (c[nn][i]))j=nn;
}
if (j !=i)
{
for (int mm = 1; mm<=n+1; mm++)
{
float temp = c[i][mm];
c[i][mm] = c[j][mm];
c[j][mm] = temp;
}}
for (k = i+1; k<= n; k++)
{
d=-c[k][i]/c[i][i];
for (p = 1; p<=n+1; p++)
c[k][p] = c[k][p] + d*c[i][p];
}}
a[n] = c[n][n+1]/c[n][n];
for (int ii = 1; ii < n; ii++)
{
int kk = n-ii;
float sum = 0.0;
int jj = n - kk;
for (int pp = 1; pp<=jj; pp++)
{
int mmm =kk + pp;
sum=sum+c[kk][mmm]*a[mmm];
}
a[kk]=(c[kk][n+1]-sum)/c[kk][kk];
}
for (k = 1; k <=m; k++)
cout<<"\nForce in member no."
<<k<<"="<<a[k]<<endl;
getch ();
for (k ; k <= m+2*ps; k++)
cout<<"\nHinge reaction no."
<<(k-m)<<" = "<<a[k]<<endl;
for (k ; k <=n; k++)
cout <<"\nRoller reaction no."
<<(k-m-2*ps)<<" = "<<a[k]<<endl;
getch();
}
```

```
Input no. of joints (less than 15) = 3
Input x and y coordinate of joint no. 1
0 0
input x and y coordinate of joint no. 2
2 2
Input x and y coordinate of joint no. 3
4 0
Input no. of members = 3
Input initial and end node no. of member no.1
1 2
Input initial and end nod no. of member no.2
2 3
Input initial and end nod no. of member no.3
1 3
How many joints have pin support reaction =1
Input the joint no. of pin support = 1
How many joints roller support = ?1
Input 1: the joint no. of roller support
2: angle (in degrees) at which the roller
reaction is inclined counterclockwise wrt
for +ve y-axis =
3 0
How many joints have external forces = ? 1
Input the joint no. = 2
Input x direction force = 5
Input y direction force = -10
Force in member no. 1 = - 3.54
Force in member no. 2 = - 10.61
Force in member no. 3 = 7.5
Hinge reaction no. 1 = -5
Hinge reaction no. 2 = 2.5
Roller reaction no. 1 = 7.5
```

Appendix E

Useful properties of plane area

Shape	Area	$I_{x,cg}$	I_x	I_y	J_z
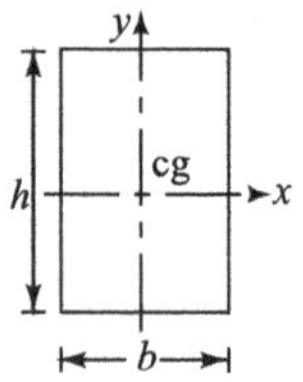 Rectangle	bh	$\dfrac{bh^3}{12}$	$\dfrac{bh^3}{12}$	$\dfrac{hb^3}{12}$	$\dfrac{bh}{12}(b^2+h^2)$
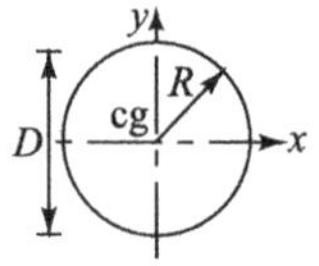 Circle	$\pi R^2 = \dfrac{\pi D^2}{4}$	$\dfrac{\pi R^4}{4} = \dfrac{\pi D^4}{64}$	$\dfrac{\pi R^4}{4}$	$\dfrac{\pi R^4}{4}$	$\dfrac{\pi R^4}{2}$
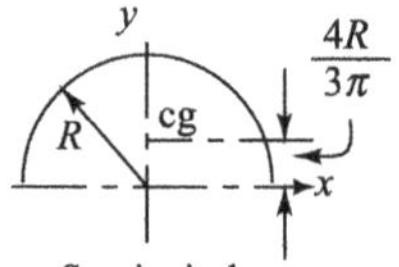 Semi–circle	$\dfrac{\pi R^2}{2}$	$\left(\dfrac{\pi}{8} - \dfrac{8}{9\pi}\right)R^4$	$\dfrac{\pi R^4}{8}$	$\dfrac{\pi R^4}{8}$	$J_{cg} = \left(\dfrac{\pi}{4} - \dfrac{8}{9\pi}\right)R^4$
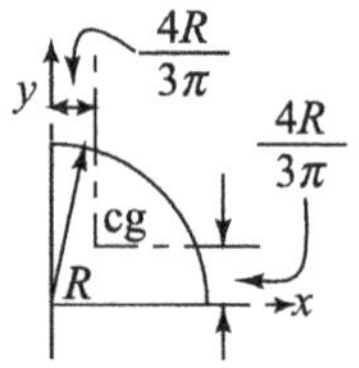 Quarter–circle	$\dfrac{\pi R^2}{4}$	$\left(\dfrac{\pi}{16} - \dfrac{4}{9\pi}\right)R^4$	$\dfrac{\pi R^4}{16}$	$\dfrac{\pi R^4}{16}$	$J_{cg} = \left(\dfrac{\pi}{8} - \dfrac{8}{9\pi}\right)R^4$
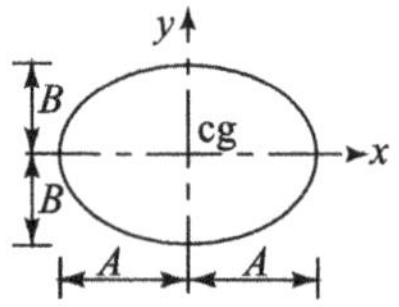 Ellipse	πAB	$\dfrac{\pi AB^3}{4}$	$\dfrac{\pi AB^3}{4}$	$\dfrac{\pi A^3 B}{4}$	$\dfrac{\pi AB}{4}\left(A^2+B^2\right)$

Appendix F

Table F.1

Beam and loading	*End rotation*	*Maximum deflection*
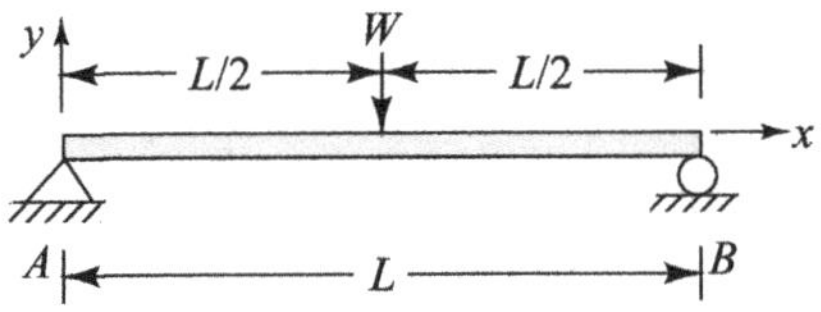	$\frac{1}{16}\cdot\frac{WL^2}{EI}$	$\frac{1}{48}\cdot\frac{WL^3}{EI}$
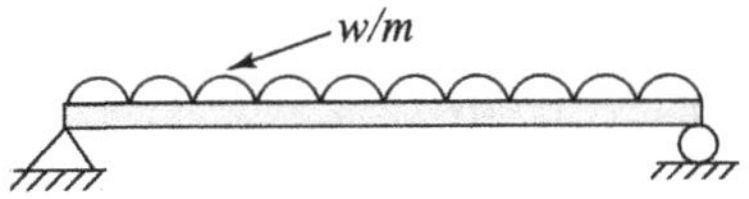	$\frac{1}{24}\cdot\frac{wL^3}{EI}$	$\frac{5}{384}\cdot\frac{wL^4}{EI}$
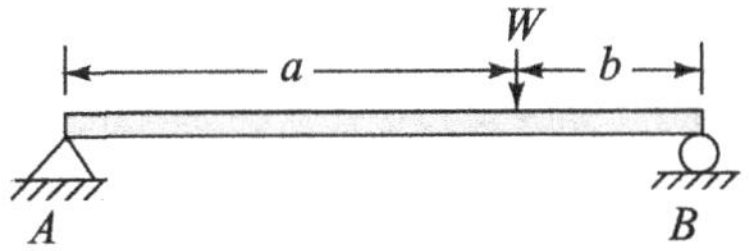	For $a > b$ $\theta_A = +\frac{Wb(L^2 - b^2)}{6EIL}$ $\theta_B = \frac{-Wa(L^2 - a^2)}{6EIL}$	$\frac{Wb(L^2 - b^2)^{3/2}}{9\sqrt{3}EIL}$ $x = \sqrt{(L^2 - b^2)/3}$ y at $L/2 = \frac{Wb\,(3L^2 - 4b^2)}{48EI}$
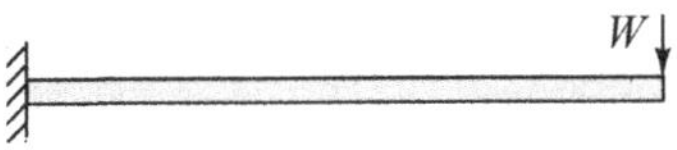	$\frac{1}{2}\,\frac{wL^2}{EI}$	$\frac{1}{3}\,\frac{WL^3}{EI}$

(*Contd*)

Beam and loading	*End rotation*	*Maximum deflection*
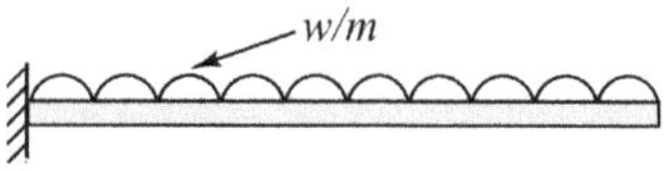	$\frac{1}{6}\frac{wL^3}{EI}$	$\frac{1}{8}\frac{wL^4}{EI}$
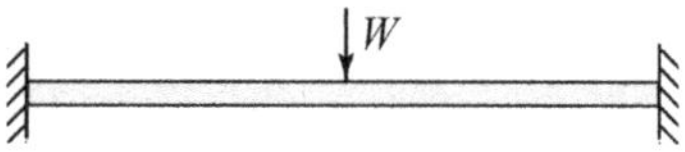	0	$\frac{1}{192}\frac{WL^3}{EI}$
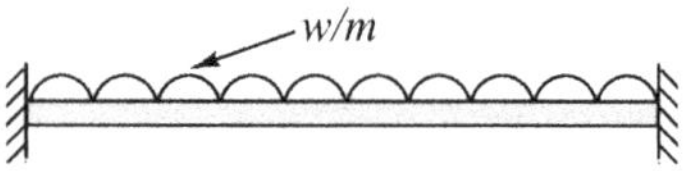	0	$\frac{1}{384}\frac{wL^4}{EI}$
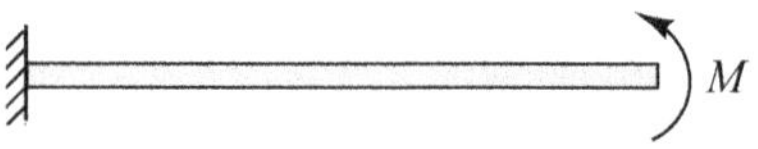	$\frac{ML}{EI}$	$\frac{1}{2}\cdot\frac{ML^2}{EI}$
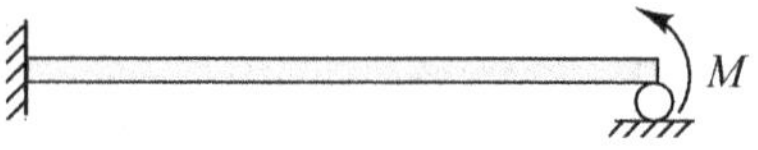	$\frac{1}{4}\cdot\frac{ML}{EI}$	$\frac{1}{27}\frac{ML^2}{EI}$ at $x = 2L/3$
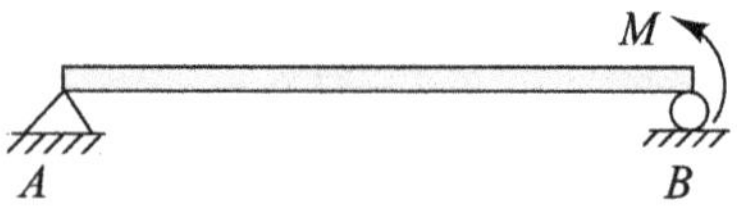	$\theta_A = \frac{1}{6}\frac{ML}{EI}$ $\theta_B = \frac{1}{3}\cdot\frac{ML}{EI}$	$\frac{1}{9\sqrt{3}}\cdot\frac{ML^2}{EI}$ at $x = L/\sqrt{3}$

Appendix G

Let l_1, m_1 and n_1 = direction cosines of x' axis wrt x, y and z axes

l_2, m_2 and n_2 = direction cosines of y' axis wrt x, y and z axes

l_3, m_3 and n_3 = direction cosines of z' axis wrt x, y and z axes

These nine cosines are related as follow.

$$l_1^2 + m_1^2 + n_1^2 = 1 \quad (1)$$

$$l_2^2 + m_2^2 + n_3^2 = 1 \quad (2)$$

$$l_3^2 + m_3^2 + n_3^2 = 1 \quad (3)$$

$$l_1 l_2 + m_1 m_2 + n_1 n_2 = 0 \quad (4)$$

$$l_2 l_3 + m_2 m_3 + n_2 n_3 = 0 \quad (5)$$

$$l_3 l_1 + m_3 m_1 + n_3 n_1 = 0 \quad (6)$$

Solution for these direction cosines is possible if 3 direction cosines say l_1, m_1 and l_2 are known. The computer program for stress/strain analysis requires inclination of x' axis wrt x and y axes and inclination of y' axis wrt x axis. Location of x'-y'-z axes is unique with reference to x - y - z axes if these three direction cosines l_1, m_1 and l_2 are specified. The computer programme will evaluate the rest of the direction cosines using six equations as above.

The cubic stress equation (2.13b) is solved by standard procedure as follow.

$$x^3 + bx^2 + cx + d = 0$$

Its solution is expressed in the form

$$x_1 = g \cos(\phi/3) - b/3$$

$$x_2 = g \cos(120° + \phi/3) - b/3$$

$$x_3 = g \cos(240° + \phi/3) - b/3$$

where, $g = 2\sqrt{-p/3}$

$$p = (3c - b^2)/3$$

$$\phi = \frac{q}{2(-p^3/27)^{1/2}}$$

$$q = (2b^3 - 3bc + 27d)/27$$

```
#include<iostream.h>
#include<math.h>
#include<conio.h>
#include<fstream.h>
#include<stdlib.h> \\for exit fuction
void main ()
{       clrscr ();
float cxx, cxy, cyx, c[3][3], ct[3][3],
        s[3][3], cs[3][3], st[3][3], pr[3],
        A,B,C,D,E,g,p,q,fi,u[3],v[3],
        w[3],z[3],l[3],m[3],n[3];
long double pie = acos (-1);
int i,j,k;
cout.precision (2);
cout<<"Given stresses wrt x,y,z axes"
<<"\nCalculate stresses on x', y' and"
<<"z' axes\n\nInput inclilnation of x'"
<<"axis wrt x and y axes";
cin>>cxx>>cxy;
cout<<"\nInput inclination of y' axis"
<<"wrt x axis";
cin>>cyx;
c[1][1] = cos(cxx*pie/180.);
c[1][2] = cos(cxy*pie/180.);
c[1][3] = sqrt(1-c[1][1]*c[1][1])
              -c[1][2]*c[1][2]);
c[2][1] = cos(cyx*pie/180.);
```

```
A = c[1][3]*c[1][3]+c[1][2];
B = 2*c[1][1]*c[2] [1]*c[1][3];
C = pow(c[1][2]*c[2][1],2)+
   pow(c[1][1]*c[2][1],2)-
   c[1][2]*c[1][2];
D = B*B-4*A*C;
if (D<0)
{cout<<"\nThe roots are imaginary";
exit(1);}else
c[2][3] = (-B+sqrt(D))/(2.*A);
c[2][2] = -(c[1][1]*c[2][1]+
         c[1][3]*c[2][3])/c[1][2];
E = (c[1][3]/c[1][1] -c[2][3]/c[2][1])/
   (c[2][2]/c[2][1]-c[1][2]/c[1][1]);
c[3][3] = sqrt(c[1][1]*c[1][1]/
(c[1][2]*c[1][2]*E*E+c[1][3]*c[1][3]+
c[1][1]*c[1][1]*E*E+c[1][1]*c[1][1]+
                2*c[1][2]*c[1][3]*E));
c[3][2] = E*c[3][3];
c[3][1] = -(c[1][2]*c[3][2]+
         c[1][3]*c[3][3])/c[1][1];
cout<<"\nDirection cosines"
<<"\nc11,    c12,    c13 = "
<<c[1][1]<<"\t"<<c[1][2]<<"\t"<<c[1][3]
<<"\nc21,    c22,    c23 ="
<<c[2][1]<<"\t"<<c[2][2]<<"\t"<<c[2][3]
<<"\nc31,    c32,    c33 ="
<<c[3][1]<<"\t"<<c[3][2]<<"\t"<<c[3][3];
for(i=1;i<=3;i++)
for(j=1;j<=3;j++)
ct[j][i] = c[i][j];
cout<<"\n\nInput axial stress in"
   <<"x, y, z directions";
cin>>s[1][1]>>s[2][2]>>s[3][3];
cout<<"\nInput shear stress in"
   <<"xy, xz and yz directions";
cin>>s[1][2]>>s[1][3]>>s[2][3];
s[2][1] = s[1][2];
s[3][1] = s[1][3];
s[3][2] = s[2][3];
for(i=1;i<=3;i++)
for(j=1;j<=3;j++){
cs[i][j] = 0.0;
for(k=1;k<=3;k++)
cs[i][j] = cs[i][j] + c[i][k]*s[k][j];
   }
for(i=1;i<=3;i++)
for(j=1;j<=3;j++){
st[i][j] = 0.0;
for(k=1;k<=3;k++)
st[i][j] = st[i][j]+ cs[i][k]*ct[k][j];
   }
cout<<"\nTransformed stressess"
<<"\nst11,   st12,   st13 = "
<<st[1][1]<<",  " <<st[1][2]
<<","<<st[1][3]
<<"\nst21,   st22,   st23 = "
<<st[2][1]<<",  "<<st[2][2]
<<"  "<<st[2][3]
<<"\nst31,   st32,   st33 = "
<<st[3][1]<<",  "<<st[3][2]
<<",           " <<st[3][3]<<endl;
B = -(st[1][1]+st[2][2]+st[3][3]);
C = st[1][1]*st[2][2]+st[1][1]*st[3][3]
   +st[2][2]*st[3][3]-st[1][2]*st[1][2]
   -st[1][3]*st[1][3]-st[2][3]*st[2][3];
D = -(st[1][1]*st[2][2]*st[3][3]+
   2*st[1][2]*st[1][3]*st[2][3]-
   st[1][1]*st[2][3]*st[2][3]-
   st[2][2]*st[1][3]*st[1][3]-
   st[3][3]*st[1][2]*st[1][2]);
p = C - B*B/3;
q = (2*B*B*B - 9*B*C + 27*D)/27;
g = 2*sqrt(-p/3);
fi = acos(-q/(2*sqrt(-p*p*p/27)));
pr[1] = g*cos(fi/3) -B/3;
pr[2] = g*cos((fi/3)+pie*120/180)-B/3;
pr[3] = g*cos((fi/3)+pie*240/180)-B/3;
cout<<"\nPrincipal stressess"
<<"\npr1 = "<<pr[1]<<"\npr2 = "<<pr[2]
<<"\npr3 = "<<pr[3]<<endl;
cout<<"\nDirection Cosine of Principal"
     <<"Stressess/n";
for(i=1;i<=3;i++){
u[i] = ((s[2][2] - pr[i])*(s[3][3]-
      pr[i])-s[2][3]*s[2][3]);
v[i] = -(s[1][2]*(s[3][3]-pr[i])-
      (s[1][3]*s[2][3]));
w[i] = ((s[1][2]*s[2][3])-
      s[1][3]*(s[2][2]-pr[i]));
z[i] = 1/(sqrt(u[i]*u[i]+
      v[i]*v[i]+w[i]*w[i]));
```

```
l[i] = u[i]*z[i];
m[i] = v[i]*z[i];
n[i] = w[i]*z[i];
cout<<"\nl"<<i<<"\tm"<<i<<"\tn"
<<i<<" = "<<l[i]<<",\t"<<m[i]
<<",\t"<<n[i];}getch();}
```

Example 2.18 is solved by this programme and its input/output are as follow. Readers may check the validity of the programme as its output show some variation with the results of Example 2.18.

```
Given stresses wrt x, y, z axes
Calculate stresses on x', y' and z' axes

Input inclination of x' axis wrt x and y axes 54.7356 54.7356
Input inclination of y' axis wrt x axis 60
Direction cosines
c11,        c12,        c13 = 0.57735         0.57735        0.57735
c21,        c22,        c23 = 0.5            -0.78229        0.28229
c31,        c32,        c33 = -0.633387      -0.12953        0.762917

Input axial stress in x, y, z directions 3 4 2
Input shear stress in xy, xz and yz direction 2 0 1
Transformed stresses
st11,       st12,       st13 = 5.000001,     -1.22927,       -1.030508
st21,       st22,       st23 = -1.22927,      1.35104,        0.114033
st31,       st32,       st33 = -1.032584,     0.114294,       2.563591

Principal stresses
pr1 = 5.706592
pr2 = 0.948855
pr3 = 2.259185

Direction Cosine of Principal Stresses
l1          m1          n1 = 0.569929,        0.793329,       0.214032
l2          m2          n2 = 0.60542,        -0.576645,       0.548587
l3          m3          n3 = -0.574773,       0.20531,        0.792139
```

Index

SI units

Basic units	*Name*	*Symbol*
Length	meter	m
Mass	kilogram	kg
Time	Second	s
Temperature	kelvin	k

Derived units	*Name*	*Symbol*	*In terms of*	
			Basic units	*Other units*
Force	Newton	N	$kg.m/s^2$	–
Pressure, stress	Pascal	Pa	$kg/m.s^2$	N/m^2
Energy, work	Joule	J	$kg.m^2/s^2$	N.m
Power	Watt	W	$kg.m^2/s^3$	J/s
Moment of force	Newton meter	N.m	kgm^2/s^2	–
Frequency	Hertz	Hz	cycles/s	–

In many problems these are not convenient units. For stress MPa = MN/m^2 = N/mm^2 is generally used. For modulus of elasticity GPa = 1000 N/mm^2 is used. For conversion, 1 horsepower = 550 foot pound/s = 746 W, Frequency $f = \frac{\omega}{2\pi}$ cycles where ω = angular velocity or circular frequency measured in rad/s.

Roman Script

α	alpha	μ	mu
β	beta	π	pi
γ	gamma	ρ	row
δ	delta	σ	sigma
Δ	delta	τ	tau
ε	epsilon	ϕ	phi
η	eta	ψ	psi
θ	theeta	ω	omega
ν	nu		

Made in the USA
Monee, IL
05 November 2026

67807947R10188